AF411439

Transcriptional Regulation and Its Misregulation in Human Diseases

Transcriptional Regulation and Its Misregulation in Human Diseases

Editors

Amelia Casamassimi
Alfredo Ciccodicola
Monica Rienzo

MDPI • Basel • Beijing • Wuhan • Barcelona • Belgrade • Manchester • Tokyo • Cluj • Tianjin

Editors

Amelia Casamassimi
Department of Precision
Medicine
University of Campania
"Luigi Vanvitelli"
Naples
Italy

Alfredo Ciccodicola
Institute of Genetics and
Biophysics "Adriano
Buzzati-Traverso"
CNR
Naples
Italy

Monica Rienzo
Department of
Environmental, Biological,
and Pharmaceutical Sciences
and Technologies
University of Campania
"Luigi Vanvitelli"
Caserta
Italy

Editorial Office
MDPI
St. Alban-Anlage 66
4052 Basel, Switzerland

This is a reprint of articles from the Special Issue published online in the open access journal *International Journal of Molecular Sciences* (ISSN 1422-0067) (available at: www.mdpi.com/journal/ijms/special_issues/Regulation_Misregulation).

For citation purposes, cite each article independently as indicated on the article page online and as indicated below:

LastName, A.A.; LastName, B.B.; LastName, C.C. Article Title. *Journal Name* **Year**, *Volume Number*, Page Range.

ISBN 978-3-0365-7737-1 (Hbk)
ISBN 978-3-0365-7736-4 (PDF)

Contents

About the Editors . . . . **vii**

Preface to "Transcriptional Regulation and Its Misregulation in Human Diseases" . . . . **ix**

Amelia Casamassimi, Alfredo Ciccodicola and Monica Rienzo
Transcriptional Regulation and Its Misregulation in Human Diseases
Reprinted from: *Int. J. Mol. Sci.* **2023**, *24*, 8640, doi:10.3390/ijms24108640 . . . . **1**

Stefan Nagel, Claudia Pommerenke, Corinna Meyer and Roderick A. F. MacLeod
The Hematopoietic TALE-Code Shows Normal Activity of IRX1 in Myeloid Progenitors and
Reveals Ectopic Expression of IRX3 and IRX5 in Acute Myeloid Leukemia
Reprinted from: *Int. J. Mol. Sci.* **2022**, *23*, 3192, doi:10.3390/ijms23063192 . . . . **9**

Stefan Nagel, Claudia Pommerenke, Corinna Meyer and Roderick A. F. MacLeod
NKL Homeobox Genes NKX2-3 and NKX2-4 Deregulate Megakaryocytic-Erythroid Cell
Differentiation in AML
Reprinted from: *Int. J. Mol. Sci.* **2021**, *22*, 11434, doi:10.3390/ijms222111434 . . . . **29**

**Gabriela Antonio-Andres, Gustavo U. Martinez-Ruiz, Mario Morales-Martinez, Elva
Jiménez-Hernandez, Estefany Martinez-Torres and Tania V. Lopez-Perez et al.**
Transcriptional Regulation of Yin-Yang 1 Expression through the Hypoxia Inducible Factor-1 in
Pediatric Acute Lymphoblastic Leukemia
Reprinted from: *Int. J. Mol. Sci.* **2022**, *23*, 1728, doi:10.3390/ijms23031728 . . . . **47**

**Chiung-Min Wang, William Harry Yang, Leticia Cardoso, Ninoska Gutierrez, Richard Henry
Yang and Wei-Hsiung Yang**
Forkhead Box Protein P3 (FOXP3) Represses ATF3 Transcriptional Activity
Reprinted from: *Int. J. Mol. Sci.* **2021**, *22*, 11400, doi:10.3390/ijms222111400 . . . . **67**

Bo Dong and Yadi Wu
Epigenetic Regulation and Post-Translational Modifications of SNAI1 in Cancer Metastasis
Reprinted from: *Int. J. Mol. Sci.* **2021**, *22*, 11062, doi:10.3390/ijms222011062 . . . . **81**

Tsuyoshi Waku and Akira Kobayashi
Pathophysiological Potentials of NRF3-Regulated Transcriptional Axes in Protein and Lipid
Homeostasis
Reprinted from: *Int. J. Mol. Sci.* **2021**, *22*, 12686, doi:10.3390/ijms222312686 . . . . **95**

Vikrant Rai, Shu Tu, Joseph R. Frank and Jian Zuo
Molecular Pathways Modulating Sensory Hair Cell Regeneration in Adult Mammalian
Cochleae: Progress and Perspectives
Reprinted from: *Int. J. Mol. Sci.* **2021**, *23*, 66, doi:10.3390/ijms23010066 . . . . **105**

**Mareike Polenkowski, Sebastian Burbano de Lara, Aldrige Bernardus Allister, Thi Nhu
Quynh Nguyen, Teruko Tamura and Doan Duy Hai Tran**
Identification of Novel Micropeptides Derived from Hepatocellular Carcinoma-Specific Long
Noncoding RNA
Reprinted from: *Int. J. Mol. Sci.* **2021**, *23*, 58, doi:10.3390/ijms23010058 . . . . **121**

**Bijay P. Dhungel, Geoffray Monteuuis, Caroline Giardina, Mehdi S. Tabar, Yue Feng and
Cynthia Metierre et al.**
The Fusion of *CLEC12A* and *MIR223HG* Arises from a *trans*-Splicing Event in Normal and
Transformed Human Cells
Reprinted from: *Int. J. Mol. Sci.* **2021**, *22*, 12178, doi:10.3390/ijms222212178 . . . . **135**

Naixia Ren, Qingqing Liu, Lingjie Yan and Qilai Huang
Parallel Reporter Assays Identify Altered Regulatory Role of rs684232 in Leading to Prostate Cancer Predisposition
Reprinted from: *Int. J. Mol. Sci.* **2021**, 22, 8792, doi:10.3390/ijms22168792 . . . . **151**

Qian Zhang, Juan Pan, Yusheng Cong and Jian Mao
Transcriptional Regulation of Endogenous Retroviruses and Their Misregulation in Human Diseases
Reprinted from: *Int. J. Mol. Sci.* **2022**, 23, 10112, doi:10.3390/ijms231710112 . . . . **171**

Jelena Pozojevic, Shela Marie Algodon, Joseph Neos Cruz, Joanne Trinh, Norbert Brüggemann and Joshua Laß et al.
Transcriptional Alterations in X-Linked Dystonia–Parkinsonism Caused by the SVA Retrotransposon
Reprinted from: *Int. J. Mol. Sci.* **2022**, 23, 2231, doi:10.3390/ijms23042231 . . . . **191**

Yumin Kim, Baeki E. Kang, Dongryeol Ryu, So Won Oh and Chang-Myung Oh
Comparative Transcriptome Profiling of Young and Old Brown Adipose Tissue Thermogenesis
Reprinted from: *Int. J. Mol. Sci.* **2021**, 22, 13143, doi:10.3390/ijms222313143 . . . . **205**

Hille Suojalehto, Joseph Ndika, Irmeli Lindström, Liisa Airaksinen, Kirsi Karvala and Paula Kauppi et al.
Transcriptomic Profiling of Adult-Onset Asthma Related to Damp and Moldy Buildings and Idiopathic Environmental Intolerance
Reprinted from: *Int. J. Mol. Sci.* **2021**, 22, 10679, doi:10.3390/ijms221910679 . . . . **219**

Hyun-Sun Park, Jongbok Lee, Hyun-Shik Lee, Seong Hoon Ahn and Hong-Yeoul Ryu
Nuclear mRNA Export and Aging
Reprinted from: *Int. J. Mol. Sci.* **2022**, 23, 5451, doi:10.3390/ijms23105451 . . . . **237**

Carolina Gutiérrez-Repiso, Teresa María Linares-Pineda, Andres Gonzalez-Jimenez, Francisca Aguilar-Lineros, Sergio Valdés and Federico Soriguer et al.
Epigenetic Biomarkers of Transition from Metabolically Healthy Obesity to Metabolically Unhealthy Obesity Phenotype: A Prospective Study
Reprinted from: *Int. J. Mol. Sci.* **2021**, 22, 10417, doi:10.3390/ijms221910417 . . . . **249**

Wen-Yuan Lin, Yuan-Ju Lee, Ping-Hung Yu, Yi-Lin Tsai, Pin-Yi She and Tzung-Shian Li et al.
The QseEF Two-Component System-GlmY Small RNA Regulatory Pathway Controls Swarming in Uropathogenic *Proteus mirabilis*
Reprinted from: *Int. J. Mol. Sci.* **2022**, 23, 487, doi:10.3390/ijms23010487 . . . . **261**

Peter Braun, Martin Duy-Thanh Nguyen, Mathias C. Walter and Gregor Grass
Ultrasensitive Detection of *Bacillus anthracis* by Real-Time PCR Targeting a Polymorphism in Multi-Copy 16S rRNA Genes and Their Transcripts
Reprinted from: *Int. J. Mol. Sci.* **2021**, 22, 12224, doi:10.3390/ijms222212224 . . . . **279**

About the Editors

Amelia Casamassimi

Amelia Casamassimi obtained her Biological Sciences degree in 1989 from the University of Naples, Federico II (Italy). She worked at IGB-CNR institute and Pascale Foundation (IRCSS) in Naples. Currently, she is working at the Department of Precision Medicine of University of Campania "Luigi Vanvitelli". She has been interested in genomics and post-genomics approaches, particularly transcriptome analysis, to study human diseases. Currently, she is studying the role of PRDMs in cancer and immunity. She is co-author of several scientific papers in these research fields.

Alfredo Ciccodicola

Alfredo Ciccodicola graduated in Biological Sciences from the Federico II University of Naples. He is currently Research Director at the Institute of Genetics and Biophysics "Adriano Buzzati-Traverso" of the National Research Council of Naples and Professor of Molecular Biology at the Department of Science and Technology of the "Parthenope" University of Naples. His main scientific interests concern the study of the molecular mechanisms underlying complex human diseases (type 2 diabetes, obesity, and cancer) using 'omics' approaches including the analysis of Next Generation Sequencing (NGS) data. Furthermore, his studies are also aimed at understanding the mechanisms of transcriptional regulation, epigenetic regulation, alternative splicing, and their deregulation in human diseases.

Monica Rienzo

Monica Rienzo obtained her Biological Sciences degree in 2003 from the University of Campania "Luigi Vanvitelli" in Caserta (Italy) and her Phd in Neuroscience at the University of Naples, Federico II. She worked at IGB-CNR institute and Department of Precision Medicine of University of Campania, in Naples. Currently, she is working at the Department of Environmental, Biological, and Pharmaceutical Sciences and Technologies, University of Campania "Luigi Vanvitelli" in Caserta. She has been interested in post-genomics approaches, particularly transcriptome analysis, to identify differential expression of alternative transcripts in human diseases. Currently, she is studying the role of PRDMs and their transcripts in cancer and immunity.

Preface to "Transcriptional Regulation and Its Misregulation in Human Diseases"

Transcriptional regulation is a critical biological process that allows the cell or organism to respond to a variety of intra- and extracellular signals, to define cell identity during development, to maintain it throughout its lifetime, and to coordinate cellular activity. This control involves multiple temporal and functional steps, as well as innumerable molecules, including transcription factors, cofactors, and chromatin regulators. It is well known that many human disorders are characterized by global transcriptional dysregulation, since most of the signaling pathways ultimately target transcription machinery. Indeed, many syndromes and genetic and complex diseases, including cancer, autoimmunity, neurological and developmental disorders, and metabolic and cardiovascular diseases, can be caused by mutations/alterations in regulatory sequences, transcription factors, splicing regulators, cofactors, chromatin regulators, ncRNAs, and other components of transcription apparatus. It is worth noting that advances in our understanding of molecules and mechanisms involved in the transcriptional circuitry and apparatus lead to new insights into the pathogenetic mechanisms of various human diseases and disorders. Thus, this Special Issue is focused on molecular genetics and genomics studies, exploring the effects of transcriptional misregulation on human diseases.

Amelia Casamassimi, Alfredo Ciccodicola, and Monica Rienzo
Editors

International Journal of
Molecular Sciences

Editorial

Transcriptional Regulation and Its Misregulation in Human Diseases

Amelia Casamassimi [1], Alfredo Ciccodicola [2,3,*] and Monica Rienzo [4]

1   Department of Precision Medicine, University of Campania "Luigi Vanvitelli", Via L. De Crecchio, 80138 Naples, Italy; amelia.casamassimi@unicampania.it

2   Institute of Genetics and Biophysics "Adriano Buzzati-Traverso", CNR, Via P. Castellino 111, 80131 Naples, Italy

3   Department of Science and Technology, University of Naples "Parthenope", 80143 Naples, Italy

4   Department of Environmental, Biological, and Pharmaceutical Sciences and Technologies, University of Campania "Luigi Vanvitelli", 81100 Caserta, Italy; monica.rienzo@unicampania.it

*   Correspondence: alfredo.ciccodicola@igb.cnr.it or alfredo.ciccodicola@uniparthenope.it

Citation: Casamassimi, A.; Ciccodicola, A.; Rienzo, M. Transcriptional Regulation and Its Misregulation in Human Diseases. *Int. J. Mol. Sci.* **2023**, 24, 8640. https://doi.org/10.3390/ijms24108640

Received: 4 May 2023

Accepted: 9 May 2023

Published: 12 May 2023

Transcriptional regulation is a critical biological process that allows the cell or an organism to respond to a variety of intra- and extracellular signals, to define cell identity during development, to maintain it throughout its lifetime, and to coordinate cellular activity. This control involves multiple temporal and functional steps, as well as innumerable molecules, including transcription factors, cofactors, and chromatin regulators. It is well known that many human disorders are characterized by global transcriptional dysregulation, since most of the signaling pathways ultimately target transcription machinery. Indeed, many syndromes and genetic and complex diseases, including cancer, autoimmunity, neurological and developmental disorders, and metabolic and cardiovascular diseases, can be caused by mutations/alterations in regulatory sequences, transcription factors, splicing regulators, cofactors, chromatin regulators, ncRNAs, and other components of transcription apparatus. It is worth noting that advances in our understanding of molecules and mechanisms involved in the transcriptional circuitry and apparatus lead to new insights into the pathogenetic mechanisms of various human diseases and disorders. Thus, this Special Issue is focused on molecular genetics and genomics studies, exploring the effects of transcriptional misregulation on human diseases [1,2].

In this Special Issue, a total of 18 excellent and interesting papers [3–20], consisting of 13 original research studies [3–6,10–12,14–16,18–20], as well as five reviews, are published [7–9,13,17]. They cover many subjects of transcriptional regulation, including studies on cis-regulatory elements, transcription factors (TF), chromatin regulators, and ncRNAs. As expected, a substantial number of transcriptome studies and computational analyses are also included in this issue.

Significantly, all the published papers have provided novel insights on the knowledge of human pathophysiological mechanisms, having the purpose of proposing novel biomarkers and/or therapeutic targets for the diagnosis and treatment of many diseases, especially cancer. In the field of TFs, two papers by Nagel et al. [3,4] are focused on the class of homeobox genes, encoding developmental factors containing a homeodomain with three helices, which interact with DNA, cofactors, and chromatin, thus allowing gene-regulating activities essential for the control of basic cell and tissue differentiation decisions. In accordance with their normal functions, these genes, when deregulated, contribute to carcinogenesis along with hematopoietic malignancies. In one of the manuscripts [3], they established the so-called myeloid TALE-code, representing a TALE homeobox gene expression pattern in normal myelopoiesis. The class of TALE homeobox genes comprises specific homeodomain factors that share a three-amino-acid residue loop extension (abbreviated as TALE) between helix 1 and helix 2. These transcription factors control basic developmental decisions. The same authors had previously constructed the lymphoid TALE-code that codifies expression patterns of all active TALE class homeobox genes in early hematopoiesis and lymphopoiesis, thus extending the TALE code to the entire hematopoietic system. Collectively, data showed expression patterns for eleven TALE homeobox genes and highlighted the exclusive expression of IRX1 in megakaryocyte-erythroid progenitors, suggesting that this TALE class member is involved in a specific myeloid differentiation route. Interestingly, the analysis of public transcription profiles from acute myeloid leukemia (AML) patients revealed aberrant expression of IRX1, IRX3, and IRX5, indicating an oncogenic role for these TALE homeobox genes when deregulated in AML [3].

In the second work [4], Nagel and colleagues investigated the subclass of NKL homeobox genes that also function in normal development and are often deregulated in hematopoietic malignancies; indeed, a previous systematic analysis revealed 18 deregulated NKL homeobox genes in AML, underlining the relevance of these developmental oncogenes in driving this cancer type. In this newly published study, the authors also identified aberrantly activated NKL genes, NKX2-3 and NKX2-4, in cell lines derived from two different AML subtypes, where they deregulate target genes involved in megakaryocytic and erythroid differentiation, thus providing the molecular basis to the classification of specific AML subtypes [4].

Another TF, which is involved in tumor progression and metastasis is Yin-Yang transcription factor 1 (YY1); indeed, it also overexpressed in different cancers, including leukemia. Antonio-Andres, G et al. [5] observed that the expression of *YY1* in patients with pediatric acute lymphoblastic leukemia (ALL) positively correlates with *HIF1A* transcription. Besides, their findings clearly indicate, for the first time, that *YY1* is transcriptionally regulated by HIF-1α and suggest that both *HIF1A* and *YY1* transcription factors could be possible therapeutic targets and/or biomarkers of ALL [5].

A further regulatory mechanism that may play a role in tumor development and progression involves two additional TFs, Forkhead Box Protein P3 (*FOXP3*) and activating transcription factor 3 (*ATF3*) [6]. FOXP3 has an essential and critical role in autoimmunity, cancer development, and Treg development, with hundreds of target genes already identified in both cancer cells and Treg cells. Additionally, *FOXP3* is a recognized breast and prostate tumor suppressor gene from the X chromosome, acting as a transcriptional repressor for several oncogenes. Using several human cell lines, Chiung-Min Wang et al. [6] assessed the function of *FOXP3* in the transcriptional activity of *ATF*, which binds several promoters of key regulatory proteins that determine cell fate, circadian signaling, and homeostasis, and it is rapidly induced by many pathophysiological signals and is essential in cellular stress response. Overall, their findings suggest that *FOXP3*, through FOX protein response element, functions as a novel repressor of *ATF3* and that phosphorylation at Y342 plays a critical role for *FOXP3* transcriptional activity [6].

Besides, three reviews of this Special Issue are mainly focused on TFs. The first one discusses the regulation of *SNAI1* (Snail Family Transcriptional Repressor 1), a zinc finger transcription factor, which acts as a master regulator of epithelial–mesenchymal transition (EMT). Noteworthy, *SNAI1* is involved in the formation of cancer metastases by epigenetic regulation and post-translational modifications [7]. The Waku and Kobayashi [8] review describes the pathophysiological aspects of the biomolecular pathways regulated by NRF3 (NFE2L3; NFE2-like BZIP Transcription Factor 3), a transcription factor belonging to the cap'n'collar (CNC)-based leucine zipper family and functioning through proteasome regulation. The NRF3 factor and its regulated axes are involved in cancer cell growth and have anti-obesity potential, thus suggesting a possible role in the development of obesity-induced cancer [8]. Finally, Rai V. et al. critically summarized the studies performed on the regeneration of sensory hair cells (HCs) in adult mammalian cochleas to elucidate the molecular pathways, and particularly transcription factors, involved in the regeneration of cochlear HCs, which aids in proposing a biological approach for better therapeutics to treat hearing loss and to restore hearing [9].

In the last years, the old annotation of protein-coding genes, based on the presence of an open reading frame (ORF) with minimal lengths for translated proteins, has significantly changed. Indeed, the recent literature indicates that the proteome is more complex than previously estimated, since RNAs previously considered noncoding, such as long non-coding RNAs (lncRNAs) and circular RNAs, are instead translated into functional small proteins [10]. In an interesting study of this Special Issue, the authors utilized transcriptome and polysome profiling to identify novel micropeptides that originate from lncRNAs that are expressed exclusively in hepatocellular carcinoma (HCC) cells, but not in the liver or other normal tissues. Specifically, they found three HCC-specific lncRNAs, containing at least one ORF longer than 50 amino acid (aa) and enriched in the polysome fraction. Besides, through a peptide specific antibody, they characterized one lncRNA candidate, NONHSAT013026.2/Linc013026-68AA, which is translated into a 68 aa micropeptide. This small protein is mainly localized at the perinuclear region and is mainly expressed in moderately—but not well-differentiated—HCC cells, and it plays a role in cell proliferation, suggesting that it could be used as an HCC-specific target molecule. This finding is noteworthy, since it represents an important advance in the study of the previously overlooked "dark proteome" and its role in human pathologies, particularly in cancer [10].

Another emerging field, especially in cancer, is represented by chimeric RNAs, which are transcripts consisting of exons from different parental genes. They can be produced by several mechanisms mostly involving chromosomal rearrangements; besides, they can also be generated by intergenic splicing, cis-splicing from two same-strand adjacent genes, and trans-splicing from two separate RNA transcripts [11]. Although these events were initially considered rare, human transcriptome profiles have revealed that a huge amount of chimeric RNAs develop from intergenic splicing and can be also detected in normal tissues, thus contributing to transcriptomic complexity. An interesting study has analyzed the genetic structure and biological roles of *CLEC12A-MIR223HG*, a novel chimeric transcript produced through trans-splicing by the fusion of the cell surface receptor CLEC12A (C-Type Lectin Domain Family 12 Member A) and the miRNA-223 host gene (*MIR223HG*), first identified through transcriptome profiling of chronic myeloid leukemia (CML) patients. Unexpectedly, *CLEC12A-MIR223HG* was detected not only in CML, but also in a variety of normal tissues and cell lines as pro-monocytic cells resistant to chemotherapy or during monocyte-to-macrophage differentiation [11]. Transcriptional activation of CLEC12A increased *CLEC12A-MIR223HG* expression. This chimeric RNA also translates into a chimeric protein, which largely resembles CLEC12A, but contains a modified C-type lectin domain, altering key disulphide bonds. Consequently, differences in post-translational modifications, cellular localization, and protein–protein interactions occur. These findings not only support a possible involvement of *CLEC12A-MIR223HG* in the regulation of CLEC12A function, but they could also provide a roadmap to study the other uncharacterized chimeric RNAs that are continuously recognized by RNA-Seq analyses [11].

In the last decades, next generation sequencing (NGS) strategies have been greatly applied in a huge number of cancer studies with different purposes. Among the NGS applications, several DNA barcode-based parallel reporter methods have been implemented for the screening of regulatory risk sites. Among them, the dinucleotide reporter system (DiR)-seq screening system was developed to investigate the gene regulatory effect from the risk single nucleotide polymorphisms (SNPs) that have a modest impact. In their paper, Ren and coworkers [12] applied the DiR system in prostate cancer cells (22Rv1) to screen the regulatory risk SNPs, leading to transcriptional misregulation, and they identified 32 regulatory SNPs that exhibited different regulatory activities with two alleles. Among them, fourteen SNPs exhibited decreased expression levels for the risk alleles, whereas eighteen SNPs showed increased expression. Particularly, they discovered that the rs684232 T allele altered chromatin binding of transcription factor FOXA1 on the DNA region and led to aberrant gene expression of *VPS53*, *FAM57A*, and *GEMIN4*, which are often upregulated in prostate cancer patients. Thus, these findings provide novel insights to further elucidate the basis mechanism of the functional prostate cancer risk SNPs [12].

An important role in the mechanisms of transcriptional regulation is also played by transposable elements, repetitive genetic sequences with the ability (sometimes lost during evolution) to transpose elsewhere in the genome. A class of these elements is represented by the endogenous retroviruses (ERVs), which represent about 8% of the sequences present in the human genome. An interesting overview of this Special Issue [13] describes the mechanisms underlying their transcriptional regulation. During the evolution of the human genome, the accumulation of mutations, insertions, deletions, and/or truncations has rendered these elements inactive. However, it is increasingly evident that, under the influence of genetic and epigenetic mechanisms, they can be involved in some physiological and pathological conditions; examples are their function in embryonic development, or, even more importantly, their reactivation in the development of human diseases, such as cancer and neurodegenerative disorders [13]. Besides, a remarkable paper [14] shows that transcriptional alterations observed in X-linked dystonia–parkinsonism (XDP) are caused by the insertion of a SINE-VNTR-Alu (SVA) retrotransposon in an intron of the *TAF1* (TATA-Box Binding Protein Associated Factor 1) gene, encoding for the largest subunit of TFIID; as an interesting effect, increased levels of the *TAF1* intron retention transcript *TAF1-32i* can be found in XDP cells, as compared to healthy controls. Overall, the results of this study provide further evidence that transposable elements affect gene expression and suggest that a mechanism of splicing alteration occurs in XDP patients, probably caused by binding sites for transcription factors and splicing regulators present within this retrotransposon and that need to be exactly proved through additional experiments [14].

Currently, transcriptome profiling is one of the most utilized approaches to investigate human diseases at the molecular level [2]. Here, Kim and colleagues [15] compare the transcriptomic profiles, extracted from the NCBI Gene Expression Omnibus (GEO) database, of brown adipose tissue (BAT) of young and elderly subjects in response to thermogenic stimuli. Interestingly, they observe that aging does not cause transcriptional changes in thermogenic genes, but it upregulates several pathways related to the immune response and downregulates metabolic pathways. Furthermore, they note that acute severe cold exposure (CE) upregulates several pathways related to protein folding, whereas chronic mild CE upregulates metabolic pathways, mostly related to carbohydrate metabolism [15].

Furthermore, in the study of Suojalehto et al. [16], transcriptome profiling was assessed to investigate whether a distinct clinical subtype of adult-onset asthma could be related to damp and moldy buildings, which are symptoms of idiopathic environmental intolerance, thus identifying potential molecular similarities with this disease. To this purpose, fifty female adult-onset asthma patients were categorized based on their exposure to building dampness and molds and other clinical parameters (inflammation, cytokine profile, etc.), together with gene signatures of nasal biopsies and peripheral blood mononuclear cells. Overall, the results of this study revealed a greater degree of similarity between idiopathic environmental intolerance and dampness related asthma than between the same patients and those with asthma not associated to dampness and mold [16]. Transcriptome analysis revealed well defined pathological mechanisms for asthma without exposure to dampness, but not dampness-and-mold-related asthma patients. Besides, a distinct molecular pathological profile in nasal and blood immune cells of idiopathic environmental intolerance subjects was found, including several differentially expressed genes (DEGs) that were also detected in dampness-and-mold-related asthma samples, thereby suggesting idiopathic environmental intolerance-type mechanisms [16].

It is well known that eukaryotic transcription is a complex, biological, and stepwise process, ranging from initiation, elongation, and termination to the process of pre-mRNA with $5'$-end capping, splicing, $3'$-end cleavage, and polyadenylation; subsequently, the mature mRNA is exported from the nucleus to the cytoplasm to undergo protein translation, whereas the aberrantly processed pre-mRNAs and mRNAs are removed via the RNA surveillance system. All these steps are also interconnected with each other, and with chromatin accessibility and additional epigenetic mechanisms [1]. Of note, alteration of any step may constitute the basis of a disease. For instance, Park HS. et al. [17] illustrate recent findings on the role of the nuclear mRNAs export in cellular aging and age-related neurodegenerative disorders. Indeed, it is now established that there is a close relationship between transcription and aging; however, the role of the nuclear mRNAs export in these issues is still poorly characterized. Of note, disruption of the regulation of factors mediating mRNA export from the nucleus (namely, TREX, TREX-2, and nuclear pore complex) has been shown to result in the accumulation of aberrant nuclear mRNAs, with the consequent alteration of normal lifespan and the development of neurodegenerative diseases [17].

Concerning epigenetics, an interesting paper focused on obesity without metabolic complications, a phenotype defined as metabolically healthy obesity (MHO), which can progress to an unhealthy state known as metabolically unhealthy obesity (MUO), although a relevant percentage of MHO individuals are likely to maintain their status over time [18]. The authors aimed to analyze the long-term evolution of DNA methylation patterns in a subset of MHO subjects in order to search for epigenetic markers that could predict the progression of MHO to MUO. As a result, twenty-six CpG sites were significantly differentially methylated, both at baseline and after eleven years of follow-up. Two potential biomarkers of the transition to an unhealthy state were identified: specifically, higher methylation in cg20707527 (*ZFPM2*) and lower methylation in cg11445109 (*CYP2E1*) could impact the stability of a healthy phenotype in obesity [18].

Finally, the study of transcriptional regulation is also useful to elucidate the pathogenetic mechanisms of human pathogens. Bacterial sensing of environmental signals has an essential role in modulating virulence and bacterium–host interactions. Generally, bacteria utilize the two-component system (TCS) method, such as QseEF, to control gene expression in response to rapid environmental changes. Of note, many recognized mechanisms function through the post-transcriptional control of small non-coding RNAs (sRNAs), and their identification is rapidly increasing in number and variety in the context of bacteria regulatory functions [19]. Interestingly, in a study in the present Special Issue, the authors identified the QseEF homologue of *Proteus mirabilis*, an Important pathogen of the urinary tract, principally in patients with indwelling urinary catheters, and they found it is involved in the modulation of swarming motility through the sRNA *GlmY*. This is the first study investigating a pathway mediated by a two-component system through a sRNA as an underlying pathogenetic mechanism of *Proteus mirabilis* swarming migration, during which expression of several virulence genes is increased. Since it is assumed that *P. mirabilis* swarming up catheters is primed to infect the urinary tract, clarifying the swarming mechanisms could provide new approaches in the development of intervention strategies and facilitate the discovery of novel therapeutics [19].

Again, in the field of human pathogens, Braun et al. [20] focused on the molecular diagnosis of the anthrax pathogen *Bacillus anthracis*, which is challenging because its identification is complicated by the close relationship with other bacteria of the group species. The authors have designed and validated an ultrasensitive detection method that can be run as a real-time PCR with solely DNA as a template or as a RT-real-time version using both cellular nucleic acid pools (DNA and RNA) as a template without requirement of DNase treatment. This assay was found to be highly species specific, yielding no false positives, and it was highly sensitive targeting a unique single nucleotide polymorphism within a variable number of loci of the multi-copy 16S rRNA gene and related transcripts. With the high abundance of 16S rRNA moieties in cells, it is expected to facilitate the detection of *B. anthracis* by PCR. While standard PCR assays are well established for the identification of *B. anthracis* from pure culture, the exceptional sensitivity of the new 16S rRNA-based test might excel in clinical and public health laboratories when detection of minute residues of the pathogen is required [20].

In conclusion, we hope the readers enjoy this Special Issue of IJMS and the effort to present the current advances and promising results in the field of transcriptional regulation and its involvement in all the relevant biological processes and in pathophysiology.

Funding: This research received no external funding.

Institutional Review Board Statement: Not applicable.

Informed Consent Statement: Not applicable.

Data Availability Statement: Not applicable.

Acknowledgments: We thank Mariagiovanna Tramontano for her helpful assistance in drafting and writing our editorial. We also thank Jerry Whang and all the participating assistant editors and reviewers for their important contribution to this Special Issue.

Conflicts of Interest: The authors declare no conflict of interest.

References

1. Casamassimi, A.; Ciccodicola, A. Transcriptional Regulation: Molecules, Involved Mechanisms, and Misregulation. *Int. J. Mol. Sci.* **2019**, *20*, 1281. [CrossRef] [PubMed]
2. Casamassimi, A.; Federico, A.; Rienzo, M.; Esposito, S.; Ciccodicola, A. Transcriptome Profiling in Human Diseases: New Advances and Perspectives. *Int. J. Mol. Sci.* **2017**, *18*, 1652. [CrossRef] [PubMed]
3. Nagel, S.; Pommerenke, C.; Meyer, C.; MacLeod, R.A.F. The Hematopoietic TALE-Code Shows Normal Activity of IRX1 in Myeloid Progenitors and Reveals Ectopic Expression of IRX3 and IRX5 in Acute Myeloid Leukemia. *Int. J. Mol. Sci.* **2022**, *23*, 3192. [CrossRef] [PubMed]
4. Nagel, S.; Pommerenke, C.; Meyer, C.; MacLeod, R.A.F. NKL Homeobox Genes NKX2-3 and NKX2-4 Deregulate Megakaryocytic-Erythroid Cell Differentiation in AML. *Int. J. Mol. Sci.* **2021**, *22*, 11434. [CrossRef] [PubMed]
5. Antonio-Andres, G.; Martinez-Ruiz, G.U.; Morales-Martinez, M.; Jiménez-Hernandez, E.; Martinez-Torres, E.; Lopez-Perez, T.V.; Estrada-Abreo, L.A.; Patino-Lopez, G.; Juarez-Mendez, S.; Davila-Borja, V.M.; et al. Transcriptional Regulation of Yin-Yang 1 Expression through the Hypoxia Inducible Factor-1 in Pediatric Acute Lymphoblastic Leukemia. *Int. J. Mol. Sci.* **2022**, *23*, 1728. [CrossRef] [PubMed]
6. Wang, C.-M.; Yang, W.H.; Cardoso, L.; Gutierrez, N.; Yang, R.H.; Yang, W.-H. Forkhead Box Protein P3 (FOXP3) Represses ATF3 Transcriptional Activity. *Int. J. Mol. Sci.* **2021**, *22*, 11400. [CrossRef] [PubMed]
7. Dong, B.; Wu, Y. Epigenetic Regulation and Post-Translational Modifications of SNAI1 in Cancer Metastasis. *Int. J. Mol. Sci.* **2021**, *22*, 11062. [CrossRef] [PubMed]
8. Waku, T.; Kobayashi, A. Pathophysiological Potentials of NRF3-Regulated Transcriptional Axes in Protein and Lipid Homeostasis. *Int. J. Mol. Sci.* **2021**, *22*, 12686. [CrossRef] [PubMed]
9. Rai, V.; Tu, S.; Frank, J.R.; Zuo, J. Molecular Pathways Modulating Sensory Hair Cell Regeneration in Adult Mammalian Cochleae: Progress and Perspectives. *Int. J. Mol. Sci.* **2022**, *23*, 66. [CrossRef] [PubMed]
10. Polenkowski, M.; Burbano de Lara, S.; Allister, A.B.; Nguyen, T.N.Q.; Tamura, T.; Tran, D.D.H. Identification of Novel Micropeptides Derived from Hepatocellular Carcinoma-Specific Long Noncoding RNA. *Int. J. Mol. Sci.* **2022**, *23*, 58. [CrossRef] [PubMed]
11. Dhungel, B.P.; Monteuuis, G.; Giardina, C.; Tabar, M.S.; Feng, Y.; Metierre, C.; Ho, S.; Nagarajah, R.; Fontaine, A.R.M.; Shah, J.S.; et al. The Fusion of CLEC12A and MIR223HG Arises from a trans-Splicing Event in Normal and Transformed Human Cells. *Int. J. Mol. Sci.* **2021**, *22*, 12178. [CrossRef] [PubMed]
12. Ren, N.; Liu, Q.; Yan, L.; Huang, Q. Parallel Reporter Assays Identify Altered Regulatory Role of rs684232 in Leading to Prostate Cancer Predisposition. *Int. J. Mol. Sci.* **2021**, *22*, 8792. [CrossRef] [PubMed]
13. Zhang, Q.; Pan, J.; Cong, Y.; Mao, J. Transcriptional Regulation of Endogenous Retroviruses and Their Misregulation in Human Diseases. *Int. J. Mol. Sci.* **2022**, *23*, 10112. [CrossRef] [PubMed]
14. Pozojevic, J.; Algodon, S.M.; Cruz, J.N.; Trinh, J.; Brüggemann, N.; Laß, J.; Grütz, K.; Schaake, S.; Tse, R.; Yumiceba, V.; et al. Transcriptional Alterations in X-Linked Dystonia–Parkinsonism Caused by the SVA Retrotransposon. *Int. J. Mol. Sci.* **2022**, *23*, 2231. [CrossRef] [PubMed]
15. Kim, Y.; Kang, B.E.; Ryu, D.; Oh, S.W.; Oh, C.-M. Comparative Transcriptome Profiling of Young and Old Brown Adipose Tissue Thermogenesis. *Int. J. Mol. Sci.* **2021**, *22*, 13143. [CrossRef] [PubMed]
16. Suojalehto, H.; Ndika, J.; Lindström, I.; Airaksinen, L.; Karvala, K.; Kauppi, P.; Lauerma, A.; Toppila-Salmi, S.; Karisola, P.; Alenius, H. Transcriptomic Profiling of Adult-Onset Asthma Related to Damp and Moldy Buildings and Idiopathic Environmental Intolerance. *Int. J. Mol. Sci.* **2021**, *22*, 10679. [CrossRef] [PubMed]
17. Park, H.-S.; Lee, J.; Lee, H.-S.; Ahn, S.H.; Ryu, H.-Y. Nuclear mRNA Export and Aging. *Int. J. Mol. Sci.* **2022**, *23*, 5451. [CrossRef] [PubMed]
18. Gutiérrez-Repiso, C.; Linares-Pineda, T.M.; Gonzalez-Jimenez, A.; Aguilar-Lineros, F.; Valdés, S.; Soriguer, F.; Rojo-Martínez, G.; Tinahones, F.J.; Morcillo, S. Epigenetic Biomarkers of Transition from Metabolically Healthy Obesity to Metabolically Unhealthy Obesity Phenotype: A Prospective Study. *Int. J. Mol. Sci.* **2021**, *22*, 10417. [CrossRef] [PubMed]

19. Lin, W.-Y.; Lee, Y.-J.; Yu, P.-H.; Tsai, Y.-L.; She, P.-Y.; Li, T.-S.; Liaw, S.-J. The QseEF Two-Component System-GlmY Small RNA Regulatory Pathway Controls Swarming in Uropathogenic Proteus mirabilis. *Int. J. Mol. Sci.* **2022**, *23*, 487. [CrossRef] [PubMed]
20. Braun, P.; Nguyen, M.D.-T.; Walter, M.C.; Grass, G. Ultrasensitive Detection of Bacillus anthracis by Real-Time PCR Targeting a Polymorphism in Multi-Copy 16S rRNA Genes and Their Transcripts. *Int. J. Mol. Sci.* **2021**, *22*, 12224. [CrossRef] [PubMed]

International Journal of
Molecular Sciences

Article

The Hematopoietic TALE-Code Shows Normal Activity of IRX1 in Myeloid Progenitors and Reveals Ectopic Expression of IRX3 and IRX5 in Acute Myeloid Leukemia

Stefan Nagel *, Claudia Pommerenke, Corinna Meyer and Roderick A. F. MacLeod

Department of Human and Animal Cell Lines, Leibniz-Institute DSMZ, German Collection of Microorganisms and Cell Cultures, 38124 Braunschweig, Germany; cpo14@dsmz.de (C.P.); cme@dsmz.de (C.M.); rafmacleod@gmail.com (R.A.F.M.)
* Correspondence: sna@dsmz.de; Tel.: +49-531-2616167

Abstract: Homeobox genes encode transcription factors that control basic developmental decisions. Knowledge of their hematopoietic activities casts light on normal and malignant immune cell development. Recently, we constructed the so-called lymphoid TALE-code that codifies expression patterns of all active TALE class homeobox genes in early hematopoiesis and lymphopoiesis. Here, we present the corresponding myeloid TALE-code to extend this gene signature, covering the entire hematopoietic system. The collective data showed expression patterns for eleven TALE homeobox genes and highlighted the exclusive expression of IRX1 in megakaryocyte-erythroid progenitors (MEPs), implicating this TALE class member in a specific myeloid differentiation process. Analysis of public profiling data from acute myeloid leukemia (AML) patients revealed aberrant activity of IRX1 in addition to IRX3 and IRX5, indicating an oncogenic role for these TALE homeobox genes when deregulated. Screening of RNA-seq data from 100 leukemia/lymphoma cell lines showed overexpression of IRX1, IRX3, and IRX5 in megakaryoblastic and myelomonocytic AML cell lines, chosen as suitable models for studying the regulation and function of these homeo-oncogenes. Genomic copy number analysis of IRX-positive cell lines demonstrated chromosomal amplification of the neighboring IRX3 and IRX5 genes at position 16q12 in MEGAL, underlying their overexpression in this cell line model. Comparative gene expression analysis of these cell lines revealed candidate upstream factors and target genes, namely the co-expression of GATA1 and GATA2 together with IRX1, and of BMP2 and HOXA10 with IRX3/IRX5. Subsequent knockdown and stimulation experiments in AML cell lines confirmed their activating impact in the corresponding IRX gene expression. Furthermore, we demonstrated that IRX1 activated KLF1 and TAL1, while IRX3 inhibited GATA1, GATA2, and FST. Accordingly, we propose that these regulatory relationships may represent major physiological and oncogenic activities of IRX factors in normal and malignant myeloid differentiation, respectively. Finally, the established myeloid TALE-code is a useful tool for evaluating TALE homeobox gene activities in AML.

Keywords: homeodomain; HOX-code; NKL-code; PBX1; TALE-code; TBX-code

Citation: Nagel, S.; Pommerenke, C.; Meyer, C.; MacLeod, R.A.F. The Hematopoietic TALE-Code Shows Normal Activity of IRX1 in Myeloid Progenitors and Reveals Ectopic Expression of IRX3 and IRX5 in Acute Myeloid Leukemia. *Int. J. Mol. Sci.* **2022**, *23*, 3192. https://doi.org/10.3390/ijms23063192

Academic Editors: Haifa Kathrin Al-Ali, Amelia Casamassimi, Alfredo Ciccodicola and Monica Rienzo

Received: 15 February 2022
Accepted: 14 March 2022
Published: 16 March 2022

Publisher's Note: MDPI stays neutral with regard to jurisdictional claims in published maps and institutional affiliations.

1. Introduction

Stem and progenitor cells expand and subsequently pass through several developmental stages to differentiate into mature cells and tissues. In the hematopoietic system, corresponding processes generate all types of blood and immune cells, starting with hematopoietic stem cells and their derived progenitors, which establish the lymphoid and myeloid lineage [1]. The first produces lymphocytes such as B- and T-cells, while the latter generates inter alia granulocytes, monocytes, megakaryocytes, and erythrocytes. The myeloid system contains several progenitors that differ in their developmental potential. The common myeloid progenitor (CMP) is able to produce all types of myeloid cells, while the megakaryocyte-erythroid progenitors (MEPs) are developmentally restricted, generating either megakaryocytes or erythrocytes.

Hematopoietic differentiation processes are typically regulated at the transcriptional level [2–4]. Therefore, the master genes controlling these operations mostly encode transcription factors (TFs). Homeobox genes encode developmental TFs, controlling basic decisions in cell and tissue differentiation. They contain a homeodomain at the protein level, which consists of three helices. This domain interacts with DNA, cofactors, and chromatin, thus representing a platform for gene-regulating activities [5]. According to sequence similarities of their conserved homeobox, these genes fall into eleven classes and several subclasses [6]. ANTP represents the largest class containing 39 clustered HOX genes and the 48-member-strong NKL homeobox gene subclass. The class of TALE homeobox genes comprises specific homeodomain factors that share a three-amino-acid residue loop extension (abbreviated as TALE) between helix 1 and helix 2. The human genome encodes 20 TALE homeodomain proteins including six IRX-factors and the well-known PBX1 and MEIS1 members, which are able to cooperate with HOX factors [7–9].

The human IRX genes are genomically arranged in two clusters, consisting of IRX1, IRX2, and IRX4 (located at chromosomal position 5p15) and of IRX3, IRX5, and IRX6 (16q12). This clustering is evolutionarily conserved and may be related to conjoint transcriptional regulation [10]. IRX genes are embryonically expressed and tasked with regulating the development of particular tissues and organs [11]. IRX1, for example, regulates the development of the limbs, gut, kidney, and lung and plays an oncogenic role in several types of cancer [12]. Aberrant expression of IRX1 and IRX3 has been reported in myeloid leukemia, supporting general commitments in carcinogenesis [13,14].

Reflecting their physiological functions in development, deregulated homeobox genes drive specific hematopoietic malignancies and are frequently targeted by chromosomal aberrations [15–17]. To evaluate the activities of major subgroups of homeobox genes in lymphoid and myeloid malignancies, we generated the "NKL-code" [18]. This gene signature describes physiological activities of NKL homeobox genes during hematopoiesis and assists the identification of deregulated NKL homeobox gene expression in patients. Accordingly, we also described the lymphoid TALE-code and applied that gene signature to identify the aberrant activity of TALE homeobox gene PBX1 in Hodgkin lymphoma [19]. Here, we extended the established lymphoid TALE-code to the entire hematopoietic system and included the myeloid lineage.

Acute myeloid leukemia (AML) is the most frequent hematopoietic malignancy in adults. The tumor cells derive from particular myeloid progenitors. Several subtypes of AML may be distinguished according to originating cells and stages, phenotypes, chromosomal aberrations, and gene mutations that differ in prognosis and outcome [20]. Knowledge of normal and abnormal activities of basic developmental regulators should serve to refine diagnostic procedures and thus help identify novel therapeutic targets. Accordingly, we describe here the physiological expression of TALE homeobox genes during myelopoiesis and investigate the aberrant activities of IRX homeobox genes in AML subsets.

2. Results

2.1. Establishment of the Myeloid TALE-Code

Recently, we delineated the lymphoid TALE-code for early hematopoiesis and lymphopoiesis [19]. Here, we extended this gene signature to include the myeloid system according to our reported approach generating the myeloid NKL-code [21]. By exploiting public gene expression profiling and RNA-seq datasets, we generated a TALE homeobox gene expression pattern for myeloid progenitors and mature cells (Figure S1). The applied cutoffs to discriminate positive and negative expression levels were adopted from our previous studies [21]. The results are depicted in Figure 1, representing the myeloid TALE-code.

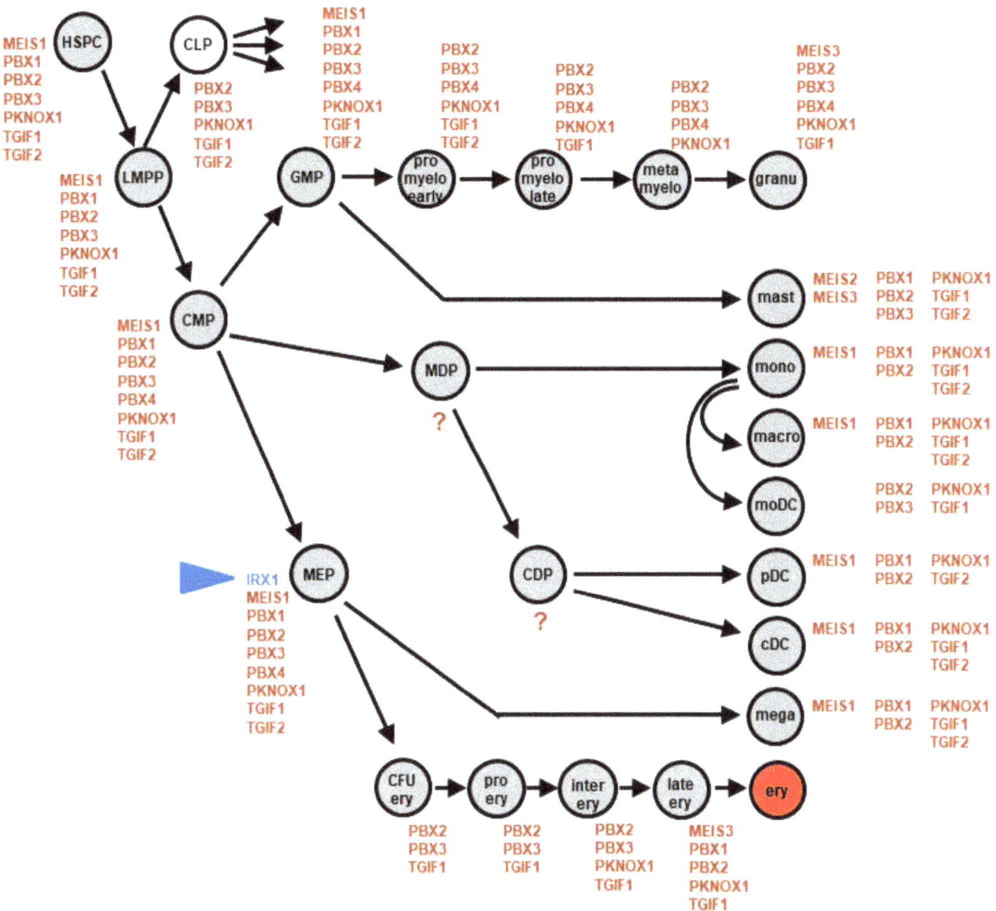

Figure 1. Myeloid TALE-code. This diagram summarizes the screening results for expression analyses of TALE homeobox genes (red) in early hematopoiesis and myelopoiesis. We have termed this expression pattern myeloid TALE-code. Expression of IRX1 is highlighted in blue (arrowhead). Abbreviations: cDC, conventional dendritic cell; CDP, common dendritic progenitor; CFU ery, erythroid colony forming unit; CLP, common lymphoid progenitor; CMP, common myeloid progenitor; ery, erythrocyte; GMP, granulo-myeloid progenitor; granu, granulocyte; HSPC, hematopoietic stem and progenitor cell; inter ery, intermediate stage erythroblast; late ery, pyknotio-stage erythroblast; LMPP, lymphomyelo-primed progenitor; macro, macrophage; mast, mast cell; MDP, monocyte dendritic cell progenitor; mega, megakaryocyte; MEP, megakaryocytic-erythroid progenitor; metamyelo, metamyelocyte; moDC, monocyte-derived dendritic cell; mono, monocyte; pDC, plasmacytoid dendritic cell; pro ery, pro-erythroblast; pro myelo early/late, early/late promyelocyte.

The established myeloid TALE-code comprises eleven genes: IRX1, MEIS1, MEIS2, MEIS3, PBX1, PBX2, PBX3, PBX4, PKNOX1, TGIF1, and TGFI2. We found TALE homeobox gene activities in all cell types analyzed. The numbers of expressed genes in single entities ranged from three (pro-erythroblast) to nine (MEP). TGIF genes were expressed in all but one entity (metamyelocyte). PBX1 was also active in most mature cell types, contrasting with the findings for this gene in lymphopoiesis [19]. Interestingly, IRX1 expression was only detected in MEPs, while the remaining IRX genes remained silent in the complete hematopoietic compartment [19]. This observation may indicate that IRX1 plays a major role in early differentiation processes of this particular progenitor, which generates both megakaryocytes and erythrocytes. In accordance with the TALE-code, RNA-seq gene expression data from the Human Protein Atlas showed repression of all six IRX genes in the myeloid entities represented, despite expression in cells and tissues of other compartments (Figure S2). In the following, we examined a potential oncogenic role for IRX genes in AML.

2.2. Aberrant Expression of IRX1, IRX3, and IRX5 in AML

To scrutinize the aberrant expression of IRX genes in AML patients, we screened three public expression profiling datasets. Dataset GSE19577 covers 42 AML patients with MLL-rearrangements, GSE15434 contains 251 patients with normal karyotypes, and GSE6891 contains 537 patients with unknown karyotypes and specific recurrent chromosomal translocations and/or gene mutations. Our results are shown in Figure S3. In all datasets, we detected aberrant expressions of IRX1, IRX3, and IRX5, while IRX2 and IRX4 showed only inconsistent or low expression levels. Of note, IRX6 was not represented by the applied arrays for these datasets. Thus, we detected the aberrant overexpression of IRX1 and ectopic activity of IRX3 and IRX5 in AML patient subsets.

IRX1 is located at chromosomal locus 5p15, while IRX3 and IRX5 are neighbors at 16q12, indicating possible coregulation. The analysis of co-expression for IRX1, IRX3, and IRX5 in AML patients revealed significant correlations for IRX3 and IRX5 using datasets GSE15434 and GSE6891 (Figure S4). Thus, the combined expression of IRX3 and IRX5 may be mediated by their genomic proximity, albeit lacking in patients with MLL-rearrangements. Taken together, TALE homeobox genes IRX1, IRX3, and IRX5 are aberrantly expressed in AML patient subsets, which differ in karyotypes and gene mutations.

To find suitable models for functional studies, we screened IRX gene activities using our published RNA-seq dataset LL-100, which contains 34 myeloid and 66 lymphoid leukemia/lymphoma cell lines [22]. Significant RNA expression levels in myeloid cell lines were detected just for IRX1, IRX3, and IRX5 (Figure S5), corresponding to the AML patient data. Interestingly, the IRX1 expression was restricted to cell lines derived from megakaryoblastic AML, while expressions of both IRX3 and IRX5 were elevated in cell lines derived from myelomonocytic AML. However, megakaryoblastic cell line MEGAL expressed IRX3 and IRX5 but not IRX1, representing an exception to that rule. RQ-PCR analysis of selected AML cell lines confirmed the RNA-seq data, showing IRX1 expressions in CMK, M07e, MKPL1, and UT-7 and IRX3/IRX5 expressions in MEGAL and OCI-AML3 (Figure 2). An additional comparison with primary cells from the cerebellum, kidney, lung, and salivary gland showed similar or even higher RNA expression levels in the cell lines, demonstrating significant overexpression of these genes in AML (see also Figure S2). Finally, Western blot analysis of IRX1 and IRX3 confirmed the transcript data at the protein level (Figure 2), endorsing the analyzed cell lines as models for functional studies.

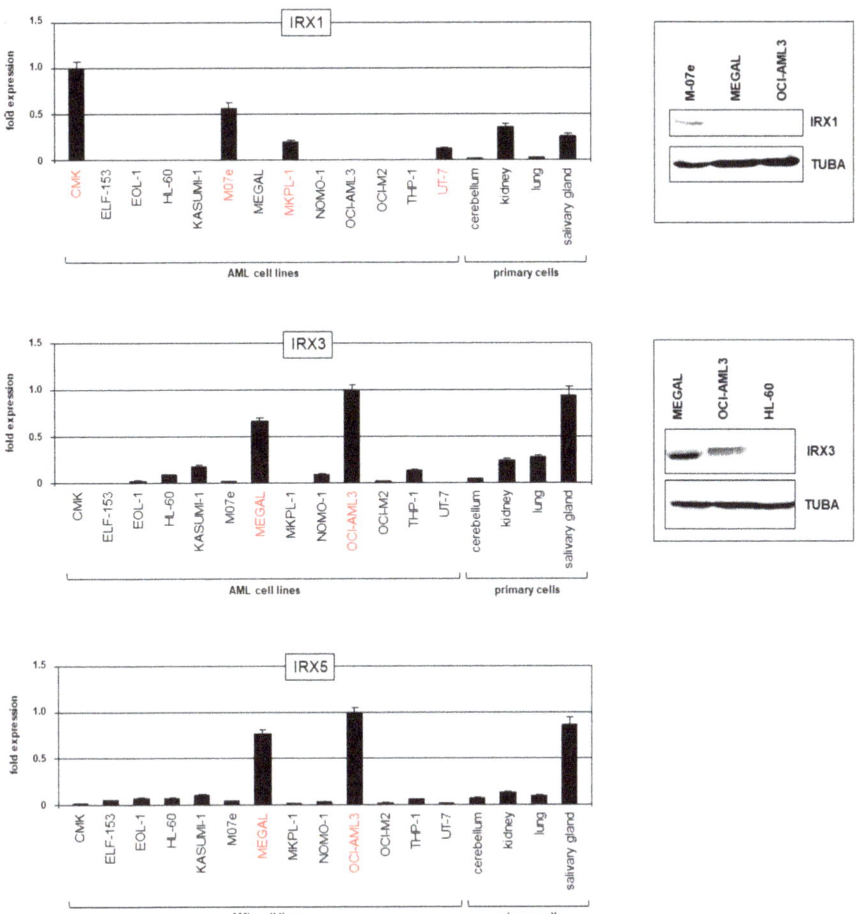

Figure 2. IRX gene activities in AML cell lines. Expression analyses of TALE homeobox genes IRX1, IRX3, and IRX5 in AML cell lines and primary controls as performed by RQ-PCR (**left**). Cell lines showing significant expression levels of the corresponding IRX genes are shown in red. P-values are indicated by asterisks. Western blot analyses (**right**) were performed for IRX1 and IRX3 in selected AML cell lines. TUBA served as loading control.

2.3. Chromosomal and Genomic Analyses of the Gene Loci for IRX1, IRX3, and IRX5 in AML

In hematopoietic malignancies, aberrantly activated oncogenes including homeobox genes are frequently targeted by chromosomal rearrangements [15,16]. To analyze if this deregulating mechanism underlies the activation of IRX genes in AML, we inspected the published karyotypes of the corresponding cell lines (www.DSMZ.de, accessed on 15 February 2022) and shortlisted i(5)(p10) in OCI-AML3 and MKPL-1 and del(16)(q13q23) in MEGAL, which may have activating impacts. In addition, we performed genomic profiling analysis of AML cell lines CMK, M-07e, MKPL-1, UT-7, MEGAL, and OCI-AML3. The data for chromosomes 5 and 16 are shown in Figure 3. IRX1 (5p15) is duplicated together with the whole short arm in OCI-AML3 and MKPL-1, consistent with the karyotype data showing the formation of isochromosomes for the short arms. However, OCI-AML3 is IRX1-negative, discounting the likely contribution of this aberration to its activation. In contrast, IRX3 and IRX5 are located at 16q12 and considerably amplified in MEGAL. The data showed several amplicons covering nearly the complete long arm of chromosome 16.

Moreover, RNA-seq data for the transcribed enhancer locus FTO, which is located adjacent to IRX3 and regulates IRX3/IRX5 activity [23,24], showed enhanced expression restricted to MEGAL (Figure S5). Thus, genomic amplification of IRX3/IRX5 together with FTO may underlie their activation in megakaryoblastic AML cell line MEGAL. Despite these substantial rearrangements of chromosome 16 in MEGAL, RT-PCR excluded the generation of AML-specific fusion gene CBFb-MYH11 via inv(16)(p13q22), confirming the cytogenetic findings (Figure S6A).

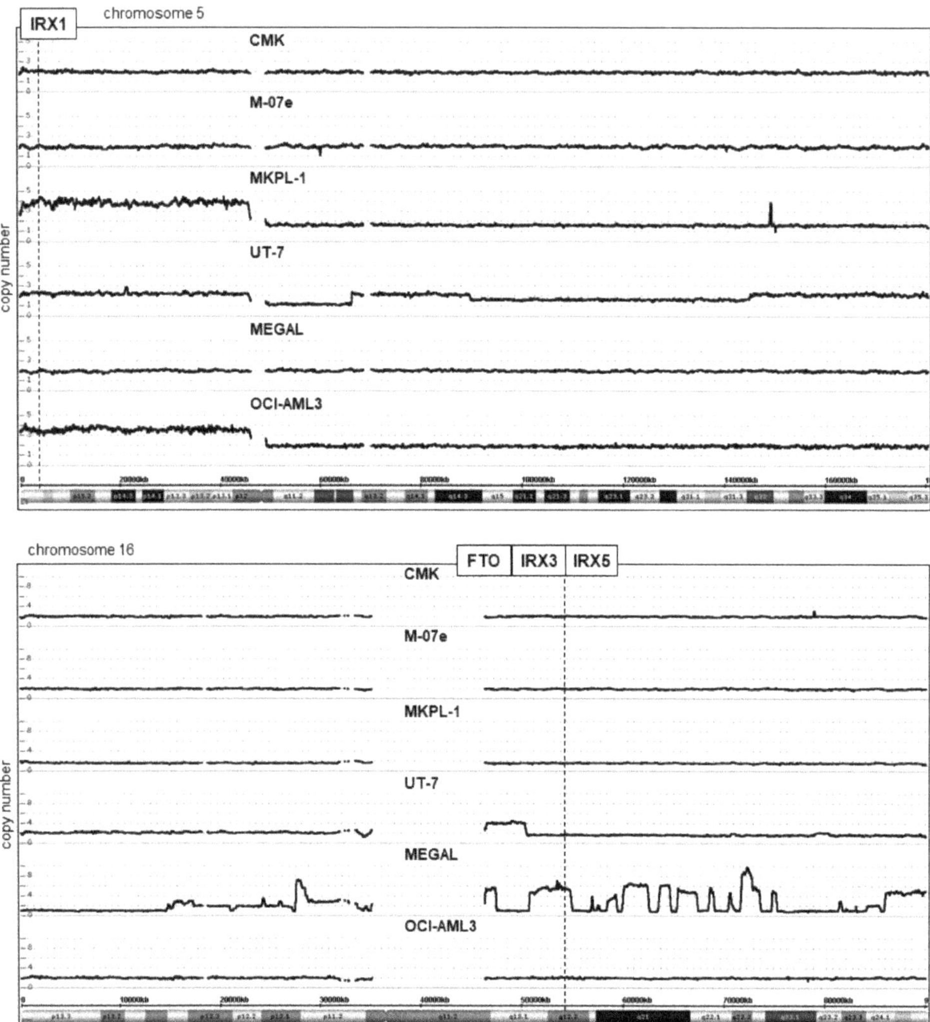

Figure 3. Genomic analysis of AML cell lines. Copy number states for chromosome 5 (**above**) and chromosome 16 (**below**) of IRX1-positive cell lines CMK, M-07e, MKPL-1, and UT-7, and of IRX3/IRX5-positive AML cell lines MEGAL and OCI-AML3 were determined by genomic profiling analysis. A copy number gain for IRX1 (located at 5p15) was detected in MKPL-1 and OCI-AML3. Amplification of FTO, IRX3, and IRX5 (16q12) was detected among multiple complex chromosome 16 rearrangements in MEGAL.

2.4. GATA1 and GATA2 Activate IRX1 in AML Cell Lines

To identify potential regulators and target genes of aberrantly expressed IRX1, we compared LL-100 RNA-seq expression data from IRX1-positive (CMK, M07e, MKPL-1, and UT-7) with IRX3/IRX5-positive (MEGAL and OCI-AML3) AML cell lines. The results are shown in Table S1. Gene set annotation analysis of the top-1000 upregulated genes in IRX1-positve cell lines revealed the statistically significant GO-term erythrocyte differentiation ($p = 0.0052$) and the associated master genes GATA1, GATA2, KLF1, and TAL1, which were chosen for analysis in more detail.

In developing imaginal discs of the fruit fly Drosophila melanogaster, GATA factor pannier regulates the IRX-cluster Iro-C [25], spotlighting this relationship for consideration in humans. RQ-PCR analysis of GATA1 and GATA2 confirmed the higher expression and thus correlating IRX1 levels in cell lines CMK, M07e, MKPL-1, and UT-7. In addition, the siRNA-mediated knockdown of GATA1 and GATA2 in M-07e and MKPL-1 cells resulted in the concomitant downregulation of IRX1 (Figure 4), demonstrating an activating impact of both factors.

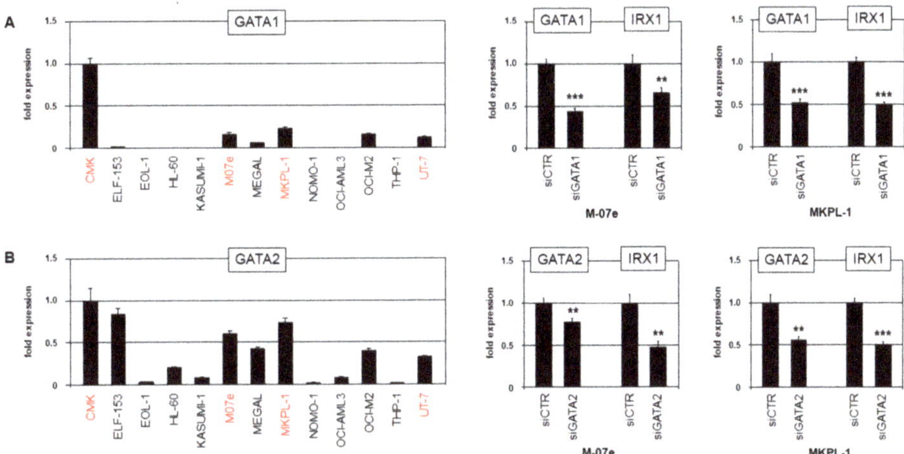

Figure 4. Transcriptional regulation of IRX1. Expression analysis of (**A**) GATA1 and (**B**) GATA2 by RQ-PCR in selected AML cell lines (**left**). Transcript levels are indicated in relation to CMK. Cell lines expressing significant levels of IRX1 are highlighted in red. SiRNA-mediated knockdown of (**A**) GATA1 and (**B**) GATA2 demonstrates activations of IRX1 and IRX5 in AML cell lines M-07e and MKPL-1 (**right**). *p*-values are indicated by asterisks (** $p < 0.01$, *** $p < 0.001$).

Our genomic profiling data showed for cell line CMK a duplication of the short arm of chromosome X, which includes the GATA1 locus at Xp11 (Figure S7). CMK expressed the highest GATA1 RNA levels, indicating that this genomic aberration may contribute indirectly to IRX1 activation. An additional chromosomal aberration revealed by genomic profiling was an amplification at 19p13 in cell lines MKPL-1 and UT-7 (Figure S7). This amplicon includes the erythropoietin receptor encoding gene EPOR and may underlie its overexpression, as shown by RNA-seq data (Figure S5). EPOR and GATA1 are mutual activators and thus may also support IRX1 expression [26]. Finally, expression profiling data from various myelopoietic stages (dataset GSE42519) showed elevations of GATA1, GATA2, and EPOR in IRX1-positive MEPs, while earlier progenitors and later myeloid stages expressed decreased levels (Figure S8). Thus, master factors GATA1 and GATA2 are prominently co-expressed along with IRX1 in MEPs and megakaryoblastic AML cell lines, and activates IRX1 expression.

2.5. HOXA10 and BMP2-Signaling Activate IRX3 and IRX5 in AML

To identify potential regulators and target genes of aberrantly expressed IRX3 and IRX5, we then compared RNA-seq data from IRX3/IRX5-positive with IRX1-positive AML cell lines (Table S2). Statistically significant GO-terms identified by gene set annotation analysis of the top-1000 upregulated genes in IRX3/IRX5-positive cell lines included the negative regulation of myeloid cell differentiation ($p = 0.0029$), SMAD protein signal transduction ($p = 0.05$), and negative regulation of apoptotic processes ($p = 0.00007$). Among the top-1000 upregulated genes, we chose the following candidates for further studies: BCL2, BMP2, HOXA10, and NUP214.

Elevated NUP214 expression correlated with the presence of fusion gene SET-NUP214 in MEGAL, which reportedly activates HOXA genes in T-cell leukemia [27,28]. Accordingly, we confirmed the chromosomal deletion at 9q34 by genomic profiling and the generated fusion gene by RT-PCR in this cell line (Figure S6B).

In AMLs containing MLL-AF4 rearrangements, HOXA genes have been shown to correlate with the expression of IRX3 but not with IRX1 [13,14,29–31], suggesting that particular HOXA-members may support IRX3/IRX5 activation. RQ-PCR analysis of HOXA10 showed low and high RNA expression levels in IRX3/IRX5-positive cell lines MEGAL and OCI-AML3, respectively, while IRX1-positive cell lines CMK, M07e, MKPL-1, and UT-7 tested negative (Figure 5A). Furthermore, siRNA-mediated knockdown of HOXA10 in OCI-AML3 and MEGAL resulted in the concomitant downregulation of IRX3 and IRX5, demonstrating an activating input (Figure 5A). Expression profiling data covering multiple myelopoietic stages (dataset GSE42519) showed an elevated HOXA10 expression in early stages, which decreased after the MEP-stage (Figure S8). Thus, HOXA10 was highly expressed in myeloid progenitors including MEPs and activated the expression of IRX3/IRX5 in AML ectopically.

As BMP2-signaling has been shown to regulate IRX gene IRO3 in the organizer region during gastrulation in zebrafish [32], we decided to explore this relationship in humans. RQ-PCR analysis of BMP2 confirmed the conspicuous expression in OCI-AML3 (Figure 5B). Furthermore, quantification of the BMP2 protein in cell culture supernatants revealed significant levels in OCI-AML3, while M-07e and MEGAL tested negative. Additional treatment of OCI-AML3 with recombinant BMP2 upregulated both IRX3 and IRX5 (Figure 5B), demonstrating an activating input. The TFs JUNB and SMAD4 operate downstream in BMP2-signaling [33,34]. Accordingly, the siRNA-mediated knockdown of JUNB and SMAD4 in OCI-AML3 resulted in the concomitant downregulation of IRX3 and IRX5, showing an activating role of these factors (Figure 5C). However, RQ-PCR analysis of JUNB indicated no correlation with IRX3/IRX5 expression (Figure 5D). However, the suppression of BMP2-activity by the treatment of OCI-AML3 with an inhibitory antibody resulted in the concomitant downregulation of IRX3, IRX5, and JUNB, while HOXA10 showed no significant alteration (Figure 5D).

Taken together, we identified HOXA10 and BMP2-signaling via JUNB and SMAD4 as activating factors for IRX3 and IRX5 in AML. The correlation of SET-NUP214, HOXA, and BMP2-signaling with IRX3/IRX5 expression in MEGAL and OCI-AML3, respectively, indicated that diverse aberrant mechanisms may underlie the ectopic activation of these TALE homeobox genes.

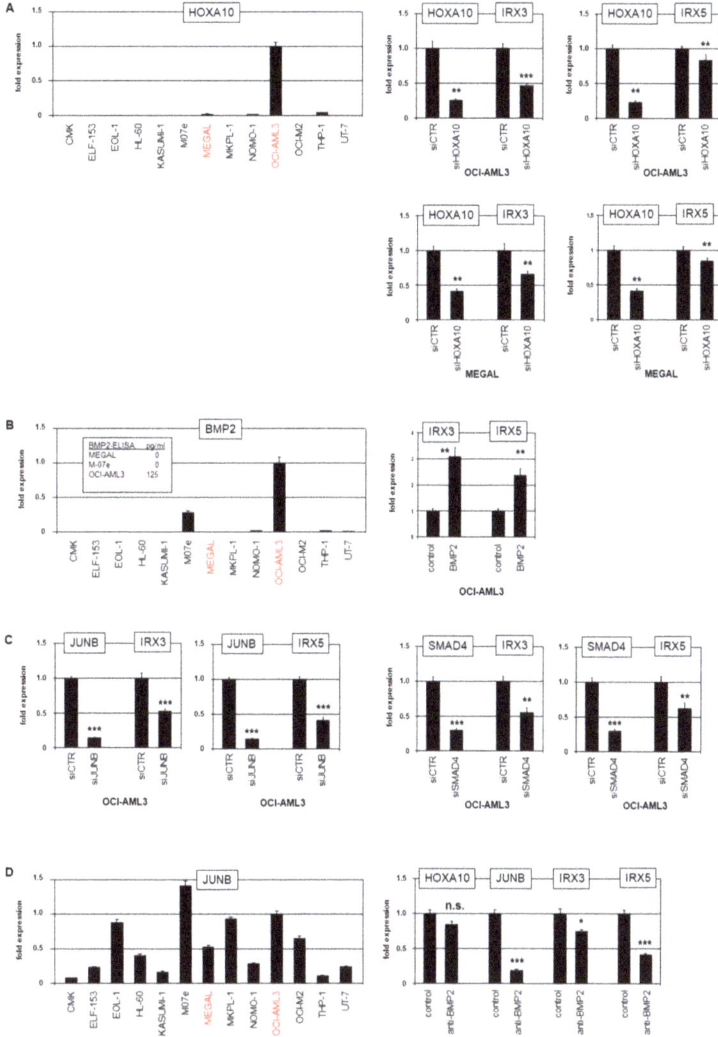

Figure 5. Transcriptional regulation of IRX3 and IRX5. (**A**) RQ-PCR analysis of HOXA10 in AML cell lines (**left**). Transcript levels are indicated in relation to OCI-AML3, and IRX3/IRX5-positive cell lines are indicated in red. RQ-PCR analysis of OCI-AML3 (**above**) and MEGAL (**below**) showed downregulation of IRX3 and IRX5 after siRNA-mediated knockdown of HOXA10 (**right**). (**B**) RQ-PCR analysis of BMP2 in AML cell lines (**left**). Transcript levels are indicated in relation to OCI-AML3. ELISA results for BMP2 protein levels in AML cell line supernatants are inserted. RQ-PCR analysis of OCI-AML3 treated with BMP2 resulted in the upregulation of IRX3 and IRX5 (**right**). (**C**) RQ-PCR analysis of OCI-AML3 after siRNA-mediated knockdown of JUNB (**left**) and SMAD4 (**right**) showed downregulation of IRX3 and IRX5. (**D**) RQ-PCR analysis of JUNB in AML cell lines (**left**). Transcript levels are indicated in relation to OCI-AML3. RQ-PCR analysis of OCI-AML3 treated with inhibitory BMP2-antibody resulted in downregulation of JUNB, IRX3, and IRX5, while HOXA10 was not significantly affected (**right**). p-values are indicated by asterisks (* $p < 0.05$, ** $p < 0.01$, *** $p < 0.001$, n.s. not significant).

2.6. IRX1 and IRX3 Differ in Target Gene Regulation

For target gene analysis of IRX1 in AML, we tested the erythropoietic differentiation genes identified in megakaryoblastic cell line MKPL-1. The SiRNA-mediated knockdown of IRX1 effected the significant downregulation of KLF1 and TAL1 while sparing GATA1 and GATA2 (Figure 6A). Thus, IRX1 activated the erythroid master TFs KLF1 and TAL1, which may disturb megakaryopoiesis if aberrantly activated. Interestingly, both genes are also prominently expressed in MEPs (Figure S8), suggesting that this regulatory relationship may play a physiological role in these progenitor cells.

To examine if the differentiation factors GATA1, GATA2, KLF1, and TAL1 are regulated by IRX3, we used myelomonocytic cell line OCI-AML3. The SiRNA-mediated knockdown of IRX3 resulted in the elevated expression of GATA1 and GATA2, while TAL1 remained unaffected and KLF1 silent (Figure 6B). Thus, IRX3 suppressed the expression of the myeloid differentiation factors GATA1 and GATA2, unlike erythroid factors TAL1 and KLF1. A potential impact of IRX3 in myeloid development was examined by the quantification of differentiation markers. The knockdown of IRX3 in OCI-AML3 resulted in the significant upregulation of CD11b/ITGAM and CD14 (Figure 6C). Furthermore, morphological inspection of these treated cells showed nuclear alterations indicative for myeloid differentiation (Figure 6D), collectively demonstrating an inhibitory role of IRX3 for these processes.

Functional live-cell imaging analysis of OCI-AML3 cells treated for IRX3-knockdown indicated that IRX3 inhibited apoptosis while sparing proliferation (Figure 6E). However, IRX1 and IRX3 knockdown experiments indicated that this anti-apoptotic effect is not mediated by the transcriptional activation of BCL2 (Figure 6A,B). Of note, BCL2 showed the conspicuous downregulation in MEPs (Figure S8), indicating potential sensitivity to apoptosis of these progenitors and conceivably of any derived malignant cells.

The observed suppressive activity of IRX3 prompted scanning downregulated genes in IRX3/IRX5-positive cell lines as well. This exercise revealed BMP2-inhibitor FST, which was downregulated in OCI-AML3 and MEGAL (Table S1, Figure S5). Accordingly, siRNA-mediated knockdown of IRX3 in OCI-AML3 boosted FST expression (Figure 6B), showing that IRX3 inhibited FST. Thus, IRX3 operated in AML as both gene activator and suppressor.

Recently, we reported that NKL homeodomain factors NKX2-3 and NKX2-4 deregulate megakaryocytic-erythroid differentiation factor FLI1 in AML, which shows conspicuous downregulation in MEPs (Figure S8) [35]. However, knockdown experiments for IRX1 and IRX3 excluded FLI1 deregulation (Figure 6A,B), demonstrating functional differences in downstream activities of NKL and TALE homeo-oncogenes in AML.

Taken together, IRX1 and IRX3 showed significant differences in target gene regulation: IRX1 activated KLF1 and TAL1, while IRX3 inhibited GATA1 and GATA2. However, all represent master TFs involved in megakaryocytic-erythroid differentiation. Furthermore, the IRX3-mediated suppression of FST may create an activating feedback loop as BMP2-signaling supported IRX3 expression.

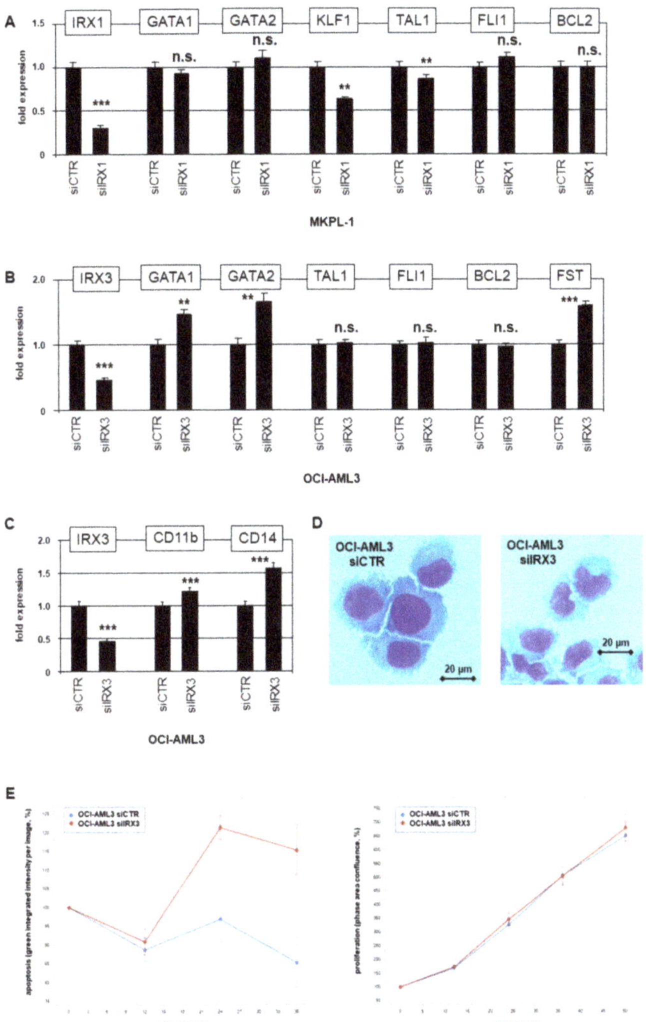

Figure 6. Target gene analyses for IRX1 and IRX3 in AML cell lines. (**A**) RQ-PCR analysis of MKPL-1 treated for siRNA-mediated knockdown of IRX1 resulted in downregulation of KLF1 and TAL1, while GATA1, GATA2, FLI1, and BCL2 remained unaffected. Thus, IRX1 activates KLF1 and TAL1. (**B**) RQ-PCR analysis of OCI-AML3 treated for siRNA-mediated knockdown of IRX3 resulted in upregulation of GATA1, GATA2, and FST, while TAL1, FLI1 and BCL2 remained unaffected. Thus, IRX3 inhibits GATA1, GATA2, and FST. (**C**) RQ-PCR analysis of OCI-AML3 treated for siRNA-mediated knockdown of IRX3 resulted in upregulation of myeloid differentiation marker CD11b/ITGAM and CD14. (**D**) SiRNA-mediated knockdown of IRX3 in OCI-AML3 cells induced morphological alterations of their nuclei, as shown by Giemsa–May–Grünwald staining. (**E**) Live-cell imaging analyses of OCI-AML3 treated for siRNA-mediated knockdown of IRX3 showed increased levels of apoptosis (**left**), while proliferation remained unaffected (**right**). Standard deviations are indicated as bars. *p*-values are indicated by asterisks (** *p* < 0.01, *** *p* < 0.001, n.s. not significant).

3. Discussion

In this study, we established the myeloid TALE-code, representing a TALE homeobox gene expression pattern in normal myelopoiesis. Thus, we completed this signature for the entire hematopoietic system [19]. According to our previously generated NKL-code, we used the same approach to analyze public expression data for TALE homeobox genes in hematopoiesis [18]. These codes show (i) physiological activities of selected homeobox gene groups in hematopoietic entities, and (ii) a blueprint permitting the evaluation of homeobox gene expressions in corresponding malignancies of patients.

The myeloid TALE-code comprises eleven TALE homeobox genes (Figure 1). Interestingly, one of these, PBX1, was expressed in most progenitors and terminal differentiated myeloid cells and only silenced in granulopoiesis. This expression pattern contrasts with that reported in lymphopoiesis, in which all terminal-differentiated lymphocytes downregulate PBX1 [19]. In accordance with these physiological data, the aberrant maintenance of PBX1 activity in developing B-cells contributes to the generation of pre-B-cell leukemia or Hodgkin lymphoma [19,36–38]. In the myeloid lineage, the aberrant expression of PBX1 has been associated with severe congenital neutropenia. Its repression is mediated by GFI1, whose loss reactivates PBX1 in addition to MEIS1 and HOXA genes [39], providing a mechanistic explanation for the aberrant expression of PBX1 in granulopoiesis. Thus, TALE-code data serve to elucidate clinical findings in various hematopoietic malignancies.

In addition, our myeloid TALE-code data revealed a conspicuous expression pattern for IRX1, which was restricted to the MEP stage. This TALE homeobox gene represented the only IRX gene active in normal hematopoiesis. Furthermore, the aberrant expression of IRX genes in AML patients and cell lines was detected for IRX1, IRX3, and IRX5. The deregulated activity of IRX genes in AML has been reported in previous studies as well. Accordingly, the aberrant expression of IRX1 plays an oncogenic role in cases with particular MLL rearrangements, and IRX3 inhibits myelomonocytic differentiation, while IRX2 supports myeloid differentiation [13,14,40]. Thus, the published results together with our data indicate functional differences between IRX factors in AML.

The human genome encodes six IRX genes that are arranged in two clusters. Clustering of the homeobox genes is evolutionarily conserved and often connected with the coregulation of gene neighbors, as described for HOX, DLX, and IRX genes [10,41,42]. They share regulatory elements as demonstrated for IRX3 and IRX5, which are controlled by the transcribed enhancer locus FTO [23,24]. AML cell line MEGAL expressed IRX3 and IRX5 ectopically and showed an amplification of these genes together with FTO at 16q12. This chromosomal aberration was part of a chain of consecutive amplicons at 16q, signifying chromothripsis. This type of cataclysmic genomic rearrangement affects single chromosomes or chromosomal arms and has been described in myeloid and lymphoid malignancies [43,44]. Previously reported aberrations altering chromosome 16 in AML include inv(16)(p13q22) and del(16)(q11) [45–47]. However, our data excluded for MEGAL the presence of both del(16)(q11) and inv(16)(p13q22), which generates the fusion gene CBFb-MYH11.

Our additional findings concerning the regulation and function of IRX genes in AML are summarized in Figure 7. The data include chromosomal aberrations and involvement of the myeloid master factors GATA1, GATA2, HOXA10, KLF1, and TAL1, generating normal and aberrant gene networks, which control differentiation processes. We showed that GATA1 and GATA2 performed an activating role in IRX1 expression, which, in turn, activated KLF1 and TAL1. In contrast, HOXA10 activated IRX3 and IRX5, while IRX3 inhibited GATA1 and GATA2 expression.

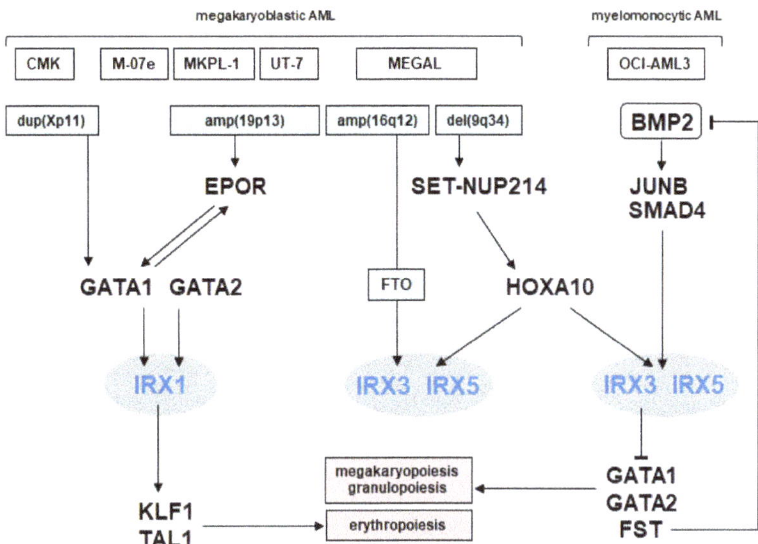

Figure 7. Gene regulatory network of IRX1 and IRX3/IRX5 in AML. The figure summarizes the results of this study. TALE homeobox genes IRX1, IRX3, and IRX5 are located centrally, chromosomal aberrations lie upstream, and developmental TFs and BMP2-signaling components both upstream and downstream. Thus, these IRX genes are part of a regulatory gene network, controlling myeloid differentiation.

GATA1 and GATA2 are fundamental regulators in myelopoiesis and interact with TAL1 in TF-complexes [48–50]. GATA1 drives granulopoiesis, mast cell differentiation, and megakaryopoiesis [51–53]. GATA2 plays basic roles in stem and progenitor cells, including erythroid progenitors, and becomes substituted by GATA1 during development [48]. The mutation or downregulation of GATA2 contributes to the generation of myeloid malignancies including AML [48,54,55]. Thus, the downregulation of GATA1 and GATA2 by IRX3 may represent a novel type of oncogenic alteration in subsets of myelomonocytic AML. Consistent with our data, IRX3 blocks myeloid differentiation [14].

HOXA genes including HOXA10 represent key developmental regulators in myelopoiesis, and their deregulation plays a role in acute leukemias containing MLL-rearrangements and SET-NUP214 fusions [27,56]. We showed that HOXA10 activated ectopic expressions of IRX3 and IRX5 in AML. This finding corresponds to reported correlations of IRX3 to elevated and of IRX1 to decreased HOXA levels in AML [14,29–31]. Furthermore, we showed that IRX3 and IRX5 were activated by BMP2-signaling via JUNB and SMAD4, thus representing an additional mechanism of IRX gene deregulation. FST inhibits the activity of BMP proteins and thus BMP-signaling [34,57]. BMP4 supports megakaryopoiesis and BMP2 erythropoiesis [58,59], demonstrating the developmental impact of this signaling pathway. In T-cell leukemia, the ectopic activity of CHRDL1 has been shown to inhibit BMP-signaling, which results in the aberrant expression of NKL homeobox gene MSX1 [60]. Thus, aberrant deregulation of the BMP pathway plays a common oncogenic role in leukemia.

KLF1 and TAL1 are master genes for erythroid development and suppressors of megakaryopoiesis [61,62]. Our data showed that both genes were activated by IRX1, which may, thus, represent a physiological relationship. On the other hand, the aberrant activation of KLF1 and TAL1 may illustrate an oncogenic function of IRX1 in megakaryoblastic AML. The CBFb-MYH11 fusion gene reportedly deregulates the activity of GATA2 and KLF1, and thus inhibits megakaryopoiesis and erythropoiesis [63]. Here, while excluding the presence of CBFb-MYH11 fusion, we detected the aberrant activity of IRX3/IRX5, an alternative mechanism to perturb myeloid differentiation via KLF1 deregulation.

FLI1 represents an additional myeloid master factor conspicuously downregulated in MEPs, which activates megakaryopoiesis and inhibits erythropoiesis [35,64]. Here, we excluded a regulatory impact of IRX factors in FLI1 expression. However, in our previous study of deregulated NKL homeobox genes, we showed that FLI1 is aberrantly activated by NKX2-3 in megakaryoblastic AML and aberrantly inhibited by NKX2-4 in erythroblastic AML [35]. These observations suggest that MEPs are developmentally susceptible to generate AML and that aberrantly activated homeodomain factors deregulate components of gene regulatory networks, controlling basic myeloid differentiation processes.

Finally, we showed that IRX3 inhibited apoptosis in AML. This effect was not performed by transcriptional activation of BCL2, although BCL2 is physiologically downregulated in MEPs. However, in lung cancer, IRX1 inhibits expression of the pro-apoptotic gene BAX [65], supporting a role of IRX factors in the (de)regulation of cell survival.

Taken together, our study revealed basic impacts of IRX factors in normal and aberrant myelopoiesis. IRX1 is a novel physiological player in hematopoiesis and part of a gene regulatory network controlling the differentiation of megakaryocytes, erythrocytes, and granulocytes. Aberrantly activated IRX1, IRX3, and IRX5 may disturb or deregulate developmental processes in myelopoiesis, driving the generation of AML subsets. Thus, our study contributes to understanding normal and abnormal processes in myelopoiesis.

4. Materials and Methods

4.1. Bioinformatic Analyses of Expression Profiling and RNA-Seq Data

Expression data for normal cell types were obtained from Gene Expression Omnibus (GEO, www.ncbi.nlm.nih.gov, accessed on 15 February 2022), using expression profiling datasets GSE42519, GSE22552, GSE109348, and GSE24759 [66–69], in addition to RNA-seq data from The Human Protein Atlas (www.proteinatlas.org, accessed on 15 February 2022) [70]. For screening of cell lines, we exploited RNA-sequencing data from 100 leukemia/lymphoma cell lines (termed LL-100), available at ArrayExpress (www.ebi.ac.uk/arrayexpress, accessed on 15 February 2022) via E-MTAB-7721 [22]. Gene expression profiling data from AML patients were examined using datasets GSE19577, GSE15434, and GSE6891 [71–73]. To parse biological functions of 1000 shortlisted genes, gene set annotation enrichment analysis was performed using DAVID bioinformatics resources (www.david.ncifcrf.gov, accessed on 15 February 2022) [74]. Analysis of gene co-expression in profiling datasets was performed by Spearman correlation using R-based tools.

4.2. Cell Lines and Treatments

Cell lines are held at the DSMZ (Braunschweig, Germany) and cultivated as described (www.DSMZ.de, accessed on 15 February 2022). All cell lines had been authenticated and tested negative for mycoplasma infection. Modification of gene expression levels was performed using gene-specific siRNA oligonucleotides with reference to AllStars negative Control siRNA (siCTR) obtained from Qiagen (Hilden, Germany). SiRNAs (80 pmol) were transfected into 1×10^6 cells by electroporation using the EPI-2500 impulse generator (Fischer, Heidelberg, Germany) at 350 V for 10 ms. Electroporated cells were harvested after 20 h of cultivation. Cell treatments were performed for 20 h using 20 ng/mL of recombinant BMP2 (R & D Systems, Wiesbaden, Germany, #355-BM-010/CF) or 20 µg/mL of inhibitory monoclonal anti-BMP2 antibody (R & D Systems, #MAB3552).

For cytological analyses, cell lines were stained with Giemsa–May–Grünwald as follows: Cells were spun onto microscope slides and fixed for 5 min with methanol. Subsequently, they were stained for 3 min with May–Grünwald's eosin-methylene blue modified solution (Merck, Darmstadt, Germany) diluted in Titrisol (Merck), and for 15 min with Giemsa's azur eosin methylene blue solution (Merck). Images were captured with an Axion A1 microscope using Axiocam 208 color and software ZEN 3.3 blue edition (Zeiss, Göttingen, Germany). For functional testing, treated cells were analyzed with the

IncuCyte S3 Live-Cell Analysis System (Essen Bioscience, Hertfordshire, UK). For detection of apoptotic cells, we additionally used the IncuCyte Caspase-3/7 Green Apoptosis Assay diluted at 1:2000 (Essen Bioscience). Live-cell imaging experiments were performed twice with fourfold parallel tests.

4.3. Polymerase-Chain-Reaction (PCR) Analyses

Total RNA was extracted from cultivated cell lines using TRIzol reagent (Invitrogen, Darmstadt, Germany). Primary human total RNA derived from the cerebellum, kidney, lung, and salivary gland was purchased from Biochain/BioCat (Heidelberg, Germany). cDNA was synthesized using 1 µg of RNA, random priming, and Superscript II (Invitrogen). Real-time quantitative (RQ)-PCR analysis was performed using the 7500 Real-time System and commercial buffer and primer sets (Applied Biosystems/Life Technologies, Darmstadt, Germany). For normalization of expression levels, we quantified the transcripts of TATA box binding protein (TBP). Quantitative analyses were performed as biological replicates and measured in triplicate. Standard deviations are presented in the figures as error bars. Statistical significance was assessed by Student's *t*-test (two-tailed) and the calculated p-values indicated by asterisks (* $p < 0.05$, ** $p < 0.01$, *** $p < 0.001$, n.s. not significant).

For detection of CBFb-MYH11 and SET-NUP214 fusion transcripts, we performed reverse transcription (RT)-PCR, using the following oligonucleotides as reported previously: CBFb-for 5′-GGGCTGTCTGGAGTTTGATG-3′, and MYH11-rev 5′-CTTGAGCGCCTGCA TGTT-3′, SET-for 5′-TGACGAAGAAGGGGATGAGGAT-3′, NUP214-rev 5′-ATCATTCACA TCTTGGACAGCA-3′ [28,44]. As a control, we analyzed ETV6, using ETV6-for 5′-AGGCC AATTGACAGCAACAC-3′ and ETV6-rev 5′-TGCACATTATCCACGGATGG-3′. All oligonucleotides were purchased from Eurofins MWG (Ebersberg, Germany). PCR products were generated using taqpol (Qiagen) and thermocycler TGradient (Biometra, Göttingen, Germany), analyzed by gel electrophoresis, and documented with the Azure c200 Gel Imaging System (Azure Biosystems, Dublin, CA, USA).

4.4. Protein Analysis

Western blots were generated by the semi-dry method. Protein lysates from cell lines were prepared using SIGMAFast protease inhibitor cocktail (Sigma, Taufkirchen, Germany). Proteins were transferred onto nitrocellulose membranes (Bio-Rad, München, Germany) and blocked with 5% dry milk powder dissolved in phosphate-buffered-saline buffer (PBS). The following antibodies were used: alpha-Tubulin (Sigma, #T6199), IRX1 (Biozol, Eching, Germany, #DF3225), and IRX3 (Biozol, #MBS8223417). For loading control, blots were reversibly stained with Poinceau (Sigma) and the detection of alpha-Tubulin (TUBA) was performed thereafter. Secondary antibodies were linked to peroxidase for detection by Western-Lightning-ECL (Perkin Elmer, Waltham, MA, USA). Documentation was performed using the digital system ChemoStar Imager (INTAS, Göttingen, Germany).

The enzyme-linked immunosorbent assay (ELISA) was used to quantify BMP2 protein levels in the supernatant of cell cultures. In addition, 2×10^6 cells were cultured in 2 mL of fresh medium in a 24-well plate. After 24 h, 1 of mL medium was harvested and frozen in aliquots. Quantification was performed using the Quantikine ELISA BMP-2 kit (R & D Systems, #DBP200), as described by the company. Two biological replicates were analyzed in triplicate.

4.5. Karyotyping and Genomic Profiling Analysis

Karyotyping was performed as described previously [75]. For genomic profiling, genomic DNA of AML cell lines was prepared by the Qiagen Gentra Puregene Kit (Qiagen). Labeling, hybridization, and scanning of Cytoscan HD arrays were performed by the Genome Analytics Facility located at the Helmholtz Centre for Infection Research (Braunschweig, Germany), using the manufacturer's protocols (Affymetrix, High Wycombe, UK). Data were interpreted using the Chromosome Analysis Suite software version 3.1.0.15 (Affymetrix, High Wycombe, UK) and copy number alterations were determined accordingly.

Supplementary Materials: The following supporting information can be downloaded at: https://www.mdpi.com/article/10.3390/ijms23063192/s1.

Author Contributions: Conceptualization, S.N.; formal analysis, S.N., C.P., C.M. and R.A.F.M.; investigation, S.N.; writing—original draft preparation, S.N.; writing—review and editing, R.A.F.M. All authors have read and agreed to the published version of the manuscript.

Funding: This research received no external funding.

Institutional Review Board Statement: Not applicable.

Informed Consent Statement: Not applicable.

Data Availability Statement: Data is contained within the article or Supplementary Materials.

Conflicts of Interest: The authors declare no conflict of interest.

References

1. Liggett, L.A.; Sankaran, V.G. Unraveling Hematopoiesis through the Lens of Genomics. *Cell* **2020**, *182*, 1384–1400. [CrossRef] [PubMed]
2. Wilson, N.K.; Foster, S.D.; Wang, X.; Knezevic, K.; Schütte, J.; Kaimakis, P.; Chilarska, P.M.; Kinston, S.; Ouwehand, W.H.; Dzierzak, E.; et al. Combinatorial transcriptional control in blood stem/progenitor cells: Genome-wide analysis of ten major transcriptional regulators. *Cell Stem Cell* **2010**, *7*, 532–544. [CrossRef] [PubMed]
3. Kucinski, I.; Wilson, N.K.; Hannah, R.; Kinston, S.J.; Cauchy, P.; Lenaerts, A.; Grosschedl, R.; Göttgens, B. Interactions between lineage-associated transcription factors govern haematopoietic progenitor states. *EMBO. J.* **2020**, *39*, e104983. [CrossRef] [PubMed]
4. Rothenberg, E.V. Transcriptional Control of Early T and B Cell Developmental Choices. *Annu. Rev. Immunol.* **2014**, *32*, 283–321. [CrossRef] [PubMed]
5. Bürglin, T.R.; Affolter, M. Homeodomain proteins: An update. *Chromosoma* **2015**, *125*, 497–521. [CrossRef] [PubMed]
6. Holland, P.W.H.; Booth, H.A.F.; Bruford, E. Classification and nomenclature of all human homeobox genes. *BMC Biol.* **2007**, *5*, 47. [CrossRef]
7. Bürglin, T.R. Analysis of TALE superclass homeobox genes (MEIS, PBC, KNOX, Iroquois, TGIF) reveals a novel domain conserved between plants and animals. *Nucleic Acids Res.* **1997**, *25*, 4173–4180. [CrossRef]
8. Mukherjee, K.; Bürglin, T.R. Comprehensive analysis of animal tale homeobox genes: New conserved motifs and cases of accelerated evolution. *J. Mol. Evol.* **2007**, *65*, 137–153. [CrossRef]
9. Dard, A.; Reboulet, J.; Jia, Y.; Bleicher, F.; Duffraisse, M.; Vanaker, J.-M.; Forcet, C.; Merabet, S. Human hox proteins use diverse and context-dependent motifs to interact with tale class cofactors. *Cell Rep.* **2018**, *22*, 3058–3071. [CrossRef]
10. Kerner, P.; Ikmi, A.; Coen, D.; Vervoort, M. Evolutionary history of the iroquois/Irx genes in metazoans. *BMC Evol. Biol.* **2009**, *9*, 74. [CrossRef]
11. Bosse, A.; Zülch, A.; Becker, M.B.; Torres, M.; Gómez-Skarmeta, J.L.; Modolell, J.; Gruss, P. Identification of the vertebrate Iroquois homeobox gene family with overlapping expression during early development of the nervous system. *Mech Dev.* **1997**, *69*, 169–181. [CrossRef]
12. Zhang, P.; Yang, H.; Wang, X.; Wang, L.; Cheng, Y.; Zhang, Y.; Tu, Y. The genomic organization and function of IRX1 in tumorigenesis and development. *Cancer Transl. Med.* **2017**, *3*, 29–33.
13. Kühn, A.; Löscher, D.; Marschalek, R. The IRX1/HOXA connection: Insights into a novel t(4;11)-specific cancer mechanism. *Oncotarget* **2016**, *7*, 35341–35352. [CrossRef] [PubMed]
14. Somerville, T.D.; Simeoni, F.; Chadwick, J.A.; Williams, E.L.; Spencer, G.J.; Boros, K.; Wirth, C.; Tholouli, E.; Byers, R.J.; Somervaille, T.C. Derepression of the Iroquois Homeodomain Transcription Factor Gene IRX3 Confers Differentiation Block in Acute Leukemia. *Cell Rep.* **2018**, *22*, 638–652. [CrossRef]
15. Dash, A.; Gilliland, D. Molecular genetics of acute myeloid leukaemia. *Best Pr. Res. Clin. Haematol.* **2001**, *14*, 49–64. [CrossRef]
16. Graux, C.; Cools, J.; Michaux, L.; Vandenberghe, P.; Hagemeijer-Hausman, A. Cytogenetics and molecular genetics of T-cell acute lymphoblastic leukemia: From thymocyte to lymphoblast. *Leukemia* **2006**, *20*, 1496–1510. [CrossRef]
17. Alharbi, R.; Pettengell, R.; Pandha, H.S.; Morgan, R. The role of HOX genes in normal hematopoiesis and acute leukemia. *Leukemia* **2012**, *27*, 1000–1008. [CrossRef]
18. Nagel, S. NKL-Code in Normal and Aberrant Hematopoiesis. *Cancers* **2021**, *13*, 1961. [CrossRef]
19. Nagel, S.; Pommerenke, C.; Meyer, C.; MacLeod, R.A.F.; Drexler, H.G. Establishment of the TALE-code reveals aberrantly activated homeobox gene PBX1 in Hodgkin lymphoma. *PLoS ONE* **2021**, *16*, e0246603. [CrossRef]
20. Arber, D.A.; Orazi, A.; Hasserjian, R.; Thiele, J.; Borowitz, M.J.; Le Beau, M.M.; Bloomfield, C.D.; Cazzola, M.; Vardiman, J.W. The 2016 revision to the World Health Organization classification of myeloid neoplasms and acute leukemia. *Blood* **2016**, *127*, 2391–2405. [CrossRef]
21. Nagel, S.; Scherr, M.; MacLeod, R.A.F.; Pommerenke, C.; Koeppel, M.; Meyer, C.; Kaufmann, M.; Dallmann, I.; Drexler, H.G. NKL homeobox gene activities in normal and malignant myeloid cells. *PLoS ONE* **2019**, *14*, e0226212. [CrossRef] [PubMed]

22. Quentmeier, H.; Pommerenke, C.; Dirks, W.G.; Eberth, S.; Koeppel, M.; MacLeod, R.A.F.; Nagel, S.; Steube, K.; Uphoff, C.C.; Drexler, H.G. The LL-100 panel: 100 cell lines for blood cancer studies. *Sci. Rep.* **2019**, *9*, 8218. [CrossRef] [PubMed]

23. Claussnitzer, M.; Dankel, S.N.; Kim, K.-H.; Quon, G.; Meuleman, W.; Haugen, C.; Glunk, V.; Sousa, I.S.; Beaudry, J.L.; Puviindran, V.; et al. FTO Obesity Variant Circuitry and Adipocyte Browning in Humans. *N. Engl. J. Med.* **2015**, *373*, 895–907. [CrossRef] [PubMed]

24. Rask-Andersen, M.; Almén, M.S.; Schiöth, H.B. Scrutinizing the FTO locus: Compelling evidence for a complex, long-range regulatory context. *Qual. Life Res.* **2015**, *134*, 1183–1193. [CrossRef] [PubMed]

25. Calleja, M.; Herranz, H.; Estella, C.; Casal, J.; Lawrence, P.; Simpson, P.; Morata, G. Generation of medial and lateral dorsal body domains by the pannier gene of Drosophila. *Development* **2000**, *127*, 3971–3980. [CrossRef]

26. Chiba, T.; Ikawa, Y.; Todokoro, K. GATA-1 transactivates erythropoietin receptor gene, and erythropoietin receptor-mediated signals enhance GATA-1 gene expression. *Nucleic Acids Res.* **1991**, *19*, 3843–3848. [CrossRef]

27. Van Vlierberghe, P.; van Grotel, M.; Tchinda, J.; Lee, C.; Beverloo, H.B.; van der Spek, P.J.; Stubbs, A.; Cools, J.; Nagata, K.; Fornerod, M.; et al. The recurrent SET-NUP214 fusion as a new HOXA activation mechanism in pediatric T-cell acute lymphoblastic leukemia. *Blood* **2008**, *111*, 4668–4680. [CrossRef]

28. Quentmeier, H.; Schneider, B.; Röhrs, S.; Romani, J.; Zaborski, M.; Macleod, R.A.; Drexler, H.G. SET-NUP214 fusion in acute myeloid leukemia- and T-cell acute lymphoblastic leukemia-derived cell lines. *J. Hematol. Oncol.* **2009**, *2*, 3. [CrossRef]

29. Trentin, L.; Giordan, M.; Dingermann, T.; Basso, G.; Te Kronnie, G.; Marschalek, R. Two independent gene signatures in pediatric t(4;11) acute lymphoblastic leukemia patients. *Eur. J. Haematol.* **2009**, *83*, 406–419. [CrossRef]

30. Stam, R.W.; Schneider, P.; Hagelstein, J.A.; van der Linden, M.H.; Stumpel, D.J.; de Menezes, R.X.; de Lorenzo, P.; Valsecchi, M.G.; Pieters, R. Gene expression profiling-based dissection of MLL translocated and MLL germline acute lymphoblastic leukemia in infants. *Blood* **2010**, *115*, 2835–2844. [CrossRef]

31. Symeonidou, V.; Ottersbach, K. HOXA9/IRX1 expression pattern defines two subgroups of infant MLL-AF4-driven acute lymphoblastic leukemia. *Exp. Hematol.* **2021**, *93*, 38–43.e5. [CrossRef] [PubMed]

32. Kudoh, T.; Dawid, I.B. Role of the *iroquois3* homeobox gene in organizer formation. *Proc. Natl. Acad. Sci. USA* **2001**, *98*, 7852–7857. [CrossRef] [PubMed]

33. Chalaux, E.; López-Rovira, T.; Rosa, J.L.; Bartrons, R.; Ventura, F. JunB is involved in the inhibition of myogenic differentiation by bone morphogenetic protein-2. *J. Biol. Chem.* **1998**, *273*, 537–543. [CrossRef]

34. Ebara, S.; Nakayama, K. Mechanism for the action of bone morphogenetic proteins and regulation of their activity. *Spine* **2002**, *27* (Suppl. 1), S10–S15. [CrossRef] [PubMed]

35. Nagel, S.; Pommerenke, C.; Meyer, C.; MacLeod, R.A.F. NKL Homeobox genes NKX2-3 and NKX2-4 deregulate megakaryocytic-erythroid cell differentiation in AML. *Int. J. Mol. Sci.* **2021**, *22*, 11434. [CrossRef]

36. Nagel, S. The NKL-and TALE-codes represent hematopoietic gene signatures to evaluate deregulated homeobox genes in hodgkin lymphoma. *Hemato* **2022**, *3*, 122–130. [CrossRef]

37. Kamps, M.P. E2A-Pbx1 induces growth, blocks differentiation, and interacts with other homeodomain proteins regulating normal differentiation. *Curr. Top. Microbiol. Immunol.* **1997**, *220*, 25–43.

38. Sanyal, M.; Tung, J.W.; Karsunky, H.; Zeng, H.; Selleri, L.; Weissman, I.L.; Herzenberg, L.A.; Cleary, M.L. B-cell development fails in the absence of the Pbx1 proto-oncogene. *Blood* **2007**, *109*, 4191–4199. [CrossRef]

39. Horman, S.R.; Velu, C.S.; Chaubey, A.; Bourdeau, T.; Zhu, J.; Paul, W.E.; Gebelein, B.; Grimes, H.L. Gfi1 integrates progenitor versus granulocytic transcriptional programming. *Blood* **2009**, *113*, 5466–5475. [CrossRef]

40. Scalea, S.; Maresca, C.; Catalanotto, C.; Marino, R.; Cogoni, C.; Reale, A.; Zampieri, M.; Zardo, G. Modifications of H3K4 methylation levels are associated with DNA hypermethylation in acute myeloid leukemia. *FEBS J.* **2020**, *287*, 1155–1175. [CrossRef]

41. Gaunt, S.J. The significance of Hox gene collinearity. *Int. J. Dev. Biol.* **2015**, *59*, 159–170. [CrossRef] [PubMed]

42. Sumiyama, K.; Tanave, A. The regulatory landscape of the Dlx gene system in branchial arches: Shared characteristics among Dlx bigene clusters and evolution. *Dev. Growth Differ.* **2020**, *62*, 355–362. [CrossRef] [PubMed]

43. Fontana, M.C.; Marconi, G.; Feenstra, J.D.M.; Fonzi, E.; Papayannidis, C.; Ghelli Luserna di Rorá, A.; Padella, A.; Solli, V.; Franchini, E.; Ottaviani, E.; et al. Chromothripsis in acute myeloid leukemia: Biological features and impact on survival. *Leukemia* **2018**, *32*, 1609–1620. [CrossRef] [PubMed]

44. Nagel, S.; Meyer, C.; Quentmeier, H.; Kaufmann, M.; Drexler, H.G.; MacLeod, R.A.F. Chromothripsis in Hodgkin lymphoma. *Genes Chromosom. Cancer* **2013**, *52*, 741–747. [CrossRef]

45. Liu, P.; Tarlé, S.A.; Hajra, A.; Claxton, D.F.; Marlton, P.; Freedman, M.; Siciliano, M.J.; Collins, F.S. Fusion between transcription factor CBF beta/PEBP2 beta and a myosin heavy chain in acute myeloid leukemia. *Science* **1993**, *261*, 1041–1044. [CrossRef]

46. van Dongen, J.J.; Macintyre, E.A.; Gabert, J.A.; Delabesse, E.; Rossi, V.; Saglio, G.; Gottardi, E.; Rambaldi, A.; Dotti, G.; Griesinger, F.; et al. Standardized RT-PCR analysis of fusion gene transcripts from chromosome aberrations in acute leukemia for detection of minimal residual disease. Report of the BIOMED-1 Concerted Action: Investigation of minimal residual disease in acute leukemia. *Leukemia* **1999**, *13*, 1901–1928. [CrossRef]

47. Yamamoto, K.; Nagata, K.; Kida, A.; Hamaguchi, H. Deletion of 16q11 is a recurrent cytogenetic aberration in acute myeloblastic leukemia during disease progression. *Cancer Genet. Cytogenet.* **2001**, *131*, 65–68. [CrossRef]

48. Fujiwara, T. GATA Transcription factors: Basic principles and related human disorders. *Tohoku J. Exp. Med.* **2017**, *242*, 83–91. [CrossRef]
49. Wadman, I.A.; Osada, H.; Grütz, G.G.; Agulnick, A.D.; Westphal, H.; Forster, A.; Rabbitts, T. The LIM-only protein Lmo2 is a bridging molecule assembling an erythroid, DNA-binding complex which includes the TAL1, E47, GATA-1 and Ldb1/NLI proteins. *EMBO J.* **1997**, *16*, 3145–3157. [CrossRef]
50. Thoms, J.A.I.; Truong, P.; Subramanian, S.; Knezevic, K.; Harvey, G.; Huang, Y.; Seneviratne, J.A.; Carter, D.R.; Joshi, S.; Skhinas, J.; et al. Disruption of a GA-TA2-TAL1-ERG regulatory circuit promotes erythroid transition in healthy and leukemic stem cells. *Blood* **2021**, *138*, 1441–1455. [CrossRef]
51. Vyas, P.; Ault, K.; Jackson, C.W.; Orkin, S.H.; Shivdasani, R.A. Consequences of GATA-1 deficiency in megakaryocytes and platelets. *Blood* **1999**, *93*, 2867–2875. [CrossRef] [PubMed]
52. Yu, C.; Cantor, A.B.; Yang, H.; Browne, C.; Wells, R.A.; Fujiwara, Y.; Orkin, S.H. Targeted deletion of a high-affinity GA-TA-binding site in the GATA-1 promoter leads to selective loss of the eosinophil lineage in vivo. *J. Exp. Med.* **2002**, *195*, 1387–1395. [CrossRef] [PubMed]
53. Nei, Y.; Obata-Ninomiya, K.; Tsutsui, H.; Ishiwata, K.; Miyasaka, M.; Matsumoto, K.; Nakae, S.; Kanuka, H.; Inase, N.; Karasuyama, H. GATA-1 regulates the generation and function of basophils. *Proc. Natl. Acad. Sci. USA* **2013**, *110*, 18620–18625. [CrossRef] [PubMed]
54. Hahn, C.N.; Chong, C.E.; Carmichael, C.L.; Wilkins, E.J.; Brautigan, P.J.; Li, X.C.; Babic, M.; Lin, M.; Carmagnac, A.; Lee, Y.K.; et al. Heritable GATA2 mutations associated with familial myelodysplastic syndrome and acute myeloid leukemia. *Nat. Genet.* **2011**, *43*, 1012–1017. [CrossRef] [PubMed]
55. Shimizu, R.; Yamamoto, M. Quantitative and qualitative impairments in GATA2 and myeloid neoplasms. *IUBMB Life* **2019**, *72*, 142–150. [CrossRef]
56. Eklund, E.A. The role of HOX genes in malignant myeloid disease. *Curr. Opin. Hematol.* **2007**, *14*, 85–89. [CrossRef]
57. Tsuchida, K.; Arai, K.Y.; Kuramoto, Y.; Yamakawa, N.; Hasegawa, Y.; Sugino, H. Identification and characterization of a novel folistatin-like protein as a binding protein for the TGF-beta family. *J. Biol. Chem.* **2000**, *275*, 40788–40796. [CrossRef]
58. Jeanpierre, S.; Nicolini, F.E.; Kaniewski, B.; Dumontet, C.; Rimokh, R.; Puisieux, A.; Maguer-Satta, V. BMP4 regulation of human megakaryocytic differentiation is involved in thrombopoietin signaling. *Blood* **2008**, *112*, 3154–3163. [CrossRef]
59. Maguer-Satta, V.; Bartholin, L.; Jeanpierre, S.; Ffrench, M.; Martel, S.; Magaud, J.P.; Rimokh, R. Regulation of human erythropoiesis by activin A, BMP2, and BMP4, members of the TGFbeta family. *Exp. Cell Res.* **2003**, *282*, 110–120. [CrossRef]
60. Nagel, S.; Ehrentraut, S.; Meyer, C.; Kaufmann, M.; Drexler, H.G.; MacLeod, R.A.F. Repressed BMP signaling reactivates NKL homeobox geneMSX1in a T-ALL subset. *Leuk. Lymphoma* **2014**, *56*, 480–491. [CrossRef]
61. Siatecka, M.; Bieker, J.J. The multifunctional role of EKLF/KLF1 during erythropoiesis. *Blood* **2011**, *118*, 2044–2054. [CrossRef] [PubMed]
62. Porcher, C.; Chagraoui, H.; Kristiansen, M.S. SCL/TAL1: A multifaceted regulator from blood development to disease. *Blood* **2017**, *129*, 2051–2060. [CrossRef] [PubMed]
63. Yi, G.; Mandoli, A.; Jussen, L.; Tijchon, E.; van Bergen, M.G.J.M.; Cordonnier, G.; Hansen, M.; Kim, B.; Nguyen, L.N.; Jansen, P.W.T.C.; et al. CBFβ-MYH11 interferes with megakaryocyte differentiation via modulating a gene program that includes GATA2 and KLF1. *Blood Cancer J.* **2019**, *9*, 33. [CrossRef] [PubMed]
64. Li, Y.; Luo, H.; Liu, T.; Zacksenhaus, E.; Ben-David, Y. The ets transcription factor Fli-1 in development, cancer and disease. *Oncogene* **2014**, *34*, 2022–2031. [CrossRef]
65. Küster, M.M.; Schneider, M.A.; Richter, A.M.; Richtmann, S.; Winter, H.; Kriegsmann, M.; Pullamsetti, S.S.; Stiewe, T.; Savai, R.; Muley, T.; et al. Epigenetic inactivation of the tumor suppressor IRX1 occurs frequently in lung adenocarcinoma and its silencing is associated with impaired prognosis. *Cancers* **2020**, *12*, 3528. [CrossRef]
66. Rapin, N.; Bagger, F.O.; Jendholm, J.; Mora-Jensen, H.; Krogh, A.; Kohlmann, A.; Thiede, C.; Borregaard, N.; Bullinger, L.; Winther, O.; et al. Comparing cancer vs normal gene expression profiles identifies new disease entities and common transcriptional programs in AML patients. *Blood* **2014**, *123*, 894–904. [CrossRef]
67. Merryweather-Clarke, A.T.; Atzberger, A.; Soneji, S.; Gray, N.; Clark, K.; Waugh, C.; McGowan, S.J.; Taylor, S.; Nandi, A.K.; Wood, W.G.; et al. Global gene expression analysis of human erythroid progenitors. *Blood* **2011**, *117*, e96–e108. [CrossRef]
68. Lu, N.; Li, Y.; Zhang, Z.; Xing, J.; Sun, Y.; Yao, S.; Chen, L. Human Semaphorin-4A drives Th2 responses by binding to receptor ILT-4. *Nat. Commun.* **2018**, *9*, 742. [CrossRef]
69. Novershtern, N.; Subramanian, A.; Lawton, L.N.; Mak, R.H.; Haining, W.N.; McConkey, M.E.; Habib, N.; Yosef, N.; Chang, C.Y.; Shay, T.; et al. Densely Interconnected Transcriptional Circuits Control Cell States in Human Hematopoiesis. *Cell* **2011**, *144*, 296–309. [CrossRef]
70. Uhlen, M.; Karlsson, M.J.; Zhong, W.; Tebani, A.; Pou, C.; Mikes, J.; Lakshmikanth, T.; Forsström, B.; Edfors, F.; Odeberg, J.; et al. A genome-wide transcriptomic analysis of protein-coding genes in human blood cells. *Science* **2019**, *366*, eaax9198. [CrossRef]
71. Pigazzi, M.; Masetti, R.; Bresolin, S.; Beghin, A.; Di Meglio, A.; Gelain, S.; Trentin, L.; Baron, E.; Giordan, M.; Zangrando, A.; et al. MLL partner genes drive distinct gene expression profiles and genomic alterations in pediatric acute myeloid leukemia: An AIEOP study. *Leukemia* **2011**, *25*, 560–563. [CrossRef] [PubMed]
72. Klein, H.U.; Ruckert, C.; Kohlmann, A.; Bullinger, L.; Thiede, C.; Haferlach, T.; Dugas, M. Quantitative comparison of microar-ray experiments with published leukemia related gene expression signatures. *BMC Bioinform.* **2009**, *10*, 422. [CrossRef] [PubMed]

73. Verhaak, R.G.; Wouters, B.J.; Erpelinck, C.A.; Abbas, S.; Beverloo, H.B.; Lugthart, S.; Lowenberg, B.; Delwel, R.; Valk, P.J. Prediction of molecular subtypes in acute myeloid leukemia based on gene expression profiling. *Haematologica* **2009**, *94*, 131–134. [CrossRef] [PubMed]
74. Huang, D.W.; Sherman, B.T.; Tan, Q.; Collins, J.R.; Alvord, W.G.; Roayaei, J.; Stephens, R.; Baseler, M.W.; Lane, H.C.; Lempicki, R.A. The DAVID Gene Functional Classification Tool: A novel biological module-centric algorithm to functionally analyze large gene lists. *Genome Biol.* **2007**, *8*, R183. [CrossRef] [PubMed]
75. MacLeod, R.A.F.; Kaufmann, M.E.; Drexler, H.G. Cytogenetic harvesting of cancer cells and cell lines. *Methods Mol. Biol.* **2016**, *1541*, 43–58. [CrossRef]

International Journal of
Molecular Sciences

Article

NKL Homeobox Genes NKX2-3 and NKX2-4 Deregulate Megakaryocytic-Erythroid Cell Differentiation in AML

Stefan Nagel *, Claudia Pommerenke, Corinna Meyer and Roderick A. F. MacLeod

Department of Human and Animal Cell Lines, Leibniz-Institute DSMZ, German Collection of Microorganisms and Cell Cultures, 38124 Braunschweig, Germany; cpo14@dsmz.de (C.P.); cme@dsmz.de (C.M.); rafmacleod@gmail.com (R.A.F.M.)
* Correspondence: sna@dsmz.de; Tel.: +49-531-2616167

Abstract: NKL homeobox genes encode transcription factors that impact normal development and hematopoietic malignancies if deregulated. Recently, we established an NKL-code that describes the physiological expression pattern of eleven NKL homeobox genes in the course of hematopoiesis, allowing evaluation of aberrantly activated NKL genes in leukemia/lymphoma. Here, we identify ectopic expression of NKL homeobox gene NKX2-4 in an erythroblastic acute myeloid leukemia (AML) cell line OCI-M2 and describe investigation of its activating factors and target genes. Comparative expression profiling data of AML cell lines revealed in OCI-M2 an aberrantly activated program for endothelial development including master factor ETV2 and the additional endothelial signature genes HEY1, IRF6, and SOX7. Corresponding siRNA-mediated knockdown experiments showed their role in activating NKX2-4 expression. Furthermore, the ETV2 locus at 19p13 was genomically amplified, possibly underlying its aberrant expression. Target gene analyses of NKX2-4 revealed activated ETV2, HEY1, and SIX5 and suppressed FLI1. Comparative expression profiling analysis of public datasets for AML patients and primary megakaryocyte–erythroid progenitor cells showed conspicuous similarities to NKX2-4 activating factors and the target genes we identified, supporting the clinical relevance of our findings and developmental disturbance by NKX2-4. Finally, identification and target gene analysis of aberrantly expressed NKX2-3 in AML patients and a megakaryoblastic AML cell line ELF-153 showed activation of FLI1, contrasting with OCI-M2. FLI1 encodes a master factor for myelopoiesis, driving megakaryocytic differentiation and suppressing erythroid differentiation, thus representing a basic developmental target of these homeo-oncogenes. Taken together, we have identified aberrantly activated NKL homeobox genes NKX2-3 and NKX2-4 in AML, deregulating genes involved in megakaryocytic and erythroid differentiation processes, and thereby contributing to the formation of specific AML subtypes.

Keywords: HOX-code; NKL-code; TALE-code; TBX-code

Citation: Nagel, S.; Pommerenke, C.; Meyer, C.; MacLeod, R.A.F. NKL Homeobox Genes NKX2-3 and NKX2-4 Deregulate Megakaryocytic-Erythroid Cell Differentiation in AML. *Int. J. Mol. Sci.* **2021**, 22, 11434. https://doi.org/10.3390/ijms222111434

Academic Editor: Amelia Casamassimi

Received: 24 September 2021
Accepted: 18 October 2021
Published: 22 October 2021

1. Introduction

Stem and progenitor cells pass through several developmental stages and subsequently differentiate into mature cells and tissues. During early embryogenesis, hematopoietic and endothelial cells share a common progenitor, termed hemangioblast. Later in development, these cell types differentiate separately, starting from hematopoietic and endothelial stem cells, respectively. The process of hematopoiesis generates all types of blood and immune cells, split into the lymphoid and myeloid lineages. The latter produces mature cell types such as erythrocytes and megakaryocytes via the joint megakaryocyte and erythroid progenitor [1]. Megakaryocytes develop subsequently the thrombocytes.

Differentiation processes during hematopoietic and endothelial development are mainly regulated at the transcriptional level [2,3]. The master genes responsible for controlling these processes mostly encode transcription factors (TFs). E26 transformation-specific (ETS) and NKL homeodomain factors represent two types of such developmental TFs and operate at specific stages and lineages. ETS factors share the conserved ETS domain, which forms a winged helix-turn-helix structure and performs both sequence-specific DNA binding and protein–protein interactions [4]. According to sequence similarities, their genes are classified into 13 groups [5]. ETV2 and FLI1 are two basic ETS factors of the ERG group, regulating early steps in hematopoietic and endothelial development. *ETV2* is expressed in the hemangioblast and drives endothelial differentiation in the adult, while FLI1 is active both in hematopoietic stem cells (HSCs) and during megakaryopoiesis [6,7].

Homeobox genes encode a homeodomain that consists of three helices, generating a helix-turn-helix structure. The homeodomain performs an interaction with DNA, cofactors, and chromatin, forming a platform for gene regulation [8]. According to sequence similarities of their conserved homeobox, these genes are arranged into eleven classes and several subclasses [9]. Thus, for example, NKL homeobox genes represent a subclass of the Antennapedia (ANTP) class of homeobox genes. The human genome contains 49 NKL homeobox genes that play fundamental roles in embryonal development of tissues and organs and regulate cell differentiation in the adult. Examples are *NKX2-3*, which is expressed in developing spleen, intestine and in HSCs; *NKX2-4*, which is active in brain and testis development; and *NKX2-5*, which controls development of heart and spleen [10–14]. Recently, we have described the NKL-code that encompasses the gene signature of eleven NKL homeobox genes expressed in specific patterns during hematopoiesis [15]. Basic NKL-code members are *NKX2-3*, in addition to both *HHEX* and *HLX*, which orchestrate differentiation processes in myeloid and lymphoid lineages and are themselves regulated by specific hematopoietic TFs [16].

According to their functions in normal development, deregulated ETS and NKL homeobox genes promote generation of hematopoietic malignancies and are frequently targeted by chromosomal aberrations. For example, the ETS gene *ETV6* is fused with different partner genes by specific translocations in lymphoblastic and myeloid acute leukemias, and NKL homeobox gene *NKX2-5* is aberrantly activated via juxtaposition to an enhancer region of *BCL11B* in T-cell acute lymphoblastic leukemia (T-ALL) [17,18]. Other reported activating mechanisms for NKL homeobox genes are deregulated chromatin components, signaling pathways, and TFs [15]. Furthermore, aberrantly expressed ETS and NKL homeobox genes serve as diagnostic markers for certain hematopoietic cancers, such as *ETV6* in ALL and *TLX1* in T-ALL.

Acute myeloid leukemia (AML) is the most frequent acute leukemia in adults. The tumor cells derive from specific progenitor cells of the myeloid lineage. According to originating cells and stages, phenotypes, chromosomal aberrations, and gene mutations, several subtypes of AML are distinguished that differ in prognosis and treatment. The established French–American–British (FAB) system differentiates eight subtypes, called AML-M0 to -M7 [19]. Thus, for example, AML-M6 represents an erythroblastic subtype and the corresponding tumor cells express several erythropoietic genes and are blocked in development, unable to terminate differentiation. Today, additional criteria including sequencing data basically serve to classify AML [20].

Systematic analysis of NKL homeobox genes in AML has revealed 18 deregulated genes, highlighting the importance of these developmental oncogenes in driving this malignancy [21]. Focused studies in AML have investigated activating factors and downstream functions of the selected deregulated NKL homeobox genes *NANOG*, *HMX2*, and *HMX3* and have shown the oncogenic roles of these genes [21,22]. In this study, we performed detailed analyses of the oncogenically deregulated NKL homeobox gene *NKX2-4* using AML cell line OCI-M2 as model. We found that *NKX2-4* is aberrantly connected with the developmental ETS genes *ETV2* and *FLI1*, generating a leukemogenic network, which impacts myeloid differentiation.

2. Results

2.1. NKX2-4 Expression in AML Cell Line OCI-M2

Systematic examinations of aberrantly expressed NKL homeobox genes in AML patients and cell lines were performed using public gene expression profiling datasets, highlighting their frequent deregulation and oncogenic impact in this myeloid malignancy [21]. To investigate their pathological function in myeloid in vitro models, we searched for deregulated NKL homeobox genes in our published RNA-seq dataset LL-100, which covers 34 myeloid and 66 lymphoid leukemia/lymphoma cell lines. We detected *NKX2-4* expression in two AML cell lines, conspicuously high in OCI-M2 and low in THP-1 (Figure 1A). Additionally, B-cell lymphoma cell line U-2932 expressed elevated *NKX2-4* levels as well. Importantly, this NKL homeobox gene is not represented in standard expression profiling arrays, and its transcriptional deregulation has, therefore, yet to be reported in leukemia patients or cell lines.

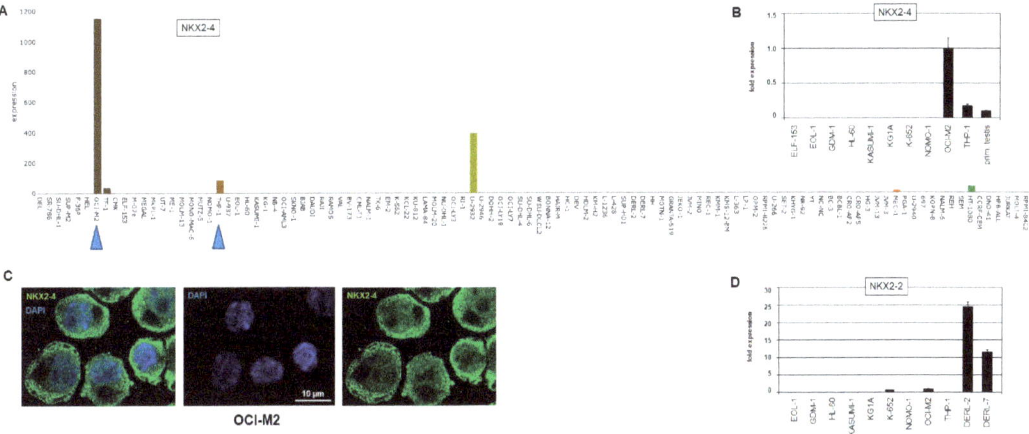

Figure 1. *NKX2-4* expression in hematopoietic cell lines. (**A**) LL-100 RNA-seq data show enhanced expression of NKL homeobox gene *NKX2-4* in AML cell line OCI-M2, while THP-1 expresses low levels (blue arrow heads). Gene expression values are given as DESeq2 normalized count data. (**B**) RQ-PCR analysis confirms high transcript levels in OCI-M2, while THP-1 and primary testis express lower levels. The expression level in OCI-M2 was set as 1. (**C**) Immunostaining of NKX2-4 protein in OCI-M2 cells, showing signals in both cytoplasm and nucleus (green). DAPI was used as nuclear counterstain (blue). (**D**) RQ-PCR analysis of *NKX2-2* in selected AML cell lines and T-cell lymphoma cell lines DERL-2 and DERL-7 shows elevated transcript levels in the latter, while THP-1 tested negative. The expression level in OCI-M2 was set as 1.

Analyses of additional public RNA-seq datasets containing samples from normal cells and tissues showed an absence of *NKX2-4* expression in developing and mature hematopoietic cells but showed its presence in the hypothalamus, pituitary gland, and testis (Figure S1). RQ-PCR and immunostaining confirmed *NKX2-4* activity in OCI-M2 cells at the RNA and protein levels, respectively (Figure 1B,C). The subcellular distribution of NKX2-4 protein showed localization in both nucleus and cytoplasm, suggesting functional regulation of this TF via nuclear import. Furthermore, combined analysis of *NKX2-4* in a primary testis sample indicated enhanced transcript levels in OCI-M2 (Figure 1B). Thus, *NKX2-4* is ectopically overexpressed in AML cell line OCI-M2, which was therefore used as a model to investigate its oncogenic role in this malignancy, including activating mechanisms and target genes.

2.2. Karyotyping and Genomic Profiling of OCI-M2

NKL homeobox genes are frequently deregulated by chromosomal aberrations [15]. To check if the *NKX2-4* locus is targeted by chromosomal rearrangements, we performed karyotyping of OCI-M2. The resultant karyotype was as follows. 51(46–51) < 2n > XX, der(X)t(X;8)(q23;q23), +6, +8, add(9)(p23), del(9)(p12p21), t(10;12)(p12;p12), del(17)(q11q21.1), +20, −21, +3 mar. However, no aberrations of the NKX2-4 locus at 20p11 were detected in OCI-M2. Nevertheless, the observed trisomy of chromosome 20 may boost *NKX2-4* expression by copy number gain.

Furthermore, we performed genomic profiling analysis of OCI-M2 to identify potential copy number alterations. The data for all chromosomes are shown in Figure 2, indicating, however, absence of a focal gain at the *NKX2-4* locus. In contrast, the results show several copy number alterations at other chromosomal positions, including duplications at 7q31-q36 and 8q22-q24, strongly amplified regions at 19p13 and 21q22, and deletions at 9p23-p24, 12p12, Xp11, and Xq12-q22. These regions may be indirectly implemented in *NKX2-4* deregulation. Interestingly, genomic profiling data of THP-1 cells that weakly transcribed *NKX2-4* showed a focal amplification at the *NKX2-4* locus (Figure S2A). However, this aberration neither enhanced the expression level of *NKX2-4* nor of the expression of its gene neighbor *NKX2-2*, as demonstrated by RQ-PCR analysis (Figure 1B,D), indicating absence of cognate activating TFs in THP-1. Of note, T-cell lymphoma cell lines DERL-2 and DERL-7 have been shown to aberrantly express *NKX2-2* and were used here as controls [23].

2.3. OCI-M2 Displays an Aberrant Program of Endothelial Development

To identify *NKX2-4* activating factors, we performed comparative expression profiling analysis of AML cell line OCI-M2 versus 31 AML control cell lines using public dataset GSE59808 and the associated online tool GEOR, which calculates the 250 most highly statistically significant differences in gene expression levels. This examination revealed for OCI-M2 differentially expressed up- or downregulated genes (Table S1). Subsequent gene set annotation analysis of these genes using the public online platform DAVID showed several significantly associated GO terms, including activated chromatin/histones and endothelial signaling (Table S2). The latter GO term was of special interest because of the close relationships between endothelial and hematopoietic development.

In support of this finding, OCI-M2 overexpressed *ETV2*, encoding a master factor for endothelial development, in addition to members of its associated gene signature, including *EPOR*, *HEY1*, *KDR*, and *SOX7* [24–26] (Table S1). RNA-seq data, RQ-PCR, and Western blot analyses of selected AML cell lines confirmed enhanced expression of these genes in OCI-M2 (Figure 3A–C and Figure S3). Furthermore, according to the comparative expression profiling data, *IRF6* and the histone genes *H1C* and *H2BB* from the HIST1-cluster were also overexpressed, while *FLI1* and *KDM6A* were downregulated in OCI-M2 (Figure 3D,E and Figure S3, Table S1). Of note, *IRF6* has been associated with endothelial development as well [27], thus representing an additional overexpressed TF in this context.

Interestingly, overexpressed *ETV2* is located at chromosomal position 19p13, which is targeted by genomic amplification in OCI-M2, while suppressed *KDM6A* is located at Xp11, which is focally deleted (Figure 2 and Figure S2B). Thus, these genomic aberrations may directly cause the deregulated activities observed for the indicated genes. Whole chromosome gains may underlie elevated expression levels of HIST1-genes at 6p22, *SOX7* at 8p23, *HEY1* at 8q21, and *NKX2-4* at 20p11 (Figure 2). However, a potential role for overexpressed histone genes in aberrant *NKX2-4* expression remained unclear, although we additionally detected elevated levels of ubiquitinated H2B in OCI-M2 (Figure 3D), which has been shown to impact NKL homeobox gene activity in B-cell lymphoma [28]. Taken together, AML cell line OCI-M2 displays an aberrantly activated program for endothelial development, which may drive its oncogenic transformation, notably including *NKX2-4* expression. Copy number alterations may underlie aberrant activation of particular endothelial signature genes.

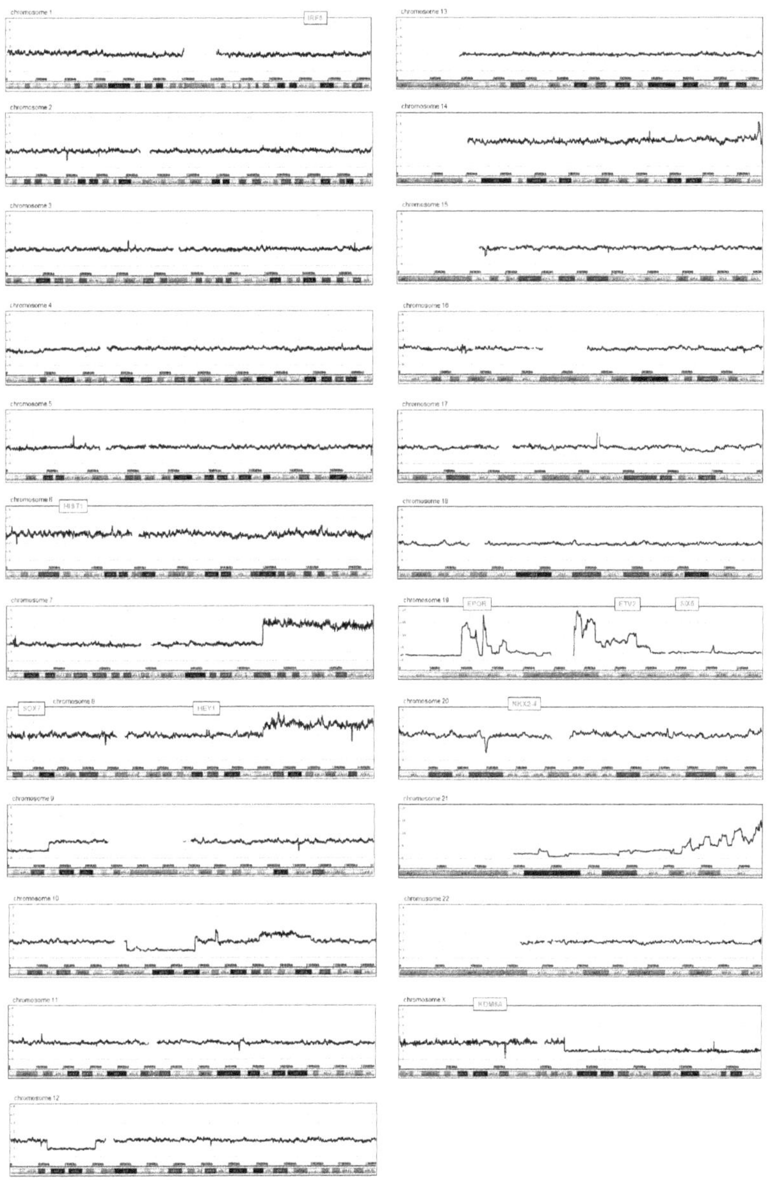

Figure 2. Genomic profiling data of OCI-M2. The data show copy number states for all chromosomes. The y-axis indicates the copy number state, the x-axis the chromosomal position. Selected gene loci are indicated, including *IRF6*, *HIST1*, *SOX7*, *HEY1*, *EPOR*, *ETV2*, *SIX5*, *NKX2-4*, and *KDM6A*.

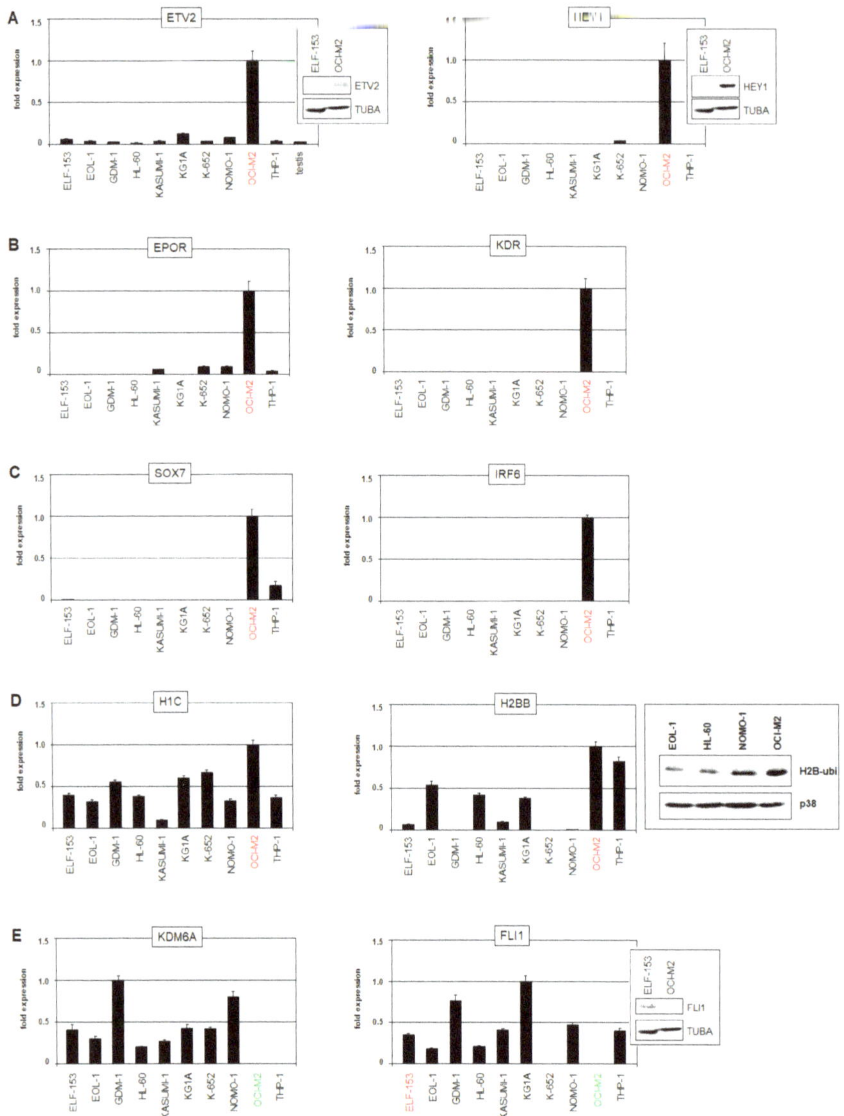

Figure 3. RQ-PCR and Western blot analysis of selected genes in AML cell lines. (**A**) RQ-PCR and Western blot analyses of ETV2 (**left**) and HEY1 (**right**) demonstrate elevated expression levels in OCI-M2. (**B**) RQ-PCR analyses of EPOR (**left**) and KDR (**right**) demonstrate increased expression levels in OCI-M2. (**C**) RQ-PCR analyses of SOX7 (**left**) and IRF6 (**right**) demonstrate elevated expression levels in OCI-M2. (**D**) RQ-PCR analyses of H1C (**left**) and H2BB (**middle**) demonstrate high expression levels in OCI-M2. Raised levels of ubiquinated H2B are shown by Western blot analysis, using p38 as loading control (**right**). (**E**) RQ-PCR and Western blot analyses of KDM6A (**left**) and FLI1 (**right**) demonstrate reduced expression levels in OCI-M2. The cell lines OCI-M2 and ELF-153 are indicated in red and green. The expression levels of OCI-M2 were set as 1, except for analyses of KDM6A (GDM-1 was set as 1) and FLI1 (KG1A was set as 1).

2.4. Endothelial Transcription Factors Activate NKX2-4 in OCI-M2

To examine whether the overexpressed endothelial TFs contribute to *NKX2-4* expression, we screened its promoter region for potential TF binding sites, using the UCSC genome browser. This approach revealed a SOX consensus site, as shown in Figure 4A. Accordingly, siRNA-mediated knockdown of overexpressed *SOX7* in OCI-M2 resulted in concomitantly reduced *NKX2-4* expression, showing that this endothelial TF activated NKL homeobox gene *NKX2-4* (Figure 4B). Then, a search for additional potential binding sites at *NKX2-4* using the CIS-BP database indicated an IRF6-site at -2648 bp, an ETV2-site at -2021 bp, and a HEY1-site within exon 2. To analyze their regulatory impact on *NKX2-4* expression, we performed corresponding siRNA-mediated knockdown experiments. The results showed that all three factors—IRF6, ETV2 and HEY1—activated expression of *NKX2-4* in OCI-M2 (Figure 4C,D).

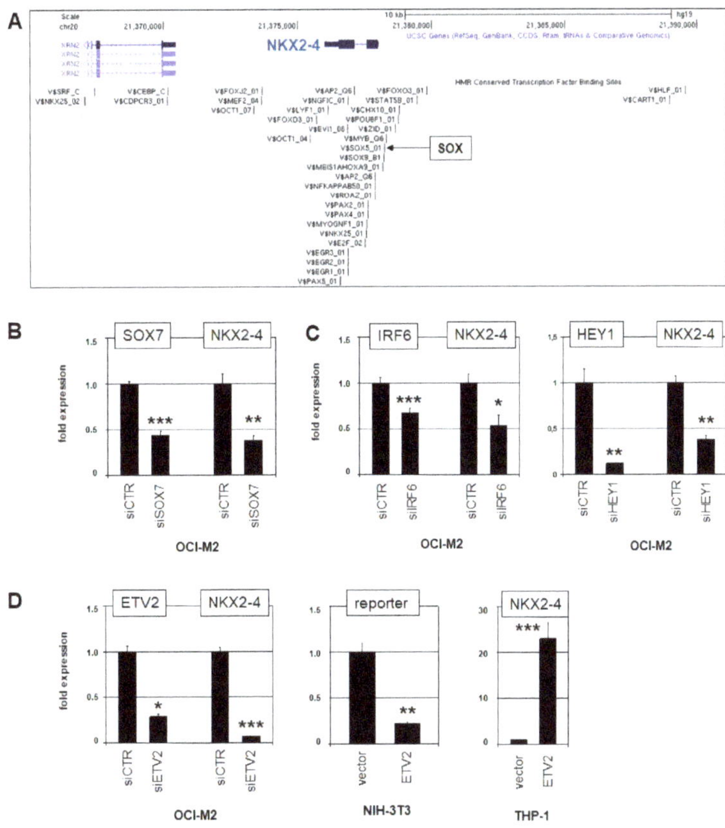

Figure 4. Endothelial TFs activate *NKX2-4* in OCI-M2. (**A**) TF binding sites at the *NKX2-4* locus were obtained from the UCSC genome browser. A consensus binding site for SOX factors is indicated. (**B**) RQ-PCR analysis of OCI-M2 after siRNA-mediated knockdown of *SOX7* shows concomitantly reduced transcript levels of *NKX2-4*. (**C**) RQ-PCR analyses of OCI-M2 after siRNA-mediated knockdown of *IRF6* (**left**) and *HEY1* (**right**) show concomitantly reduced transcript levels of *NKX2-4*. (**D**) RQ-PCR analysis of OCI-M2 after siRNA-mediated knockdown of *ETV2* shows concomitantly reduced transcript levels of *NKX2-4* (**left**). Reporter gene assay for a potential ETV2 binding site at *NKX2-4* was performed in NIH-3T3 cells, showing a suppressive effect. Nevertheless, these context-dependent results may indicate direct regulation of *NKX2-4* by ETV2 (**middle**). Forced expression of ETV2 in THP-1 cells resulted in strongly elevated *NKX2-4* expression (**right**). Statistical significance was assessed by Student's *t*-test (two-tailed), and the calculated *p*-values are indicated by asterisks (* $p < 0.05$, ** $p < 0.01$, *** $p < 0.001$, n.s. not significant). The expression levels of siCTR-treated and vector-treated cells were set as 1.

To study the role of ETV2 in more detail, we established a reporter gene assay for the identified binding site, showing that ETV2 regulated *NKX2-4* directly (Figure 4D). Furthermore, forced expression of ETV2 in THP-1 cells resulted in strongly elevated transcript levels of *NKX2-4* (Figure 4D), highlighting the activatory power of this TF in AML cells. Taken together, the endothelial TFs SOX7, IRF6, HEY1, and ETV2 are aberrantly overexpressed activators of *NKX2-4* in AML cell line OCI-M2.

2.5. NKX2-4 Impacts Erythroid Development

Functional analyses of NKX2-4 were performed by life-cell imaging. Accordingly, OCI-M2 cells were treated for *NKX2-4* knockdown and subsequently quantified for proliferation and apoptosis (Figure 5A). However, the results show no significant impact on these processes. The better to understand the potential oncogenic role of TF NKX2-4 in AML, we searched for its target genes. We postulated that the above-identified *NKX2-4* regulators ETV2 and HEY1 may simultaneously represent NKX2-4 target genes because they contain consensus binding sites for NKX2-4 at -1235 bp and -961 bp, respectively. Moreover, indeed, siRNA-mediated knockdown of *NKX2-4* resulted in concurrent downregulation of *ETV2* and *HEY1*, confirming targeting of these genes by NKX2-4 and thus their participation in an aberrant mutually activating network (Figure 5B). The known role of *ETV2* and *HEY1* in endothelial development may indicate that *NKX2-4* deregulates differentiation processes in AML.

To search for NKX2-4 target genes more systematically, we performed expression profiling analysis of OCI-M2 after *NKX2-4* knockdown in comparison with a control (Table S3). Gene set annotation analysis of the top-1000 upregulated and downregulated genes revealed several GO terms associated with tissue and organ development (Table S4). Thus, these results further support that NKX2-4 may deregulate developmental processes in AML. However, this approach showed no GO term specifically associated with myeloid differentiation. Therefore, after inspection of these differentially expressed genes, we selected six myeloid-associated candidates encoding TFs, receptors, and markers for detailed analyses—namely *FLI1, FOXA1, MAML2, SIX5, SIRPA*, and *TGFBR1* [7,29–33]. Again, siRNA-mediated knockdown of *NKX2-4* in OCI-M2 and subsequent transcript quantification by RQ-PCR confirmed that NKX2-4 activated *FOXA1, MAML2*, and *SIX5* (Figure 5C) and repressed *FLI1* and *SIRPA* (Figure 5D). The identified NKX2-4 target genes may thus play a role in deregulated myeloid development and thus in the leukemogenesis of OCI-M2.

To evaluate our findings, we analyzed gene expression data obtained from primary samples, including peripheral blood cells from AML patients as well as hematopoietic progenitor cells from healthy donors. OCI-M2 is derived from a patient with acute erythroblastic leukemia, AML-M6 [34]. To compare our cell line data with those from patients, we performed comparative expression profiling analysis of samples from six AML-M6 patients versus 429 controls (patients with AML-M0, -M1, -M2, -M3, -M4, and -M5) using public dataset GSE6891 and analysis tool GEOR. Differentially expressed genes revealed by this approach included significantly upregulated *EPOR, GATA1, HIST1, IRF6*, and *SIX5*, and significantly downregulated *FLI1* in AML-M6 (Table S5). Thus, key regulators and target genes of NKX2-4 identified in OCI-M2 are likewise expressed in AML-M6 patients, verifying the clinical relevance of our data obtained from a cell line model.

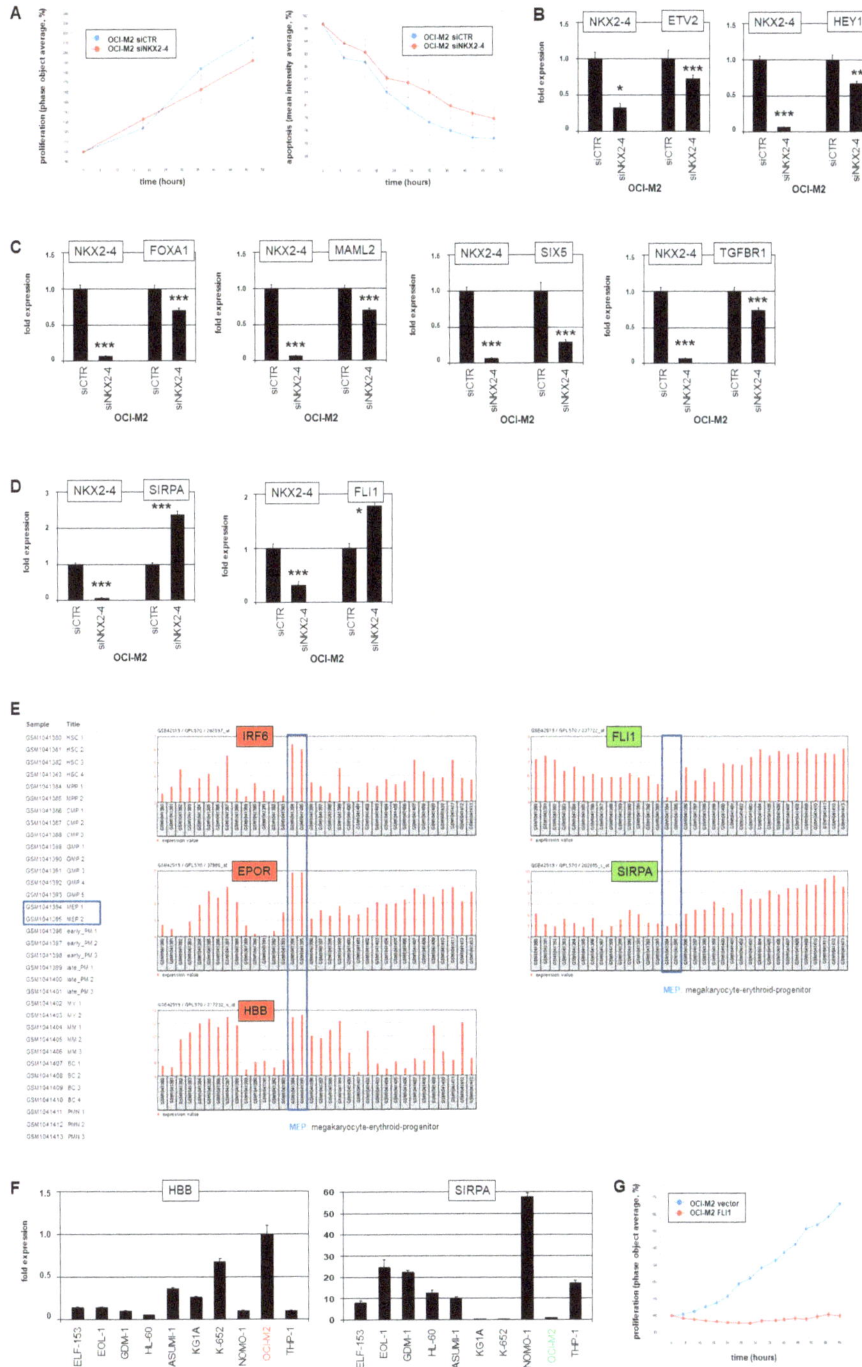

Figure 5. NKX2-4 target gene analyses. (**A**) Life-cell imaging analysis of OCI-M2 cells treated by siRNA-mediated knockdown of *NKX2-4* show no significant impact in proliferation (**left**) or apoptosis (**right**). (**B**) RQ-PCR analyses of

ETV2 (**left**) and *HEY1* (**right**) in OCI-M2 after siRNA-mediated knockdown of *NKX2-4* confirm these genes as activated targets. (**C**) RQ-PCR analyses of *FOXA1, MAML2, SIX5*, and *TGFBR1* in OCI-M2 after siRNA-mediated knockdown of *NKX2-4* confirm these genes as activated targets. (**D**) RQ-PCR analyses of *SIRPA* (**left**) and *FLI1* (**right**) in OCI-M2 after siRNA-mediated knockdown of *NKX2-4* confirm these genes as suppressed targets. (**E**) Expression profiling analysis of developing myeloid cells using public dataset GSE42519 shows elevated expression levels of *IRF6, EPOR*, and *HBB* and reduced levels of *FLI1* and *SIRPA* in megakaryocyte–erythroid progenitor cells (MEP). The y-axis represents the expression levels. (**F**) RQ-PCR analysis of selected AML cell lines for *HBB* (**left**) and *SIRPA* (**right**) show respective elevated and reduced expression levels in OCI-M2. (**G**) Life-cell imaging analysis of OCI-M2 cells treated by forced expression of FLI1 shows significant reduction in proliferation. Statistical significance was assessed by Student's *t*-test (two-tailed), and the calculated *p*-values are indicated by asterisks (* $p < 0.05$, ** $p < 0.01$, *** $p < 0.001$, n.s. not significant). The expression levels of siCTR-treated cells and of untreated OCI-M2 were set as 1.

In addition, we compared expression profiling data of samples from normal megakaryocyte and erythroid progenitors (MEPs) versus more differentiated granulopoietic progenitors, including promyelocytes, myelocytes, metamyelocytes, and band cells, using dataset GSE42519 and analysis tool GEOR. This approach revealed downregulated *FLI1* and *SIRPA* and upregulated *IRF6, EPOR, GATA1*, and *HBB* in MEPs (Table S6, Figure 5E). FLI1, SIRPA, GATA1 and HBB play basic roles in myeloid development especially in erythropoiesis [7,32,35]. Accordingly, RQ-PCR analysis demonstrated that OCI-M2 expressed elevated *GATA1* and *HBB* and reduced *SIRPA* levels as well (Figure 5F), consistent with its erythroblastic phenotype resembling MEPs.

FLI1 encodes a key myeloid TF, repressing erythroid while activating megakaryocytic differentiation [7,36]. Therefore, aberrant suppression of *FLI1* by *NKX2-4* in OCI-M2 may promote leukemogenic transformation towards erythroblastic cells. Accordingly, forced expression of FLI1 in OCI-M2 cells inhibited proliferation, suggesting that FLI1-mediated cell differentiation processes are uniformly connected with termination of cell growth (Figure 5G). Taken together, our experimental data for *NKX2-4* in acute erythroblastic leukemia cell line OCI-M2 correspond to expression data from AML-M6 patients and primary MEPs, suggesting that this aberrantly activated NKL homeobox gene may provoke a developmental defect in the process of megakaryocyte–erythroid differentiation.

2.6. NKX2-3 Impacts Megakaryocytic Development

As mentioned above, *NKX2-4* is not represented in standard expression profiling datasets. However, our comparative profiling results of AML patients (M6 versus M0 to M5) showed ectopic expression of NKL homeobox gene *NKX2-3* in AML-M6 patients (Table S5). This finding indicated that NKX2-3 and the closely related TF NKX2-4 may control similar oncogenic processes in this particular AML subtype. To compare their regulatory impacts in vitro, we searched for an *NKX2-3* expressing AML cell line model. Anew screening of RNA-seq dataset LL-100 revealed cell line ELF-153 which, however, derives from acute megakaryoblastic leukemia, AML-M7 [37]. RQ-PCR and Western blot analyses confirmed *NKX2-3* expression in ELF-153 at the RNA and protein level, respectively, validating its suitability for functional tests (Figure 6A).

Subsequent siRNA-mediated knockdown experiments in ELF-153 demonstrated that NKX2-3 activated *ETV2, SIX5*, and *FLI1* and suppressed *HEY1* (Figure 6B). Thus, NKX2-3 differs from NKX2-4 in regulation of *HEY1* and *FLI1*, while both NKL-TFs activated *ETV2* and *SIX5*. Accordingly, RQ-PCR and Western blot analysis demonstrated that ELF-153 expressed elevated levels of *FLI1*, and OCI-M2 reduced levels of *FLI1* (Figure 3E). Furthermore, siRNA-mediated knockdown of *FLI1* in ELF-153 resulted in slightly decreased proliferation, as analyzed by life-cell imaging (Figure 6C). Thus, proliferation was promoted by FLI1 in ELF-153, contrasting with OCI-M2. *NKX2-4* is aberrantly expressed in erythroblastic leukemia cell line OCI-M2 and represses *FLI1*, while *NKX2-3* is aberrantly expressed in megakaryoblastic leukemia cell line ELF-153 and activates *FLI1*. *FLI1* contains a consensus binding site for both NKX2-3 and NKX2-4 in its upstream region at −2564 bp, suggesting direct regulation. Moreover, both *NKX2-3* and *FLI1* are physiologically upregulated in HSCs (Figure 6D), indicating aberrant reactivation of their regulatory connection in ELF-153.

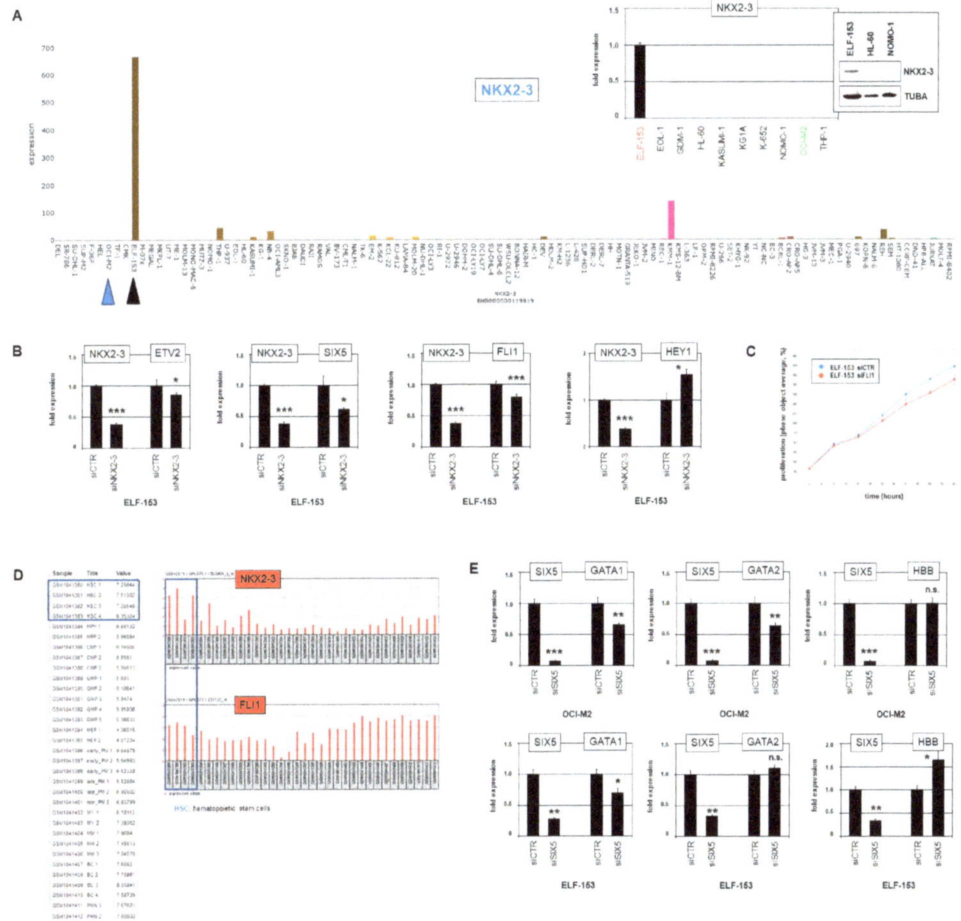

Figure 6. Expression and target gene analysis of NKX2-3. (**A**) LL-100 RNA-seq data show enhanced expression of NKL homeobox gene NKX2-3 in AML cell line ELF-153 (black arrowhead), while this gene is silent in OCI-M2 (blue arrow heads). Gene expression values are given as DESeq2 normalized count data. RQ-PCR and Western blot analyses of selected AML cell lines confirm enhanced expression of NKX2-3 in ELF-153 (insert). The expression level of ELF-153 was set as 1. (**B**) RQ-PCR analyses of ETV2 (**above left**), SIX5 (**above right**), FLI1 (**below left**), and of HEY1 (**below right**) in ELF-153 after siRNA-mediated knockdown of NKX2-3 confirm these genes as targets. Of note, FLI1 is activated by NKX2-3. (**C**) Life-cell imaging analysis of ELF-153 cells treated for knockdown of FLI1 shows reduction in proliferation (*p* = 0.021). (**D**) Expression profiling analysis of developing myeloid cells using public dataset GSE42519 shows elevated expression levels of NKX2-3 and FLI1 in hematopoietic stem cells (HSC). (**E**) RQ-PCR analyses of GATA1, GATA2, and HBB in OCI-M2 (**above**) and ELF-153 (**below**) after siRNA-mediated knockdown of SIX5 show differences in gene regulation between these AML cell lines. Statistical significance was assessed by Student's *t*-test (two-tailed), and the calculated *p*-values are indicated by asterisks (* *p* < 0.05, ** *p* < 0.01, *** *p* < 0.001, n.s. not significant). The expression levels of siCTR-treated cells were set as 1.

GATA1 and GATA2 represent additional regulators in megakaryopoiesis and erythropoiesis [35,38]. Furthermore, GATA1 has been shown to interact with the coactivating TFs

SIX1 and SIX2 in erythropoiesis, and *GATA2* is regulated by SIX1 in embryonal development of the placodes [39,40]. Hence, we speculated whether the NKX2-3 and NKX2-4 activated target SIX5 regulates or cooperates with GATA1 and/or GATA2. Thus, we performed *SIX5* knockdown experiments in OCI-M2 and ELF-153 (Figure 6E). The results show that SIX5 activated the expression of *GATA1* in both cell lines, while *GATA2* was activated by SIX5 only in OCI-M2. GATA1-target gene *HBB* was suppressed by SIX5 in ELF-513, while no impact was detectable in OCI-M2. Thus, SIX5 functionally differed in both cell lines. In OCI-M2, *SIX5* supported erythropoiesis via *GATA1* and *GATA2*, while in ELF-153, *SIX5* supported megakaryopoiesis via *GATA1* and inhibited erythropoiesis via *HBB*. Taken together, our findings highlight that the aberrantly expressed NKL homeobox genes *NKX2-3* and *NKX2-4* manipulate developmental lineage decisions in respective megakaryoblastic and erythroblastic AML via *FLI1* deregulation and *SIX5* activation.

3. Discussion

In this study, we report aberrant expression of NKL homeobox genes *NKX2-3* and *NKX2-4* in cell lines derived from two different AML subtypes. *NKX2-4* is ectopically activated in erythroblastic AML-M6 cell line OCI-M2 via the endothelial TFs ETV2, HEY1, IRF6, and SOX7, generating an aberrant developmental gene network. Prominent target genes identified are repressed *FLI1* and activated *SIX5* (Figure 7). In contrast, *NKX2-3* is aberrantly expressed in megakaryoblastic AML-M7 cell line ELF-153 by so far unknown factors and activates both *FLI1* and *SIX5* (Figure 7). In AML-M6 patients, we were also able to find gene activities of endothelial activators and target genes of NKX2-4 first identified in OCI-M2. Thus, this comparison with AML patient data revealed interesting concordances, supporting the clinical relevance of our data. Comparison with primary megakaryocyte–erythroid progenitors indicated deregulation of developmental processes by these aberrantly expressed NKL homeobox genes, operating at that developmental stage. Our data revealed deregulation of master factor FLI1. This ETS factor controls the differentiation of megakaryocytes and erythrocytes and may, therefore, represent a new key target in megakaryoblastic and erythroblastic AML.

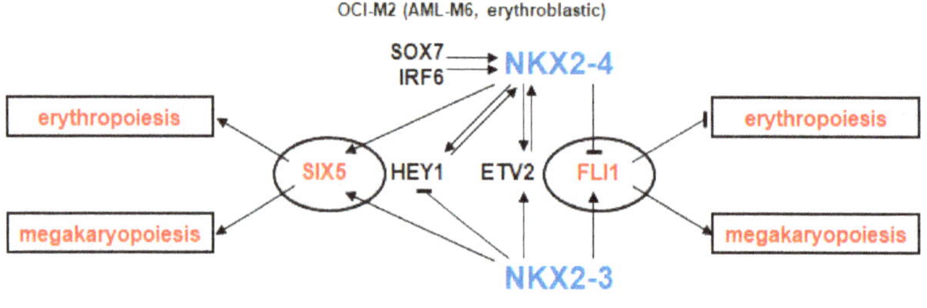

Figure 7. Aberrant gene regulatory network of *NKX2-4* in OCI-M2 (**above**) and *NKX2-3* in ELF-153 (**below**). These NKL homeobox genes promote respective erythroblastic and megakaryoblastic leukemogenesis via their target genes *FLI1* and *SIX5*. Endothelial activators of *NKX2-4* are indicated.

NKX2-3 is a member of the NKL-code and hematopoietically expressed in HSCs [11]. This restricted expression pattern may indicate a physiological role of this NKL homeobox gene in the control of stemness and lineage differentiation, while its deregulation may promote leukemogenesis by disturbing these developmental processes. Aberrant *NKX2-3* activity has been described in AML patients carrying mutations of *NPM1* or aberrations of *KMT2A* [41–43]. Enhanced *NKX2-3* expression correlates with *KMT2A*-rearrangements in T-ALL as well [44]. Moreover, *NKX2-3* is a direct target gene of *KMT2A-ENL* and impacts proliferation and cell differentiation [45]. In accordance with these published results, combined analysis of expression profiling data from normal hematopoietic cells and AML patient samples demonstrated physiological *NKX2-3* activity in stem cells and aberrant expression in patients with *KMT2A* rearrangements and complex karyotypes (Figure S4). In addition, *NKX2-3* expression has been detected in myelodysplastic syndrome, diffuse large B-cell lymphoma, T-ALL, and T-cell lymphoma, showing its oncogenic potential in both myeloid and lymphoid cell lineages [15]. Increased colony forming and replating capacity after NKX2-3 overexpression has been shown in murine hematopoietic progenitor cells [42], supporting an impact in early differentiation processes.

In contrast, *NKX2-4* lies without the NKL-code and is, therefore, normally silent throughout hematopoiesis. This gene is not represented on standard expression arrays and thus is less studied in cancer patients. Nevertheless, a screen of T-ALL patients revealed a chromosomal translocation that juxtaposes *NKX2-4* to the *TCRA* gene, supporting its oncogenic activity in hematopoietic malignancies [46].

Our results show that aberrantly expressed endothelial TFs ETV2, HEY1, IRF6, and SOX7 activate *NKX2-4* in AML cell line OCI-M2. The development of endothelial and hematopoietic cell types is closely linked during embryogenesis. The hemangioblast represents an embryonic stem cell that forms the basis of developing blood and endothelial cells. Although this stem cell is absent in the adult, several control genes are shared between these lineages [2]. *ETV2* is a master gene of the hemangioblast and regulates endothelial development in the adult [6]. In the embryo, ETV2 activates myelopoiesis via *RUNX1*, *SPI1*, and *TAL1* [47,48]. In the endothelial development of the heart, *ETV2* is a target gene of the heart master factor NKX2-5 [49]. This physiological connection may be aberrantly captured by the related *NKX2-4* in OCI-M2. Additional endothelial factors are encoded by *FLI1*, *HEY1*, *IRF6*, and *SOX7* [25,27,50,51]. Chromosomal amplifications and chromosomal gains underlie their aberrant activation in OCI-M2. Several of these chromosomal regions are rearranged in erythroleukemic AML patients, including 6p22, 19p13, 20p11, and Xp11, supporting the pathogenic basis of our findings [52].

FLI1 is a master gene in myelopoiesis, driving megakaryocyte differentiation and repressing erythropoiesis [53]. Furthermore, *FLI1* is prominently expressed in HSCs, alongside *NKX2-3* [7,11]. This coexpression, together with our data showing a regulatory connection between both genes, may indicate that *FLI1* represents a target gene of NKX2-3 in HSCs as well. B-cell lymphoma cell line U-2932 expressed elevated *NKX2-4* and reduced *FLI1* (Figure 1 and Figure S3), suggesting that the repressive impact of NKX2-4 may play a role in malignant lymphoid cells as well. *FLI1* and *IRF6* play a role in both endothelial and erythroid development [53,54]. Additional genes connected with NKL homeobox genes in our study are *HBB* and *SIRPA*. *HBB* encodes beta-globin expressed in erythroid progenitors and *SIRPA* a receptor implicated in innate immunological processes [32]. *SIRPA* is highly expressed in granulocytes and monocytes, while downregulated in megakaryocytes and erythrocytes [55]. Thus, NKX2-4-mediated inhibition of *SIRPA* may shift the myeloid differentiation of granulocytes and monocytes towards megakaryocytes and erythrocytes.

SIX1 is implicated in erythropoiesis by interaction with and activation of *GATA1* [39]. *GATA2* is suppressed by SIX1 and SIX2 during erythropoiesis but activated by SIX1 in Hodgkin lymphoma [39,56]. Furthermore, *SIX1* may promote the development of MEPs [57]. Thus, several studies show a regulatory connection between SIX and GATA factors. Here, our data indicate that aberrantly activated *SIX5* impacts *GATA1* and *GATA2* in AML. Depending on the cell type, *SIX5* can promote erythropoietic or megakaryopoietic processes.

Taken together, this study shows that *FLI1* and *SIX5* represent two target genes of NKX2-3 and NKX2-4 that disturb the differentiation of megakaryocytes and erythrocytes. Overexpression experiments of *NKX2-3* and *NKX2-4* in primary progenitor cells may support this interpretation in the future. However, a developmental impact at the same stage of differentiation was described for NKL homeobox gene *DLX4* [58], highlighting the potential of these genes in lineage decisions. Moreover, we also show that deregulated *NKX2-3* and *NKX2-4* impact developmental gene activities in AML in a context dependent manner. Detection of their aberrant expression may assist the diagnosis and prognosis of certain AML subtypes.

4. Materials and Methods

4.1. Bioinformatic Analyses of RNA-Seq and Expression Profiling Data

For screening cell lines, we exploited RNA-sequencing data from 100 leukemia/lymphoma cell lines (termed LL-100), available at ArrayExpress (www.ebi.ac.uk/arrayexpress) (accessed on 18 October 2021) via E-MTAB-7721. Gene expression values are given as DE-Seq2 normalized count data [59]. Expression data for normal cell types were obtained from Gene Expression Omnibus (GEO, www.ncbi.nlm.nih.gov) (accessed on 18 October 2021), using RNA-seq dataset GSE69239 and expression profiling dataset GSE42519 [60,61], in addition to RNA-seq data from The Human Protein Atlas (www.proteinatlas.org) (accessed on 18 October 2021). Gene expression profiling data from AML cell lines and patients were examined using datasets GSE59808 and GSE15434, respectively [62]. Combined expression analysis of normal hematopoietic cells (GSE42519) and AML patients (GSE13159) was performed using the online database BloodSpot [63]. Gene set annotation analysis was performed using the online tool DAVID (www.david.abcc.ncifcrf.gov) (accessed on 18 October 2021) [64]. Consensus binding sites for TFs were obtained from the CIS-BP database (www.cisbp.ccbr.utoronto.ca) (accessed on 18 October 2021) and were used for screening at the UCSC genome browser (www.genome.cse.ucsc.edu) (accessed on 18 October 2021). Expression profiling data from siRNA-treated OCI-M2 cells were generated at the Genome Analytics Facility (Helmholtz Centre for Infection Research, Braunschweig, Germany) using HG U133 Plus 2.0 gene chips (Affymetrix, High Wycombe, UK). The data are available from ArrayExpress via E-MTAB-10941. After RMA background correction and quantile normalization of the spot intensities, the profiling data were expressed as ratios of sample means and subsequently log2 transformed. Data processing was performed via R/Bioconductor using public limma and affy packages.

4.2. Cell Lines and Treatments

Cell lines are held by the DSMZ (Braunschweig, Germany) and cultivated as described previously [65]. All cell lines had been authenticated and tested negative for mycoplasma infection. Modification of gene expression levels was performed using gene-specific siRNA oligonucleotides with reference to AllStars negative Control siRNA (siCTR) obtained from Qiagen (Hilden, Germany). The gene expression constructs for ETV2, NKX2-4, and FLI1, in addition to an empty control-vector, were obtained from Origene (Wiesbaden, Germany). SiRNAs (80 pmol) and vector DNA (2 µg) were transfected into 1×10^6 cells by electroporation using the EPI-2500 impulse generator (Fischer, Heidelberg, Germany) at 350 V for 10 ms. Electroporated cells were harvested after 20 h cultivation.

Proliferation and apoptosis were analyzed using the IncuCyte S3 Live-Cell Analysis System (Essen Bioscience, Hertfordshire, UK). For detection of apoptotic cells, we used the IncuCyte Caspase-3/7 Green Apoptosis Assay diluted at 1:2000 (Essen Bioscience, Essen, Germany). Live-cell imaging experiments were performed twice with fourfold parallel tests.

4.3. Polymerase Chain-Reaction (PCR) Analyses

Total RNA was extracted from cultivated cell lines using TRIzol reagent (Invitrogen, Darmstadt, Germany). Primary human total RNA derived from testis was purchased from Biochain/BioCat (Heidelberg, Germany). cDNA was synthesized using 5 µg RNA, random priming, and Superscript II (Invitrogen). Real time quantitative (RQ)-PCR analysis was performed using the 7500 Real-time System and commercial buffer and primer sets (Applied Biosystems/Life Technologies, Darmstadt, Germany). For normalization of expression levels, we quantified the transcripts of TATA box binding protein (TBP). Quantitative analyses were performed as biological replicates and measured in triplicate. Standard deviations are presented in the figures as error bars. Statistical significance was assessed by

Student's *t*-test (two-tailed) and the calculated *p*-values are indicated by asterisks (* $p < 0.05$, ** $p < 0.01$, *** $p < 0.001$, n.s. not significant).

4.4. Protein Analysis

Western blots were generated by the semi-dry method. Protein lysates from cell lines were prepared using SIGMAFast protease inhibitor cocktail (Sigma, Taufkirchen, Germany). Proteins were transferred onto nitrocellulose membranes (Bio-Rad, Munich, Germany) and blocked with 5% dry milk powder dissolved in phosphate-buffered saline buffer (PBS). The following antibodies were used: alpha-Tubulin (Sigma, #T6199), NKX2-4 (Novus Biologicals, Centennial, CO, USA, #NBP1-91541), ETV2 (MyBioSource, San Diego, CA, USA, #MBS7112932), HEY1 (Novus Biologicals, #NBP2-56068), ubiquinated H2B (Cell Signaling, Frankfurt, Germany, #5546), p38 (Cell Signaling, #8690), FLI1 (Thermo Fisher Scientific, Darmstadt, Germany, #MA1-196), NKX2-3 (Abcam, Cambridge, UK, #ab66366). For loading, control blots were reversibly stained with Poinceau (Sigma) and detection of alpha-Tubulin (TUBA) performed thereafter. Secondary antibodies were linked to peroxidase for detection by Western-Lightning-ECL (Perkin Elmer, Waltham, MA, USA). Documentation was performed using the digital system ChemoStar Imager (INTAS, Göttingen, Germany).

Immuno-cytology was performed as follows: cells were spun onto slides and subsequently air-dried and fixed with methanol/acetic acid for 90 s. Antibodies were diluted 1:20 in PBS containing 5% BSA and incubated for 30 min. Washing was performed 3 times with PBS. Preparations were incubated with secondary antibody (diluted 1:100) for 20 min. After final washing, cells were mounted for nuclear in Vectashield (Vector Laboratories, Burlingame, CA, USA), containing DAPI. Documentation of subcellular protein localization was performed using an Axio-Imager microscope (Zeiss, Göttingen, Germany) configured to a dual Spectral Imaging system (Applied Spectral Imaging, Neckarhausen, Germany).

4.5. Karyotyping and Genomic Profiling Analysis

Karyotyping was performed as described previously [66]. For genomic profiling, genomic DNA of AML cell lines was prepared by the Qiagen Gentra Puregene Kit (Qiagen). Labelling, hybridization, and scanning of Cytoscan HD arrays was performed by the Genome Analytics Facility located at the Helmholtz Centre for Infection Research (Braunschweig, Germany), according to the manufacturer's protocols (Affymetrix, High Wycombe, UK). These arrays base on single nucleotide polymorphisms (SNPs) and allow the determination of copy number states of most gene loci. Data were interpreted using the Chromosome Analysis Suite software version 3.1.0.15 (Affymetrix, High Wycombe, UK) and copy number alterations were determined accordingly.

4.6. Reporter-Gene Assay

For creation of reporter gene constructs, we combined a reporter with a regulatory genomic fragment derived from the upstream region of NKX2-4. PCR products of the corresponding genomic region (regulator) and of the HOXA9 gene, comprising exon1-intron1-exon2 (reporter), were cloned into respective HindIII/BamHI and EcoRI sites of expression vector pcDNA3 downstream of the CMV enhancer. Oligonucleotides used for the amplification of the regulator were obtained from Eurofins MWG, Ebersberg, Germany. Their sequences were as follows: NKX2-4-for 5′-GATGAAGCTTATAGCCTGAAAACAGAG-3′, NKX2-4-rev 5′-AACACTTGCTGGGATCCTTCTG-3′. Introduced restriction sites used for cloning are underlined. Constructs were validated by sequence analysis (Eurofins MWG). Transfections of plasmid-DNA into NIH-3T3 cells were performed using SuperFect Transfection Reagent (Qiagen). Commercial HOXA9 and TBP assays were used for RQ-PCR to quantify the spliced reporter–transcript, corresponding to the regulator activity (Thermo Fisher Scientific).

Supplementary Materials: The following are available online at https://www.mdpi.com/article/10.3390/ijms222111434/s1.

Author Contributions: Conceptualization, S.N.; formal analysis, S.N., C.P., C.M. and R.A.F.M.; investigation, S.N.; writing—original draft preparation, S.N.; writing—review and editing, R.A.F.M. All authors have read and agreed to the published version of the manuscript.

Funding: This research received no external funding.

Institutional Review Board Statement: Not applicable.

Informed Consent Statement: Not applicable.

Data Availability Statement: Data is contained within the article or Supplementary Materials. The data presented in this study are available in [Table S3].

Acknowledgments: We thank Hans G. Drexler for critically reading the manuscript.

Conflicts of Interest: The authors declare no conflict of interest.

References

1. Xavier-Ferrucio, J.; Krause, D.S. Concise review: Bipotent megakaryocytic-erythroid progenitors: Concepts and controversies. *Stem Cells* **2018**, *36*, 1138–1145. [CrossRef]
2. Canu, G.; Ruhrberg, C. First blood: The endothelial origins of hematopoietic progenitors. *Angiogenesis* **2021**, *24*, 199–211. [CrossRef] [PubMed]
3. Wilson, N.K.; Foster, S.D.; Wang, X.; Knezevic, K.; Schütte, J.; Kaimakis, P.; Chilarska, P.M.; Kinston, S.; Ouwehand, W.H.; Dzierzak, E.; et al. Combinatorial transcriptional control in blood stem/progenitor cells: Genome-wide analysis of ten major transcriptional regulators. *Cell Stem Cell* **2010**, *7*, 532–544. [CrossRef]
4. Gallant, S.; Gilkeson, G. ETS transcription factors and regulation of immunity. *Arch. Immunol. Ther. Exp.* **2006**, *54*, 149–163. [CrossRef]
5. Laudet, V.; Hänni, C.; Stéhelin, D.; Duterque-Coquillaud, M. Molecular phylogeny of the ETS gene family. *Oncogene* **1999**, *18*, 1351–1359. [CrossRef] [PubMed]
6. Wareing, S.; Eliades, A.; Lacaud, G.; Kouskoff, V. ETV2 expression marks blood and endothelium precursors, including hemogenic endothelium, at the onset of blood development. *Dev. Dyn.* **2012**, *241*, 1454–1464. [CrossRef] [PubMed]
7. Kruse, E.A.; Loughran, S.J.; Baldwin, T.M.; Josefsson, E.C.; Ellis, S.; Watson, D.K.; Nurden, P.; Metcalf, D.; Hilton, D.J.; Alexander, W.S.; et al. Dual requirement for the ETS transcription factors Fli-1 and Erg in hematopoietic stem cells and the megakar-yocyte lineage. *Proc. Natl. Acad. Sci. USA* **2009**, *106*, 13814–13819. [CrossRef]
8. Bürglin, T.R.; Affolter, M. Homeodomain proteins: An update. *Chromosoma* **2016**, *125*, 497–521. [CrossRef] [PubMed]
9. Holland, P.W.H.; Booth, H.A.F.; Bruford, E. Classification and nomenclature of all human homeobox genes. *BMC Biol.* **2007**, *5*, 47. [CrossRef]
10. Pabst, O.; Zweigerdt, R.; Arnold, H. Targeted disruption of the homeobox transcription factor Nkx2-3 in mice results in postnatal lethality and abnormal development of small intestine and spleen. *Development* **1999**, *126*, 2215–2225. [CrossRef]
11. Nagel, S.; Pommerenke, C.; Scherr, M.; Meyer, C.; Kaufmann, M.; Battmer, K.; MacLeod, R.A.F.; Drexler, H.G. NKL homeobox gene activities in hematopoietic stem cells, T-cell development and T-cell leukemia. *PLoS ONE* **2017**, *12*, e0171164. [CrossRef]
12. Small, E.M.; Vokes, S.; Garriock, R.J.; Li, D.; Krieg, A.P. Developmental expression of the Xenopus Nkx2-1 and Nkx2-4 genes. *Mech. Dev.* **2000**, *96*, 259–262. [CrossRef]
13. Lints, T.; Parsons, L.; Hartley, L.; Lyons, I.; Harvey, R. Nkx-2.5: A novel murine homeobox gene expressed in early heart progenitor cells and their myogenic descendants. *Development* **1993**, *119*, 419–431. [CrossRef]
14. Brendolan, A.; Ferretti, E.; Salsi, V.; Moses, K.; Quaggin, S.; Blasi, F.; Cleary, M.L.; Selleri, L. A Pbx1-dependent genetic and transcriptional network regulates spleen ontogeny. *Development* **2005**, *132*, 3113–3126. [CrossRef]
15. Nagel, S. NKL-Code in Normal and Aberrant Hematopoiesis. *Cancers* **2021**, *13*, 1961. [CrossRef] [PubMed]
16. Migueles, R.P.; Shaw, L.; Rodrigues, N.P.; May, G.; Henseleit, K.; Anderson, K.G.; Goker, H.; Jones, C.M.; De Bruijn, M.F.; Brickman, J.M.; et al. Transcriptional regulation of Hhex in hematopoiesis and hematopoietic stem cell ontogeny. *Dev. Biol.* **2017**, *424*, 236–245. [CrossRef]
17. Rasighaemi, P.; Ward, A.C. ETV6 and ETV7: Siblings in hematopoiesis and its disruption in disease. *Crit. Rev. Oncol./Hematol.* **2017**, *116*, 106–115. [CrossRef] [PubMed]
18. Nagel, S.; Scherr, M.; Kel, A.; Hornischer, K.; Crawford, G.E.; Kaufmann, M.; Meyer, C.; Drexler, H.G.; MacLeod, R.A. Activation of TLX3 and NKX2-5 in t(5;14)(q35;q32) T-cell acute lymphoblastic leukemia by remote 3′-BCL11B enhancers and coregu-lation by PU.1 and HMGA1. *Cancer Res.* **2007**, *67*, 1461–1471. [CrossRef] [PubMed]
19. Arber, D.A. Realistic Pathologic Classification of Acute Myeloid Leukemias. *Am. J. Clin. Pathol.* **2001**, *115*, 552–560. [CrossRef] [PubMed]

20. Arber, D.A.; Orazi, A.; Hasserjian, R.; Thiele, J.; Borowitz, M.J.; Le Beau, M.M.; Bloomfield, C.D.; Cazzola, M.; Vardiman, J.W. The 2016 revision to the World Health Organization classification of myeloid neoplasms and acute leukemia. *Blood* **2016**, *127*, 2391–2405. [CrossRef]

21. Nagel, S.; Scherr, M.; MacLeod, R.A.F.; Pommerenke, C.; Koeppel, M.; Meyer, C.; Kaufmann, M.; Dallmann, I.; Drexler, H.G. NKL homeobox gene activities in normal and malignant myeloid cells. *PLoS ONE* **2019**, *14*, e0226212. [CrossRef]

22. Nagel, S.; Pommerenke, C.; Meyer, C.; MacLeod, R.A.F.; Drexler, H.G. Aberrant expression of NKL homeobox genes HMX2 and HMX3 interferes with cell differentiation in acute myeloid leukemia. *PLoS ONE* **2020**, *15*, e0240120. [CrossRef]

23. Nagel, S.; Pommerenke, C.; MacLeod, R.A.; Meyer, C.; Kaufmann, M.; Fähnrich, S.; Drexler, H.G. Deregulated expression of NKL homeobox genes in T-cell lymphomas. *Oncotarget* **2019**, *10*, 3227–3247. [CrossRef] [PubMed]

24. Sumanas, S.; Choi, K. ETS transcription factor ETV2/ER71/Etsrp in hematopoietic and vascular development. *Curr. Top. Dev. Biol.* **2016**, *118*, 77–111. [PubMed]

25. Kanki, Y.; Nakaki, R.; Shimamura, T.; Matsunaga, T.; Yamamizu, K.; Katayama, S.; Suehiro, J.-I.; Osawa, T.; Aburatani, H.; Kodama, T.; et al. Dynamically and epigenetically coordinated GATA/ETS/SOX transcription factor expression is indispensable for endothelial cell differentiation. *Nucleic Acids Res.* **2017**, *45*, 4344–4358. [CrossRef]

26. Liu, F.; Li, D.; Yu, Y.Y.; Kang, I.; Cha, M.J.; Kim, J.Y.; Park, C.; Watson, D.K.; Wang, T.; Choi, K. Induction of hematopoietic and en-dothelial cell program orchestrated by ETS transcription factor ER71/ETV2. *EMBO Rep.* **2015**, *16*, 654–669. [CrossRef]

27. Cheng, C.-C.; Chang, S.-J.; Chueh, Y.-N.; Huang, T.-S.; Huang, P.-H.; Cheng, S.-M.; Tsai, T.-N.; Chen, J.-W.; Wang, H.-W. Distinct angiogenesis roles and surface markers of early and late endothelial progenitor cells revealed by functional group analyses. *BMC Genom.* **2013**, *14*, 1–10. [CrossRef]

28. Nagel, S.; Ehrentraut, S.; Tomasch, J.; Quentmeier, H.; Meyer, C.; Kaufmann, M.; Drexler, H.G.; MacLeod, R.A. Ectopic expression of homeobox gene NKX2-1 in diffuse large B-cell lymphoma is mediated by aberrant chromatin modifications. *PLoS ONE.* **2013**, *8*, e61447. [CrossRef]

29. Neben, K.; Schnittger, S.; Brors, B.; Tews, B.; Kokocinski, F.; Haferlach, T.; Muller, J.; Hahn, M.; Hiddemann, W.; Eils, R.; et al. Distinct gene expression patterns associated with FLT3- and NRAS-activating mutations in acute myeloid leukemia with normal karyotype. *Oncogene* **2005**, *24*, 1580–1588. [CrossRef] [PubMed]

30. Nemoto, N.; Suzukawa, K.; Shimizu, S.; Shinagawa, A.; Takei, N.; Taki, T.; Hayashi, Y.; Kojima, H.; Kawakami, Y.; Nagasawa, T. Identification of a novel fusion geneMLL-MAML2 in secondary acute myelogenous leukemia and myelodysplastic syndrome with inv(11)(q21q23). *Genes Chromosom. Cancer* **2007**, *46*, 813–819. [CrossRef] [PubMed]

31. Chu, Y.; Chen, Y.; Li, M.; Shi, D.; Wang, B.; Lian, Y.; Cheng, X.; Wang, X.; Xu, M.; Cheng, T.; et al. Six1 regulates leukemia stem cell maintenance in acute myeloid leukemia. *Cancer Sci.* **2019**, *110*, 2200–2210. [CrossRef]

32. Barclay, A.N.; Brown, M.H. The SIRP family of receptors and immune regulation. *Nat. Rev. Immunol.* **2006**, *6*, 457–464. [CrossRef] [PubMed]

33. Otten, J.; Schmitz, L.; Vettorazzi, E.; Schultze, A.; Marx, A.H.; Simon, R.; Krauter, J.; Loges, S.; Sauter, G.; Bokemeyer, C.; et al. Expression of TGF-β receptor ALK-5 has a negative impact on outcome of patients with acute myeloid leukemia. *Leukemia* **2010**, *25*, 375–379. [CrossRef]

34. Papayannopoulou, T.; Nakamoto, B.; Kurachi, S.; Tweeddale, M.; Messner, H. Surface antigenic profile and globin phenotype of two new human erythroleukemia lines: Characterization and interpretations. *Blood* **1988**, *72*, 1029–1038. [CrossRef]

35. Moriguchi, T.; Yamamoto, M. A regulatory network governing Gata1 and Gata2 gene transcription orchestrates erythroid lineage differentiation. *Int. J. Hematol.* **2014**, *100*, 417–424. [CrossRef] [PubMed]

36. Starck, J.; Weiss-Gayet, M.; Gonnet, C.; Guyot, B.; Vicat, J.M.; Morlé, F. Inducible Fli-1 gene deletion in adult mice modifies several myeloid lineage commitment decisions and accelerates proliferation arrest and terminal erythrocytic differen-tiation. *Blood* **2010**, *116*, 4795–4805. [CrossRef]

37. Mouthon, M.A.; Freund, M.; Titeux, M.; Katz, A.; Guichard, J.; Breton-Gorius, J.; Vainchenker, W. Growth and differentiation of the human megakaryoblastic cell line (ELF-153): A model for early stages of megakaryocytopoiesis. *Blood* **1994**, *84*, 1085–1097. [CrossRef]

38. Szalai, G.; LaRue, A.C.; Watson, D.K. Molecular mechanisms of megakaryopoiesis. *Cell. Mol. Life Sci.* **2006**, *63*, 2460–2476. [CrossRef]

39. Creed, T.M.; Baldeosingh, R.; Eberly, C.L.; Schlee, C.S.; Kim, M.; Cutler, J.A.; Pandey, A.; Civin, C.I.; Fossett, N.G.; Kingsbury, T.J. PAX-SIX-EYA-DACH Network modulates GATA-FOG function in fly hematopoiesis and human erythropoiesis. *Development* **2020**, *147*, dev177022. [CrossRef]

40. Maharana, S.K.; Schlosser, G. A gene regulatory network underlying the formation of pre-placodal ectoderm in Xenopus laevis. *BMC Biol.* **2018**, *16*, 79. [CrossRef]

41. Becker, H.; Marcucci, G.; Maharry, K.; Radmacher, M.D.; Mrózek, K.; Margeson, D.; Whitman, S.P.; Wu, Y.Z.; Schwind, S.; Paschka, P.; et al. Favorable prognostic impact of NPM1 mutations in older patients with cytogenetically normal de novo acute myeloid leukemia and associated gene- and microRNA-expression signatures: A Cancer and leukemia group B study. *J. Clin. Oncol.* **2010**, *28*, 596–604. [CrossRef]

42. Dovey, O.M.; Cooper, J.L.; Mupo, A.; Grove, C.S.; Lynn, C.; Conte, N.; Andrews, R.M.; Pacharne, S.; Tzelepis, K.; Vijayabaskar, M.S.; et al. Molecular synergy underlies the co-occurrence patterns and phenotype of NPM1-mutant acute myeloid leukemia. *Blood* **2017**, *130*, 1911–1922. [CrossRef] [PubMed]

43. Al Hinai, A.S.; Pratcorona, M.; Grob, T.; Kavelaars, F.G.; Bussaglia, E.; Sanders, M.A.; Nomdedeu, J.; Valk, P.J. The Landscape of KMT2A-PTD AML: Concurrent Mutations, Gene Expression Signatures, and Clinical Outcome. *HemaSphere* **2019**, *3*, e181. [CrossRef] [PubMed]
44. Kang, H.; Sharma, N.D.; Nickl, C.K.; Devidas, M.; Loh, M.L.; Hunger, S.P.; Dunsmore, K.P.; Winter, S.S.; Matlawska-Wasowska, K. Dysregulated transcriptional networks in KMT2A- and MLLT10-rearranged T-ALL. *Biomark. Res.* **2018**, *6*, 27. [CrossRef]
45. Garcia-Cuellar, M.-P.; Büttner, C.; Bartenhagen, C.; Dugas, M.; Slany, R.K. Leukemogenic MLL-ENL Fusions Induce Alternative Chromatin States to Drive a Functionally Dichotomous Group of Target Genes. *Cell Rep.* **2016**, *15*, 310–322. [CrossRef] [PubMed]
46. Le Noir, S.; Ben Abdelali, R.; Lelorch, M.; Bergeron, J.; Sungalee, S.; Payet-Bornet, D.; Villarèse, P.; Petit, A.; Callens, C.; Lhermitte, L.; et al. Extensive molecular mapping of TCRα/δ- and TCRβ-involved chromosomal translocations reveals distinct mechanisms of oncogene activation in T-ALL. *Blood* **2012**, *120*, 3298–3309. [CrossRef]
47. Salanga, M.; Meadows, S.M.; Myers, C.T.; Krieg, P.A. ETS family protein ETV2 is required for initiation of the endothelial lineage but not the hematopoietic lineage in the Xenopus embryo. *Dev. Dyn.* **2010**, *239*, 1178–1187. [CrossRef]
48. Glenn, N.O.; Schumacher, J.A.; Kim, H.J.; Zhao, E.J.; Skerniskyte, J.; Sumanas, S. Distinct regulation of the anterior and posterior myeloperoxidase expression by Etv2 and Gata1 during primitive Granulopoiesis in zebrafish. *Dev. Biol.* **2014**, *393*, 149–159. [CrossRef]
49. Ferdous, A.; Caprioli, A.; Iacovino, M.; Martin, C.M.; Morris, J.; Richardson, J.A.; Latif, S.; Hammer, R.E.; Harvey, R.P.; Olson, E.N.; et al. Nkx2-5 transactivates theEts-related protein 71gene and specifies an endothelial/endocardial fate in the developing embryo. *Proc. Natl. Acad. Sci. USA* **2009**, *106*, 814–819. [CrossRef] [PubMed]
50. Watanabe, Y.; Seya, D.; Ihara, D.; Ishii, S.; Uemoto, T.; Kubo, A.; Arai, Y.; Isomoto, Y.; Nakano, A.; Abe, T.; et al. Importance of endothelial Hey1 expression for thoracic great vessel development and its distal enhancer for Notch-dependent endothelial transcription. *J. Biol. Chem.* **2020**, *295*, 17632–17645. [CrossRef]
51. Lilly, A.J.; Costa, G.; Largeot, A.; Fadlullah, M.Z.H.; Lie-A-Ling, M.; Lacaud, G.; Kouskoff, V. Interplay between SOX7 and RUNX1 regulates hemogenic endothelial fate in the yolk sac. *Development* **2016**, *143*, 4341–4351. [CrossRef]
52. Cigudosa, J.C.; Odero, M.D.; Calasanz, M.J.; Solé, F.; Salido, M.; Arranz, E.; Martínez-Ramirez, A.; Urioste, M.; Alvarez, S.; Cervera, J.V.; et al. De novo erythroleukemia chromosome features include multiple re-arrangements, with special involvement of chromosomes 11 and 19. *Genes Chromosom. Cancer* **2003**, *36*, 406–412. [CrossRef] [PubMed]
53. Li, Y.; Luo, H.; Liu, T.; Zacksenhaus, E.; Ben-David, Y. The ets transcription factor Fli-1 in development, cancer and disease. *Oncogene* **2015**, *34*, 2022–2031. [CrossRef]
54. Xu, J.; Shao, Z.; Glass, K.; Bauer, D.E.; Pinello, L.; Van Handel, B.; Hou, S.; Stamatoyannopoulos, J.A.; Mikkola, H.K.; Yuan, G.-C.; et al. Combinatorial Assembly of Developmental Stage-Specific Enhancers Controls Gene Expression Programs during Human Erythropoiesis. *Dev. Cell* **2012**, *23*, 796–811. [CrossRef] [PubMed]
55. Seiffert, M.; Cant, C.; Chen, Z.; Rappold, I.; Brugger, W.; Kanz, L.; Brown, E.J.; Ullrich, A.; Bühring, H.J. Human signal-regulatory protein is expressed on normal, but not on subsets of leukemic myeloid cells and mediates cellular adhesion involving its counterreceptor CD47. *Blood* **1999**, *94*, 3633–3643. [CrossRef]
56. Nagel, S.; Meyer, C.; Kaufmann, M.; Drexler, H.G.; MacLeod, R.A. Aberrant expression of homeobox gene SIX1 in Hodgkin lymphoma. *Oncotarget* **2015**, *6*, 40112–40126. [CrossRef] [PubMed]
57. Zhang, L.-S.; Kang, X.; Lu, J.; Zhang, Y.; Wu, X.; Wu, G.; Zheng, J.; Tuladhar, R.; Shi, H.; Wang, Q.; et al. Installation of a cancer promoting WNT/SIX1 signaling axis by the oncofusion protein MLL-AF9. *EBioMedicine* **2019**, *39*, 145–158. [CrossRef]
58. Trinh, B.Q.; Barengo, N.; Kim, S.B.; Lee, J.S.; Zweidler-McKay, P.A.; Naora, H. The homeobox gene DLX4 regulates erythro-megakaryocytic differentiation by stimulating IL-1β and NF-κB signaling. *J. Cell Sci.* **2015**, *128*, 3055–3067.
59. Quentmeier, H.; Pommerenke, C.; Dirks, W.G.; Eberth, S.; Koeppel, M.; MacLeod, R.A.F.; Nagel, S.; Steube, K.; Uphoff, C.C.; Drexler, H.G. The LL-100 panel: 100 cell lines for blood cancer studies. *Sci. Rep.* **2019**, *9*, 8218. [CrossRef]
60. Casero, D.; Sandoval, S.; Seet, C.S.; Scholes, J.; Zhu, Y.; Ha, V.L.; Luong, A.; Parekh, C.; Crooks, G.M. Long non-coding RNA profil-ing of human lymphoid progenitor cells reveals transcriptional divergence of B cell and T cell lineages. *Nat. Immunol.* **2015**, *16*, 1282–1291. [CrossRef] [PubMed]
61. Rapin, N.; Bagger, F.O.; Jendholm, J.; Mora-Jensen, H.; Krogh, A.; Kohlmann, A.; Thiede, C.; Borregaard, N.; Bullinger, L.; Winther, O.; et al. Comparing cancer vs normal gene expression profiles identifies new disease entities and common transcriptional programs in AML patients. *Blood* **2014**, *123*, 894–904. [CrossRef] [PubMed]
62. Klein, H.U.; Ruckert, C.; Kohlmann, A.; Bullinger, L.; Thiede, C.; Haferlach, T.; Dugas, M. Quantitative comparison of microar-ray experiments with published leukemia related gene expression signatures. *BMC Bioinform.* **2009**, *10*, 422. [CrossRef] [PubMed]
63. Bagger, F.O.; Sasivarevic, D.; Sohi, S.H.; Laursen, L.G.; Pundhir, S.; Sønderby, C.K.; Winther, O.; Rapin, N.; Porse, B.T. BloodSpot: A database of gene expression profiles and transcriptional programs for healthy and malignant haematopoiesis. *Nucleic Acids Res.* **2016**, *44*, D917–D924. [CrossRef] [PubMed]
64. Huang, D.W.; Sherman, B.T.; Tan, Q.; Collins, J.R.; Alvord, W.G.; Roayaei, J.; Stephens, R.; Baseler, M.W.; Lane, H.C.; Lempicki, R.A. DAVID Gene functional classification tool: A novel biological module-centric algorithm to functionally analyze large gene list. *Genome Biol.* **2007**, *8*, R183. [CrossRef] [PubMed]
65. Drexler, H.G. *Guide to Leukemia-Lymphoma Cell Lines*, 2nd ed.; DSMZ: Braunschweig, Germany, 2010.
66. MacLeod, R.A.F.; Kaufmann, M.E.; Drexler, H.G. Cytogenetic Harvesting of Cancer Cells and Cell Lines. *Methods Mol. Biol.* **2016**, *1541*, 43–58. [CrossRef]

International Journal of
Molecular Sciences

Article

Transcriptional Regulation of Yin-Yang 1 Expression through the Hypoxia Inducible Factor-1 in Pediatric Acute Lymphoblastic Leukemia

Gabriela Antonio-Andres [1], Gustavo U. Martinez-Ruiz [2], Mario Morales-Martinez [1], Elva Jiménez-Hernandez [3], Estefany Martinez-Torres [1], Tania V. Lopez-Perez [1,4], Laura A. Estrada-Abreo [5], Genaro Patino-Lopez [5], Sergio Juarez-Mendez [6], Víctor M. Davila-Borja [6] and Sara Huerta-Yepez [1,*]

1 Unidad de Investigación en Enfermedades Oncológicas, Hospital Infantil de México, Federico Gómez, Mexico City 06720, Mexico; gabya_24@yahoo.com.mx (G.A.-A.); ixnergal@gmail.com (M.M.-M.); stephy.mt12@gmail.com (E.M.-T.); valecita_vale@hotmail.com (T.V.L.-P.)
2 División de Investigación, Facultad de Medicina, Universidad Nacional Autónoma de México, Mexico City 04510, Mexico; gustavobambam@gmail.com
3 Servicio de Hemato-Oncología, Hospital Infantil de Moctezuma, Mexico City 15530, Mexico; elvajimenez@yahoo.com
4 Consejo Nacional de Ciencia y Tecnología (CONACYT), Mexico City 03940, Mexico
5 Laboratorio de Investigación en Inmunología y Proteómica, Hospital Infantil de México, Federico Gómez, Mexico City 06720, Mexico; lauraestrada13@yahoo.com.mx (L.A.E.-A.); gena23pat@yahoo.com (G.P.-L.)
6 Laboratorio de Oncología Experimental, Instituto Nacional de Pediatría, S.S.A., Mexico City 04530, Mexico; sjuarezm@pediatria.gob.mx (S.J.-M.); latrans86@hotmail.com (V.M.D.-B.)
* Correspondence: shuertay@gmail.com; Tel.: +52-55-52289917 (ext. 4401); Fax: +52-55-44349663

Citation: Antonio-Andres, G.; Martinez-Ruiz, G.U.; Morales-Martinez, M.; Jiménez-Hernandez, E.; Martinez-Torres, E.; Lopez-Perez, T.V.; Estrada-Abreo, L.A.; Patino-Lopez, G.; Juarez-Mendez, S.; Davila-Borja, V.M.; et al. Transcriptional Regulation of Yin-Yang 1 Expression through the Hypoxia Inducible Factor-1 in Pediatric Acute Lymphoblastic Leukemia. *Int. J. Mol. Sci.* **2022**, 23, 1728. https://doi.org/10.3390/ijms23031728

Academic Editors: Amelia Casamassimi, Alfredo Ciccodicola and Monica Rienzo

Received: 27 October 2021
Accepted: 11 January 2022
Published: 2 February 2022

Publisher's Note: MDPI stays neutral with regard to jurisdictional claims in published maps and institutional affiliations.

Abstract: Yin-Yang transcription factor 1 (YY1) is involved in tumor progression, metastasis and has been shown to be elevated in different cancers, including leukemia. The regulatory mechanism underlying YY1 expression in leukemia is still not understood. Bioinformatics analysis reveal three Hypoxia-inducible factor 1-alpha (HIF-1α) putative binding sites in the YY1 promoter region. The regulation of YY1 by HIF-1α in leukemia was analyzed. Mutation of the putative YY1 binding sites in a reporter system containing the HIF-1α promoter region and CHIP analysis confirmed that these sites are important for YY1 regulation. Leukemia cell lines showed that both proteins HIF-1α and YY1 are co-expressed under hypoxia. In addition, the expression of mRNA of YY1 was increased after 3 h of hypoxia conditions and affect several target genes expression. In contrast, chemical inhibition of HIF-1α induces downregulation of YY1 and sensitizes cells to chemotherapeutic drugs. The clinical implications of HIF-1α in the regulation of YY1 were investigated by evaluation of expression of HIF-1α and YY1 in 108 peripheral blood samples and by RT-PCR in 46 bone marrow samples of patients with pediatric acute lymphoblastic leukemia (ALL). We found that the expression of HIF-1α positively correlates with YY1 expression in those patients. This is consistent with bioinformatic analyses of several databases. Our findings demonstrate for the first time that YY1 can be transcriptionally regulated by HIF-1α, and a correlation between HIF-1α expression and YY1 was found in ALL clinical samples. Hence, HIF-1α and YY1 may be possible therapeutic target and/or biomarkers of ALL.

Keywords: HIF-1α; YY1; ALL

1. Introduction

Acute lymphoblastic leukemia (ALL) is the most common childhood malignancy worldwide, with an incidence of around 4.8 cases per 100,000 population aged 0 to 19 years. Around 2500 cases of ALL are diagnosed annually in the USA [1–4]. Advancements in treatment have been remarkable and have seen a 70–85% increase in survival rate [5,6].

Standard-risk patients have a 4-year event-free survival of 70–80%. However, the potential development of chemoresistance still represents a main obstacle for ALL treatment [7,8].

One of the mechanisms involved in the chemoresistance of tumors is the HIF-1. HIF-1 plays a very important role in cancer biology, participating in processes such as angiogenesis, maintenance of stem cells, metabolic reprogramming, epithelial-mesenchymal transition, as well as invasion, metastasis and resistance to radiation therapy and chemotherapy. HIF-1 levels correlate with tumor growth, vascularization and metastasis in both animal models and clinical studies [9–13].

HIF-1 is a heterodimeric protein complex composed of two subunits; one constitutively stable and expressed HIF1β, and the other inducible by O_2 and growth factors, HIF-1α. HIF-1α is post-translationally modified by prolyl-hydroxylases within the oxygen dependent degradation domains (ODD), which promote the binding of pVHL (von Hippel-Lindau protein) and subsequent degradation via the proteasome [14]. Multiple signaling pathways have been shown to contribute to the regulation of transcription of the inducible gene in hypoxia-inducible gene and protein stabilization of HIF-1α even in normoxia. Some of these pathways include extracellular signal-regulated Ras/kinase (ERK) and mitogen-activated p38-protein kinase (MAPK), phosphatidylinositol-3-kinase (PI3K) pathway and mTOR signaling pathway [15]. In addition to hypoxia, other oncogenic pathways including signaling pathways for growth factors or genetic loss of tumor suppressor genes, such as VHL and PTEN, over-regulate HIF-1 activity [16]. Importantly, for this work it has been shown that HIF-1α is overexpressed in various types of leukemia, including pediatric ALL and interestingly the high expression of this transcription factor correlates with poor survival [17].

Different types of resistance to cytotoxic agents have been identified, including the involvement of membrane transporters. These transporters are proteins that act as ATP-dependent expulsion pumps, causing a decrease in intracellular concentrations and drug resistance [18,19]. Among the most studied are the multi-drug resistance 1 (MDR1) protein (or gp-170). MDR1 is encoded by the *ABCB1* gene, and constitutively expressed in normal tissue [19]. The overexpression of this protein induces excessive flow and therefore an insufficient intracellular concentration of cytotoxic agents even at maximum doses, which results in a resistance to various chemotherapeutic drugs. Drugs most often used in leukemia treatment include anthracycline, vinca alkaloids, and podophyllines, which are substrates for MDR1. Expression of *ABCB1* messenger RNA is frequently detected in tumors of patients with ovarian cancer, acute myeloid leukemia, and other cancers. Some studies have proposed that expression of *ABCB1* messenger RNA correlates with tumor severity intrinsic to drug resistance after chemotherapy [8,20,21]. It is known that in general, in leukemia, the MDR1 phenotype is generally acquired after the administration of chemotherapeutic agents and is more frequent in ALL and in aggressive carcinomas (e.g., breast and ovarian) [21]. Recent studies have shown that HIF-1α positively regulates the expression of MDR1, which represents a mechanism of chemotherapy resistance in various types of tumors [22]. However, the mechanisms that regulate the expression of MDR1 are not fully elucidated.

The transcription factor Yin-Yang-1 (YY1) is known to play a fundamental role both in normal biological processes such as embryogenesis, differentiation, replication, cell proliferation, and in mechanisms of carcinogenesis, tumor progression and metastasis. It is estimated that more than 7% of vertebrate genes contain binding sites for YY1, which reflects the importance of this transcription factor [23]. YY1 participates in response events to various apoptotic stimuli and has been associated with carcinogenic processes by activating relevant proto-oncogenes such as c-Myc, and downregulating tumor suppressor genes such as p53. An increase in expression and/or activation of this transcription factor has been shown in different neoplasms, such as hematopoietic neoplasia, carcinomas, hepatocarcinoma and retinoblastoma [24–26]. Recent studies have shown that YY1 is elevated in patients with non-Hodgkin's lymphoma and leukemia and its high expression correlates with poor prognosis [27]. Our group recently reported that YY1 regulates the expression of MDR1 and its over-expression is correlated with poor prognosis in ALL pediatric patients [28]. In addition, we demonstrated that high nuclear expression of YY1 correlates with poor survival in leukemia patients. Nevertheless, the role of these transcription factors in the pathogenesis of ALL is not clear and given their possible co-expression and correlation with poor prognosis, it is plausible to think that there is a relationship between these two transcription factors. Our first approach hypothesized that HIF-1α could regulate transcriptionally the expression of YY1 since we previously showed that HIF-1α and YY1 increase their expression under hypoxic conditions [29]. Understanding the regulatory mechanism underlying YY1 expression and its implications in ALL, as well as its relationship with the transcription factor HIF-1α, is important for diagnostic and prognostic purposes.

2. Results

2.1. Transcriptional Regulation of the YY1 Protein by HIF-1α in Leukemia Cell Lines

Based on independent findings regarding the expression of HIF-1α and YY1 in lymphomas and leukemia [24,30,31], we proposed that there is a correlation between these proteins. To investigate this, we performed a bioinformatics analysis to predict HIF-1α binding sites in the YY1 promoter to described in detail material and methods section. Three putative binding sites located at nucleotides −622 bp, −592 bp and +199 bp were identified with respect to the YY1 gene transcription start site (TSS) (Figure 1A).

To determine if HIF-1α can regulate the expression of YY1 through activation of its promoter, we evaluated the role of each binding site in regulating the promoter region of the gene encoding YY1. The YY1 promoter region was cloned into the reporter plasmid pGL3 as described in the Materials and Methods Section. The reporter plasmid pGL3-YY1-pro-luc was generated. A single or double mutation of the sites in the YY1 promoter was performed. The mutants were designated as pGL3-YY1-MutA-pro-luc (site −622), pGL3-YY1-MutB-pro-luc (site −592) and pGL3-YY1-MutC-pro-luc (site +199) for the single mutants and as pGL3-YY1-MutAB-pro-luc and pGL3-YY1-MutBC-pro-luc for the double mutants. A triple mutant, pGL3-YY1-MutABC-pro-luc was also generated. Reporter plasmids containing their respective mutations were transfected into the PC3 cell line as in the transfection model previously reported [21]. Figure 1B shows the luciferase results. For plasmid pGL3-YY1-MutA-pro-luc and pGL3-YY1-MutC-pro-luc the luciferase/B-galactosidase results were significant at * $p < 0.005$ when comparing with pGL3-YY1-pro-luc plasmid, which contains the complete promotor of YY1. However, the most dramatic effect observed with the reporter gene (luciferase) was obtained with the plasmid pGL3-YY1-MutB-pro-luc (* $p < 0.001$), for which luciferase/β-galactosidase activity was very similar to the results observed with the empty plasmid. This result was corroborated by the luciferase/β-galactosidase activity observed with the plasmid containing double and triple mutant. When sites A, B and C were mutated, the activity of the reporter plasmid was affected, and the fold change with respect to control is shown (Figure 1B). These results show that the sites at −622, +199, and especially site −592 play an important role in the positive regulation of YY1 by HIF-1α.

To confirm the interaction of the transcription factor HIF-1α and the promoter region of YY1, ChIP assays were performed. Chromatin from the RS4;11 cell line was used. For immunoprecipitation, an anti-HIF-1α antibody was used, and then segments were amplified by PCR using specific oligonucleotides for each possible binding site of HIF-1α in the YY1 promoter region. The results are shown in Figure 1C, and we observed that HIF-1α binds to all three sites (+199, −592 and −622) in the YY1 promoter. Non-immunoprecipitated chromatin was used as a positive control, and control IgG was used as a negative control. ChIP is shown as efficiency of the immunoprecipitation YY1 and control (Figure 1C).

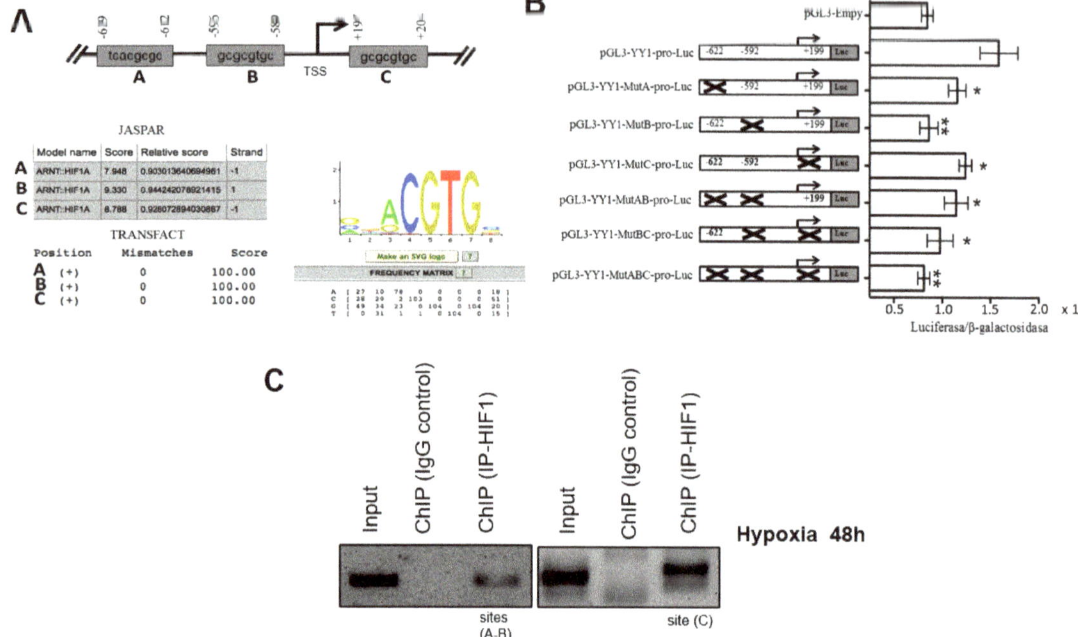

Figure 1. HIF-1α regulates YY1 transcriptional activation by direct interaction with its promotor. (**A**) Three potential binding sites for the transcription factor HIF-1α obtained after bioinformatics analysis using two online servers, JASPAR and TRANSFACT, are displayed. The region from −2000 to +350 bp in the YY1 gene was analyzed for Transcription Start Site (TSS). A weight matrix obtained from the JASPAR database for the transcription factor HIF-1α is displayed. (**B**) Putative binding HIF-1α sites in the YY1 promoter that are involved in regulating expression. Transfection assays were performed using the PC3 cell line to assess the effects of directed mutagenesis at each of the YY1 binding sequences, located at sites −622 bp, −592 bp and +199 in the promoter region of the YY1 gene. The schematic shows each of the mutated sites, and the graph indicates normalized luciferase reporter gene expression levels obtained by measuring β-galactosidase via co-transfection with a reporter gene plasmid; fold changes are reported. The results are representative of three independent experiments (one-way ANOVA, * $p < 0.005$, ** $p < 0.001$). (**C**) ChIP was conducted for each potential HIF-1α binding site in the YY1 promoter. The results show that HIF-1α binds the promoter region of YY1. Results of three independent experiments are shown.

2.2. Induction or Inhibition of HIF-1α Expression Affects in the YY1 Expression

In order to demonstrate whether the inhibition of HIF-1α can modify YY1 expression, we incubated the leukemia cell line RS4;11 in normoxia and hypoxia conditions for different times. Figure 2A, upper panel shows a representative photomicrograph of HIF-1α immunostaining in RS4;11 cells cultured under normoxia and hypoxia over time (0, 0.5, 1, 3, 6 and 9 h). The results show, as expected, that there is a gradual increase in the expression of HIF-1α over time under hypoxia. Interestingly, immunostaining was preferably observed at the nuclear level indicating activity of this transcription factor. When we quantified the expression of this transcription factor, we observed a significant gradual increase after three hours under hypoxic conditions ($p = 0.02$) until nine hours (Figure 2A, bottom panel). Remarkably, very similar results were obtained when we evaluated YY1 expression in the same experiment; Figure 2B, upper panel shows a representative photomicrograph of YY1 immunostaining. Immunostaining observed mainly at the nuclear level revealed a gradual increase in YY1 expression with respect to time under hypoxic conditions. When we performed the quantification of the expression of this transcription factor, a significant increase was observed from six hours in hypoxic conditions ($p = 0.03$) and a gradual escalation after 9 h ($p = 0.005$) (Figure 2B, bottom panel). The increased expression of YY1 under hypoxic conditions were analyzed by RT-PCR (Figure 2C). A similar significant increase in the expression of YY1 mRNA is observed at 3, 6 and 9 h under hypoxic conditions. We next evaluated the expression of genes targeting these transcription factors. The expression of MDR1 (target gene of HIF-1α) and c-Myc (target gene of YY1) significantly increased after 6 h in hypoxic conditions (Figure 3).

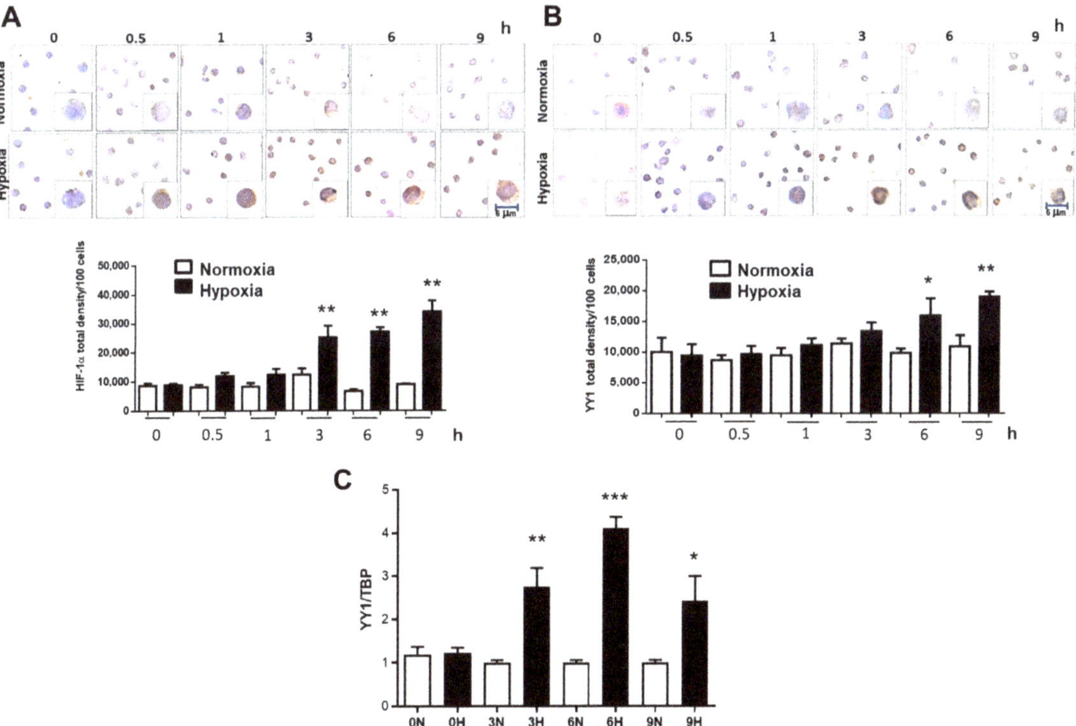

Figure 2. Hypoxia induces expression of YY1. RS4:11 were cultured under normoxia or hypoxia conditions for different times (0, 0.5, 1, 3, 6, 9 h). Slides were then prepared for ICC staining for HIF-1α, (**A**) and YY1 (**B**). The immunostaining shows significant increase of HIF-1α after two hours (* $p = 0.035$) under hypoxia compared with normoxia conditions. Interestingly, YY1 expression increases significantly later, at time 6 h (** $p \leq 0.03$), both proteins continue to increase after 9 h (*** $p = 0.005$ one-way ANOVA-test). (**C**) Analysis of YY1 expression by evaluated using real-time PCR. Bars represent the media of an assay by triplicate. Differences were analyzed by one-way ANOVA and Tukey's multiple comparisons test (** $p = 0.01$, *** $p = 0.0005$, * $p = 0.05$ one-way ANOVA-test).

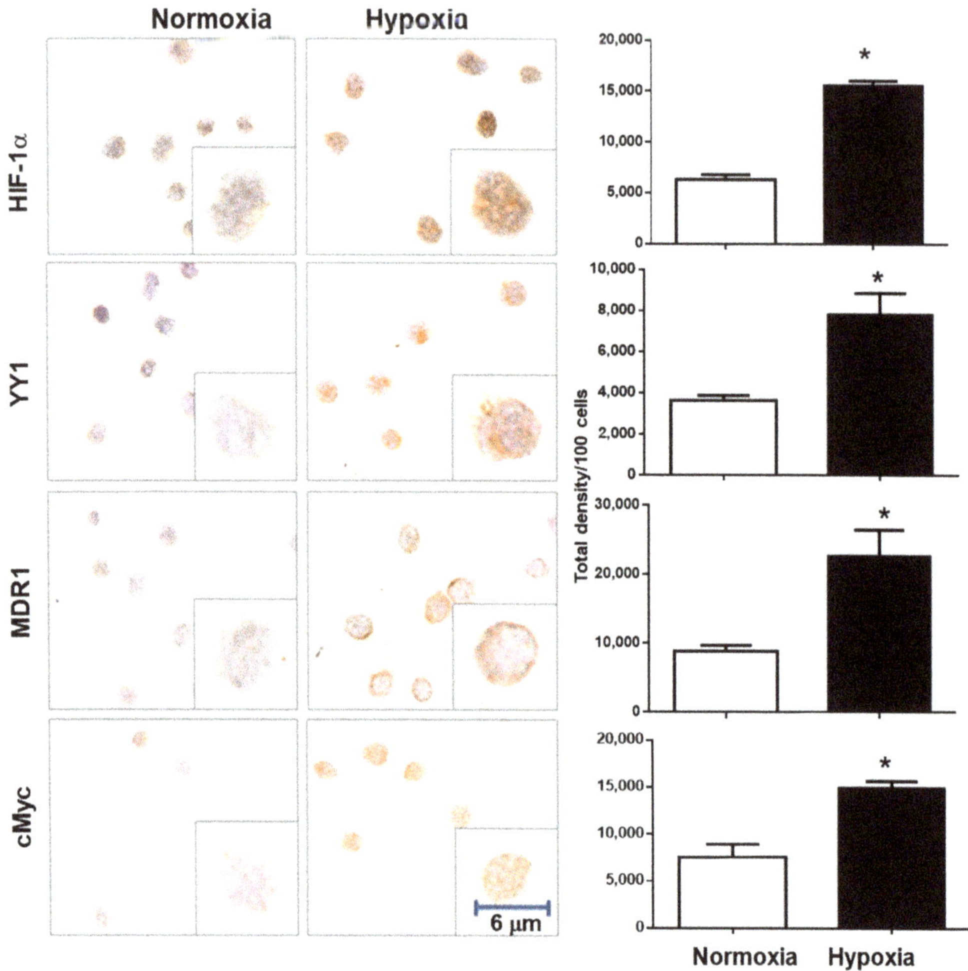

Figure 3. Hypoxia induces the expression of YY1 target genes. To evaluate the transcription factor activation of HIF-1α and YY1, ICC for target genes of each transcription factor were performed. MDR1 for HIF-1α and c-Myc for YY1. Significant increases for both target genes were observed after 6 h. (* $p < 0.05$ Student *t*-test). Representative images of a triplicate of each experiment are shown.

We next explored the effects of inhibiting HIF-1α activity using a chemical inhibitor 2-methoxyestradiol (2ME). Figure 4A shows a representative photomicrograph of HIF-1α and YY1 immunostaining in RS4;11 after a 6-h culture under hypoxia. We found that treating cells with 2ME (0.5 or 1 µM) inhibits the HIF-1α expression. As expected, the expression of YY1 also decreased proportionally with HIF-1α expression. These results were corroborated by real-time PCR analysis under normoxia and hypoxia at 3 and 6 h with 2ME (0.5 µM)(Figure 4B).

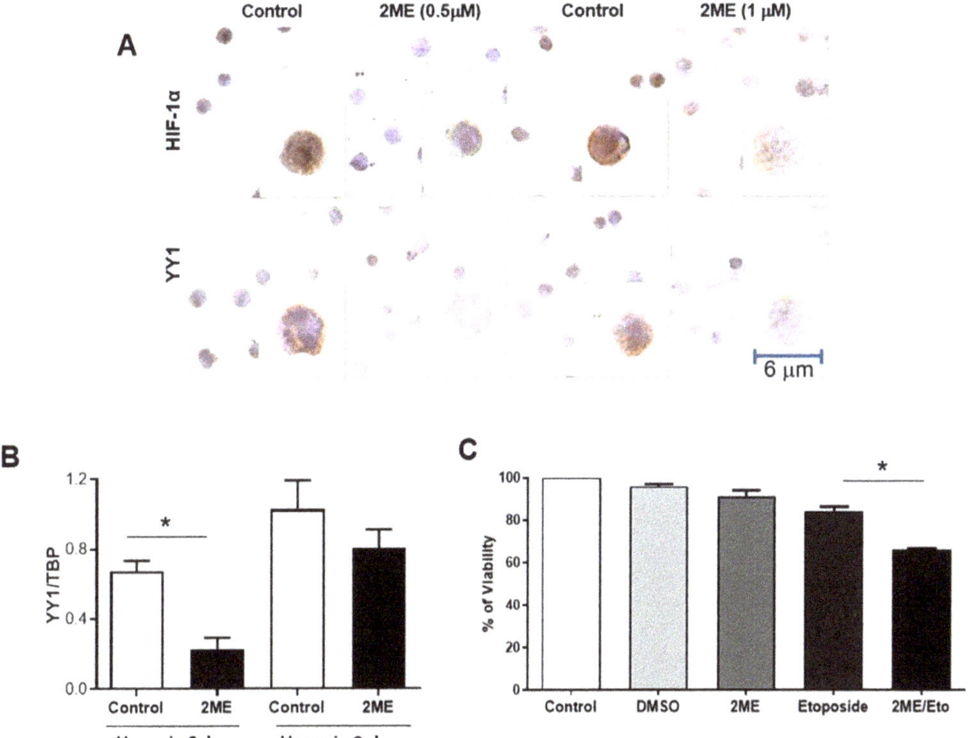

Figure 4. The abrogation of HIF-1α by a chemical inhibitor induces a reduction in YY1 expression. (**A**) is an ICC staining of HIF-1α and YY1 in RS4;11 cells, untreated and treated with 2ME (0.5 and 1 µM) for 6 h. Results show that when HIF-1α is inhibited with 2ME, YY1 decreases correspondingly. (**B**) These results were corroborated by real-time PCR, untreated and treated with 2ME (0.5 µM) for 3 or 6 h. A significant decrease was found at 3 h under hypoxia (* $p = 0.01$ one-way *t*-student). At 6 h there were no significant changes but there was a clear trend. (**C**) RS4;11 cells were treated with 2ME alongside with etoposide, a chemotherapy drug. A significant decrease in viability of RS4;11 cells is observed (* $p = 0.01$ one-way ANOVA-test) when treated with etoposide in combination with 2ME. Graph shows the results of triplicates of three independent experiments. RS4;11 cells were pre-treated for 12 h with 2ME (0.5 µM). Subsequently, the cells were treated with etoposide (0.125 µg/mL). Cell viability was determined after 50 h of co-treatment using MTT (* $p < 0.05$, vehicle or Eto vs. 2ME + Eto).

We then evaluated if 2ME sensitizes RS4;11 cells to the chemotherapeutic drug, etoposide, since it has been shown that both HIF-1α and YY1 positively regulate the expression of MDR1 [28,32]. Remarkably, we found that the combined treatment of 2ME with etoposide (2ME/Eto) significantly decreases the percent viability of RS4;11 compared with single treatment with 2ME, etoposide or control ($p = 0.01$) (Figure 4C).

2.3. Correlation of HIF-1α and YY1 Expression in Pediatric Patients with ALL

We next examined the expression of HIF-1α and YY1 in 108 patients with ALL chemotherapy, untreated, as well as 50 healthy controls; the clinical characteristics of our study population are shown in Table 1. In Figure 5A, we show a representative photomicrograph of immunostained HIF1α and YY1, where both proteins are observed in patients with ALL were compared to healthy controls. The expressions of HIF-1α and YY1 were mainly at the nuclear level. When expression of both proteins was quantified, a significant increase in their expression was observed in patients with ALL as compared to healthy controls, $p = 0.0001$ for HIF-1α and $p = 0.04$ for nuclear YY1 (Student t-test) (Figure 5B). In order to validate our results in peripheral blood an analysis of YY1 expression in patients undergoing treatment was conducted. YY1 levels were measured from patients at time of diagnosis as well as different phases of treatment (remission and consolidation) and compared with YY1 expression in healthy controls. Results show that patients at diagnosis have a significantly greater expression of YY1 compared with healthy controls ($p = 0.001$). However, after remission and consolidation phases of treatment, expression of YY1 decreases to levels similar to that of control cells (Figure 5C). We performed an analysis of mRNA expression for HIF-1α and YY1 in 46 bone marrow samples from pediatric patients with ALL. The results show that elevated expression of HIF-1 mRNA correlates with high expression of YY1 mRNA ($r = 0.35$, $p = 0.0197$) (Figure 5D).

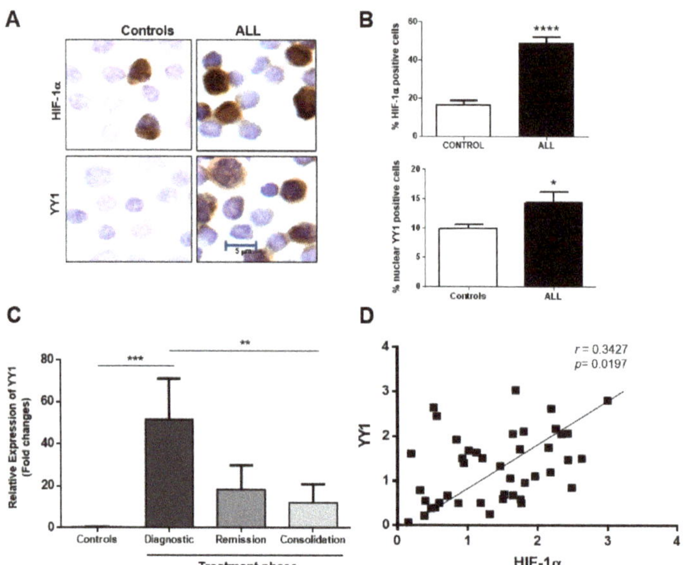

Figure 5. Overexpression of HIF-1α and YY1 in samples from ALL pediatric patients. (**A**) Representative microphotography that show the ICC staining for HIF-1α and YY1 in peripheral blood cells from ALL pediatric patients which clearly shows an overexpression of both proteins in ALL patients as compared with healthy controls. (**B**) This difference is significant when 108 ALL patients are evaluated (**** $p = 0.0001$ and * $p = 0.04$, HIF-1α and YY1, respectively; Student t-test). (**C**) RT-PCR of YY1 levels measured from patients at diagnoses, remission and consolidation shows that YY1 levels are significantly greater in 10 ALL patients at time of diagnoses compared to healthy controls (*** $p = 0.001$) but decrease to levels similar to those of healthy control during remission and consolidation phases of treatment (** $p = 0.01$ one-way ANOVA test). (**D**) To corroborate results from peripheral blood cells, RT-PCR was performed for bone marrow samples of 46 pediatric patients with ALL. The results again show a strong positive correlation between HIF-1α and YY1 (* $p = 0.0197$, $r = 0.35$, Pearson analysis).

Table 1. Clinical patient characteristics.

Total Number	108
Gender	
Female	45
Male	63
Age (years)	7.8 (0.1–16)
Phenotype	
B	32
Pre-B	40
Pro-B	13
T	11
Unknown	12
Survivor	79
Deaths	29

2.4. Network Analysis of HIF-1α/YY1 and Correlation between HIF-1α and YY1 Expression in ALL

The analysis performed showed the protein–protein interaction (PPI) relationships for YY1 and HIF-1α genes. With a combined score of >0.4 when selected and restricted to *Homo sapiens*, we found a close relationship between seven genes confirmed by curated and experimental data. The analysis identified the co-activator EP300 as likely playing a role in the regulation of HIF-1α and YY1, so YY1 probably also plays an important role in the regulation of EP300 and HIF-1α. Additionally, the meta-analysis showed an interaction with HIF1AN, the inhibitor of HIF-1α, VHL, a protein involved in the ubiquitination and degradation of HIF-1α, TCEB1, a subunit of the transcription factor B (SIII) complex closely regulated by VHL and finally with EGLN1, which codes a protein that catalyzes the post-translational formation of 4-hydroxyproline in HIF-1α proteins (Figure 6A). In order to corroborate our data, we performed bioinformatics analysis. The microarray data were from 87 ALL-B phenotype samples out of the 127 different types of leukemia samples present in a related data set (GSE7186). Figure 6B showed the selective analysis of HIF-1α and YY1 co-expression, where significant correlation was found for ALL-B samples (** $p < 0.0011$, $r = 0.3597$). These results are consistent with our results showed in Figure 1 where we demonstrated that HIF-1α expression and YY1 expression are positively correlated in samples derived from pediatric ALL patients.

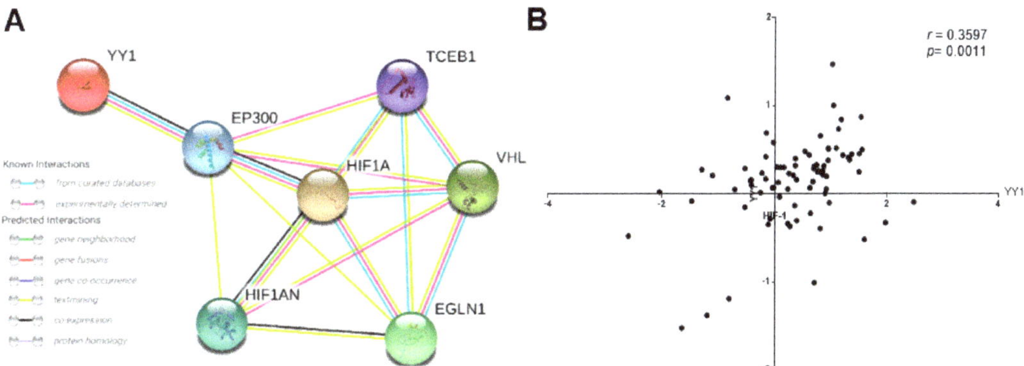

Figure 6. Gene expression and correlation of HIF-1α and YY1 in ALL. (**A**) Network analysis between

HIF-1α/YY1. The Search Tool for the Retrieval of Interacting Genes (STRING) database [32], was used to visualize biological networks and integrate data of the protein–protein interaction between HIF-1α and YY1 [33]. The version 11.0 of STRING (Academic Consortium) was employed to seek for the protein–protein interaction (PPI), data limited to *Homo Sapiens* and a confidence score >0.4. (**B**) Analysis of HIF-1α and YY1 expression levels in several subtypes of ALL was performed using a public dataset of microarrays retrieved from the Oncomine database and the Gene Expression Omnibus NCBI gene expression and hybridization array data repository, obtained from an analysis from Anderson et al. [34]. According with Oncomine^{TM} the results shown that HIF-1α expression levels correlated with YY1 expression (*** $p < 0.0001$, $r = 0.468$, Pearson Analysis).

3. Discussion

HIF-1 was initially characterized as a molecular regulator to control oxygen homeostasis [35]. HIFs have long been studied in solid tumors, where they have been shown to promote wide-ranging tumor-promoting processes, including neo-angiogenesis, cancer metabolism, maintenance of cancer stem cells, and immune evasion [36]. Tumor hypoxia strongly correlates with poor prognosis in solid cancers [37], and in the past 10 years, clear involvements of HIFs have also been demonstrated in leukemia. However, the role of HIF-1α in the transcriptional regulation of the YY1 gene so far remains largely unknown. Upon analysis of the YY1 promoter, we found three putative HIF-1α binding sites. We therefore hypothesized that HIF-1α is involved in positively regulating YY1 and is the reason for chemoresistance in tumor cells.

To demonstrate the above, we performed reporter plasmids and site-directed mutagenesis and eliminated each of the binding sites for HIF-1α. The results demonstrated that mutation of the -592 site results in a drastic decrease in the activity of the reporter gene (site with higher 9.3. JASPER), indicating that this site is the most important for the regulation of the expression of YY1 by HIF-1α. Furthermore, ChIP assays demonstrate that the binding of HIF-1α in the YY1 promoter in ALL cells. To corroborate the impact of HIF-1α inhibition of YY1 positive expression, we cultured RS4;11 leukemia cells in hypoxic conditions over time. The results of this experiment reveal that hypoxia induces significant nuclear translocation of HIF-1α after two hours of hypoxia. Importantly, in the same experiment the expression of YY1 was also upregulated under 6 h of hypoxia. This was corroborated by WB analysis. To demonstrate the activity of HIF-1α and YY1, we evaluated the expression of the target genes of these transcription factors, MDR1, regulated by HIF-1α and YY1 [22], and c-Myc regulated by YY1 [38]. The results demonstrate that hypoxic conditions induce the activity of both HIF-1α and YY1. These results are very significant, if we consider that tumor cells show an increase HIF-1α activity (due to either conditions of hypoxia or the presence of a pro-inflammatory environment) and that increase correlates with more angiogenesis, resistance to apoptosis and induction of MDR1 expression, among other mechanisms [22,39]. These findings represent a new mechanism induced by HIF-1α to increase the malignancy of tumor cells. In addition, we recently published that leukemia cell RS4;11 co-expressed both with HIF-1α and YY1 under hypoxia, which correlated with a downregulation of Fas expression. During hypoxia, the levels of apoptosis diminished after an agonist of FasL (DX2) treatment. Moreover, a bioinformatics analysis revealed that patients with high levels of HIF-1α also express high levels of YY1 and low levels of Fas. These results suggest that YY1 negatively regulates the expression of the Fas receptor, which would be involved in the escape of leukemic cell from the immune response is another mechanism to contributing to the ALL pathogenesis [29]. Once we demonstrated that HIF-1α positively regulates the expression of YY1, we then evaluated the effect of using a chemical inhibitor of HIF-1α activated activity. A natural metabolite of 17-β-estradiol that does not bind to the estrogen receptor, 2-methoxyestradiol (2ME), has anti-proliferative and anti-angiogenic activity [40,41]. Therefore, 2ME was approved several years ago by the FDA for use in humans to treat cancer, and to date it is used in patients with nasopharyngeal cancer, multiple myeloma, prostate cancer, among others [41,42]. However, its effect on ALL cells has not been studied. To evaluate the effect of 2ME in tumor cells, we performed a series of experiments using this chemical inhibitor to treat RS4;11 cells. As previously demonstrated, RS4;11 cells under 6 h of hypoxic conditions show a significant increase in HIF-1α and YY1 expression. In order to discern the localization of HIF-1α and YY1 expression in ALL cells before and after treatment with 2ME, we performed immunocytochemical assays. The results demonstrated a clear decrease in HIF-1α and YY1 expression after treatment with this chemical inhibitor, and very similar results were obtained after performing RT-PCR.

Next, it was important to evaluate whether treatment with 2ME induces reversal of chemoresistance in ALL cells, so we performed cell viability tests with RS4;11 cells after treatment with 2ME alone and in combination with different concentrations of the chemotherapeutic drug etoposide, which is used in the treatment of pediatric patients with ALL. The results are shown in Figure 4C. A sensitivity to the drug of up to 40% after treatment with 2ME is observed in contrast to cells treated with the drug or inhibitor separately. This is very relevant, as this drug is currently used in the treatment of pediatric patients with ALL, and is a substrate of MDR1 [43]. These results are consistent with those reported by other groups demonstrating that 2ME sensitizes different cancer cells to die in the presence of chemotherapeutic drugs, including acute myeloblastic leukemia [44].

To corroborate our results in vitro, we evaluated HIF-1α and YY1 expression in patients with ALL. Our results demonstrated the significant expression of both proteins in ALL pediatric patients. Furthermore, this expression was directly proportional. These results are novel, considering that no previous studies have shown the involvement of HIF-1α in the regulation of YY1 in the pathophysiology of ALL. Our results clearly show constitutive expression of both proteins, strongly suggesting the importance of YY1 protein expression in the pathogenesis of ALL pediatric patients. In this study, we determined HIF-1α and YY1 expression levels by ICC and analyzed this expression in mononuclear peripheral blood cells and bone marrow cells derived from pediatric ALL patients. Our results are consistent with previous findings which demonstrate that tumor hypoxia strongly correlates with poor prognosis of solid cancers [37]. Additionally, in the past 10 years, clear evidence of HIFs have been shown to be involved in the pathogenesis of leukemia [17,44,45]. However, some reports have suggested that HIFs may exert tumor suppressive functions in acute myeloid leukemia, albeit this may be limited to specific disease sub-contexts. For example, Vukovic et al., demonstrate that knockdown of HIF-1α or HIF-2α in human acute myeloid leukemia (AML) samples results in their apoptosis and inability to engraft. Both Hif-1α and Hif-2α synergize to suppress the development of AML. (Vukovic M, Hif-1α and Hif-2α synergize to suppress AML development but are dispensable for disease maintenance [45].

It has been shown that YY1 is an important negative regulator of the tumor suppressor factor p53 [46]. Wu S. et al. demonstrated that inhibition of YY1 reduced the accumulation of HIF-1 α and its activity under hypoxic conditions, and consequently downregulated the expression of HIF-1 α target genes. In addition, it was demonstrated that the downregulation of HIF-1 α by inhibiting YY1 is p53-independent. Therefore, YY1 inhibition could be considered as a potential tumor therapeutic strategy to give consistent clinical outcomesYY1 is associated with HIF-1 α regulation under hypoxia, and targeting YY1 might be a potential therapeutic strategy of solid cancer [47].

Very limited studies are available on the function of HIF1 factors in ALL. So far, it has been shown that HIF-1α is expressed in ALL that reside in the BM [48]. Accordingly, HIF-1α is induced by stroma-mediated AKT/mTOR signaling in pre-B-ALL, and confers resistance to chemotherapy [49]. This information is consistent with our findings which show that HIF-1α positively regulates YY1 expression, and that several transcription factors are reportedly involved in chemoresistance of different types of cancer including ALL [10,11,33,34]. These findings represent a new mechanism of chemoresistance in ALL. In addition, Zhea, N. et al. demonstrated a high expression of HIF-1α in human AML cell lines and the inhibition of HIF-1α by 2ME has potential anti-leukemia activity through activation of the mitochondrial apoptotic pathway mediated by ROS. In addition, it is not cytotoxic to normal cells. 2ME is therefore a potential candidate for the treatment of

AML [44]. Our results, demonstrating chemo-sensitization of drugs in leukemia cells using 2MF, strongly suggest that this metabolite can also be effective in ALL.

Based on data retrieved from Oncomine, we found that HIF-1α and YY1 mRNA were expressed in several leukemia subtypes, especially ALL, as shown in Figure 6. We found a positive correlation between the expression of HIF-1α and YY1 in several data sets analyzed from the leukemia study by Anderson et al. [34]. This correlation is consistent with findings from our in vivo patient samples and confirm the interaction and regulation of YY1 by HIF-1α and also are consistent with our previously reports to reveal that by a bioinformatics analysis that patients with high levels of HIF-1α also express high levels of YY1 and low levels of Fas [29]. These findings indicate that HIF-1α and YY1 might participate in the initiation and progression of ALL via positive transcriptional regulation. On the other hand it has been described that the inhibition of YY1 disrupts hypoxia-stimulated HIF-1α stabilization in a p53-independent manner, reducing the accumulation of HIF-1α and its activity under hypoxic condition, and consequently downregulated the expression of HIF-1α target genes [47]. Therefore, the transcriptional regulation of YY1 on the HIF-1a promoter is feasible, and possible participates in the regulation of its expression, which suggests a bidirectional regulation.

A bioinformatic analysis with Cytoscape permits the identification of active subsets/modules. A network was analyzed in conjunction with gene expression databases (microarray databases used in this study: ONCOMINE, GEO-NCBI) to identify sets of connecting interactions between proteins by identifying interaction subsets in which genes show particularly high levels of differential expression. The interactions contained within each subset provide hypotheses for regulatory and signaling interactions controlling observed changes in expression. One can search for groups (highly interconnected regions) and load any network in Cytoscape. Depending on the type of network, groups can have different meanings. Networks are designed with automated algorithms. Our Cytoscape analysis identified interactions between HIF-1α and YY1, and this correlation was confirmed by experimental findings obtained with ChIP and binding site mutation. The analysis identified co-activator EP300 as playing a role in the regulation of HIF-1α and probably in YY1 as well, but interestingly, HIF-1α may also play an important role in the regulation of both EP300 and YY1. Additionally, the meta-analysis showed an interaction with HIF1AN, the inhibitor of HIF-1α, VHL, a protein involved in the ubiquitination and degradation of HIF-1α, TCEB1, a subunit of the transcription factor B (SIII) complex closely regulated by VHL and finally EGLN1, which catalyzes the post-translational formation of 4-hydroxyproline in HIF-1α proteins, from data that was published already by other research groups [50].

This is the first report describing a correlation between HIF-1α and YY1 expression in pediatric ALL patients, and this study identifies HIF-1α and YY1 as potential disease markers, which could be considered biomarkers at the time of diagnosis for predicting disease behavior. We also propose that the use of pharmacological or chemical inhibitors targeting HIF-1α and YY1 could be an alternative treatment for pediatric patients with ALL that are known to be positive for HIF-1α and YY1 expression, thus offering a therapeutic alternative for this disease.

The model in Figure 7 summarizes all our findings.

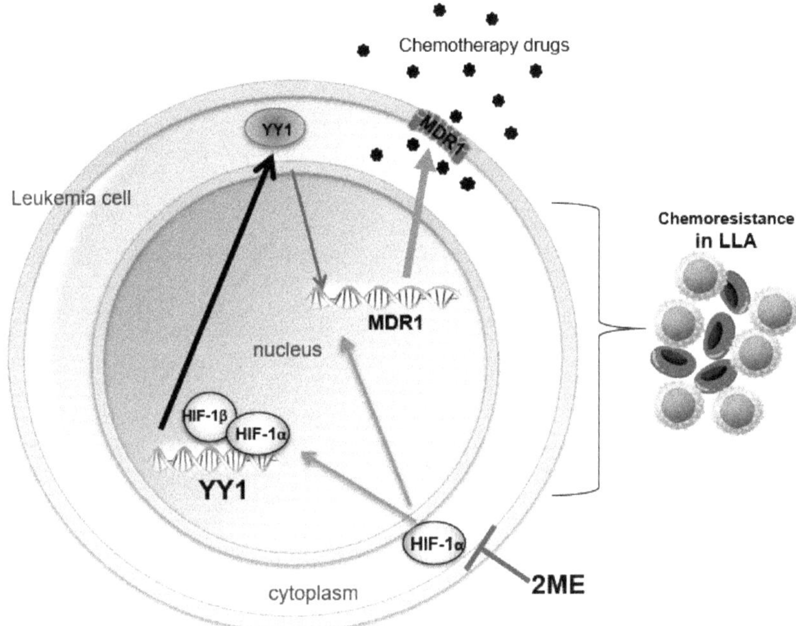

Figure 7. A schematic of HIF-1α transcriptionally regulating YY1 expression. In this study we demonstrated that HIF-1α transcriptionally regulates YY1. In addition, it has been shown that both HIF-1α and YY1 increase expression of MDR1, a membranous protein which pumps out chemotherapy drugs, inducing chemoresistance in leukemia cells. Interestingly, here we demonstrated that this pathway is inactivated by the inhibition of HIF-1α by 2ME, which induces sensitizes leukemia cells to chemotherapy.

4. Materials and Methods

4.1. Ethical and Biosecurity Aspects

This project involved the manipulation of blood of pediatric patients with ALL. Due to this, during development, the standards of good laboratory practices were followed to avoid occupational risks. The study was in accordance with the regulations of the General Health Law on Research in Mexico, as well as the research standards of the Hospital Infantil de México Federico Gómez. A letter of informed consent was signed in all cases. The data provided by the patient's clinical file were kept confidential, according to the Helsinski international standard for research.

4.2. Patients

In this study, mononuclear cells from peripheral blood (PBMC) samples were isolated from 108 patients with ALL (0–16 years of age) who were diagnosed between 2009 and 2018. 10 samples were collected during the different phases of treatment (remission and consolidation). The samples were obtained from the Oncology Pediatric Services of both the Hospital Infantil de México Federico Gomez and the Hospital Pediatrico Moctezuma. In this study, we also included 46 bone marrow (BM) samples from pediatric patients with ALL, who were diagnosed between 2014 and 2016 in the Oncology Pediatric Service of the Instituto Nacional de Pediatria, SSA. The diagnosis was established by cytological examination of bone marrow smears according to the French-American-British (FAB) group. Cytochemical tests included staining for Periodic Acid Schiff (PAS) and myeloperoxidase (MPO) [51,52]. The pediatric control group was composed of 50 children (mean age:

7.8 years; range: 5 to 15 years) who were admitted to the hospital for elective surgery and who were free of any known viral or bacterial infections.

4.3. Cell Culture

Cells were cultured from the RS4;11 (acute phenotype B lymphoblastic leukemia cell line, ATCC: CRL-1873) and PC3 cell lines (prostatic carcinoma; ATCC CRL-1435), which overexpress YY1. A 25 cm^2 box was cultured with RPMI advanced 1640 medium supplemented with 5% FBS (GIBCO-Invitrogen), 1% v/v L-glutamine (GIBCO-Invitrogen), 1% v/v sodium pyruvate (100 mM GIBCO-Invitrogen), 1% antibiotic (GIBCO-Invitrogen) and 1% non-essential amino acids (GIBCO-Invitrogen). Cells were maintained in culture at 37 °C and 5% CO_2.

4.4. Determination of the Putative Binding Sites of HIF1α in the YY1 Promoter

The prediction of the binding sites for HIF-1α in the promoter of the YY1 gene was performed by the TESS (Transcription Element Search System) program which conjugates the databases of TRANSFAC v6.0, JASPAR 20060301, IMD v1.1 and IWC/GibbsMat v1. 2000 nucleotides upstream (−2000 bp) of the gene promoter ATG sequence at 350 nucleotides downstream (+350 bp) were analyzed. Three putative binding sites located at nucleotides −622 bp, −592 bp and +199 bp were identified with respect to the YY1 gene TSS.

4.5. Cloning of the Promoter Region of YY1

Genomic DNA was extracted from a healthy volunteer using TRIzol (invitrogen) and used as template to amplify the promoter region of the YY1 (−2000 bp to +350 bp relative to the start site) by PCR using a specific set of primers. Once amplified, the YY1 promoter was purified and cloned into the vector pJet (Thermo Fisher Scientific). Then the YY1 promoter was subcloned from pJet to the pGL3 vector (Promega) generating the construct pGL3-YY1-pro with Luciferase as a reporter gene.

4.6. Site-Direct Mutagenesis in Putative Binding Sites for HIF-1α of the Promoter YY1

Once the pGL3-YY1-pro construct was obtained, a site directed mutagenesis was performed in the three putative binding sites using a system of commercial site-directed mutagenesis QuickChange Lighting Site-Directed Mutagenesis, (Agilent Technologies, Santa Clara, CA, USA) following the manufactures instructions. Importantly, the computer algorithm used provided the supplier with the primer design. The incorporation of the mutations in the putative HIF-1α sites was performed by PCR using the primers show in Table 1. After the restriction enzyme digestion of the PCR products with Dpn-1, *E. coli* strain DH5-alpha bacteria were transformed with the digested PCR products.

4.7. Transfection of Cell Lines with the Construct Generated

PC3 cells were co-transfected with 2 μg of DNA total from each of the plasmid constructs generated (pGL3-YY1-pro or the mutants generated from it) and pCMV-SPORT-β-gal vector (Invitrogen, Waltham, MA USA) at a 6:1 proportion. The pCMV-SPORT-β-gal vector was used to normalize the enzymatic activities from the reporter genes regarding transfection efficiency. Following the manufacturer's recommendations, Lipofectamine 2000 (Invitrogen) was used as a transfection reagent. After 48 h post-transfection, intracellular proteins were obtained to determine the enzymatic activity of the implicated reporters. Using commercial substrates, the luciferase (Promega, Madison, WI, USA), and β-galactosidase (Clontech, Santa Clara, CA, USA), activities were measured in the multimodal reader plates EnSpire (Perkin Elmer, Waltham, MA, USA).

4.8. Chromatin Immunoprecipitation

RS4;11 cells (1×10^6 previously stimulated using hypoxic or normoxic conditions (at the indicated time points) and then chromatin immunoprecipitation (ChIP) was performed as previously described [28]. The specific antibody against HIF1α (ab2185; Abcam) was used. The DNA recovered after the ChIP assay was used as template for PCR reactions using the specific set of primers (Table 2).

Table 2. Primer sequences.

Name	Sequence
YY1pro_Sen (BglI underlined)	CG**AGATCT**GCTTTTTTGAACAGAGAGCC
YY1Pro_Ant (BamH1 underlined)	GA**GGATCC**GGGTGCAAACCG
YY1mutAant (Mutation in bold)	GCCCGCGGCG**AAGA**ACGTCAGCGCGCCGCCGCC
YY1mutAsen (Mutation in bold)	GGCGGCGGCGCGCTGACGTT**CTT**CGCCGCGGGC
YY1mutBant (Mutation in bold)	GGCGGGGCGGCTCG**AGAA**CGCCCTGGCTGGCC
YY1mutBsen (Mutation in bold)	GGCCAGCCAGGGCG**TTCT**CGAGCCGCCCCGCC
YY1mutCant (Mutation in bold)	GTGGTGGTGGCCGG**AAGA**CCCGTGGCCGCCCC
YY1mutCsen (Mutation in bold)	GGGGCGGCCACGGG**TCTT**CCGGCCACCACCAC
Ch(PromYY)_HIFABsen	TTTTGTGGCTGTTGCACCG
Ch(PromYY)_HIFABant	AATCGATCTGTCCGCTGGC
Ch(PromYY)_HIFCsen	AGACCATCGAGACCACAGTGGTGG
Ch(PromYY)_HIFant	CTGCAGAGCGATCATGGGCG
Ch(PromTel)_HIFposSen	AGCGCTGCGTCCTGCT
Ch(PromTel)_HIFposAnt	AGCACCTCGCGGTAGTGG

4.9. Treatments and Exposure of ALL Cells to Hypoxic Conditions

RS4;11 (5×10^5) cells in culture were exposed to low oxygen conditions (1% of O_2) in a hypoxia chamber Bactrox (Scientific Biogen, Madrid, Spain), during different periods. Subsequently the cells were harvested to perform immunocytochemistry and RT-PCR assays, cells were treated or untreated for 3 or 6 h with 2ME (0.5 or 1 μM) under normoxia or hypoxia. The RS4;11 cells were pretreated 12 h with 2ME (0.5 μM) and then treated or untreated for 18 h with etoposide (0.0625 μM). The viability of RS4;11cells were evaluated after treatment as describe below.

4.10. Immunocytochemistry

Immunostaining was performed as previously describe [28]. Anti-HIF-1α, anti-c-Myc, anti-YY1 or anti-MDR1 (Novus Biological, Litletown, CO, USA).

4.11. Real-Time RT-PCR Assays

For YY1 mRNA expression, the total RNA was obtained from RS4;11, 5×10^6 cells were cultured under normoxic or hypoxic conditions. After treatment, cells were collected and RNA was purified using the miRNeasy Mini Kit (QIAGEN, Germantown, MD, USA) according to manufacture instructions. cDNA was performed using 1 μg of RNA of each sample according to TaqMan Reverse Transcription kit (Applied Biosystems, Foster City, CA, USA). Gene expression was analyzed by real-time PCR using Maxima SYBR Green/Fluorescein qPCR Master Mix (Thermo Scientific, Waltham, MA, USA) according to manufacture instructions. In addition, the total RNA was obtained from bone marrow aspirate samples from patients with ALL and healthy controls as previously described [52]. Subsequently, the expression of HIF-1α and YY1 was determined using the Universal Probe Library Set, Human (Roche Diagnostics GmbH, Mannheim, Germany). The expression levels were then in a computer attached to the thermocycler StepOne™ Real-Time PCR System (Applied Biosystems Foster City, CA, USA). Data capture and analysis was carried out with the thermal cycler program. The PCR conditions were 1 cycle at 95 °C/10 min, 45 cycles of 95 °C/15 s and 60 °C/1 min. Using specific primers for HIF1α, and YY1. To normalize the amount of cDNA the OAZ gene (ornithine decarboxylase antizyme) was used. The sequence of the primers for each gene are shown in Table 2.

4.12. Cell Viability Assays

Cell viability assays were performed via MTT colorimetric assays following the manufacture's instruction (Roche Diagnostics Co., (Basel, Switzerland). Briefly, after RS4;11 cells treatment the MTT reagent was added, after 4 h the solubilization of the salts was carried out by means of the solubilizing agent included in the MTT kit to later incubate for one night. Once the incubation occurred, the plate is revealed by quantifying the absorbance in a multi-reader EnSpire (PerkinElmer) plate of PE at 620 nm.

4.13. Network Analysis of HIF-1α/YY1 and the Construction of Gene Networks Related to Function

To demonstrate the functional interactions between HIF-1α and YY1 and other transcriptional factors, a bioinformatics analysis was performed with GeneMANIA using the free software Cytoscape 2.8, which visualizes biological networks and integrates data, and the database Oncomine [53]. Typically, annotations used by Cytoscape correspond to the GEO database (Gene Ontology Database) [54]. Both databases permit free access to microarray banks and data meta-analyses and networks, which allow the prediction of interactions between genes or proteins of interest.

To elucidate the interactions between YY1/HIF-1α and other proteins, a bioinformatics analysis was performed with The Search Tool for the Retrieval of Interacting Genes (STRING) database (https://string-db.org/), which visualizes biological networks and integrates data from curated and experimental determinations [32].

4.14. Bioinformatics Analysis of and Correlation between HIF-1α and YY1 Gene Expression in ALL

An analysis of HIF-1α and YY1 expression levels in ALL was performed using a public data set of microarrays retrieved from the Oncomine and Gene Expression Omnibus databases, derived from a published analysis reported by Andersson, A. et al. [48].

4.15. Statistical Analysis

A database was analyzed, and the information was processed using a statistical analysis program (Prism 4® from GraphPad Software, Inc., San Diego, CA, USA), and the evaluation of the difference in the number of positive cells from the immunocytochemical reactions was performed by analysis of variance (ANOVA). The correlation analysis was performed using Pearson analysis. All the data in the graphs are represented as a mean ± SEM. A value less than or equal to 0.05 was considered significant. For statistical analysis we used the STATA program (version 11.00).

5. Conclusions

In conclusion, our findings demonstrate for the first time that HIF-1α regulates transcriptional YY1 and increasing or inhibiting HIF-1α expression directly affects YY1 expression in ALL cells. Furthermore, HIF-1α and YY1 expression are increased in pediatric patients with ALL, and high expression of HIF-1α correlates with the presence of YY1. Therefore, both HIF-1α and YY1 may be possible therapeutic biomarkers in ALL. The clinical significance of this finding across different cancer types has yet to be determined.

Author Contributions: Conceptualization, S.H.-Y. and G.A.-A.; Methodology, G.A.-A., G.U.M.-R., M.M.-M., E.M.-T., T.V.L.-P., L.A.E.-A., S.J.-M. and E.J.-H.; Software, M.M.-M.; Validation, G.P.-L., S.J.-M. and V.M.D.-B.; Formal Analysis, G.A.-A., G.U.M.-R., M.M.-M. and S.H.-Y.; Investigation, S.H.-Y. and G.A.-A.; Resources, S.H.-Y. and G.A.-A.; Data Curation, G.A.-A., G.U.M.-R. and M.M.-M.; Writing—Original Draft Preparation, S.H.-Y. and G.A.-A.; Writing—Review & Editing, T.V.L.-P., S.H.-Y. and G.A.-A.; Supervision, S.H.-Y. and G.A.-A.; Project Administration, S.H.-Y.; Funding Acquisition, S.H.-Y. All authors have read and agreed to the published version of the manuscript.

Funding: The study was funded by the National Council of Science and Technology (2007-C01-69789 CONACYT and 2019-CONACYT 302142 S.H.-Y.), and the Mexico Federal Funds (Grant HIM/2014/018/SSA1140, G.A.-A. and S.H.-Y.).

Institutional Review Board Statement: The study was conducted according to the guidelines of the Declaration of Helsinki and approved by the Ethics Committee of Hospital Infantil de México Federico Goméz (HIM/2014/018/SSA1140).

Informed Consent Statement: Informed consent was obtained from all subjects involved in the study to publish this paper.

Data Availability Statement: The data underlying this article will be shared on reasonable request from the corresponding author.

Acknowledgments: The authors would like to thank the National Council of Science and Technology, the Mexico Federal Funds of Hospital Infantil de Mexico Federico Gomez and the National Institute of Pediatrics for the facilities provided.

Conflicts of Interest: The authors declare no conflict of interest.

References

1. Fajardo-Gutierrez, A.; Juarez-Ocana, S.; Gonzalez-Miranda, G.; Palma-Padilla, V.; Carreon-Cruz, R.; Ortega-Alvarez, M.C.; Mejia-Arangure, J.M. Incidence of cancer in children residing in ten jurisdictions of the Mexican Republic: Importance of the Cancer registry (a population-based study). *BMC Cancer* **2007**, *7*, 68. [CrossRef] [PubMed]
2. Fajardo-Gutierrez, A.; Rendon-Macias, M.E.; Mejia-Arangure, J.M. Cancer epidemiology in Mexican children. Overall results. *Rev. Med. Del Inst. Mex. Seguro Soc.* **2011**, *49* (Suppl. 1), S43–S70.
3. Mejia-Arangure, J.M.; Fajardo-Gutierrez, A.; Bernaldez-Rios, R.; Paredes-Aguilera, R.; Flores-Aguilar, H.; Martinez-Garcia, M.C. Incidence of acute leukemia in children in Mexico City, from 1982 to 1991. *Salud Publica Mex* **2000**, *42*, 431–437. [PubMed]
4. Perez-Saldivar, M.L.; Ortega-Alvarez, M.C.; Fajardo-Gutierrez, A.; Bernaldez-Rios, R.; Del Campo-Martinez Mde, L.; Medina-Sanson, A.; Palomo-Colli, M.A.; Paredes-Aguilera, R.; Martinez-Avalos, A.; Borja-Aburto, V.H.; et al. Father's occupational exposure to carcinogenic agents and childhood acute leukemia: A new method to assess exposure (a case-control study). *BMC Cancer* **2008**, *8*, 7. [CrossRef] [PubMed]
5. Pui, C.H. Update on Pediatric ALL. *Clin. Adv. Hematol. Oncol.* **2006**, *4*, 884–886. [PubMed]
6. Pui, C.H. Central nervous system disease in acute lymphoblastic leukemia: Prophylaxis and treatment. *Hematol. Am. Soc. Hematol. Educ. Progr.* **2006**, 142–146. [CrossRef]
7. Kourti, M.; Vavatsi, N.; Gombakis, N.; Sidi, V.; Tzimagiorgis, G.; Papageorgiou, T.; Koliouskas, D.; Athanassiadou, F. Expression of multidrug resistance 1 (MDR1), multidrug resistance-related protein 1 (MRP1), lung resistance protein (LRP), and breast cancer resistance protein (BCRP) genes and clinical outcome in childhood acute lymphoblastic leukemia. *Int. J. Hematol.* **2007**, *86*, 166–173. [CrossRef]
8. Kourti, M.; Vavatsi, N.; Gombakis, N.; Tzimagiorgis, G.; Sidi, V.; Koliouskas, D.; Athanassiadou, F. Increased expression of multidrug resistance gene (MDR1) at relapse in a child with acute lymphoblastic leukemia. *Pediatr. Hematol. Oncol.* **2006**, *23*, 489–494. [CrossRef]
9. Li, Y.P.; Tian, F.G.; Shi, P.C.; Guo, L.Y.; Wu, H.M.; Chen, R.Q.; Xue, J.M. 4-Hydroxynonenal promotes growth and angiogenesis of breast cancer cells through HIF-1alpha stabilization. *Asian Pac. J. Cancer Prev.* **2014**, *15*, 10151–10156. [CrossRef]
10. Lu, G.W.; Shao, G. Hypoxic preconditioning: Effect, mechanism and clinical implication (Part 1). *Zhongguo Ying Yong Sheng Li Xue Za Zhi* **2014**, *30*, 489–501.
11. Niu, F.; Li, Y.; Lai, F.F.; Chen, X.G. Research progress of hypoxia-inducible factor 1 inhibitors against tumors. *Yao Xue Xue Bao* **2014**, *49*, 832–836.
12. Ye, Y.; Wang, M.; Li, J.; Shi, Y.; Zhang, X.; Zhou, Y.; Zhao, C.; Wen, J. Hypoxia-inducible factor-1alpha G polymorphism and the risk of cancer: A meta-analysis. *Tumori* **2014**, *100*, e257-65. [CrossRef]
13. Zhu, J.; Cheng, X.; Xie, R.; Chen, Z.; Li, Y.; Lin, G.; Liu, J.; Yang, Y. Genetic association between the HIF-1alpha P582S polymorphism and cervical cancer risk: A meta analysis. *Int. J. Clin. Exp. Pathol.* **2014**, *7*, 6085–6090. [PubMed]
14. Moroz, E.; Carlin, S.; Dyomina, K.; Burke, S.; Thaler, H.T.; Blasberg, R.; Serganova, I. Real-time imaging of HIF-1alpha stabilization and degradation. *PLoS ONE* **2009**, *4*, e5077. [CrossRef] [PubMed]
15. Brader, S.; Eccles, S.A. Phosphoinositide 3-kinase signalling pathways in tumor progression, invasion and angiogenesis. *Tumori* **2004**, *90*, 2–8. [CrossRef]
16. Maxwell, P.H.; Pugh, C.W.; Ratcliffe, P.J. Activation of the HIF pathway in cancer. *Curr. Opin. Genet. Dev.* **2001**, *11*, 293–299. [CrossRef]
17. Wellmann, S.; Guschmann, M.; Griethe, W.; Eckert, C.; von Stackelberg, A.; Lottaz, C.; Moderegger, E.; Einsiedel, H.G.; Eckardt, K.U.; Henze, G.; et al. Activation of the HIF pathway in childhood ALL, prognostic implications of VEGF. *Leukemia* **2004**, *18*, 926–933. [CrossRef] [PubMed]

18. Sonneveld, P. Multidrug resistance in haematological malignancies. *J. Intern. Med.* **2000**, *247*, 521–534. [CrossRef]
19. van den Heuvel-Eibrink, M.M.; Sonneveld, P.; Pieters, R. The prognostic significance of membrane transport-associated multidrug resistance (MDR) proteins in leukemia. *Int. J. Clin. Pharmacol. Ther.* **2000**, *38*, 94–110. [CrossRef]
20. Fardel, O.; Lecureur, V.; Guillouzo, A. The P-glycoprotein multidrug transporter. *Gen. Pharmacol.* **1996**, *27*, 1283–1291. [CrossRef]
21. Sugawara, I. Expression and functions of P-glycoprotein (mdr1 gene product) in normal and malignant tissues. *Acta Pathol. Jpn.* **1990**, *40*, 545–553.
22. Comerford, K.M.; Wallace, T.J.; Karhausen, J.; Louis, N.A.; Montalto, M.C.; Colgan, S.P. Hypoxia-inducible factor-1-dependent regulation of the multidrug resistance (MDR1) gene. *Cancer Res.* **2002**, *62*, 3387–3394.
23. Gordon, S.; Akopyan, G.; Garban, H.; Bonavida, B. Transcription factor YY1: Structure, function, and therapeutic implications in cancer biology. *Oncogene* **2006**, *25*, 1125–1142. [CrossRef]
24. Nicholson, S.; Whitehouse, H.; Naidoo, K.; Byers, R.J. Yin Yang 1 in human cancer. *Crit. Rev. Oncog.* **2011**, *16*, 245–260. [CrossRef] [PubMed]
25. Zhang, Q.; Stovall, D.B.; Inoue, K.; Sui, G. The oncogenic role of Yin Yang 1. *Crit. Rev. Oncog.* **2011**, *16*, 163–197. [CrossRef]
26. Deng, Z.; Cao, P.; Wan, M.M.; Sui, G. Yin Yang 1: A multifaceted protein beyond a transcription factor. *Transcription* **2010**, *1*, 81–84. [CrossRef] [PubMed]
27. Castellano, G.; Torrisi, E.; Ligresti, G.; Nicoletti, F.; Malaponte, G.; Traval, S.; McCubrey, J.A.; Canevari, S.; Libra, M. Yin Yang 1 overexpression in diffuse large B-cell lymphoma is associated with B-cell transformation and tumor progression. *Cell Cycle* **2010**, *9*, 557–563. [CrossRef]
28. Antonio-Andrés, G.; Rangel-Santiago, J.; Tirado-Rodríguez, B.; Martinez-Ruiz, G.U.; Klunder-Klunder, M.; Vega, M.I.; Lopez-Martinez, B.; Jiménez-Hernández, E.; Torres Nava, J.; Medina-Sanson, A.; et al. Role of Yin Yang-1 (YY1) in the transcription regulation of the multi-drug resistance (MDR1) gene. *Leuk. Lymphoma* **2018**, *59*, 2628–2638. [CrossRef] [PubMed]
29. Martínez-Torres, E.; López-Pérez, T.V.; Morales-Martínez, M.; Huerta-Yepez, S. Yin-yang-1 decreases fas-induced apoptosis in acute lymphoblastic leukemia under hypoxic conditions: Its implications in immune evasion. *Bol. Med. Hosp. Infant. Mex.* **2020**, *77*, 186–194. [CrossRef] [PubMed]
30. Bonavida, B. Therapeutic YY1 Inhibitors in Cancer: ALL in ONE. *Crit. Rev. Oncog.* **2017**, *22*, 37–47. [CrossRef]
31. Bonavida, B.; Huerta-Yepez, S.; Baritaki, S.; Vega, M.; Liu, H.; Chen, H.; Berenson, J. Overexpression of Yin Yang 1 in the pathogenesis of human hematopoietic malignancies. *Crit. Rev. Oncog.* **2011**, *16*, 261–267. [CrossRef]
32. STRING. Available online: https://string-db.org (accessed on 12 December 2021).
33. Szklarczyk, D.; Gable, A.L.; Lyon, D.; Junge, A.; Wyder, S.; Huerta-Cepas, J.; Simonovic, M.; Doncheva, N.T.; Morris, J.H.; Bork, P.; et al. STRING v11: Protein–protein association networks with increased coverage, supporting functional discovery in genome-wide experimental datasets. *Nucleic Acids Res.* **2019**, *47*, D607–D613. [CrossRef]
34. Andersson, A.; Ritz, C.; Lindgren, D.; Eden, P.; Lassen, C.; Heldrup, J.; Olofsson, T.; Rade, J.; Fontes, M.; Porwit-Macdonald, A.; et al. Microarray-based classification of a consecutive series of 121 childhood acute leukemias: Prediction of leukemic and genetic subtype as well as of minimal residual disease status. *Leukemia* **2007**, *21*, 1198–1203. [CrossRef] [PubMed]
35. Semenza, G.L. Oxygen homeostasis. *Wiley Interdiscip. Rev. Syst. Biol. Med.* **2010**, *2*, 336–361. [CrossRef]
36. Schito, L.; Semenza, G.L. Hypoxia-Inducible Factors: Master Regulators of Cancer Progression. *Trends Cancer* **2016**, *2*, 758–770. [CrossRef]
37. Wigerup, C.; Pahlman, S.; Bexell, D. Therapeutic targeting of hypoxia and hypoxia-inducible factors in cancer. *Pharmacol. Ther.* **2016**, *164*, 152–169. [CrossRef]
38. Austen, M.; Cerni, C.; Luscher-Firzlaff, J.M.; Luscher, B. YY1 can inhibit c-Myc function through a mechanism requiring DNA binding of YY1 but neither its transactivation domain nor direct interaction with c-Myc. *Oncogene* **1998**, *17*, 511–520. [CrossRef]
39. Eltzschig, H.K.; Bratton, D.L.; Colgan, S.P. Targeting hypoxia signalling for the treatment of ischaemic and inflammatory diseases. *Nat. Rev. Drug Discov.* **2014**, *13*, 852–869. [CrossRef] [PubMed]
40. Chua, Y.S.; Chua, Y.L.; Hagen, T. Structure activity analysis of 2-methoxyestradiol analogues reveals targeting of microtubules as the major mechanism of antiproliferative and proapoptotic activity. *Mol. Cancer Ther.* **2010**, *9*, 224–235. [CrossRef] [PubMed]
41. Fotsis, T.; Zhang, Y.; Pepper, M.S.; Adlercreutz, H.; Montesano, R.; Nawroth, P.P.; Schweigerer, L. The endogenous oestrogen metabolite 2-methoxyoestradiol inhibits angiogenesis and suppresses tumour growth. *Nature* **1994**, *368*, 237–239. [CrossRef]
42. Zhou, N.N.; Zhu, X.F.; Zhou, J.M.; Li, M.Z.; Zhang, X.S.; Huang, P.; Jiang, W.Q. 2-Methoxyestradiol induces cell cycle arrest and apoptosis of nasopharyngeal carcinoma cells. *Acta Pharmacol. Sin.* **2004**, *25*, 1515–1520.
43. Hijiya, N.; Thomson, B.; Isakoff, M.S.; Silverman, L.B.; Steinherz, P.G.; Borowitz, M.J.; Kadota, R.; Cooper, T.; Shen, V.; Dahl, G.; et al. Phase 2 trial of clofarabine in combination with etoposide and cyclophosphamide in pediatric patients with refractory or relapsed acute lymphoblastic leukemia. *Blood* **2011**, *118*, 6043–6049. [CrossRef]
44. Zhe, N.; Chen, S.; Zhou, Z.; Liu, P.; Lin, X.; Yu, M.; Cheng, B.; Zhang, Y.; Wang, J. HIF-1alpha inhibition by 2-methoxyestradiol induces cell death via activation of the mitochondrial apoptotic pathway in acute myeloid leukemia. *Cancer Biol. Ther.* **2016**, *17*, 625–634. [CrossRef] [PubMed]
45. Vukovic, M.; Guitart, A.V.; Sepulveda, C.; Villacreces, A.; O'Duibhir, E.; Panagopoulou, T.I.; Ivens, A.; Menendez-Gonzalez, J.; Iglesias, J.M.; Allen, L.; et al. Hif-1α and Hif-2α synergize to suppress AML development but are dispensable for disease maintenance. *J. Exp. Med.* **2015**, *212*, 2223–2234. [CrossRef]

46. Sui, G.; El Bachir, A.; Shi, Y.; Brignone, C.; Wall, N.R.; Yin, P.; Donohoe, M.; Luke, M.P.; Calvo, D.; Grossman, S.R.; et al. Yin Yang 1 Is a Negative Regulator of p53. *Cell* **2004**, *117*, 859–872. [CrossRef] [PubMed]
47. Wu, S.; Kasim, V.; Kano, M.R.; Tanaka, S.; Ohba, S.; Miura, Y.; Miyata, K.; Liu, X.; Matsuhashi, A.; Chung, U.-I.; et al. Transcription Factor YY1 Contributes to Tumor Growth by Stabilizing Hypoxia Factor HIF-1 in a p53-Independent Manner. *Cancer Res.* **2013**, *73*, 1787–1799. [CrossRef]
48. Forristal, C.E.; Brown, A.L.; Helwani, F.M.; Winkler, I.G.; Nowlan, B.; Barbier, V.; Powell, R.J.; Engler, G.A.; Diakiw, S.M.; Zannettino, A.C.; et al. Hypoxia inducible factor (HIF)-2alpha accelerates disease progression in mouse models of leukemia and lymphoma but is not a poor prognosis factor in human AML. *Leukemia* **2015**, *29*, 2075–2085. [CrossRef] [PubMed]
49. Frolova, O.; Samudio, I.; Benito, J.M.; Jacamo, R.; Kornblau, S.M.; Markovic, A.; Schober, W.; Lu, H.; Qiu, Y.H.; Buglio, D.; et al. Regulation of HIF-1alpha signaling and chemoresistance in acute lymphocytic leukemia under hypoxic conditions of the bone marrow microenvironment. *Cancer Biol. Ther.* **2012**, *13*, 858–870. [CrossRef] [PubMed]
50. To, K.K.; Huang, L.E. Suppression of hypoxia-inducible factor 1alpha (HIF-1alpha) transcriptional activity by the HIF prolyl hydroxylase EGLN1. *J. Biol. Chem.* **2005**, *280*, 38102–38107. [CrossRef]
51. Meier, E.L.K. Nonparametric Estimation from Incomplete Observations. *J. Am. Stat. Assoc.* **1958**, *53*, 457–481.
52. Ranstam, J.; Cook, J.A. Kaplan-Meier curve. *Br. J. Surg.* **2017**, *104*, 442. [CrossRef] [PubMed]
53. Thermo Fisher Scientific Inc. Oncomine Research Edition.
54. The Gene Ontology Resource. Available online: http://geneontology.org (accessed on 12 December 2021).

International Journal of
Molecular Sciences

Article

Forkhead Box Protein P3 (FOXP3) Represses ATF3 Transcriptional Activity

Chiung-Min Wang, William Harry Yang, Leticia Cardoso, Ninoska Gutierrez, Richard Henry Yang and Wei-Hsiung Yang *

Department of Biomedical Sciences, Mercer University School of Medicine, Savannah, GA 31404, USA; chiungminw@gmail.com (C.-M.W.); theyangbossman@gmail.com (W.H.Y.); leticiacdn@gmail.com (L.C.); NinMGutierrez@stu.southuniversity.edu (N.G.); thelegoman700@gmail.com (R.H.Y.)
* Correspondence: yang_w@mercer.edu; Tel.: +1-912-721-8203; Fax: +1-912-721-8268

Abstract: Activating transcription factor 3 (ATF3), a transcription factor and acute stress sensor, is rapidly induced by a variety of pathophysiological signals and is essential in the complex processes in cellular stress response. FOXP3, a well-known breast and prostate tumor suppressor from the X chromosome, is a novel transcriptional repressor for several oncogenes. However, it remains unknown whether ATF3 is the target protein of FOXP3. Herein, we demonstrate that ATF3 expression is regulated by FOXP3. Firstly, we observed that overexpression of FOXP3 reduced ATF3 protein level. Moreover, knockdown FOXP3 by siRNA increased ATF3 expression. Secondly, FOXP3 dose-dependently reduced ATF3 promoter activity in the luciferase reporter assay. Since FOXP3 is regulated by post-translational modifications (PTMs), we next investigated whether PTMs affect FOXP3-mediated ATF3 expression. Interestingly, we observed that phosphorylation mutation on FOXP3 (Y342F) significantly abolished FOXP3-mediated ATF3 expression. However, other PTM mutations on FOXP3, including S418 phosphorylation, K263 acetylation and ubiquitination, and K268 acetylation and ubiquitination, did not alter FOXP3-mediated ATF3 expression. Finally, the FOXP3 binding site was found on ATF3 promoter region by deletion and mutagenesis analysis. Taken together, our results suggest that FOXP3 functions as a novel regulator of ATF3 and that this novel event may be involved in tumor development and progression.

Keywords: FOXP3; ATF3; post-translational modification; transcriptional activity

Citation: Wang, C.-M.; Yang, W.H.; Cardoso, L.; Gutierrez, N.; Yang, R.H.; Yang, W.-H. Forkhead Box Protein P3 (FOXP3) Represses ATF3 Transcriptional Activity. *Int. J. Mol. Sci.* **2021**, *22*, 11400. https://doi.org/10.3390/ijms222111400

Academic Editors: Amelia Casamassimi, Apostolos Zaravinos, Alfredo Ciccodicola and Monica Rienzo

Received: 24 August 2021
Accepted: 20 October 2021
Published: 22 October 2021

Publisher's Note: MDPI stays neutral with regard to jurisdictional claims in published maps and institutional affiliations.

1. Introduction

Transcription factor Forkhead Box Protein P3 (FOXP3), encoded from the X chromosome (Xp11.23 in human), was originally identified as the causative mutation for lethal X-linked autoimmune/poly-endocrine dysregulation syndrome [1–4]. Unlike other FOX proteins, FOXP3 is mainly expressed in a subset of CD4+ T cells which function as suppressors in the immune system. In the T cell lineage, FOXP3 is essential for regulatory T (Treg) cell development and maintenance of immune homeostasis, as evidence shows that mutations of FOXP3 results in defective development of CD4+ CD25+ Treg cells [5]. Therefore, FOXP3 has become the most specific biomarker of Treg cells in the immunosuppressive system and homeostasis. Animal model studies have concluded that deficient and/or truncated FOXP3 is lethal due to Treg cell deficiency [5,6]. In humans, many mutations and/or alternatively spliced variants of the FOXP3 gene are strongly associated with an extremely rare and severe autoimmune disorder termed immunodysregulation, poly-endocrinopathy and enteropathy, X-linked syndrome (IPEX) [7]. Currently, the only curative treatment for IPEX is an allogeneic stem cell transplant from a healthy donor.

In addition to CD4+ T cells, FOXP3 also expresses in a variety of normal tissues, such as breast, liver, lung, prostate, spleen, pituitary, testis, and thymus [8–11], suggesting that FOXP3 might have broad biological and physiological functions. Extensive studies in the past 20 years have strongly suggested that FOXP3 is a novel tumor suppressor, especially in breast, colon, and prostate cancers. For example, deletions and mutations of FOXP3 have been found in human breast and prostate cancer samples, and germline mutations of FOXP3 result in a high rate of spontaneous breast cancer and fatal autoimmunity in mice [12]. Secondly, as a tumor suppressor, FOXP3 represses many key target genes in cancer development and progression, such as HER2/ERBB2 [12], BRCA1 [13], SKP2 [14], and CD44 [15], providing a strong link between FOXP3 and cell cycle regulation as well as FOXP3 and DNA repair system. Thirdly, the FOXP3-miR-146-NF-κB as an oncotarget and its axis has a functional role during tumor initiation in both prostate and breast cancers [16,17]. Moreover, miR-155, induced by FOXP3 through transcriptional repression of BRCA1, is associated with tumor initiation in human breast cancer, suggesting that plasma miR-155 may serve as a non-invasive biomarker for detection of early-stage breast cancer [18]. Furthermore, miR-141 and miR-200c are regulated by a FOXP3-KAT2B axis and are associated with tumor metastasis in breast cancer, suggesting that circulating levels of miR-141 and miR-200c are also potential biomarkers for early detection of breast cancer metastases [19]. Finally, loss of FOXP3 and TSC1 accelerates prostate cancer progression through synergistic regulation of c-MYC [20]. However, FOXP3 has been shown to promote cancer growth and metastasis in non-small cell lung cancer [21,22]. Recently, tumor CD274 (PD-L1) expression is inversely associated with FOXP3+ cell density in colorectal cancer tissues [23] and FOXP3 is expressed significantly higher in cytolytic-high colorectal tumors [24]. Together, these results demonstrate that FOXP3 has broad functions as a tumor suppressor in breast and prostate cancers and a tumor promoter in non-small cell lung cancer, suggesting that the regulatory machinery associating with FOXP3 in each cancer type might be critical for FOXP3 function.

Activating transcription factor 3 (ATF3), which is a member of the basic leucine zipper family of transcription factors, acts through binding to the ATF/cAMP response element (CRE) found in a number of promoters of key regulatory proteins that determine cell fate, circadian signaling, and homeostasis [25]. ATF3, as an immediate gene, is rapidly induced in cells once exposed to stress stimuli, including those initiated by cytokines, genotoxic agents, infections, reactive oxygen species, nerve injury, tissue damage, inflammation, and/or essential physiological stresses [25,26]. In addition, some evidence has implicated that ATF3 is up-regulated in many cancers, suggesting that ATF3 is an oncogene [27,28]. However, other evidence has indicated that ATF3 is able to suppress cell proliferation and inhibit the development of tumors [29–31]. Moreover, hepatocyte ATF3 is a key regulator of high-density lipoprotein and bile acid metabolism in the development of atherosclerosis [32]. Importantly, our group is the first to demonstrate that SUMOylation of ATF3 alters its transcriptional activity on regulation of TP53 gene [33], and that loss of SUMOylation on ATF3 inhibits proliferation of prostate cancer cells by modulating CCND1/2 activity [34]. These results imply that ATF3 has broad biological and physiological functions including cancer development and metabolism.

Previously, our group demonstrated that FOXP3 directly regulates UBC9 (the only E2 enzyme for SUMOylation) expression, suggesting that FOXP3 has a potential effect on regulating the global protein SUMOylation process [35]. Though FOXP3 has an essential and critical role in autoimmunity, cancer development, and Treg development, with hundreds of FOXP3 target genes already identified in both cancer cells and Treg cells, the functional role of FOXP3 in regulating ATF3 is largely unknown. Therefore, since FOXP3 is a transcription factor and has broad biological and physiological effects in cells and organs, we assessed the function of FOXP3 in ATF3 transcriptional activity using several human cell lines in the present study.

2. Results

2.1. FOXP3 Decreases ATF3 Protein Level

Since FOXP3 is a transcription factor, we first investigated the role of FOXP3 on ATF3 expression. We used FOXP3-Tet-off MCF7 cells to evaluate whether FOXP3 affects ATF3 protein expression since MCF7 (human breast cancer) cells express very small amounts of FOXP3. As shown in Figure 1A, FOXP3 induction by doxycycline removal decreased the expression levels of ATF3. We next over-expressed FOXP3 in MCF7 cells to evaluate whether FOXP3 affects ATF3 protein expression. As shown in Figure 1B, FOXP3 dose-dependently decreased the expression levels of ATF3. In order to confirm the results of Figure 1A,B, we next knockdown FOXP3 by siRNA system in HEK293 (human embryonic kidney) cells, which express more endogenous FOXP3 than MCF7 cells, to evaluate whether FOXP3 affects ATF3 protein expression. As shown in Figure 1C, reduction of FOXP3 by siRNA increased the expression levels of ATF3. In order to confirm the previous findings, we next performed qRT-PCR experiments. The expression vectors encoding wild-type FOXP3 or empty vectors were transfected into MCF7 human breast cancer cells. As shown in Figure 1D, when wild-type FOXP3 was transfected, the level of ATF3 mRNA was significantly decreased (approximately 50% reduction). Overall, these findings indicate that FOXP3 has the potential to down-regulate ATF3 expression.

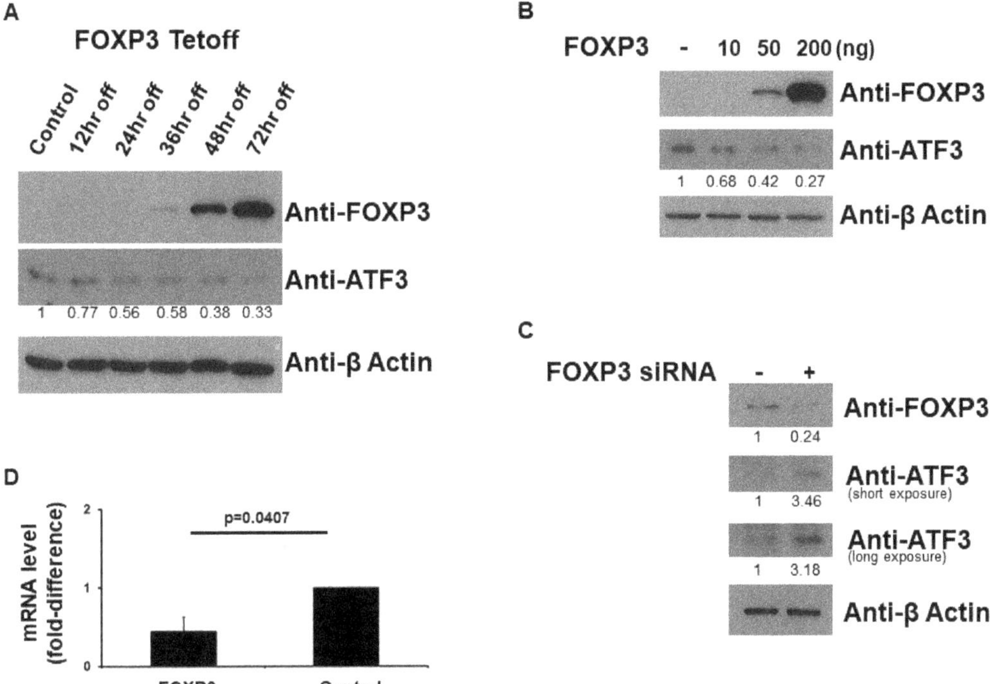

Figure 1. FOXP3 decreases ATF3 protein level. (**A**) Western blot analysis of ATF3 expression from FOXP3-Tet-off MCF7 cells. (**B**) Western blot analysis of ATF3 expression from FOXP3 over-expressed MCF7 cells. (**C**) Western blot analysis of ATF3 expression from HEK293 cells treated with FOXP3 siRNA. The expression levels of FOXP3 and ATF3 were determined using anti-FOXP3 and anti-ATF3 immunoblotting, respectively. The β-Actin levels were also determined for equal loading. (**D**) Real-time RT-PCR analysis of ATF3 expression by FOXP3 from MCF7 human breast cancer cells. Total RNA was extracted from cells and then reverse transcribed to cDNA followed by qPCR analysis with glyceraldehyde-3-phosphate dehydrogenase (GAPDH) as an internal control. The experiments were performed two times, each with triplicate samples. Error bars indicate standard errors.

2.2. FOXP3 Is a Repressor of the ATF3 Promoter

As FOXP3 decreases ATF3 protein expression, we next investigated the role of FOXP3 on ATF3 promoter activation. The −1372 bp ATF3 promoter-LUC reporter plasmid was co-transfected with FOXP3 expression plasmid into several different cell lines (MCF7, MDAMB231, H1299, or HEK293 cells) and ATF3 promoter activity was determined by measuring the LUC activity in cell lysates 48 h after transfection. As shown in Figure 2A, expression of FOXP3 generated a dose-dependent decrease in the activity of ATF3 gene transcription in MCF7 cells. Similar results were observed in MDAMB231 (triple negative human breast cancer), H1299 (human lung cancer), and HEK293 cells. This finding indicates that FOXP3 is a repressor of ATF3 transcription independent on cell types.

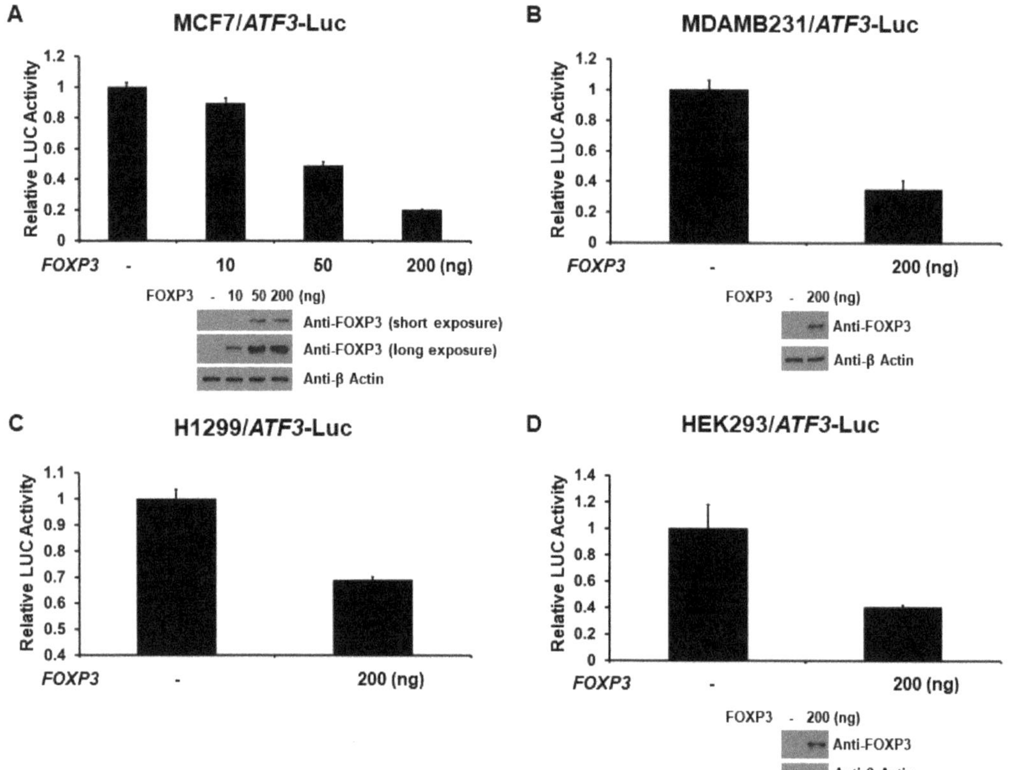

Figure 2. FOXP3 represses ATF3 transcription. (**A**) MCF7, (**B**) MDAMB231, (**C**) H1299, and (**D**) HEK293 cells were co-transfected, where indicated, with different amount of FOXP3 expression plasmid and ATF3 promoter-LUC reporter plasmid. Luciferase activities were measured 48 h after transfection and normalized with *Renilla* activity. Relative LUC activity was calculated and plotted. The protein levels of FOXP3 and β Actin in the cells from the reporter assays were confirmed using anti-FOXP3 and anti-β Actin immunoblotting, respectively.

2.3. Minimal ATF3 Promoter Region Responsive to FOXP3 Repression

To determine whether the FOXP3 response elements (REs) are required for FOXP3-mediated ATF3 expression, we first searched for potential FOXP3 binding site(s) on the ATF3 promoter region using the ALGGEN-PROMO website (http://alggen.lsi.upc.es/cgi-bin/promo_v3/promo/promoinit.cgi?dirDB=TF_8.3, access on: 1 August 2021) and rVista 2.0 website (https://rvista.dcode.org, access on: 1 August 2021). We identified one potential FOXP3 binding site on the human ATF3 promoter region. The potential

FOXP3 binding site is located 870 bp (AAAAA<u>AAAATCGAA</u>CCGATAC) upstream of the transcription start site, suggesting that FOXP3 may regulate ATF3 transcription directly.

Because the −1372 bp ATF3 promoter contains one potential candidate of FOXP3 RE, the ATF3 promoter was truncated to determine whether FOXP3 RE is important for transcriptional repression of ATF3 by FOXP3 (Figure 3A). Deletion of the FOXP3 RE (−870 bp) resulted in a major loss of FOXP3-mediated ATF3 transcriptional repression.

We next generated −870 bp mutant (AAAATCGAA→AAAAAAAAA) ATF3 promoter-LUC reporter plasmids. As shown in Figure 3B, mutation of the −870 bp FOXP3 RE resulted in approximately 75% loss of FOXP3-mediated ATF3 promoter repression. Together, these results indicate that the −870 bp RE is essential for the FOXP3 action on the ATF3 promoter. To confirm FOXP3 directly binds to the human ATF3 promoter region, we next performed FOXP3 or immunoglobin (IgG) chromotin immunoprecipitation (ChIP) assay with qPCR analysis. As shown in Figure 3C, FOXP3 strongly binds to the −870 RE region of the ATF3 promoter but not the −200 RE and −1300 RE regions. These results indicate that the −870 RE is the major FOXP3-binding site on the human ATF3 promoter region.

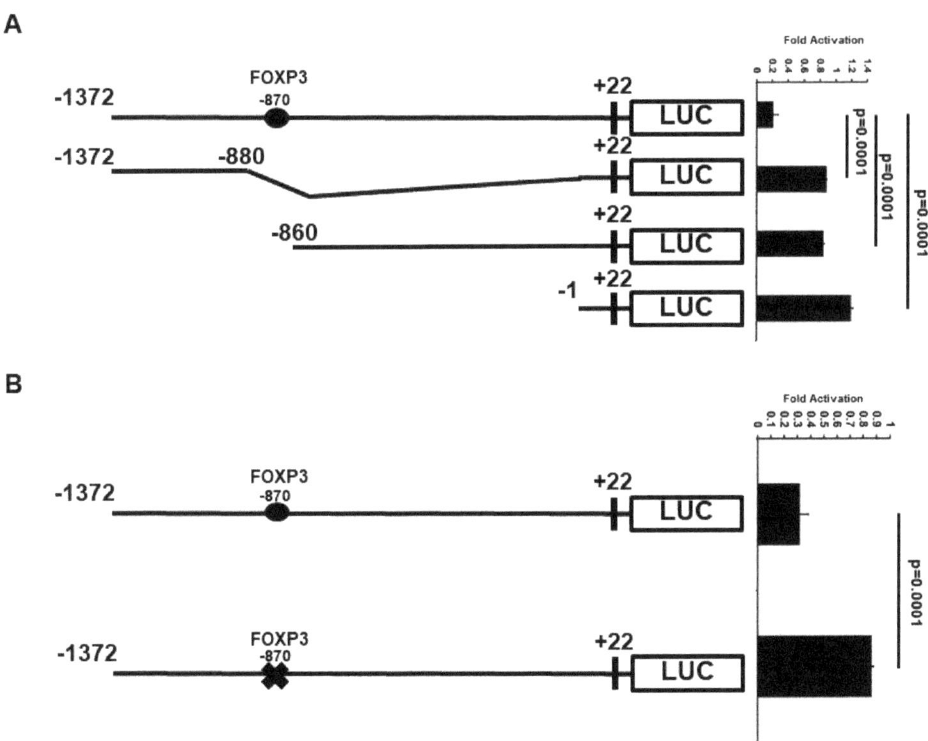

Figure 3. *Cont.*

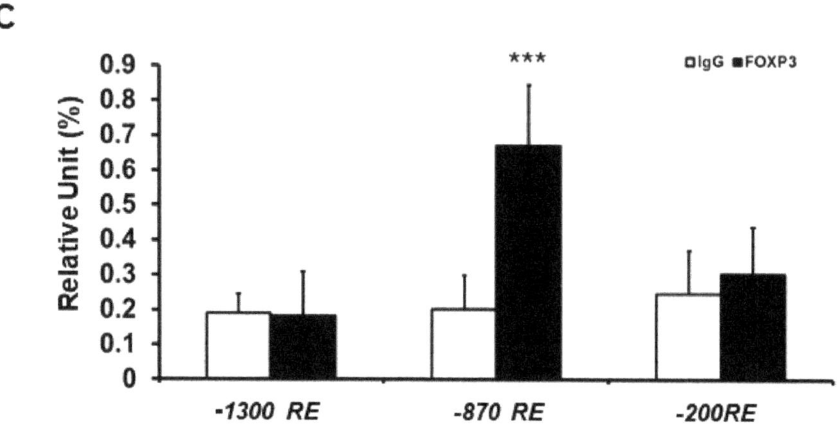

Figure 3. Regions of ATF3 promoter important for transcriptional down-regulation by FOXP3. (**A**) MCF7 cells were co-transfected with ATF3 promoter deletion constructs and FOXP3 expression plasmids. Luciferase activities were measured 48 h after transfection and normalized with *Renilla* activity. Relative LUC activity was calculated and plotted. (**B**) MCF7 cells were co-transfected with FOXP3 expression plasmids and with either wild-type (with −870 FOXP3 RE) or mutant (with −870 FOXP3 RE mutated) ATF3 promoter constructs. Luciferase activities were measured 48 h after transfection and normalized with *Renilla* activity. Relative LUC activity was calculated and plotted. (**C**) The quantification of the amount of DNA fragment precipitated (expressed as relative unit as a percentage of the total input DNA) in chromotin immunoprecipitation (ChIP) assay and qPCR analysis of possible FOXP3-binding sites (−200 RE, −870 RE, and −1300 RE) in the ATF3 promoter region in MCF7 cells. The experiments were performed two times. *** indicates $p < 0.001$ vs. IgG.

2.4. Phosphorylation at Tyr342 of FOXP3 Is Required for Full FOXP3-Mediated ATF3 Transcriptional Activity

Because the function of FOXP3 has been shown to be affected by post-translational modifications such as phosphorylation and acetylation [20,35], we next examined the effect of the post-translational modifications of FOXP3 on its transcriptional activity of the ATF3 promoter. MCF7 cells were co-transfected with the ATF3 promoter-LUC reporter plasmid and with either wild-type (WT), K31R (mimicking de-acetylated at K31), K263R (mimicking de-acetylated at K263), K263RK268R (mimicking de-acetylated at both K263 and K268), S418A (mimicking de-phosphorylated at S418), or Y342F (mimicking de-phosphorylated at Y342) FOXP3 expression plasmid. As shown in Figure 4A, in MCF7 cells, while the WT FOXP3 repressed ATF3 promoter activity as expected, K31R, K263R, K263RK268R, and S418A FOXP3 also reduced this effect (similar to WT level). Interestingly, loss of phosphorylation on Y342 abolished FOXP3-mediated ATF3 promoter repression, suggesting that phosphorylation at Y342 is extremely important for FOXP3 on ATF3 promoter activity. Similar results were observed in MDAMB231 and H1299 cells, suggesting that phosphorylation at Y342 is critical for FOXP3 function. Overall, these results suggest that phosphorylation is essential for FOXP3-mediated ATF3 promoter activity.

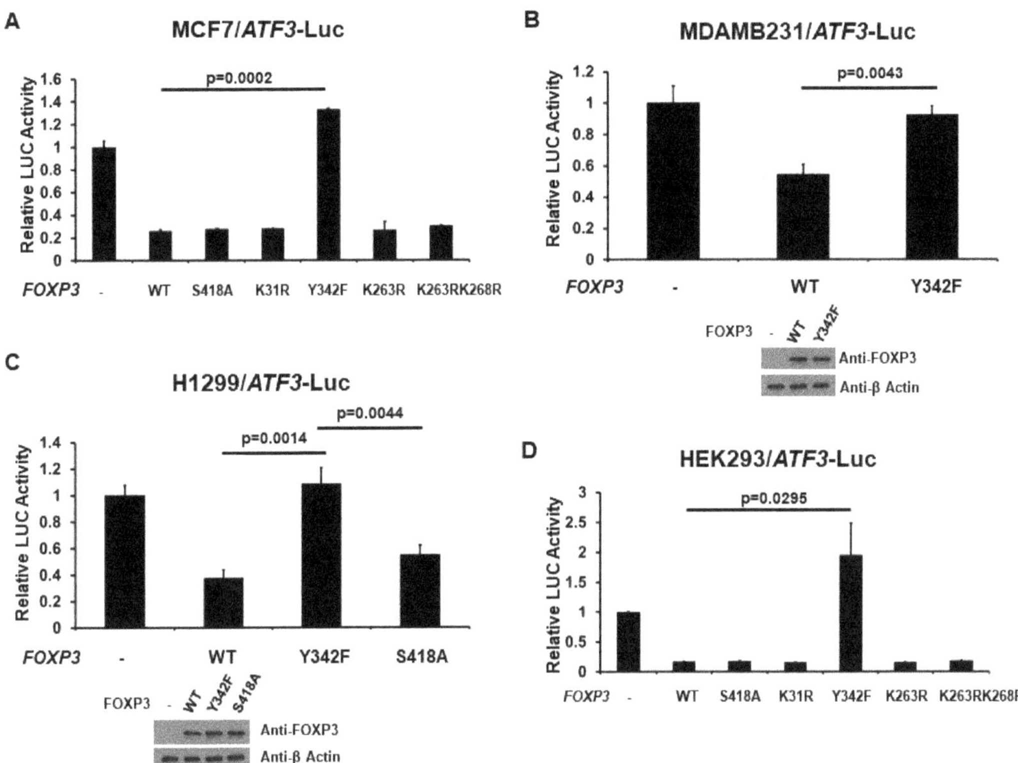

Figure 4. Post-translational modifications of FOXP3 regulate the repression of the ATF3 promoter. (**A**) MCF7, (**B**) MDAMB231, (**C**) H1299, and (**D**) HEK293 cells were co-transfected with the ATF3 promoter-LUC reporter plasmid and either wild-type (WT), S418A, K31R, Y342F, K263R, or K263RK268R FOXP3 expression plasmid. Luciferase activities were measured 48 h after transfection and normalized with *Renilla* activity. Relative LUC activity was calculated and plotted. The protein levels of FOXP3 and β Actin in the cells from the reporter assays were confirmed using anti-FOXP3 and anti-β Actin immunoblotting, respectively.

3. Discussion

Within eukaryotic cells, transcription factors (representing the connection of multiple signaling pathways for the growth and development of tissue and organisms) regulate downstream target genes by responding to a wide diversity of physiological and pathological stimuli, and therefore play essential roles in apoptosis, cell growth, developmental control, metabolism, pathogenesis, reproduction, and response to intercellular signals and environment insults [36–38]. Deregulation of transcription factors results in many human diseases, such as autoimmune dysfunctions, cardiovascular diseases, metabolic disorders, platelet disorders, and cancers [39,40]. FOXP3, a member of the transcription factor family of FOX proteins, has main functions in autoimmune homeostasis and cancer development and prevention. Herein, we show for the first time that FOXP3 acts as a transcriptional repressor of the human ATF3 gene in human cells.

It is well documented that FOXP3, as a critical master transcription regulator for Treg cell development and function, helps control the activities of various genes (such as oncogenes SKP2 and HER-2/ErbB2) related to the cancer development of the breast and prostate [12,14]. A previous report has shown that FOXP3 maintains Treg unresponsiveness by selectively inhibiting the promoter binding activity of c-Jun-based AP-1 [41]. The structure of AP-1 consists of heterodimers of families of c-Fos, c-Jun, ATF, and JDP.

Interestingly, in the present work, we showed that FOXP3 down-regulates ATF3 protein level and decreases ATF3 promoter activity. Our promoter analysis and ChIP assays further supports that FOXP3 response element −870 bp on the ATF3 promoter region is critical for regulating ATF3 gene expression. Our data highlights that ATF3 is the novel target gene for FOXP3.

FOXP3 can cooperate with several transcription factors, and among them is tumor suppressor p53. A previous report has demonstrated that FOXP3 is a key downstream regulator of p53 that is sufficient to induce p21 expression, ROS production and p53-mediated senescence [42]. Many reports, including ours, have demonstrated that ATF3 regulates the stability of p53 and the expression of the TP53 gene [33,43]. The evidence from our group has also highlighted that SUMOylation plays an important role for ATF3-mediated TP53 gene expression [33]. Interestingly, a study has demonstrated that the human ATF3 gene is one of the target genes directly activated by p53 [44], suggesting a functional link between stress-inducible transcriptional repressor ATF3 and p53. However, whether FOXP3 directly involves in p53-ATF3 interaction in a feedback loop remains unknown. Further future studies are indeed necessary to dissect this potential mechanism and regulatory loop.

Post-translational modifications such as acetylation, phosphorylation, methylation, and SUMOylation influence a wide range of cellular activities, including metabolism and cancer development [45,46]. Accumulating evidence indicates that human FOXP3 can be modified by phosphorylation (S418 and Y342), acetylation (K31, K263, and K268), ubiquitylation (K263, K268, and K31), and methylation (R48, R51) [47]. In the present work, we demonstrated that replacement of Y342 by a phenylalanine (F) residue in FOXP3 leads to significant loss of the repression of ATF3's transcriptional activity. This result is consistent with our previous report that phosphorylation at Y342 on FOXP3 is essential for UBC9 expression [35]. Both reports further highlight the crucial role on FOXP3 as a transcription factor. A previous study has shown that phosphorylation at Y342 of FOXP3 by lymphocyte-specific protein tyrosine kinase (LCK) represses cell invasion [48], suggesting that phosphorylation at Y342 of FOXP3 by LCK plays an important role for ATF3 expression. Interestingly, our current results suggest that another major phosphorylation site (S418) on FOXP3 does not play an important role in ATF3 repression. Since Y342 is modified by LCK and S418 is modified by PP1 from the previous reports, this suggests that LCK-FOXP3 pathway may be a major regulatory pathway for ATF3 regulation. Future studies are indeed necessary to dissect this potential regulatory mechanism.

The activatory/repressing role of splicing events on the neoplastic development/progression has been studied recently in different neoplasms, including alternative splicing of FOXP3. For example, immune responses may be manipulated by modulating the expression of FOXP3 isoforms, which has broad implications for the treatment of autoimmune diseases [49]. Moreover, alternative splicing of FOXP3 controls regulatory T cell effector functions and is associated with human atherosclerotic plaque stability [50]. Accumulated data also suggest that IPEX syndrome may be a consequence of alternatively spliced FOXP3 [51]. Finally, FOXP3 isoform profile has been linked to cardiovascular diseases [52]. Interestingly, SRSF1 has been shown to regulate alternative splicing events, especially in breast cancer [53] and lung cancer [54]. SRSF1 has also been found to be overexpressed in brain glioblastoma [55,56] and to be potentially used as a diagnostic marker of gliomas. With our current results of FOXP3-mediated ATF3 regulation, future studies are indeed necessary to explore the FOXP3 spliced isoforms in ATF3 regulation as well as the role of SRSF1 in FOXP3 isoform regulation, especially in the cancer field.

In conclusion, this study demonstrates that FOXP3, through FOX protein response element, is a novel repressor of ATF3 promoter, and that phosphorylation at Y342 plays a critical role for FOXP3's transcriptional activity.

4. Materials and Methods

4.1. Chemicals and Reagents

Both cell culture reagents and cell culture medium were purchased from Thermo Fisher Scientific (Waltham, MA, USA). Antibodies against ATF3, FOXP3, and β-Actin (Santa Cruz Biotechnology Inc., Santa Cruz, CA, USA). Luciferase activity was measured using the Dual-Luciferase Reporter Assay System (Promega, Madison, WI, USA).

4.2. DNA Constructs

Human FOXP3-pcDNA6 expression plasmid (with myc and HIS tags) was described previously [35]. S418A, Y342F, K31R, K263R, K263RK268R FOXP3 expression plasmids were created by PCR-based mutagenesis from WT FOXP3-pcDNA6 expression plasmid (QuikChange Lightning site-directed mutagenesis kit, Agilent/Strategene, La Jolla, CA, USA). S418A and Y342F represent the loss of phosphorylation of FOXP3. K31R, K263R, and K263RK268R represent the loss of acetylation of FOXP3. All these mutations have been confirmed by previous studies, including ours. The human ATF3 promoter ($-1372/+22$ bp) pGL2 plasmid was kindly provided by Dr. Aronheim (Technion-Israel Institute of Technology, Haifa, Israel). The human ATF3 promoter deletion constructs were then generated by removal of specific fragments of DNA sequence in Yang lab. The human ATF3 promoter with FOXP3 RE mutant (-870 bp) plasmids were created by PCR-based mutagenesis (QuikChange Lightning site-directed mutagenesis kit, Agilent/Strategene, La Jolla, CA, USA). All DNA plasmid constructs were verified by Sanger nucleotide sequencing.

4.3. Cell Culture and Transfection

H1299, HEK293, MCF7, and MDAMB231 cells were purchased from the American Type Culture Collection (Manassas, VA, USA). The cells were maintained in Dulbecco's Modified Eagle Medium (DMEM) in the presence of 10% fetal bovine serum and Pen/Strep antibiotics (GIBCO/Life Technologies, Grand Island, NY, USA) in a humidified incubator (5% CO_2 at 37 °C) and cultured for less than six months. We also routinely checked for Mycoplasma contamination using a PCR-based kit (Millipore-Sigma, Burlington, MA, USA). We chose MCF7 cells for overexpression study and HEK293 cells for siRNA experiments because MCF7 cells have the lowest FOXP3 expression and HEK293 cells have the highest FOXP3 expression among cells we tested. After incubation, the cells were transfected with specific expression plasmids described in each assay using Fugene HD Transfection Reagent (Roche, Madison, WI, USA). Forty-eight hours after transfection, the cells were harvested and lysed and ready for promoter luciferase reporter assays or Western blot analysis.

4.4. ATF3 Promoter Luciferase Reporter Assays

Cells were cultured in 24-well plates overnight and then transiently transfected with ATF3 promoter-firefly luciferase plasmid and internal control pRL-TK plasmid (which encodes *Renilla* luciferase activity) in the presence of Fugene HD Transfection Reagent (Roche, Madison, WI, USA). At 48 h after transfection, the cells were harvested and lysed in passive lysis buffer (Promega, Madison, WI, USA). Luminescence was detected with the Dual-Luciferase Reporter Assay System (Promega, Madison, WI, USA) using a luminometer (Turner Designs, Sunnyvale, CA, USA) according to the manufacturer's instructions. For each reporter assay, we also had a group of three wells without any plasmid transfected to make sure the background signals picked up by a luminometer were low. The firefly luciferase activity was normalized by calculating the ratio to *Renilla* luciferase activity. The relative luciferase activity was calculated as a fold change to the control groups. All experiments were performed three times in a triplicate setting.

4.5. Western Blot Analysis

Cells were lysed with RIPA buffer supplemented with protease and phosphatase, and protein contents of the high-speed supernatant were determined using the BCA^{TM} Protein Assay kit assay (Pierced/Thermo Scientific, Rockford, IL, USA). Equivalent quantities of protein (approximately 40 µg) were resolved on 10% polyacrylamide-SDS gels and transferred to polyvinylidene difluoride (PVDF) membrane (Bio-Rad, Hercules, CA, USA) by wet electrophoretic transfer. The membranes were probed with specific primary first and then with specific secondary antibodies. Blots were visualized using the Supersignal West Dura Extended Duration Substrate kit (Pierce Chemical Co., Rockford, IL, USA). The intensity of the protein band was quantified by ImageJ program.

4.6. RT-PCR and Real-Time ChIP

Total RNA from MCF7 cells was extracted using the TRIzol reagent (Thermo Fisher Scientific, Waltham, MA, USA) and treated with DNase (Ambion, Austin, TX, USA) to remove genomic DNA. The RNA concentration was quantified by ultraviolet spectrometry before being reverse-transcribed to cDNA. One microgram of total RNA was converted into cDNA using the iScript cDNA synthesis kit (Bio-Rad, Hercules, CA, USA) according to the manufacturer's instructions. The final cDNA product was purified and eluted in Tris-EDTA buffer using QIAquick PCR purification kits (QIAGEN, Germantown, MD, USA) according to the manufacturer's instructions. Quantitative PCR (using 5 ng cDNA per µL) was performed on an ABI 7500 qPCR system (Applied Biosystems, Foster City, CA, USA) using TaqMan Universal PCR Master Mix Kit (Applied Biosystems, Foster City, CA, USA) according to the manufacturer's instructions. Two primers (5′- CCC TTT GTC AAG TGC CG -3′ and 5′- GGCA CTT TGC AGC TGC -3′) were used to amplify 241-bp human ATF3 fragments. Two primers (5′- CAT CAC CAT CTT CCA GGA GCG AG -3′ and 5′- GTC TTC TGG GTG GCA GTG ATG G -3′) were used to amplify 341-bp human glyceraldehydes-3-phosphate dehydrogenase (GAPDH) fragments. For real-time ChIP assays, the extracted DNA fragments were quantified by real-time qPCR using pairs of primers that covered the FOXP3 response region within the human ATF3 promoter. The primers used for −1300 RE PCR were: CAAGAAGGTTCC (forward) and CCTTAAAAACG (reverse). The primers used for −870 RE PCR were: CTTGTCAATTTC (forward) and CTCCGGGCTCC (reverse). The primers used for −200 RE PCR were: GGAACACGCAG (forward) and CTGAGACACACAC (reverse).

4.7. Statistical Analysis

Statistical comparisons were performed by using the Student's *t*-test to determine the statistical significance between groups. A $p < 0.05$ was considered statistically significant between groups.

5. Conclusions

In summary, we have shown a novel relationship between FOXP3 and ATF3 for the first time. Our studies suggest FOXP3 is a novel repressor of the ATF3 promoter, and the FOXP3-mediated ATF3 transcription is critically regulated by the phosphorylation at Y342. Overall, our findings add a new layer of information to the previous understanding of FOXP3 functions.

Supplementary Materials: The following are available online at https://www.mdpi.com/article/10.3390/ijms222111400/s1.

Author Contributions: Conceptualization, C.-M.W. and W.-H.Y.; methodology, C.-M.W., W.H.Y., L.C., N.G., R.H.Y. and W.-H.Y.; validation, C.-M.W., W.H.Y. and W.-H.Y.; formal analysis, C.-M.W., W.H.Y. and W.-H.Y.; investigation, W.-H.Y.; resources, W.-H.Y.; data curation, C.-M.W., W.H.Y., W.-H.Y.; writing—original draft preparation, C.-M.W. and W.-H.Y.; writing—review and editing, W.H.Y., L.C., N.G., R.H.Y. and W.-H.Y.; visualization, W.-H.Y.; supervision, W.-H.Y.; project administration, W.-H.Y.; funding acquisition, W.-H.Y. All authors have read and agreed to the published version of the manuscript.

Funding: This work was supported by grants from a Mercer University Provost Seed Grant (W.-H.Y.) and a Rubye Smith Research Grant (W.-H.Y.).

Data Availability Statement: Data is contained within the article or Supplementary Material.

Acknowledgments: The human ATF3 promoter (−1372/+22 bp) pGL2 plasmid was kindly provided by Aronheim (Technion-Israel Institute of Technology, Haifa, Israel).

Conflicts of Interest: The authors declare no conflict of interest.

Abbreviations

FOXP3	Forkhead box protein P3
ATF3	Activating transcription factor 3
SUMO	Small ubiquitin-like modifier

References

1. Chatila, T.A.; Blaeser, F.; Ho, N.; Lederman, H.M.; Voulgaropoulos, C.; Helms, C.; Bowcock, A.M. JM2, encoding a fork head-related protein, is mutated in X-linked autoimmunity-allergic disregulation syndrome. *J. Clin. Investig.* **2000**, *106*, R75–R81. [CrossRef]
2. Bennett, C.L.; Christie, J.; Ramsdell, F.; Brunkow, M.E.; Ferguson, P.J.; Whitesell, L.; Kelly, T.E.; Saulsbury, F.T.; Chance, P.F.; Ochs, H.D. The immune dysregulation, polyendocrinopathy, enteropathy, X-linked syndrome (IPEX) is caused by mutations of FOXP3. *Nat. Genet.* **2001**, *27*, 20–21. [CrossRef]
3. Brunkow, M.E.; Jeffery, E.W.; Hjerrild, K.A.; Paeper, B.; Clark, L.B.; Yasayko, S.A.; Wilkinson, J.E.; Galas, D.; Ziegler, S.F.; Ramsdell, F. Disruption of a new forkhead/winged-helix protein, scurfin, results in the fatal lymphoproliferative disorder of the scurfy mouse. *Nat. Genet.* **2001**, *27*, 68–73. [CrossRef]
4. Wildin, R.S.; Ramsdell, F.; Peake, J.; Faravelli, F.; Casanova, J.L.; Buist, N.; Levy-Lahad, E.; Mazzella, M.; Goulet, O.; Perroni, L.; et al. X-linked neonatal diabetes mellitus, enteropathy and endocrinopathy syndrome is the human equivalent of mouse scurfy. *Nat. Genet.* **2001**, *27*, 18–20. [CrossRef] [PubMed]
5. Fontenot, J.D.; Gavin, M.A.; Rudensky, A.Y. Foxp3 programs the development and function of CD4+CD25+ regulatory T cells. *Nat. Immunol.* **2003**, *4*, 330–336. [CrossRef]
6. Khattri, R.; Cox, T.; Yasayko, S.A.; Ramsdell, F. An essential role for Scurfin in CD4+CD25+ T regulatory cells. *Nat. Immunol.* **2003**, *4*, 337–342. [CrossRef]
7. Bennett, C.L.; Ochs, H.D. IPEX is a unique X-linked syndrome characterized by immune dysfunction, polyendocrinopathy, enteropathy, and a variety of autoimmune phenomena. *Curr. Opin. Pediatr.* **2001**, *13*, 533–538. [CrossRef]
8. Martin, F.; Ladoire, S.; Mignot, G.; Apetoh, L.; Ghiringhelli, F. Human FOXP3 and cancer. *Oncogene* **2010**, *29*, 4121–4129. [CrossRef]
9. Redpath, M.; Xu, B.; van Kempen, L.C.; Spatz, A. The dual role of the X-linked FoxP3 gene in human cancers. *Mol. Oncol.* **2011**, *5*, 156–163. [CrossRef] [PubMed]
10. Jung, D.O.; Jasurda, J.S.; Egashira, N.; Ellsworth, B.S. The forkhead transcription factor, FOXP3, is required for normal pituitary gonadotropin expression in mice. *Biol. Reprod.* **2012**, *86*, 1–9. [CrossRef] [PubMed]
11. Jiang, L.L.; Ruan, L.W. Association between FOXP3 promoter polymorphisms and cancer risk: A meta-analysis. *Oncol. Lett.* **2014**, *8*, 2795–2799. [CrossRef]
12. Zuo, T.; Wang, L.; Morrison, C.; Chang, X.; Zhang, H.; Li, W.; Liu, Y.; Wang, Y.; Liu, X.; Chan, M.W.; et al. FOXP3 is an X-linked breast cancer suppressor gene and an important repressor of the HER-2/ErbB2 oncogene. *Cell* **2007**, *129*, 1275–1286. [CrossRef]
13. Li, W.; Katoh, H.; Wang, L.; Yu, X.; Du, Z.; Yan, X.; Zheng, P.; Liu, Y. FOXP3 regulates sensitivity of cancer cells to irradiation by transcriptional repression of BRCA1. *Cancer Res.* **2013**, *73*, 2170–2180. [CrossRef]
14. Zuo, T.; Liu, R.; Zhang, H.; Chang, X.; Liu, Y.; Wang, L.; Zheng, P.; Liu, Y. FOXP3 is a novel transcriptional repressor for the breast cancer oncogene SKP2. *J. Clin. Investig.* **2007**, *117*, 3765–3773. [CrossRef] [PubMed]
15. Zhang, C.; Xu, Y.; Hao, Q.; Wang, S.; Li, H.; Li, J.; Gao, Y.; Li, M.; Li, W.; Xue, X.; et al. FOXP3 suppresses breast cancer metastasis through downregulation of CD44. *Int. J. Cancer* **2015**, *137*, 1279–1290. [CrossRef] [PubMed]
16. Liu, R.; Yi, B.; Wei, S.; Yang, W.H.; Hart, K.M.; Chauhan, P.; Zhang, W.; Mao, X.; Liu, X.; Liu, C.G.; et al. FOXP3-miR-146-NF-κB Axis and Therapy for Precancerous Lesions in Prostate. *Cancer Res.* **2015**, *75*, 1714–1724. [CrossRef] [PubMed]
17. Liu, R.; Liu, C.; Chen, D.; Yang, W.H.; Liu, X.; Liu, C.G.; Dugas, C.M.; Tang, F.; Zheng, P.; Liu, Y.; et al. FOXP3 Controls an miR-146/NF-κB Negative Feedback Loop That Inhibits Apoptosis in Breast Cancer Cells. *Cancer Res.* **2015**, *75*, 1703–1713. [CrossRef] [PubMed]
18. Gao, S.; Wang, Y.; Wang, M.; Li, Z.; Zhao, Z.; Wang, R.X.; Wu, R.; Yuan, Z.; Cui, R.; Jiao, K.; et al. MicroRNA-155, induced by FOXP3 through transcriptional repression of BRCA1, is associated with tumor initiation in human breast cancer. *Oncotarget* **2017**, *8*, 41451–41464. [CrossRef]
19. Zhang, G.; Zhang, W.; Li, B.; Stringer-Reasor, E.; Chu, C.; Sun, L.; Bae, S.; Chen, D.; Wei, S.; Jiao, K.; et al. MicroRNA-200c and microRNA- 141 are regulated by a FOXP3-KAT2B axis and associated with tumor metastasis in breast cancer. *Breast Cancer Res.* **2017**, *19*, 73. [CrossRef] [PubMed]
20. Wu, L.; Yi, B.; Wei, S.; Rao, D.; He, Y.; Naik, G.; Bae, S.; Liu, X.M.; Yang, W.H.; Sonpavde, G.; et al. Loss of FOXP3 and TSC1 Accelerates Prostate Cancer Progression through Synergistic Transcriptional and Posttranslational Regulation of c-MYC. *Cancer Res.* **2019**, *79*, 1413–1425. [CrossRef]

21.	Yang, S.; Liu, Y.; Li, M.Y.; Ng, C.S.H.; Yang, S.L.; Wang, S.; Zou, C.; Dong, Y.; Du, J.; Long, X.; et al. FOXP3 promotes tumor growth and metastasis by activating Wnt/beta-catenin signaling pathway and EMT in non-small cell lung cancer. *Mol. Cancer.* **2017**, *16*, 124. [CrossRef]

22.	Qi, H.; Li, W.; Zhang, J.; Chen, J.; Peng, J.; Liu, Y.; Yang, S.; Du, J.; Long, X.; Ng, C.S.; et al. Glioma-associated oncogene homolog 1 stimulates FOXP3 to promote non-small cell lung cancer stemness. *Am. J. Transl. Res.* **2020**, *12*, 1839–1850. [PubMed]

23.	Masugi, Y.; Nishihara, R.; Yang, J.; Mima, K.; da Silva, A.; Shi, Y.; Inamura, K.; Cao, Y.; Song, M.; Nowak, J.A.; et al. Tumour CD274 (PD-L1) expression and T cells in colorectal cancer. *Gut* **2017**, *66*, 1463–1473. [CrossRef] [PubMed]

24.	Zaravinos, A.; Roufas, C.; Nagara, M.; de Lucas Moreno, B.; Oblovatskaya, M.; Efstathiades, C.; Dimopoulos, C.; Ayiomamitis, G.D. Cytolytic activity correlates with the mutational burden and deregulated expression of immune checkpoints in colorectal cancer. *J. Exp. Clin. Cancer Res.* **2019**, *38*, 364. [CrossRef]

25.	Hai, T.; Hartman, M.G. The molecular biology and nomenclature of the activating transcription factor/cAMP responsive element binding family of transcription factors: Activating transcription factor proteins and homeostasis. *Gene* **2001**, *273*, 1–11. [CrossRef]

26.	Hai, T.; Wolfgang, C.D.; Marsee, D.K.; Allen, A.E.; Sivaprasad, U. ATF3 and stress responses. *Gene Expr.* **1999**, *7*, 321–335.

27.	Yin, X.; Dewille, J.W.; Hai, T. A potential dichotomous role of ATF3, an adaptive-response gene, in cancer development. *Oncogene* **2008**, *27*, 2118–2127. [CrossRef]

28.	Pelzer, A.E.; Bektic, J.; Haag, P.; Berger, A.P.; Pycha, A.; Schafer, G.; Rogatsch, H.; Horninger, W.; Bartsch, G.; Klocker, H. The expression of transcription factor activating transcription factor 3 in the human prostate and its regulation by androgen in prostate cancer. *J. Urol.* **2006**, *175*, 1517–1522. [CrossRef]

29.	Lu, D.; Wolfgang, C.D.; Hai, T. Activating transcription factor 3, a stress-inducible gene, suppresses Ras-stimulated tumorigenesis. *J. Biol. Chem.* **2006**, *281*, 10473–10481. [CrossRef]

30.	Huang, X.; Li, X.; Guo, B. KLF6 induces apoptosis in prostate cancer cells through up-regulation of ATF3. *J. Biol. Chem.* **2008**, *283*, 29795–29801. [CrossRef]

31.	Fan, F.; Jin, S.; Amundson, S.A.; Tong, T.; Fan, W.; Zhao, H.; Zhu, X.; Mazzacurati, L.; Li, X.; Petrik, K.L.; et al. ATF3 induction following DNA damage is regulated by distinct signaling pathways and over-expression of ATF3 protein suppresses cells growth. *Oncogene* **2002**, *21*, 7488–7496. [CrossRef] [PubMed]

32.	Xu, Y.; Li, Y.; Jadhav, K.; Pan, X.; Zhu, Y.; Hu, S.; Chen, S.; Chen, L.; Tang, Y.; Wang, H.H.; et al. Hepatocyte ATF3 protects against atherosclerosis by regulating HDL and bile acid metabolism. *Nat. Metab.* **2021**, *3*, 59–74. [CrossRef] [PubMed]

33.	Wang, C.M.; Brennan, V.C.; Gutierrez, N.M.; Wang, X.; Wang, L.; Yang, W.H. SUMOylation of ATF3 alters its transcriptional activity on regulation of TP53 gene. *J. Cell Biochem.* **2013**, *114*, 589–598. [CrossRef] [PubMed]

34.	Wang, C.M.; Yang, W.H. Loss of SUMOylation on ATF3 inhibits proliferation of prostate cancer cells by modulating CCND1/2 activity. *Int. J. Mol. Sci.* **2013**, *14*, 8367–8380. [CrossRef]

35.	Wang, C.M.; Yang, W.H.; Liu, R.; Wang, L.; Yang, W.H. FOXP3 Activates SUMO-Conjugating UBC9 Gene in MCF7 Breast Cancer Cells. *Int. J. Mol. Sci.* **2018**, *19*, 2036. [CrossRef] [PubMed]

36.	Spitz, F.; Furlong, E.E. Transcription factors: From enhancer binding to developmental control. *Nat. Rev. Genet.* **2012**, *13*, 613–626. [CrossRef]

37.	Francois, M.; Donovan, P.; Fontaine, F. Modulating transcription factor activity: Interfering with protein-protein interaction networks. *Semin. Cell Dev. Biol.* **2020**, *99*, 12–19. [CrossRef]

38.	Karagianni, P.; Talianidis, I. Transcription factor networks regulating hepatic fatty acid metabolism. *Biochim. Biophys. Acta* **2015**, *1851*, 2–8. [CrossRef]

39.	Daly, M.E. Transcription factor defects causing platelet disorders. *Blood Rev.* **2017**, *31*, 1–10. [CrossRef]

40.	Bushweller, J.H. Targeting transcription factors in cancer—From undruggable to reality. *Nat. Rev. Cancer* **2019**, *19*, 611–624. [CrossRef]

41.	Lee, S.M.; Gao, B.; Fang, D. FoxP3 maintains Treg unresponsiveness by selectively inhibiting the promoter DNA-binding activity of AP-1. *Blood* **2008**, *111*, 3599–3606. [CrossRef]

42.	Kim, J.E.; Shin, J.S.; Moon, J.H.; Hong, S.W.; Jung, D.J.; Kim, J.H.; Hwang, I.Y.; Shin, Y.J.; Gong, E.Y.; Lee, D.H.; et al. Foxp3 is a key downstream regulator of p53-mediated cellular senescence. *Oncogene* **2017**, *36*, 219–230. [CrossRef]

43.	Yan, C.; Boyd, D.D. ATF3 regulates the stability of p53: A link to cancer. *Cell Cycle* **2006**, *5*, 926–929. [CrossRef] [PubMed]

44.	Zhang, C.; Gao, C.; Kawauchi, J.; Hashimoto, Y.; Tsuchida, N.; Kitajima, S. Transcriptional activation of the human stress-inducible transcriptional repressor ATF3 gene promoter by p53. *Biochem. Biophys. Res. Commun.* **2002**, *297*, 1302–1310. [CrossRef]

45.	Dou, L.; Yang, F.; Xu, L.; Zou, Q. A comprehensive review of the imbalance classification of protein post-translational modifications. *Brief. Bioinform.* **2021**, *8*, bbab089. [CrossRef]

46.	Keenan, E.K.; Zachman, D.K.; Hirschey, M.D. Discovering the landscape of protein modifications. *Mol. Cell* **2021**, *81*, 1868–1878. [CrossRef]

47.	Deng, G.; Song, X.; Fujimoto, S.; Piccirillo, C.A.; Nagai, Y.; Greene, M.I. Foxp3 Post-translational Modifications and Treg Suppressive Activity. *Front. Immunol.* **2019**, *10*, 2486. [CrossRef]

48.	Nakahira, K.; Morita, A.; Kim, N.S.; Yanagihara, I. Phosphorylation of FOXP3 by LCK downregulates MMP9 expression and represses cell invasion. *PLoS ONE* **2013**, *8*, e77099. [CrossRef]

49.	Mailer, R.K.; Joly, A.L.; Liu, S.; Elias, S.; Tegner, J.; Andersson, J. IL-1beta promotes Th17 differentiation by inducing alternative splicing of FOXP3. *Sci. Rep.* **2015**, *5*, 14674. [CrossRef]

50. Joly, A.L.; Seitz, C.; Liu, S.; Kuznetsov, N.V.; Gertow, K.; Westerberg, L.S.; Paulsson-Berne, G.; Hansson, G.K.; Andersson, J. Alternative Splicing of FOXP3 Controls Regulatory T Cell Effector Functions and Is Associated with Human Atherosclerotic Plaque Stability. *Circ. Res.* **2018**, *122*, 1385–1394. [CrossRef]

51. Mailer, R.K. IPEX as a Consequence of Alternatively Spliced FOXP3. *Front. Pediatr.* **2020**, *8*, 594375. [CrossRef] [PubMed]

52. Joly, A.L.; Andersson, J. Alternative splicing, FOXP3 and cardiovascular disease. *Aging* **2019**, *11*, 1905–1906. [CrossRef] [PubMed]

53. Anczuków, O.; Akerman, M.; Cléry, A.; Wu, J.; Shen, C.; Shirole, N.H.; Raimer, A.; Sun, S.; Jensen, M.A.; Hua, Y.; et al. SRSF1-Regulated Alternative Splicing in Breast Cancer. *Mol. Cell* **2015**, *60*, 105–117. [CrossRef]

54. Sheng, J.; Zhao, Q.; Zhao, J.; Zhang, W.; Sun, Y.; Qin, P.; Lv, Y.; Bai, L.; Yang, Q.; Chen, L.; et al. SRSF1 modulates PTPMT1 alternative splicing to regulate lung cancer cell radioresistance. *EBioMedicine* **2018**, *38*, 113–126. [CrossRef] [PubMed]

55. Barbagallo, D.; Caponnetto, A.; Cirnigliaro, M.; Brex, D.; Barbagallo, C.; D'Angeli, F.; Morrone, A.; Caltabiano, R.; Barbagallo, G.M.; Ragusa, M.; et al. CircSMARCA5 Inhibits Migration of Glioblastoma Multiforme Cells by Regulating a Molecular Axis Involving Splicing Factors SRSF1/SRSF3/PTB. *Int. J. Mol. Sci.* **2018**, *19*, 480. [CrossRef]

56. Zhou, X.; Wang, R.; Li, X.; Yu, L.; Hua, D.; Sun, C.; Shi, C.; Luo, W.; Rao, C.; Jiang, Z.; et al. Splicing factor SRSF1 promotes gliomagenesis via oncogenic splice-switching of MYO1B. *J. Clin. Investig.* **2019**, *129*, 676–693. [CrossRef] [PubMed]

International Journal of
Molecular Sciences

Review

Epigenetic Regulation and Post-Translational Modifications of SNAI1 in Cancer Metastasis

Bo Dong and Yadi Wu *

Department of Pharmacology & Nutritional Sciences and Markey Cancer Center, University of Kentucky School of Medicine, Lexington, KY 40506, USA; b.dong@uky.edu
* Correspondence: yadi.wu@uky.edu

Abstract: SNAI1, a zinc finger transcription factor, not only acts as the master regulator of epithelial-mesenchymal transition (EMT) but also functions as a driver of cancer progression, including cell invasion, survival, immune regulation, stem cell properties, and metabolic regulation. The regulation of SNAI1 occurs at the transcriptional, translational, and predominant post-translational levels including phosphorylation, acetylation, and ubiquitination. Here, we discuss the regulation and role of SNAI1 in cancer metastasis, with a particular emphasis on epigenetic regulation and post-translational modifications. Understanding how signaling networks integrate with SNAI1 in cancer progression will shed new light on the mechanism of tumor metastasis and help develop novel therapeutic strategies against cancer metastasis.

Keywords: SNAI1; metastasis; post-translational modifications; epigenetic; EMT

Citation: Dong, B.; Wu, Y. Epigenetic Regulation and Post-Translational Modifications of SNAI1 in Cancer Metastasis. *Int. J. Mol. Sci.* **2021**, 22, 11062. https://doi.org/10.3390/ijms222011062

Academic Editors: Amelia Casamassimi, Alfredo Ciccodicola and Monica Rienzo

Received: 31 August 2021
Accepted: 11 October 2021
Published: 14 October 2021

Publisher's Note: MDPI stays neutral with regard to jurisdictional claims in published maps and institutional affiliations.

1. Introduction

Tumor metastasis, the spreading of cancer cells from original tumor sites to distant organs followed by development of secondary tumors, is the foremost cause of cancer-related deaths [1]. Initiation of the metastatic program is often followed by exploitation of an embryonic development process referred to as epithelial-mesenchymal transition (EMT) [2]. During EMT, epithelial cells attain mesenchymal phenotypes such as increased motility and invasiveness by dissolving cell–cell junctions and rebuilding cell–matrix connections, accompanied by loss of epithelial markers and a gain of mesenchymal markers [3]. EMT is activated by a plethora of EMT-activating transcription factors (EMT-TFs), such as those from the SNAIL, zinc finger E-box binding homeobox (ZEB), and TWIST families [4].

SNAI1 was the first discovered and most intensively studied transcription repressor of E-cadherin, a hallmark of EMT encoded by the epithelial gene *CDH1*. SNAI1 directly binds to E-boxes present in the *CDH1* promoter to transcriptionally repress its expression. On the other hand, SNAI1 also acts as a transcriptional activator. SNAI1 not only enhances mesenchymal markers including fibronectin, collagens, and the matrix degradation enzyme matrix metalloproteinases 2 and 9 (MMP2 and MMP9), it also increases other EMT transcription factors such as TWIST and ZEB1 [5,6]. In addition, SNAI1 positively regulates transcriptional activation of target genes involved in *Drosophila* development through direct binding to the promoters [7]. In collaboration with early growth response 1(EGR1) and SP1, SNAI1 may directly activate transcription of p15^{INK4b}, lymphoid enhancer-binding factor (LEF), and cyclooxygenase 2 (COX2) by directly binding on a consensus motif in HepG2 cells stimulated by the phorbol ester tumor promoter 12-O-tetradecanoyl-phorbol 13-acetate (TPA) [5,8–10]. SNAI1 induces resistance to apoptosis, confers tumor recurrence and drug resistance, generates breast cancer stem cell (CSC)-like properties, and induces aerobic glycolysis [11–14]. Interestingly, SNAI1 is tightly controlled at both transcriptional and protein levels. Many growth factors and cytokines can transcriptionally regulate SNAI1 expression [15]. In addition, SNAI1 protein levels are regulated by post-translational modifications (PTMs). These PTMs have diverse effects on the function of SNAI1.

Because of the reversible plasticity of EMT, epigenetic alternations are required in the EMT process. In eukaryotic cells, genomic DNA interacts with histone proteins and RNA to form chromatin, which holds epigenetic information independent of the DNA genetic data [16]. Alteration of chromatin occurs through regulators responsible for DNA methylation, post-translational modifications of nucleosomal histone tails, and/or non-coding RNA modulation; these epigenetic modifications play a key role in regulating gene expression by defining whether chromatins at a given genomic locus will be transcriptionally active or inactive [17]. For EMT, a variety of epigenetic regulators are critical requirements that interpret signals passed from stimulators to transcription factors [18]. Indeed, the expression of CDH1 is regulated by multiple enzymes involving epigenetic modification. SNAI1 collaborates with multiple epigenetic enzyme complexes, such as DNA methyltransferases, histone deacetylases, and histone methyltransferase and demethylase, in the transcriptional regulation of CDH1. Recent studies suggest a crucial role of epigenetic alterations in the regulation of SNAI1 and EMT markers.

Here, we summarize the regulation of SNAI1 with an emphasis on PTMs. Moreover, we describe recent insights into the epigenetic mechanisms of SNAI1-induced cancer metastasis, focusing on the cooperation of SNAI1 with epigenetic regulators.

2. Regulation of SNAI1

Expression of SNAI1 is governed at multiple levels from gene transcription, post-transcriptional regulation, and translation to PTMs such as phosphorylation, ubiquitination, acetylation, and sumoylation.

2.1. Structure of SNAI1

SNAI1 belongs to the SNAIL family which consists of SNAI1 (Snail), SNAI2 (Slug), and SNAI3 (Smuc) [19]. The amino termini of SNAI1 contains the evolutionarily conserved SNAI1/Gfi (SNAG) domain, which interacts with several co-repressor complexes or epigenetic remodeling complexes (Figure 1). *Drosophila* SNAIL lacks the SNAG domain but has a consensus PxDLSx motif and exerts their repressive function through the interaction with the co-repressor c-terminal binding protein (CtBP) [19]. In the central region, a serine-rich domain (SRD) is adjacent to the nuclear export sequence (NES) (Scheme 1). SRD controls ubiquitination and proteasome degradation while NES is involved in the regulation of its protein stability and subcellular translocation. The c-terminal zinc finger domain with four C2H2-type zinc fingers is highly conserved. This domain mediates sequence-specific interactions with their target DNA promoters containing an E-box sequence (CAGGTG).

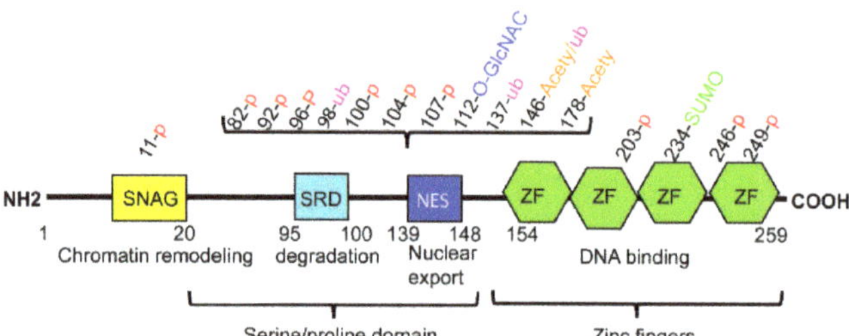

Figure 1. The structure of SNAI1 and potential post-translational modification sites. SNAI1 contains four main domains: Scheme 1. Gfi-1 (SNAG) domain, serine-rich domain (SRD), nuclear export sequence (NES) and zinc-finger (ZF) domain. P: phosphorylation. Acety: Acetylation. Ub: Ubiquitination. SUMO: Sumoylation.

2.2. Transcriptional and Post-Transcriptional Regulation

A diverse repertoire of molecular mechanisms regulating SNAI1 at the transcriptional level have been documented in a variety of organisms. Cytokines, chemokines, and growth factors, such as tumor necrosis factor (TNFα), transforming growth factor (TGFβ), interleukin-6 (IL-6), fibroblast growth factors (FGF), epidermal growth factor (EGF), and hepatocyte growth factor (HGF) [20], trigger an intracellular signaling cascade that leads to the binding of a transcription factor to the *SNAI1* promoter to regulate its expression. Extensive evaluation of this regulation has been covered in other excellent reviews [21,22]. Interestingly, the expression of SNAI1 can also be regulated by other EMT-TFs. For example, both SNAI1 and SNAI2 were upregulated under TGFβ stimulation [23]. Depletion of SNAI2 increases SNAI1 expression and vice versa; this compensatory regulation could be indispensable for EMT and cancer progression [24]. In addition, TWIST induces SNAI1 and the Twist-SNAI1 axis is critically involved in EMT and tumor metastasis [25].

Post-transcriptional control provides a fundamental regulatory mechanism for gene expression. Besides regulation by microRNAs [26], SNAI1 transcript stability is also regulated extensively. For example, recent work showed that upon activation of EGF receptor, UDP-glucose 6-dehydrogenase (UGDH) is phosphorylated in human lung cancer cells. Phosphorylated UGDH not only converts UDP-glucose to UDP-glucuronic acid but also interacts with Hu antigen R (an RNA-binding protein that binds to short-lived mRNAs to increase their stability). This interaction attenuates the UDP-glucose-mediated inhibition and therefore enhances the stability of SNAI1 mRNA [27]. In addition, the mRNA of SNAI1 can be modified with N6-Methyladenosine (m6A) by the methyltransferase-like 3 (METTL3) and YTH N6-methyladenosine RNA binding protein 1 (YTHDF1) (m6A readers). m6A in the coding sequence of SNAI1 triggers polysome-mediated translation of SNAI1 mRNA in cancer cells [28]. The stability of SNAI1 mRNA is also enhanced by heterogeneous nuclear ribonucleoprotein, which thus promotes invasion, metastasis, and EMT in breast cancer [29].

2.3. Post-Translational Regulation

Because of their critical roles in cancer metastasis, much attention has focused on the PTMs of SNAI1. PTMs function in the regulating protein stability, transcriptional activity, and intracellular localization of SNAI1. Among the number of modifications, phosphorylation and ubiquitination represent the best characterized and control a variety of biological activities, such as apoptosis, transcription, metabolism, and stem cell properties. Therefore, gaining deeper insight into the PTMs may help elucidate important steps in cancer metastasis.

2.3.1. Phosphorylation Regulation

SNAI1 stability is extensively regulated by phosphorylation (Table 1). On one hand, phosphorylation of SNAI1 promotes its proteasomal-mediated ubiquitination degradation. Both casein kinase 1(CK1) and dual specificity tyrosine phosphorylation regulated kinase 2 (DYRK2)-mediated SNAI1 phosphorylation at serine (Ser) 104 act to prime phosphorylations that allow glycogen synthase kinase 3 β (GSK3β)-mediated phosphorylation at Ser96 and Ser100, leading to β-TRCP-induced poly-ubiquitination and degradation [30,31]. Protein kinase D1 (PKD1)-mediated phosphorylation at Ser11 of SNAI1 facilitates F-box protein 11 (FBXO11)-mediated SNAI1 degradation [32]. Under intact apical-basal polarity, α protein kinase C (PKC) kinases promote degradation through phosphorylation of SNAI1 S249 [33]. On the other hand, some SNAI1 phosphorylations prevent its degradation. Most commonly, the main mechanism that regulates SNAI1 stability is phosphorylation at specific sites that reduce its affinity for GSK3β, thus blocking ubiquitination. For example, phosphorylation of SNAI1 at Ser100 by ataxia-telangiectasia mutated (ATM) and DNA-PKCs inhibits SNAI1 ubiquitination by reducing interaction with GSK3β [34,35]. Recently, it was shown that p38 stabilizes SNAI1 through phosphorylation at Ser107, which suppresses DYRK2-mediated Ser104 phosphorylation and subsequent GSK3β-mediated SNAI1 degradation [36]. However, stabilization of SNAI1 also occurs independent of GSK3β. Protein kinase A (PKA) and CK2 have been characterized as the main kinases responsible for in vitro SNAI1 phosphorylation at Ser11 and 92, respectively [37]. Phosphorylation of these two sites control SNAI1 stability and positively regulate SNAI1 repressive function and its interaction with the mSin3A corepressor. Alternatively, confinement of SNAI1 to the nucleus prevents degradation. ERK2-mediated Ser82/Ser104 phosphorylation of SNAI1 leads to nuclear SNAI1 accumulation [38]. P21 (RAC1) activated kinase 1 (PAK1) and GRO-α phosphorylate SNAI1 on Ser246 and increase SNAI1's accumulation in the nucleus, which thus promotes transcriptional activity of SNAI1 [39–41]. Large tumor suppressor kinase 2 (Lats2) phosphorylates SNAI1 at threonine (Thr)203 in the nucleus, which prevents nuclear export, thereby supporting stabilization [42]. Recently, we also found that serine/threonine kinase 39 (STK39) enhances SNAI1 stability by phosphorylation at Thr203 [43]. Notably, phosphorylation can be reversed by phosphatases. We identified c-terminal domain phosphatase (SCP) as a specific phosphatase for SNAI1 [44]. SCP physically interacts with and stabilizes SNAI1 by direct dephosphorylation [44,45].

Table 1. Phosphorylation regulators involved in SNAI1.

Function	Regulation Factor	Phosphorylation Sites
Degradation	PKD1	Ser11
	αPKC	Ser249
	CK1	Ser104/Ser107
	GSK3β	Ser96/Ser100/S104/Ser107
	DYRK2	Ser104
Stabilization	PTK6	Tyr342
	ATM	Ser100
	CK2	Ser92
	p38	Ser107
Nuclear accumulation and stabilization	ERK	Ser82/Ser104
	PAK1	Ser246
	GROα	
	LATS2	Thr203
	STK39	

2.3.2. Ubiquitination and Deubiquitination

SNAI1's ubiquitination and degradation are controlled by a number of F-box ligases, including β-TRCP1/FBXW1, FBXL14, FBXL5, FBXO11, and FBXO45 [46]. Recently, more E3 ligases have been discovered (Figure 2). F-box E3 ubiquitin ligase FBXO22 elicits antimetastatic effects by targeting SNAI1 ubiquitin-mediated proteasomal degradation in a GSK3β phosphorylation-dependent manner [47]. Through a luciferase-based genome-wide screening using small interfering RNA library against ~200 of E3 ligases and ubiquitin-related genes, SOCS box protein SplA/ryanodine receptor domain and SOCS box containing 3 (SPSB3) was identified as a novel E3 ligase component [48]. SPSB3 targets SNAI1 to promote polyubiquitination and degradation in response to GSK3β phosphorylation of SNAI1. Through yeast two-hybrid screening, the carboxyl terminus of Hsc70-interacting protein (CHIP) was identified as a novel SNAI1 ubiquitin ligase that interacts with SNAI1 to induce ubiquitin-mediated proteasomal degradation [49]. Recently, it was reported that SNAI1 was monoubiquitinated by the ubiquitin-editing enzyme A20. This monoubiquitylation of SNAI1 reduces the affinity of SNAI1 for GSK3β, and thus SNAI1 is stabilized in the nucleus [50].

Deubiquitinases (DUBs) counteract the SNAI1 degradation process to maintain a high level of SNAI1 protein in cancer cells. We recently identified DUB3 as a SNAI1 deubiquitinase that interacts with and stabilizes SNAI1 [51]. Independent research indicated that DUB3 is a target of cyclin-dependent kinase (CDK)4/6, and CDK4/6-mediated activation of DUB3 is essential to deubiquitinate and stabilize SNAI1 [52]. Resistance to platinum-based chemotherapy is a common event associated with tumor dissemination and metastasis in cancer patients. Upon platinum treatment, the ubiquitin-specific protease 1 (USP1) is phosphorylated by ATM and RAD3-related (ATR) and binds to SNAI1. Then, USP1 de-ubiquitinates and stabilizes SNAI1 expression, conferring resistance to platinum, increased stem cell-like features, and metastatic ability [53]. USP29 can be induced by major EMT and metastatic-inducing factors such as TGFβ, TNFα, and hypoxia. This protease enhances the interaction of SNAI1 and SCP1, and results in simultaneous dephosphorylation and de-ubiquitination of SNAI1 and thereafter cooperative prevention of SNAI1 degradation [54]. TGFβ also induces USP27X expression, which increases SNAI1 stability by deubiquitination [55]. Recently, more deubiquitinases have been identified. Eukaryotic translation initiation factor 3 subunit H (EIF3H), OTU deubiquitinase, ubiquitin aldehyde binding 1(OTUB1), USP3, proteasome 26S subunit, Non-ATPase 14 (PSMD14), USP26, USP36, USP37 also target SNAI1 for de-ubiquitination and stabilization (Figure 2) [56–61].

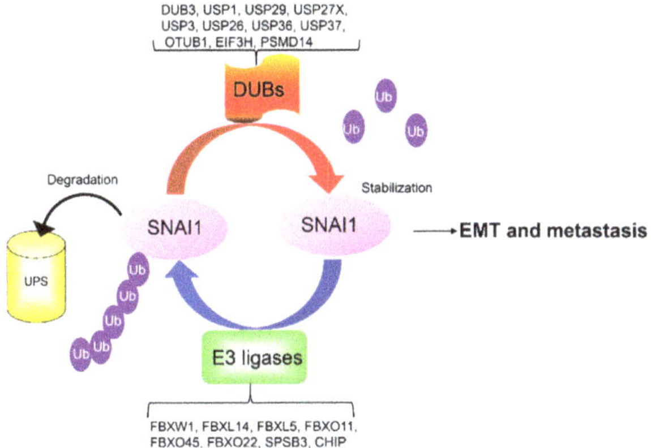

Figure 2. The ubiquitination and de-ubiquitination of SNAI1. SNAI1 is degraded by multiple E3 ligases. By contrast, de-ubiquitinases counteract E3 ligase activity and prevent SNAI1 degradation. USP: ubiquitin-specific protease; OTUB1: OTU deubiquitinase, ubiquitin aldehyde binding 1; PSMD14: proteasome 26S subunit, Non-ATPase 14; EIF3H: eukaryotic translation initiation factor 3 subunit H; DUB: deubiquitinase; UPS: ubiquitin/proteasome system; FBXW1: F-box/WD repeat-containing protein 1; FBXL: F-box and leucine rich repeat protein; FBXO: F-boxes other; SPSB3: SplA/ryanodine receptor domain and SOCS box containing 3; CHIP: carboxy-terminus of Hsc70 interacting protein.

2.3.3. Other Post-Translational Regulation

Beyond the well-characterized PTMs of phosphorylation and ubiquitination, at least three other PTMs are involved in regulating SNAI1 protein abundance and activity. First, the sumoylation pathway is very similar to its biochemical analog, ubiquitylation, and regulates diverse cellular processes including transcription and protein stability, chromosome organization, DNA repair, and other cellular processes. TGFβ induces sumoylation of SNAI1 at its lysine (K) 234 residue, which is critical for the EMT-activating function of SNAI1 [62]. Second, the O-linked β-*N*-acetylglucosamine (O-GlcNAc) modification is a monosaccharide addition. SNAI1 is subject to O-GlcNAc at Ser112 under hyperglycemic conditions [63]. This modification leads to stabilization of SNAI1 by inhibition of GSK3β-mediated phosphorylation. Consequently, the O-GlcNAc SNAI1 promotes EMT. Finally, SNAI1 is also acetylated by the histone acetyltransferase adenovirus E1A-associated protein (p300) and CREB binding protein (CBP), two key transcriptional coactivators implicated in a multitude of cellular processes including cancer progression. CBP and p300 interact with SNAI1 to acetylate SNAI1 at K146 and K187, which consequently reduces SNAI1 ubiquitination and thus enhances its protein stability [64] (Figure 3).

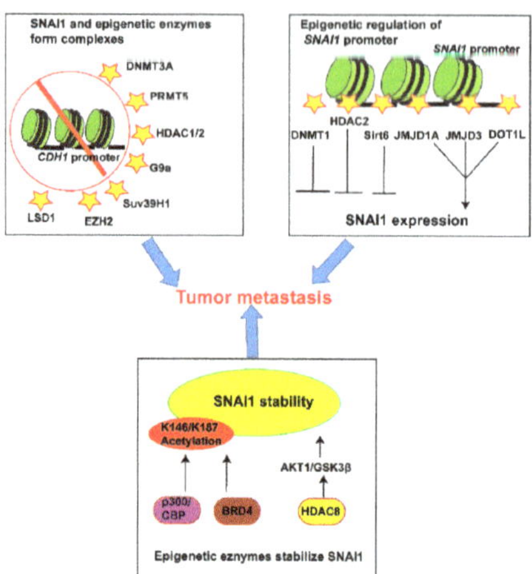

Figure 3. The interplay between SNAI1 and epigenetic regulators. SNAI1 collaborates with epigenetic regulators to repress CDH1 expression. Epigenetic regulators are recruited to the *SNAI1* promoter, leading to transcriptional activation or repression of SNAI1. In addition, epigenetic regulators regulate SNAI1 stability by post-translational modifications. The increase in SNAI1 expression via multiple epigenetic mechanisms leads to the cancer metastasis that is accompanied by the loss of CDH1. DNMT: DNA methyltransferases; PRMT: protein arginine methyltransferases; HDAC: histone deacetylases; Suv39H1: suppressor of variegation 3-9 homolog 1; EZH2: enhancer of zeste 2 polycomb repressive complex 2 subunit; LSD1: lysine-specific demethylase 1; JMJD: Jumonji C domain-containing; DOT1L: DOT1 like histone lysine methyltransferase; BRD4: bromodomain-containing protein 4; CBP: CREB binding protein.

3. The Interplay between SNAI1 and Epigenetic Regulators in Tumor Metastasis

Because EMT is a reversible and transient process, as well as having reversibility of the epigenetic marks and the enzymatic nature of the regulators, EMT-TFs and chromatin-remodeling enzymes are intimately connected (Figure 3). During tumor metastasis, SNAI1 recruits epigenetic regulators to the *CDH1* promoter, thus repressing its expression. Epigenetic alterations also play a crucial role in SNAI1 expression. The interplay between SNAI1 and epigenetic regulators indicate the complexity of epigenetic mechanisms and the potentially crucial role of histone modifications for regulating SNAI1.

3.1. SNAI1 and DNA Methylation

DNA methylation involves a covalent attachment of a methyl group to cytosine residues at CpG-rich dinucleotide sequences through DNA methyltransferases (DNMTs). Upon induction of EMT, hypermethylation of the *CDH1* promoter through DNMTs, which are recruited by EMT-TFs, is constantly observed in a wide variety of cancer cells. For example, SNAI1 interacts with DNMT3A to repress CDH1 expression via DNA hypermethylation and histone modifications of H3K9me2 and H3K27me3 in gastric cancer [65]. Previous research also indicated that DNMT1 was implicated in cell metastasis, such that downregulation or inhibition of DNMT1 could facilitate the metastasis of cancer cells [66]. DNMT1 can decrease the expression of CDH1 by increasing promoter methylation. Interestingly, DNMT1 can also act on CDH1 expression independent of its catalytic activity [67]. DNMT1 interacts with SNAI1 to prevent its interaction with the *CDH1* promoter; this interaction leads to full CDH1 expression. Furthermore, DNMT1 is recruited to the *SNAI1* promoter by AT-rich interactive domain-containing protein 2 (ARID2), a subunit of SWI/SNF chromatin remodeling complex. This complex increases the DNA methylation and suppresses SNAI1 transcription, leading to a repression of EMT. During hepatocellular carcinoma progression, loss/mutation of ARID2 impairs recruitment of DNMT1 to the *SNAI1* promoter. As a result of decreased methylation at the *SNAI1* promoter, there is an upregulation of SNAI1 expression that ultimately promotes EMT [68]. These results suggest that DNMT1 plays a cellular context-dependent role in tumor metastasis. Protein arginine methyltransferase (PRMT) 5 is a type II protein arginine methyltransferase. PRMT5 physically associates with SNAI1 and the NuRD (MAT1) complex to form a transcriptionally repressive complex that catalyzes a simultaneous histone demethylation and deacetylation. In addition, this complex also inhibits tet methylcytosine dioxygenase 1 (TET1) and contributes to DNA hypermethylation [69].

3.2. SNAI1 and Histone Modification

3.2.1. Acetylation

A variety of transcriptional co-activating complexes, which contain lysine acetyltransferase, catalyze lysine acetylation of histone tails. Because acetylation masks the positive charge on lysine residues and weakens the DNA–histone association and relaxes the chromatin structure, histone acetylation is often associated with gene activation. SNAI1 recruits the p300 activator complex to the *VEGF* and *Sox2* promoters to stimulate their expression, leading to endothelium generation and tumor growth [70].

3.2.2. Deacetylation

Histone deacetylation by histone deacetylase (HDAC) is believed to restrict gene transcription because it reveals the positive charge of lysine and permits the DNA–histone interaction. HDACs, in particular HDAC1 and HDAC2, are often recruited by EMT-TFs to gene promoter regions and form protein complexes to deacetylate histones and silence expression of epithelial gene factors. For instance, SNAI1 mediates recruitment of the HDAC1/2 that contain Sin3A or NuRD repressor complexes to inhibit CDH1 expression by deacetylation of histones H3 and H4. This effect was abolished by treatment with the HDAC inhibitor Trichostatin A (TSA) [71,72]. Interestingly, HDAC2 can also be recruited by the HOP homeobox to epigenetically inhibit SNAI1 transcription, leading to the enhanced histone H3K9 deacetylation, which subsequently suppresses tumor progression [73]. Similarly, HDAC1 can be recruited by SATB homeobox 2 (SATB2) to the *SNAI1* promoter, repressing *SNAI1* transcription and inhibiting EMT [74]. Recently, it has been reported that HDAC8 increases the protein stability of SNAI1 via AKT/GSK3β signals [75]. HDAC8 interacts with AKT1 to decrease acetylation while increasing its phosphorylation, which further increases Ser9-phosphorylation of GSK3β. Sirt6, the class III histone deacetylates, functions as an NAD$^+$-dependent histone deacetylase. Sirt6 interacts with p65 and attenuates NF-kB regulated SNAI1 expression by removing acetyl residues of histone H3K9 and H3K56 in the promoter regions of *SNAI1* [76].

3.2.3. Acetylation Readers

The bromodomain-containing proteins (BRDs) are acetylation readers that bind to ε-*N*-aminoacetyl groups of nucleosomal histone lysines and recruit histone modifiers and transcriptional/remodeling factors to gene promoters; these processes promote upregulation or repression of gene expression. Ever increasing studies in different cancer cells have demonstrated the contribution of BRDs to cancer progression [77]. For instance, BRD4 interacts with SNAI1 if certain K146 and K187 are acetylated. This interaction prevents recognition of SNAI1 by its E3 ubiquitin ligases FBXL14 and β-TrCP1, thereby inhibiting SNAI1 polyubiquitination and proteasomal degradation [78]. In addition, BRD4 increases SNAI1 expression by diminishing the PKD1-mediated proteasome degradation pathway. BRD4 inhibition suppresses the expression of Gli1, which is required for transcriptional activation of SNAI1, indicating that BRD4 controls malignancy of breast cancer cells via both transcriptional and post-translational regulation of SNAI1 [79]. Therefore, inhibition of BRD4 is a promising therapeutic approach for cancer patients with metastatic lesions.

3.2.4. Methylation

Histone lysine methylation is catalyzed by lysine methyltransferases, which directly recruit or inhibit the recruitment of histone-binding proteins. Usually, H3K9 and H3K27 methylation is associated with transcriptional repression, while H3K79 is often linked with gene activation. G9a is responsible for the transcriptionally repressive modification of H3K9. In aggressive lung cancer cells, G9a is preferentially expressed, and its elevated expression correlates with poor prognosis. G9a represses a cell adhesion molecule EPCAM, which stimulates EMT and cancer metastasis by catalyzing H3K9me2 on its promoter [80]. In breast cancer cells, SNAI1 recruits G9a to the *CDH1* promoter for transcription silencing. Therefore, inhibition of G9a reduces promoter H3K9me2 as well as DNA methylation which abrogates EMT and tumor metastasis [81]. Meanwhile, SNAI1 also interacts with Suv39H1, a histone methyltransferase for the trimethylation of histone H3 at lysine K9 (H3K9me3) to the *CDH1* promoter to repress its transcription. EZH2, the catalytic subunit of the polycomb repressive complex 2 (PRC2), promotes transcriptional silencing of CDH1 by H3K27me3 [82,83]. EZH2 can interact with HDAC1/HDAC2 in association with SNAI1 to form a complex that represses CDH1 expression [84]. DOT1L catalyzes the methylation of an active transcription mark histone H3K79, which is crucial for tumor development [85]. In breast cancers, DOT1L forms a transcriptionally active complex with c-Myc and p300 to facilitate H3K79 methylation and acetylation in the promoter regions of *SNAI1* that enhances SNAI1 de-repression, consequently promoting EMT [86].

3.2.5. Demethylation

Histone lysine-specific demethylase 1 (LSD1) functions as an epigenetic regulator by removing methyl groups on the transcription-activating H3K4 or repressing H3K9 residues through an amine oxidase reaction [87,88]. LSD1 takes part in a variety of chromatin-remodeling protein complexes to regulate tumor progression. We found that the amine oxidase domain of LSD1 interacts with the SNAG domain of SNAI1 [89]. SNAI1 recruits LSD1 and forms the SNAI1-LSD1-CoREST complex to repress CDH1 expression and enhance cell migration [89]. Another study indicated that SNAI1 recruits LSD1 on epithelial gene promoters for H3K4me2 demethylation, thereby silencing their expression and promoting EMT [90]. The chromatin remodeling factor Jumonji C (JmjC) domain-containing protein 3 (JMJD3, also known as KDM6B) is a α-ketoglutarate-dependent demethylase which is responsible for the demethylation of di- and trimethyllysine 27 (H3K27m2/3) on histone H3. JMJD3 demethylates H3K27m3 at the *SNAI1* promoter to activate the transcription of SNAI1 during TGFβ-induced EMT [91]. In addition, JMJD1A also transcriptionally activates SNAI1 expression via H3K9me1 and H3K9me2 demethylation at its promoter [92].

4. Potential Pharmacological Inhibitors of SNAI1

Given the important role of SNAI1 in driving cancer progression, targeting SNAI1 would be an attractive anticancer therapeutic approach. However, the development of small molecules to inhibit SNAI1's functions is hindered as there is no clear "ligand-binding domain" for targeting SNAI1. However, other strategies have been successfully attempted. First, the E-box, a SNAI1-binding site, was chosen as a target. A Co(III) complex conjugated to a CAGGTG hexanucleotide was synthesized. This complex binds to SNAI1 and prevents any interaction with DNA, thus reducing the invasive potential of tumor cells [93]. Second, the SNAI1-p53 complex acts as a target. Two leader compounds, GN25 and GN29, increase the expression of p53 and uncouple it from SNAI1. These two compounds selectively inhibit K-ras mutated cells [94]. Third, the LSD1-SNAI1 complex was chosen as a target. Inhibiting its interactions blocks cancer cell invasion [95,96]. Fourth, CYD19, a small-molecule compound, binds to SNAI1 and disrupts the SNAI1 interaction with p300, leading to SNAI1 degradation [97]. CYD19 impairs EMT-associated tumor invasion and metastasis by reversing SNAI1-driven EMT; this finding provides evidence that pharmacologic interference with SNAI1 acetylation may exert potent therapeutic effects in patients with cancer. Finally, chemical classes of synthetic and natural compounds affecting the transcriptional activity and expression of SNAI1 have already been characterized. For example, disulfiram inhibits cell migration, invasion, and growth of tumor grafts through the ERK/NF-κB/SNAI1 signaling pathway [98]. The proteasome inhibitor, NPI-0052, also inhibits SNAI1 expression via inhibition of NF-kB [99]. In all, targeting the SNAI1 complex or suppression of SNAI1 expression is one major approach to specifically inhibit SNAI1 activity.

Proteolysis-targeting chimeras (PROTACs) that hijack the ubiquitin-proteasome system for targeted protein degradation have expanded significantly in years [100]. This technology circumvents some of the limitations associated with traditional small-molecule therapeutics. PROTAC consists of a ligand for an E3 ligase and a ligand for a protein of interest (POI) connected by a chemical linker to form a ternary complex. In 2021, TRAnscription Factor Targeting Chimeras (TRAFTACs) technology was developed. The TRAFTAC system is composed of a HaloTag-fused dCas9 protein and a chimeric oligonucleotide that can bind transcription factor of interest (TOI) and dCas9 simultaneously [101]. This system labels the TOI with ubiquitin which then degrades the TOI by proteasomal machinery. This strategy was applied to target several transcription factors including E2F1 and NF-kB [102]. It will be attractive to design a TRAFTAC targeting SNAI1.

5. Conclusions and Perspective

SNAI1 as the key EMT regulator plays important roles in invasion and metastasis. The molecular events mediated by SNAI1 are of interest as therapeutic targets, in particular for resistant metastatic tumors. Although direct targeting of SNAI1 is unsuccessful, identifying inhibitors for PTMs of SNAI1 hold significant potential, and thus are a high priority in the development of future cancer treatments. Indeed, many pharmacological approaches, including chemical inhibitors and monoclonal antibodies that target these modification enzymes including deubiquitinase and kinase, have been devised and show promise for the treatment of tumor metastasis [103]. Furthermore, identification of SNAI1's post-transcriptional and PTMs is crucial given that these changes could be identified in the primary tumor before metastasis occurs. Such knowledge would facilitate better prediction of patients who have genotypes that are more likely to follow an aggressive clinical course and who are prone to development of metastases.

In addition, because of the intimate connection between SNAI1 and chromatin-remodeling enzymes, targeting the epigenetic enzymes to reverse the EMT process is also an efficient and promising approach [104]. Indeed, abundant pre-clinical and clinical studies examining the effects of these epigenetic enzyme inhibitors alone or in combination with other anti-cancer agents are under development [105]. However, the impact of these epigenetic alternations on tumor metastasis differs greatly in various types of cancers. Therefore, it is urgent to comprehensively understand the mechanisms of action and roles of epigenetic modulations on EMT in different cancer types. These detailed mechanisms of epigenetic regulation in tumor metastasis will provide a bright future for the use of an efficient and specific "epigendrug" as one of the important therapeutic strategies in the fight against tumor metastasis.

Author Contributions: All authors contributed to the manuscript content and editing for this review. B.D. wrote the manuscript. Y.W. wrote the manuscript, provided supervision and financial support. All authors have read and agreed to the published version of the manuscript.

Funding: This research was supported by the Shared Resources of the University of Kentucky Markey Cancer Center (P30CA177558). This research was also supported by grants from American Cancer Society Research Scholar Award (RSG13187) and NIH (P20GM121327 and CA230758) to Y.W.

Institutional Review Board Statement: Not applicable.

Informed Consent Statement: Not applicable.

Data Availability Statement: Not applicable.

Acknowledgments: We sincerely apologize to colleagues whose studies have been omitted from this review owing to space limitations. We thank Cathy Anthony for the critical editing of this manuscript.

Conflicts of Interest: The authors have declared that no conflict of interest exists.

References

1. Anderson, R.L.; Balasas, T.; Callaghan, J.; Coombes, R.C.; Evans, J.; Hall, J.A.; Kinrade, S.; Jones, D.; Jones, P.S.; Jones, R.; et al. A framework for the development of effective anti-metastatic agents. *Nat. Rev. Clin. Oncol.* **2019**, *16*, 185–204. [CrossRef]
2. Brabletz, T.; Kalluri, R.; Nieto, M.A.; Weinberg, R.A. EMT in cancer. *Nat. Rev. Cancer* **2018**, *18*, 128–134. [CrossRef] [PubMed]
3. Yuan, S.; Norgard, R.J.; Stanger, B.Z. Cellular Plasticity in Cancer. *Cancer Discov.* **2019**, *9*, 837–851. [CrossRef]
4. Puisieux, A.; Brabletz, T.; Caramel, J. Oncogenic roles of EMT-inducing transcription factors. *Nat. Cell Biol.* **2014**, *16*, 488–494. [CrossRef]
5. Hu, C.-T.; Chang, T.-Y.; Cheng, C.-C.; Liu, C.-S.; Wu, J.-R.; Li, M.-C.; Wu, W.-S. Snail associates with EGR-1 and SP-1 to upregulate transcriptional activation of p15INK4b. *FEBS J.* **2010**, *277*, 1202–1218. [CrossRef] [PubMed]
6. Guaita, S.; Puig, I.; Francí, C.; Garrido, M.; Domínguez, D.; Batlle, E.; Sancho, E.; Dedhar, S.; de Herreros, A.G.; Baulida, J. Snail Induction of Epithelial to Mesenchymal Transition in Tumor Cells Is Accompanied by MUC1 Repression andZEB1 Expression. *J. Biol. Chem.* **2002**, *277*, 39209–39216. [CrossRef]
7. Rembold, M.; Ciglar, L.; Yáñez-Cuna, J.O.; Zinzen, R.; Girardot, C.; Jain, A.; Welte, M.; Stark, A.; Leptin, M.; Furlong, E.E. A conserved role for Snail as a potentiator of active transcription. *Genes Dev.* **2014**, *28*, 167–181. [CrossRef]
8. Hu, C.-T.; Wu, J.-R.; Chang, T.Y.; Cheng, C.-C.; Wu, W.-S. The transcriptional factor Snail simultaneously triggers cell cycle arrest and migration of human hepatoma HepG2. *J. Biomed. Sci.* **2008**, *15*, 343–355. [CrossRef]
9. Ly, T.M.; Chen, Y.-C.; Lee, M.-C.; Hu, C.-T.; Cheng, C.-C.; Chang, H.-H.; You, R.-I.; Wu, W.-S. Snail Upregulates Transcription of FN, LEF, COX2, and COL1A1 in Hepatocellular Carcinoma: A General Model Established for Snail to Transactivate Mesenchymal Genes. *Cells* **2021**, *10*, 2202. [CrossRef] [PubMed]
10. Wu, W.-S.; You, R.-I.; Cheng, C.-C.; Lee, M.-C.; Lin, T.-Y.; Hu, C.-T. Snail collaborates with EGR-1 and SP-1 to directly activate transcription of MMP 9 and ZEB1. *Sci. Rep.* **2017**, *7*, 17753. [CrossRef]
11. Kajita, M.; McClinic, K.N.; Wade, P.A. Aberrant Expression of the Transcription Factors Snail and Slug Alters the Response to Genotoxic Stress. *Mol. Cell. Biol.* **2004**, *24*, 7559–7566. [CrossRef]
12. Mani, S.A.; Guo, W.; Liao, M.-J.; Eaton, E.N.; Ayyanan, A.; Zhou, A.Y.; Brooks, M.; Reinhard, F.; Zhang, C.C.; Shipitsin, M.; et al. The Epithelial-Mesenchymal Transition Generates Cells with Properties of Stem Cells. *Cell* **2008**, *133*, 704–715. [CrossRef]
13. Dong, C.; Yuan, T.; Wu, Y.; Wang, Y.; Fan, T.W.; Miriyala, S.; Lin, Y.; Yao, J.; Shi, J.; Kang, T.; et al. Loss of FBP1 by Snail-Mediated Repression Provides Metabolic Advantages in Basal-like Breast Cancer. *Cancer Cell* **2013**, *23*, 316–331. [CrossRef] [PubMed]
14. Wu, Y.; Zhou, B.P. Snail: More than Emt. *Cell Adhes. Migr.* **2010**, *4*, 199–203. [CrossRef]
15. Barrallo-Gimeno, A.; Nieto, M.A. The Snail genes as inducers of cell movement and survival: Implications in development and cancer. *Development* **2005**, *132*, 3151–3161. [CrossRef]
16. Greally, J.M. A user's guide to the ambiguous word 'epigenetics'. Nature reviews. *Mol. Cell Biol.* **2018**, *19*, 207–208.
17. Wen, B.; Wu, H.; Shinkai, Y.; Irizarry, R.A.; Feinberg, A.P. Large histone H3 lysine 9 dimethylated chromatin blocks distinguish differentiated from embryonic stem cells. *Nat. Genet.* **2009**, *41*, 246–250. [CrossRef] [PubMed]
18. Tam, W.L.; Weinberg, R.A. The epigenetics of epithelial-mesenchymal plasticity in cancer. *Nat. Med.* **2013**, *19*, 1438–1449. [CrossRef] [PubMed]
19. Nieto, M.A. The snail superfamily of zinc-finger transcription factors. *Nat. Rev. Mol. Cell Biol.* **2002**, *3*, 155–166. [CrossRef] [PubMed]
20. Joyce, J.A.; Pollard, J.W. Microenvironmental regulation of metastasis. *Nat. Rev. Cancer* **2008**, *9*, 239–252. [CrossRef] [PubMed]
21. Wang, Y.; Shi, J.; Chai, K.; Ying, X.; Zhou, B.P. The Role of Snail in EMT and Tumorigenesis. *Curr. Cancer Drug Targets* **2013**, *13*, 963–972. [CrossRef]
22. Kaufhold, S.; Bonavida, B. Central role of snail1 in the regulation of emt and resistance in cancer: A target for therapeutic intervention. *J. Exp. Clin. Cancer Res.* **2014**, *33*, 1–19. [CrossRef] [PubMed]
23. Saito, D.; Kyakumoto, S.; Chosa, N.; Ibi, M.; Takahashi, N.; Okubo, N.; Sawada, S.; Ishisaki, A.; Kamo, M. Transforming growth factor-beta1 induces epithelial-mesenchymal transition and integrin alpha3beta1-mediated cell migration of hsc-4 human squamous cell carcinoma cells through slug. *J. Biochem.* **2013**, *153*, 303–315. [CrossRef]
24. Nakamura, R.; Ishii, H.; Endo, K.; Hotta, A.; Fujii, E.; Miyazawa, K.; Saitoh, M. Reciprocal expression of Slug and Snail in human oral cancer cells. *PLoS ONE* **2018**, *13*, e0199442. [CrossRef] [PubMed]
25. Smit, M.A.; Geiger, T.R.; Song, J.Y.; Gitelman, I.; Peeper, D.S. A twist-snail axis critical for trkb-induced epithelial-mesenchymal transition-like transformation, anoikis resistance, and metastasis. *Mol. Cell Biol.* **2009**, *29*, 3722–3737. [CrossRef] [PubMed]
26. Skrzypek, K.; Majka, M. Interplay among SNAIL Transcription Factor, MicroRNAs, Long Non-Coding RNAs, and Circular RNAs in the Regulation of Tumor Growth and Metastasis. *Cancers* **2020**, *12*, 209. [CrossRef]
27. Wang, X.; Liu, R.; Zhu, W.; Chu, H.; Yu, H.; Wei, P.; Wu, X.; Zhu, H.; Gao, H.; Liang, J.; et al. UDP-glucose accelerates SNAI1 mRNA decay and impairs lung cancer metastasis. *Nature* **2019**, *571*, 127–131. [CrossRef] [PubMed]
28. Lin, X.; Chai, G.; Wu, Y.; Li, J.; Chen, F.; Liu, J.; Luo, G.; Tauler, J.; Du, J.; Lin, S.; et al. RNA m6A methylation regulates the epithelial mesenchymal transition of cancer cells and translation of Snail. *Nat. Commun.* **2019**, *10*, 2065. [CrossRef]

29. Li, F.; Zhao, H.; Su, M.; Xie, W.; Fang, Y.; Du, Y.; Yu, Z.; Hou, L.; Tan, W. Hnrnp-f regulates emt in bladder cancer by mediating the stabilization of snail1 mrna by binding to its 3′ UTR. *EBioMedicine* **2019**, *45*, 208–219. [CrossRef] [PubMed]

30. Zhou, B.P.; Deng, J.; Xia, W.; Xu, J.; Li, Y.M.; Gunduz, M.; Hung, M.C. Dual regulation of snail by gsk-3beta-mediated phosphorylation in control of epithelial-mesenchymal transition. *Nat. Cell Biol.* **2004**, *6*, 931–940. [CrossRef]

31. Mimoto, R.; Taira, N.; Takahashi, H.; Yamaguchi, T.; Okabe, M.; Uchida, K.; Miki, Y.; Yoshida, K. Dyrk2 controls the epithe-lial-mesenchymal transition in breast cancer by degrading snail. *Cancer Lett.* **2013**, *339*, 214–225. [CrossRef]

32. Zheng, H.; Shen, M.; Zha, Y.L.; Li, W.; Wei, Y.; Blanco, M.A.; Ren, G.; Zhou, T.; Storz, P.; Wang, H.Y.; et al. Pkd1 phosphor-ylation-dependent degradation of snail by scf-fbxo11 regulates epithelial-mesenchymal transition and metastasis. *Cancer Cell* **2014**, *26*, 358–373. [CrossRef] [PubMed]

33. Jung, H.Y.; Fattet, L.; Tsai, J.H.; Kajimoto, T.; Chang, Q.; Newton, A.C.; Yang, J. Apical-basal polarity inhibits epitheli-al-mesenchymal transition and tumour metastasis by par-complex-mediated snail degradation. *Nat. Cell Biol.* **2019**, *21*, 359–371. [CrossRef] [PubMed]

34. Sun, M.; Guo, X.; Qian, X.; Wang, H.; Yang, C.; Brinkman, K.L.; Serrano-Gonzalez, M.; Jope, R.S.; Zhou, B.; Engler, D.A.; et al. Activation of the ATM-Snail pathway promotes breast cancer metastasis. *J. Mol. Cell Biol.* **2012**, *4*, 304–315. [CrossRef] [PubMed]

35. Pyun, B.-J.; Seo, H.R.; Lee, H.-J.; Jin, Y.B.; Kim, E.-J.; Kim, N.H.; Kim, H.S.; Nam, H.W.; Yook, J.I.; Lee, Y.-S. Mutual regulation between DNA-PKcs and snail1 leads to increased genomic instability and aggressive tumor characteristics. *Cell Death Dis.* **2013**, *4*, e517. [CrossRef] [PubMed]

36. Ryu, K.J.; Park, S.M.; Park, S.H.; Kim, I.K.; Han, H.; Kim, H.J.; Kim, S.H.; Hong, K.S.; Kim, H.; Kim, M.; et al. P38 stabilizes snail by suppressing dyrk2-mediated phosphorylation that is required for gsk3beta-betatrcp-induced snail degradation. *Cancer Res.* **2019**, *79*, 4135–4148. [CrossRef]

37. MacPherson, M.R.; Molina, P.; Souchelnytskyi, S.; Wernstedt, C.; Martin-Pérez, J.; Portillo, F.; Cano, A. Phosphorylation of serine 11 and serine 92 as new positive regulators of human snail1 function: Potential involvement of casein kinase-2 and the camp-activated kinase protein kinase a. *Mol. Biol. Cell* **2010**, *21*, 244–253. [CrossRef]

38. Zhang, K.; Corsa, C.; Ponik, S.; Prior, J.L.; Piwnica-Worms, D.; Eliceiri, K.; Keely, P.J.; Longmore, G.D. The collagen receptor discoidin domain receptor 2 stabilizes SNAIL1 to facilitate breast cancer metastasis. *Nat. Cell Biol.* **2013**, *15*, 677–687. [CrossRef]

39. Yang, Z.; Rayala, S.; Nguyen, D.; Vadlamudi, R.K.; Chen, S.; Kumar, R. Pak1 Phosphorylation of Snail, a Master Regulator of Epithelial-to-Mesenchyme Transition, Modulates Snail's Subcellular Localization and Functions. *Cancer Res.* **2005**, *65*, 3179–3184. [CrossRef]

40. Chen, L.; Pan, X.-W.; Huang, H.; Gao, Y.; Yang, Q.-W.; Wang, L.-H.; Cui, X.-G.; Xu, D.-F. Epithelial-mesenchymal transition induced by GRO-α-CXCR2 promotes bladder cancer recurrence after intravesical chemotherapy. *Oncotarget* **2017**, *8*, 45274–45285. [CrossRef]

41. Escriva, M.; Peiro, S.; Herranz, N.; Villagrasa, P.; Dave, N.; Montserrat-Sentis, B.; Murray, S.A.; Franci, C.; Gridley, T.; Virtanen, I.; et al. Repression of pten phosphatase by snail1 transcriptional factor during gamma radiation-induced apop-tosis. *Mol. Cell Biol.* **2008**, *28*, 1528–1540. [CrossRef] [PubMed]

42. Zhang, K.; Rodriguez-Aznar, E.; Yabuta, N.; Owen, R.J.; Mingot, J.M.; Nojima, H.; Nieto, M.A.; Longmore, G.D. Lats2 kinase potentiates Snail1 activity by promoting nuclear retention upon phosphorylation. *EMBO J.* **2012**, *31*, 29–43. [CrossRef] [PubMed]

43. Qiu, Z.; Dong, B.; Guo, W.; Piotr, R.; Longmore, G.; Yang, X.; Yu, Z.; Deng, J.; Evers, B.M.; Wu, Y. STK39 promotes breast cancer invasion and metastasis by increasing SNAI1 activity upon phosphorylation. *Theranostics* **2021**, *11*, 7658–7670. [CrossRef] [PubMed]

44. Wu, Y.; Evers, B.M.; Zhou, B.P. Small C-terminal Domain Phosphatase Enhances Snail Activity through Dephosphorylation. *J. Biol. Chem.* **2009**, *284*, 640–648. [CrossRef] [PubMed]

45. Zhao, Y.; Liu, J.; Chen, F.; Feng, X.H. C-terminal domain small phosphatase-like 2 promotes epithelial-to-mesenchymal tran-sition via snail dephosphorylation and stabilization. *Open Biol.* **2018**, *8*, 170274. [CrossRef]

46. Yu, Q.; Zhou, B.P.; Wu, Y. The regulation of snail: On the ubiquitin edge. *Cancer Cell Microenviron.* **2017**, *4*, e1567. [CrossRef]

47. Sun, R.; Xie, H.-Y.; Qian, J.-X.; Huang, Y.-N.; Yang, F.; Zhang, F.-L.; Shao, Z.-M.; Li, D.-Q. Fbxo22 possesses both protumor-igenic and antimetastatic roles in breast cancer progression. *Cancer Res.* **2018**, *78*, 5274–5286. [CrossRef]

48. Liu, Y.; Zhou, H.; Zhu, R.; Ding, F.; Li, Y.; Cao, X.; Liu, Z. Spsb3 targets snail for degradation in gsk-3beta phosphoryla-tion-dependent manner and regulates metastasis. *Oncogene* **2018**, *37*, 768–776. [CrossRef]

49. Park, S.; Park, S.; Ryu, K.; Kim, I.-K.; Han, H.; Kim, H.; Kim, S.; Hong, K.; Kim, H.; Kim, M.; et al. Downregulation of CHIP promotes ovarian cancer metastasis by inducing Snail-mediated epithelial–mesenchymal transition. *Mol. Oncol.* **2019**, *13*, 1280–1295. [CrossRef]

50. Lee, J.-H.; Jung, S.M.; Yang, K.-M.; Bae, E.; Ahn, S.G.; Park, J.S.; Seo, D.; Kim, M.; Ha, J.; Lee, J.; et al. A20 promotes metastasis of aggressive basal-like breast cancers through multi-monoubiquitylation of Snail1. *Nat. Cell Biol.* **2017**, *19*, 1260–1273. [CrossRef] [PubMed]

51. Wu, Y.; Wang, Y.; Lin, Y.; Liu, Y.; Wang, Y.; Jia, J.; Singh, P.; Chi, Y.-I.; Wang, C.; Dong, C.; et al. Dub3 inhibition suppresses breast cancer invasion and metastasis by promoting Snail1 degradation. *Nat. Commun.* **2017**, *8*, 14228. [CrossRef] [PubMed]

52. Liu, T.; Yu, J.; Deng, M.; Yin, Y.; Zhang, H.; Luo, K.; Qin, B.; Li, Y.; Wu, C.; Ren, T. Cdk4/6-dependent activation of dub3 regulates cancer metastasis through snail1. *Nat. Commun.* **2017**, *8*, 1–12. [CrossRef]

53. Sonego, M.; Pellarin, I.; Costa, A.; Vinciguerra, G.L.R.; Coan, M.; Kraut, A.; D'Andrea, S.; Dall'Acqua, A.; Castillo-Tong, D.C.; Califano, D.; et al. USP1 links platinum resistance to cancer cell dissemination by regulating Snail stability. *Sci. Adv.* **2019**, *5*, eaav3235. [CrossRef]

54. Qian, W.; Li, Q.; Wu, X.; Li, W.; Li, Q.; Zhang, J.; Li, M.; Zhang, D.; Zhao, H.; Zou, X.; et al. Deubiquitinase USP29 promotes gastric cancer cell migration by cooperating with phosphatase SCP1 to stabilize Snail protein. *Oncogene* **2020**, *39*, 6802–6815. [CrossRef]

55. Lambies, G.; Miceli, M.; Martinez-Guillamon, C.; Olivera-Salguero, R.; Pena, R.; Frias, C.P.; Calderon, I.; Atanassov, B.S.; Dent, S.Y.R.; Arribas, J.; et al. Tgfbeta-activated usp27x deubiquitinase regulates cell migration and chemoresistance via stabilization of snail1. *Cancer Res.* **2019**, *79*, 33–46. [CrossRef]

56. Fan, L.; Chen, Z.; Wu, X.; Cai, X.; Feng, S.; Lu, J.; Wang, H.; Liu, N. Ubiquitin-Specific Protease 3 Promotes Glioblastoma Cell Invasion and Epithelial–Mesenchymal Transition via Stabilizing Snail. *Mol. Cancer Res.* **2019**, *17*, 1975–1984. [CrossRef] [PubMed]

57. Zhou, H.; Liu, Y.; Zhu, R.; Ding, F.; Cao, X.; Lin, D.; Liu, Z. OTUB1 promotes esophageal squamous cell carcinoma metastasis through modulating Snail stability. *Oncogene* **2018**, *37*, 3356–3368. [CrossRef]

58. Zhu, R.; Liu, Y.; Zhou, H.; Li, L.; Li, Y.; Ding, F.; Cao, X.; Liu, Z. Deubiquitinating enzyme PSMD14 promotes tumor metastasis through stabilizing SNAIL in human esophageal squamous cell carcinoma. *Cancer Lett.* **2018**, *418*, 125–134. [CrossRef]

59. Guo, X.; Zhu, R.; Luo, A.; Zhou, H.; Ding, F.; Yang, H.; Liu, Z. EIF3H promotes aggressiveness of esophageal squamous cell carcinoma by modulating Snail stability. *J. Exp. Clin. Cancer Res.* **2020**, *39*, 175. [CrossRef] [PubMed]

60. Cai, J.; Li, M.; Wang, X.; Li, L.; Li, Q.; Hou, Z.; Jia, H.; Liu, S. USP37 Promotes Lung Cancer Cell Migration by Stabilizing Snail Protein via Deubiquitination. *Front. Genet.* **2020**, *10*, 1324. [CrossRef]

61. Li, L.; Zhou, H.; Zhu, R.; Liu, Z. USP26 promotes esophageal squamous cell carcinoma metastasis through stabilizing Snail. *Cancer Lett.* **2019**, *448*, 52–60. [CrossRef]

62. Gudey, S.K.; Sundar, R.; Heldin, C.H.; Bergh, A.; Landstrom, M. Pro-invasive properties of snail1 are regulated by sumoylation in response to tgfbeta stimulation in cancer. *Oncotarget* **2017**, *8*, 97703–97726. [CrossRef] [PubMed]

63. Park, S.Y.; Kim, H.S.; Kim, N.H.; Ji, S.; Cha, S.Y.; Kang, J.G.; Ota, I.; Shimada, K.; Konishi, N.; Nam, H.W.; et al. Snail1 is stabilized by o-glcnac modification in hyperglycaemic condition. *EMBO J.* **2010**, *29*, 3787–3796. [CrossRef] [PubMed]

64. Hsu, D.S.-S.; Wang, H.-J.; Tai, S.-K.; Chou, C.-H.; Hsieh, C.-H.; Chiu, P.-H.; Chen, N.-J.; Yang, M.-H. Acetylation of Snail Modulates the Cytokinome of Cancer Cells to Enhance the Recruitment of Macrophages. *Cancer Cell* **2014**, *26*, 534–548. [CrossRef] [PubMed]

65. Cui, H.; Hu, Y.; Guo, D.; Zhang, A.; Gu, Y.; Zhang, S.; Zhao, C.; Gong, P.; Shen, X.; Li, Y.; et al. DNA methyltransferase 3A isoform b contributes to repressing E-cadherin through cooperation of DNA methylation and H3K27/H3K9 methylation in EMT-related metastasis of gastric cancer. *Oncogene* **2018**, *37*, 4358–4371. [CrossRef] [PubMed]

66. Lee, E.; Wang, J.; Yumoto, K.; Jung, Y.; Cackowski, F.C.; Decker, A.M.; Li, Y.; Franceschi, R.T.; Pienta, K.J.; Taichman, R.S. DNMT1 Regulates Epithelial-Mesenchymal Transition and Cancer Stem Cells, Which Promotes Prostate Cancer Metastasis. *Neoplasia* **2016**, *18*, 553–566. [CrossRef]

67. Espada, J.; Peinado, H.; Lopez-Serra, L.; Setién, F.; Lopez-Serra, P.; Portela, A.; Renart, J.; Carrasco, E.; Calvo, M.; Juarranz, A.; et al. Regulation of SNAIL1 and E-cadherin function by DNMT1 in a DNA methylation-independent context. *Nucleic Acids Res.* **2011**, *39*, 9194–9205. [CrossRef]

68. Jiang, H.; Cao, H.-J.; Ma, N.; Bao, W.-D.; Wang, J.-J.; Chen, T.-W.; Zhang, E.-B.; Yuan, Y.-M.; Ni, Q.-Z.; Zhang, F.-K.; et al. Chromatin remodeling factor ARID2 suppresses hepatocellular carcinoma metastasis via DNMT1-Snail axis. *Proc. Natl. Acad. Sci. USA* **2020**, *117*, 4770–4780. [CrossRef] [PubMed]

69. Gao, J.; Liu, R.; Feng, D.; Huang, W.; Huo, M.; Zhang, J.; Leng, S.; Yang, Y.; Yang, T.; Yin, X.; et al. Snail/PRMT5/NuRD complex contributes to DNA hypermethylation in cervical cancer by TET1 inhibition. *Cell Death Differ.* **2021**, *28*, 2818–2836. [CrossRef]

70. Chang, Z.; Zhang, Y.; Liu, J.; Zheng, Y.; Li, H.; Kong, Y.; Li, P.; Peng, H.; Shi, Y.; Cao, B.; et al. Snail promotes the generation of vascular endothelium by breast cancer cells. *Cell Death Dis.* **2020**, *11*, 457. [CrossRef]

71. Peinado, H.; Ballestar, E.; Esteller, M.; Cano, A. Snail Mediates E-Cadherin Repression by the Recruitment of the Sin3A/Histone Deacetylase 1 (HDAC1)/HDAC2 Complex. *Mol. Cell. Biol.* **2004**, *24*, 306–319. [CrossRef] [PubMed]

72. Fujita, N.; Jaye, D.; Kajita, M.; Geigerman, C.; Moreno, C.S.; Wade, P.A. MTA3, a Mi-2/NuRD Complex Subunit, Regulates an Invasive Growth Pathway in Breast Cancer. *Cell* **2003**, *113*, 207–219. [CrossRef]

73. Ren, X.; Yang, X.; Cheng, B.; Chen, X.; Zhang, T.; He, Q.; Li, B.; Li, Y.; Tang, X.; Wen, X.; et al. HOPX hypermethylation promotes metastasis via activating SNAIL transcription in nasopharyngeal carcinoma. *Nat. Commun.* **2017**, *8*, 14053. [CrossRef] [PubMed]

74. Wang, Y.-Q.; Jiang, D.-M.; Hu, S.-S.; Zhao, L.; Wang, L.; Yang, M.-H.; Ai, M.-L.; Jiang, H.-J.; Han, Y.; Ding, Y.-Q.; et al. SATB2-AS1 Suppresses Colorectal Carcinoma Aggressiveness by Inhibiting SATB2-Dependent Snail Transcription and Epithelial–Mesenchymal Transition. *Cancer Res.* **2019**, *79*, 3542–3556. [CrossRef] [PubMed]

75. An, P.; Chen, F.; Li, Z.; Ling, Y.; Peng, Y.; Zhang, H.; Li, J.; Chen, Z.; Wang, H. Hdac8 promotes the dissemination of breast cancer cells via akt/gsk-3beta/snail signals. *Oncogene* **2020**, *39*, 4956–4969. [CrossRef]

76. Chen, F.; Ma, X.; Liu, Y.; Ma, D.; Gao, X.; Qian, X. SIRT6 inhibits metastasis by suppressing SNAIL expression in nasopharyngeal carcinoma cells. *Int. J. Clin. Exp. Pathol.* **2021**, *14*, 63–74.

77. Belkina, A.; Denis, G.V. BET domain co-regulators in obesity, inflammation and cancer. *Nat. Rev. Cancer* **2012**, *12*, 465–477. [CrossRef]

78. Qin, Z.-Y.; Wang, T.; Su, S.; Shen, L.-T.; Zhu, G.-X.; Liu, Q.; Zhang, L.; Liu, K.-W.; Zhang, Y.; Zhou, Z.-H.; et al. BRD4 promotes gastric cancer progression and metastasis through acetylation-dependent stabilization of Snail. *Cancer Res.* **2019**, *79*, 4869–4881. [CrossRef]

79. Lu, L.; Chen, Z.; Lin, X.; Tian, L.; Su, Q.; An, P.; Li, W.; Wu, Y.; Du, J.; Shan, H.; et al. Inhibition of brd4 suppresses the ma-lignancy of breast cancer cells via regulation of snail. *Cell Death Differ.* **2020**, *27*, 255–268. [CrossRef]

80. Chen, M.-W.; Hua, K.-T.; Kao, H.-J.; Chi, C.-C.; Wei, L.-H.; Johansson, G.; Shiah, S.-G.; Chen, P.-S.; Jeng, Y.-M.; Cheng, T.-Y.; et al. H3K9 Histone Methyltransferase G9a Promotes Lung Cancer Invasion and Metastasis by Silencing the Cell Adhesion Molecule Ep-CAM. *Cancer Res.* **2010**, *70*, 7830–7840. [CrossRef]

81. Dong, C.; Wu, Y.; Yao, J.; Wang, Y.; Yu, Y.; Rychahou, P.; Evers, B.M.; Zhou, B.P. G9a interacts with Snail and is critical for Snail-mediated E-cadherin repression in human breast cancer. *J. Clin. Investig.* **2012**, *122*, 1469–1486. [CrossRef]

82. Herranz, N.; Pasini, D.; Díaz, V.M.; Francí, C.; Gutierrez, A.; Dave, N.; Escrivà, M.; Hernandez-Muñoz, I.; Di Croce, L.; Helin, K.; et al. Polycomb Complex 2 Is Required for E-cadherin Repression by the Snail1 Transcription Factor. *Mol. Cell. Biol.* **2008**, *28*, 4772–4781. [CrossRef] [PubMed]

83. Cao, Q.; Yu, J.; Dhanasekaran, S.M.; Kim, J.H.; Mani, R.; Tomlins, S.; Mehra, R.; Laxman, B.; Cao, X.; Kleer, C.G.; et al. Repression of E-cadherin by the polycomb group protein EZH2 in cancer. *Oncogene* **2008**, *27*, 7274–7284. [CrossRef]

84. Tong, Z.-T.; Cai, M.-Y.; Wang, X.-G.; Kong, L.-L.; Mai, S.-J.; Liu, Y.-H.; Zhang, H.-B.; Liao, Y.-J.; Zheng, F.; Zhu, W.-G.; et al. EZH2 supports nasopharyngeal carcinoma cell aggressiveness by forming a co-repressor complex with HDAC1/HDAC2 and Snail to inhibit E-cadherin. *Oncogene* **2011**, *31*, 583–594. [CrossRef]

85. Okada, Y.; Feng, Q.; Lin, Y.; Jiang, Q.; Li, Y.; Coffield, V.M.; Su, L.; Xu, G.; Zhang, Y. hDOT1L Links Histone Methylation to Leukemogenesis. *Cell* **2005**, *121*, 167–178. [CrossRef] [PubMed]

86. Cho, M.H.; Park, J.-H.; Choi, H.-J.; Park, M.-K.; Won, H.-Y.; Park, Y.-J.; Lee, C.H.; Oh, S.-H.; Song, Y.-S.; Kim, H.S.; et al. DOT1L cooperates with the c-Myc-p300 complex to epigenetically derepress CDH1 transcription factors in breast cancer progression. *Nat. Commun.* **2015**, *6*, 7821. [CrossRef] [PubMed]

87. Metzger, E.; Wissmann, M.; Yin, N.; Müller, J.M.; Schneider, R.; Peters, A.H.; Günther, T.; Buettner, R.; Schüle, R. Lsd1 de-methylates repressive histone marks to promote androgen-receptor-dependent transcription. *Nature* **2005**, *437*, 436–439. [CrossRef] [PubMed]

88. Shi, Y.; Lan, F.; Matson, C.; Mulligan, P.; Whetstine, J.R.; Cole, P.A.; Casero, R.; Shi, Y. Histone Demethylation Mediated by the Nuclear Amine Oxidase Homolog LSD1. *Cell* **2004**, *119*, 941–953. [CrossRef]

89. Lin, Y.; Wu, Y.; Li, J.; Dong, C.; Ye, X.; Chi, Y.-I.; Evers, B.M.; Zhou, B.P. The SNAG domain of Snail1 functions as a molecular hook for recruiting lysine-specific demethylase 1. *EMBO J.* **2010**, *29*, 1803–1816. [CrossRef]

90. Lin, T.; Ponn, A.; Hu, X.; Law, B.K.; Lu, J. Requirement of the histone demethylase LSD1 in Snai1-mediated transcriptional repression during epithelial-mesenchymal transition. *Oncogene* **2010**, *29*, 4896–4904. [CrossRef]

91. Ramadoss, S.; Chen, X.; Wang, C.-Y. Histone Demethylase KDM6B Promotes Epithelial-Mesenchymal Transition. *J. Biol. Chem.* **2012**, *287*, 44508–44517. [CrossRef]

92. Tang, D.E.; Dai, Y.; Fan, L.L.; Geng, X.Y.; Fu, D.X.; Jiang, H.W.; Xu, S.H. Histone demethylase jmjd1a promotes tumor pro-gression via activating snail in prostate cancer. *Mol. Cancer Res.* **2020**, *18*, 698–708. [CrossRef]

93. Vistain, L.F.; Yamamoto, N.; Rathore, R.; Cha, P.; Meade, T.J. Targeted Inhibition of Snail Activity in Breast Cancer Cells by Using a CoIII-Ebox Conjugate. *ChemBioChem* **2015**, *16*, 2065–2072. [CrossRef] [PubMed]

94. Lee, S.-H.; Shen, G.-N.; Jung, Y.S.; Lee, S.-J.; Chung, J.-Y.; Kim, H.-S.; Xu, Y.; Choi, Y.; Lee, J.-W.; Ha, N.-C.; et al. Antitumor effect of novel small chemical inhibitors of Snail-p53 binding in K-Ras-mutated cancer cells. *Oncogene* **2010**, *29*, 4576–4587. [CrossRef] [PubMed]

95. Baron, R.; Binda, C.; Tortorici, M.; McCammon, J.A.; Mattevi, A. Molecular Mimicry and Ligand Recognition in Binding and Catalysis by the Histone Demethylase LSD1-CoREST Complex. *Structure* **2011**, *19*, 212–220. [CrossRef] [PubMed]

96. Ferrari-Amorotti, G.; Fragliasso, V.; Esteki, R.; Prudente, Z.; Soliera, A.R.; Cattelani, S.; Manzotti, G.; Grisendi, G.; Dominici, M.; Pieraccioli, M.; et al. Inhibiting Interactions of Lysine Demethylase LSD1 with Snail/Slug Blocks Cancer Cell Invasion. *Cancer Res.* **2013**, *73*, 235–245. [CrossRef]

97. Li, H.M.; Bi, Y.R.; Li, Y.; Fu, R.; Lv, W.C.; Jiang, N.; Xu, Y.; Ren, B.X.; Chen, Y.D.; Xie, H.; et al. A potent cbp/p300-snail in-teraction inhibitor suppresses tumor growth and metastasis in wild-type p53-expressing cancer. *Sci. Adv.* **2020**, *6*, eaaw8500. [CrossRef]

98. Han, D.; Wu, G.; Chang, C.; Zhu, F.; Xiao, Y.; Li, Q.; Zhang, T.; Zhang, L. Disulfiram inhibits tgf-β-induced epitheli-al-mesenchymal transition and stem-like features in breast cancer via erk/nf-κb/snail pathway. *Oncotarget* **2015**, *6*, 40907–40919. [CrossRef]

99. Baritaki, S.; Chapman, A.; Yeung, K.; Spandidos, D.; Palladino, M.; Bonavida, B. Inhibition of epithelial to mesenchymal transition in metastatic prostate cancer cells by the novel proteasome inhibitor, NPI-0052: Pivotal roles of Snail repression and RKIP induction. *Oncogene* **2009**, *28*, 3573–3585. [CrossRef]

100. Burslem, G.; Crews, C.M. Proteolysis-Targeting Chimeras as Therapeutics and Tools for Biological Discovery. *Cell* **2020**, *181*, 102–114. [CrossRef]

101. Samarasinghe, K.T.; Jaime-Figueroa, S.; Burgess, M.; Nalawansha, D.A.; Dai, K.; Hu, Z.; Bebenek, A.; Holley, S.A.; Crews, C.M. Targeted degradation of transcription factors by TRAFTACs: TRAnscription Factor TArgeting Chimeras. *Cell Chem. Biol.* **2021**, *28*, 648–661.e5. [CrossRef]

102. Hu, Z.; Crews, C.M. Recent Developments in PROTAC-Mediated Protein Degradation: From Bench to Clinic. *ChemBioChem* **2021**. [CrossRef] [PubMed]
103. Lai, K.P.; Chen, J.; Tse, W.K.F. Role of Deubiquitinases in Human Cancers: Potential Targeted Therapy. *Int. J. Mol. Sci.* **2020**, *21*, 2548. [CrossRef]
104. Dong, B.; Qiu, Z.; Wu, Y. Tackle Epithelial-Mesenchymal Transition with Epigenetic Drugs in Cancer. *Front. Pharmacol.* **2020**, *11*, 596239. [CrossRef] [PubMed]
105. Park, J.W.; Han, J.-W. Targeting epigenetics for cancer therapy. *Arch. Pharmacal Res.* **2019**, *42*, 159–170. [CrossRef] [PubMed]

International Journal of
Molecular Sciences

Review

Pathophysiological Potentials of NRF3-Regulated Transcriptional Axes in Protein and Lipid Homeostasis

Tsuyoshi Waku [1] **and Akira Kobayashi** [2,*]

[1] Laboratory for Genetic Code, Department of Medical Life Systems, Faculty of Life and Medical Sciences, Doshisha University, Kyotanabe 610-0394, Japan; twaku@mail.doshisha.ac.jp
[2] Laboratory for Genetic Code, Graduate School of Life and Medical Sciences, Doshisha University, Kyotanabe 610-0394, Japan
* Correspondence: akobayas@mail.doshisha.ac.jp

Abstract: NRF3 (NFE2L3) belongs to the CNC-basic leucine zipper transcription factor family. An NRF3 homolog, NRF1 (NFE2L1), induces the expression of proteasome-related genes in response to proteasome inhibition. Another homolog, NRF2 (NFE2L2), induces the expression of genes related to antioxidant responses and encodes metabolic enzymes in response to oxidative stress. Dysfunction of each homolog causes several diseases, such as neurodegenerative diseases and cancer development. However, NRF3 target genes and their biological roles remain unknown. This review summarizes our recent reports that showed NRF3-regulated transcriptional axes for protein and lipid homeostasis. NRF3 induces the gene expression of *POMP* for 20S proteasome assembly and *CPEB3* for NRF1 translational repression, inhibiting tumor suppression responses, including cell-cycle arrest and apoptosis, with resistance to a proteasome inhibitor anticancer agent bortezomib. NRF3 also promotes mevalonate biosynthesis by inducing *SREBP2* and *HMGCR* gene expression, and reduces the intracellular levels of neural fatty acids by inducing *GGPS1* gene expression. In parallel, NRF3 induces macropinocytosis for cholesterol uptake by inducing *RAB5* gene expression. Finally, this review mentions not only the pathophysiological aspects of these NRF3-regulated axes for cancer cell growth and anti-obesity potential but also their possible role in obesity-induced cancer development.

Keywords: NRF3; protein homeostasis; lipid homeostasis; proteasome; translation; GGPP; macropinocytosis; cancer; obesity

Citation: Waku, T.; Kobayashi, A. Pathophysiological Potentials of NRF3-Regulated Transcriptional Axes in Protein and Lipid Homeostasis. *Int. J. Mol. Sci.* **2021**, *22*, 12686. https://doi.org/10.3390/ijms222312686

Academic Editors:
Amelia Casamassimi,
Alfredo Ciccodicola and
Monica Rienzo

Received: 4 November 2021
Accepted: 22 November 2021
Published: 24 November 2021

Publisher's Note: MDPI stays neutral with regard to jurisdictional claims in published maps and institutional affiliations.

1. Introduction

Protein and lipid homeostasis is crucial for cell survival and proliferation, and the defects interfere with several diseases, such as neurodegeneration, cancer development, metabolic disorder, and obesity [1–4]. NRF3 (nuclear factor erythroid 2-like 3; NFE2L3) belongs to the cap'n'collar (CNC)-basic leucine zipper transcription factor family, and has two homologs: NRF1 (nuclear factor erythroid 2-like 1; NFE2L1) and NRF2 (nuclear factor erythroid 2-like 2; NFE2L2) [5,6]. NRF1 mainly maintains the proteasome activity by comprehensively inducing the expression of most proteasome-related genes [7]. *Nrf1*-null mice suffer from embryonic lethality [8]. Thus, neuron, liver, or osteoblast-specific *Nrf1* knockout mice have been generated and show tissue defects, such as neurodegeneration, nonalcoholic steatohepatitis, and bone loss [9–12]. NRF2 is crucial for the cytoprotective mechanisms against xenobiotic and oxidative stress [13]. NRF2 also activates genes encoding enzymes for glutaminolysis, shifting the metabolic flux of glutamine to the glutathione synthesis pathway [14,15]. *Nrf2*-null mice do not respond to oxidative stress [16], whereas *Kelch-like ECH-associated protein 1 (Keap1)*-null mice demonstrate postnatal lethality by the constitutive activation of *Nrf2* [17]. Meanwhile, the biological function of NRF3 has long remained unclear because *Nrf3*-null mice develop and grow normally under physiological conditions [18–20]. However, studies on NRF3 have recently increased [21–23]. For example, we reported that the *NRF3* gene is highly expressed in several cancers [24].

NRF3 and NRF1 proteins are anchored to the endoplasmic reticulum (ER), and are degraded through ER-associated degradation. Proteotoxic stress, such as proteasome inhibition, leads to the cleavage of these proteins by the aspartic protease DNA damage-inducible 1 homolog 2 (DDI2), resulting in the nuclear translocation of cleaved NRF3 and NRF1 proteins for transcription activation [25–27]. Meanwhile, NRF2 proteins are negatively regulated by a cytosolic E3 ligase adaptor protein KEAP1, and are activated in response to oxidative stress [28]. In the nucleus, activated NRF proteins heterodimerize with small musculoaponeurotic fibrosarcoma (sMAF) proteins, including MAFF, MAFG, and MAFK, and bind to a consensus sequence called antioxidant response element (ARE; TGA[G/C]NNNGC) [29,30]. To date, we have identified several NRF3 target genes that coordinate protein and lipid homeostasis by gene expression analysis based on DNA microarray, real-time quantitative PCR, and chromatin immunoprecipitation (ChIP) experiments.

This review first introduces that NRF3 promotes cancer development through proteasome regulation, by inducing the gene regulation of *proteasome maturation protein (POMP)* [31] and *cytoplasmic polyadenylation element-binding protein 3 (CPEB3)* [32]. Then, this review describes the gene expression network of NRF3-regulated lipid metabolism, including *sterol regulatory element-binding protein 2 (SREBP2)* and *hydroxy-methylglutaryl-CoA reductase (HMGCR)* [33]. NRF3 also induces the gene expression of *GGP synthase 1 (GGPS1)* for geranylgeranyl pyrophosphate (GGPP)-mediated lipogenesis inhibition and *ras-related small GTPase protein (RAB5)* for macropinocytic cholesterol uptake [33]. Finally, this review remarks on the pathophysiological potential of these NRF3-regulated axes for cancer and obesity.

2. Assembly of the Ubiquitin-Independent 20S Proteasome

2.1. POMP, a 20S Proteasome Assembly Factor

The 26S proteasome is essential for ubiquitin-dependent protein degradation and consists of two subcomplexes: a 20S proteasome and a 19S-regulatory particle (RP) [34]. Several chaperones strictly coordinate the assembly of 20S proteasome and 19S-RP [35,36]. NRF1 induces the expression of almost all proteasome-related genes required for 26S proteasome [7]. Meanwhile, NRF3 does not affect the expression of almost all proteasome-related genes, but induces the gene expression of *POMP* [31], a chaperone of the 20S proteasome assembly [37]. ChIP experiments showed an ARE-like sequence (TGAGCG-GCG) near the transcription start site of the *POMP* gene as the NRF3 binding region [31] (Figure 1A). Furthermore, *POMP*-ARE mutations using CRISPR/Cas9-based genome editing reduced not only NRF3 recruitment on *POMP*-ARE but also *POMP* gene expression induced by NRF3 [31]. Proteasome activity assays using a fluorogenic substrate showed that NRF3 increases the amount and activity of 20S proteasome [31] (Figure 1B). These results provided direct evidence that *POMP* is an NRF3 target gene for enhancing the 20S proteasome activity.

2.2. NRF3-POMP-20S Proteasome Assembly Axis for Cancer Development

The 20S proteasome, a homodimer of a half-mer proteasome composed of an outer α-ring and an inner β-ring, contains proteolytic sites with different specificities: chymotrypsin-, caspase-, and trypsin-like activities. Meanwhile, the 20S proteasome lacks the 19S-RP that selects and unfolds ubiquitin substrates. Previous studies suggested that the 20S proteasome contributes to the ubiquitin-independent degradation of several tumor suppressor proteins, such as p53 and retinoblastoma (Rb) [38]. Surprisingly, NRF3 decreases p53 and Rb proteins without alteration of their mRNA levels under treatment with a ubiquitin-activating enzyme E1 inhibitor TAK-243, which inhibits 26S proteasome-mediated protein degradation by covalently binding with ubiquitin proteins [39] (Figure 1B). Furthermore, *POMP*-ARE mutation impairs the NRF3-mediated reduction in p53 and Rb protein, irrespective of TAK-243 treatment. More importantly, p53 and Rb inhibit cancer cell proliferation by inducing cell-cycle arrest or apoptosis in response to DNA damage [40]. NRF3 suppresses the expression of p53 target genes, including the cell-cycle inhibitory effector gene *p21* [41] and the proapoptotic gene *PUMA* (*p53 upregulated modulator of apoptosis*) [42]. NRF3 further inhibits p53-dependent cell-cycle arrest and apoptosis induction, leading to continuous cancer cell growth [31] (Figure 1B). These results indicated that the NRF3-POMP-20S proteasome assembly axis affects the ubiquitin-independent degradation of endogenous p53 and Rb proteins.

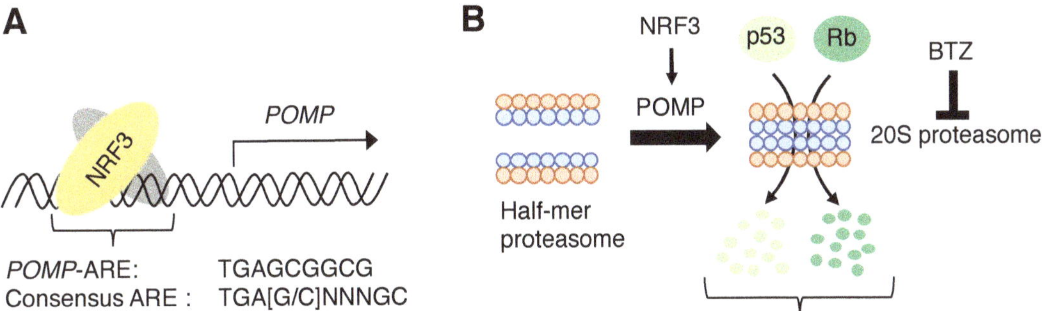

Figure 1. NRF3-POMP-20S proteasome assembly axis. (**A**) NRF3 directly induces *POMP* expression by binding to *POMP*-ARE, which is slightly different from consensus ARE. (**B**) Upregulation of the NRF3-POMP axis enhances a dimerization of a half-mer proteasome (known as the 20S proteasome assembly). The increased 20S proteasome confers the ubiquitin-independent degradation of p53 and Rb proteins, resulting in the rapid and continuous growth of cancer cells. Aberrant upregulation of the axis also confers resistance to a BTZ-type anticancer agent.

The proteasome is a target for cancer chemotherapy, and several proteasome inhibitors have been developed as anticancer agents. Among proteasome inhibitor anticancer agents, bortezomib (BTZ) inhibits both 20S and 26S proteolytic activities by binding to catalytic sites within the 20S proteasome [43]. Expectedly, upregulation of the NRF3-POMP-20S proteasome axis confers resistance to BTZ [31] (Figure 1B). Furthermore, xenograft and hepatic metastatic mouse models showed that NRF3 increases tumorigenesis and metastasis, whereas *POMP*-ARE mutation inhibits this tumor burden [31]. More importantly, clinical analyses indicated a negative correlation between *POMP/NRF3* mRNA levels and the survival rates of patients with colorectal adenocarcinoma, where *NRF3* is highly expressed [31]. These insights shed light on the crucial function of the NRF3-POMP-20S proteasome axis on cancer development, by inhibiting tumor suppression signals of p53 and Rb through ubiquitin-independent degradation. The upregulation of the axis also confers resistance to a BTZ-type proteasome inhibitor [44].

3. Complementary Maintenance of Proteasome with NRF1

3.1. CPEB3, a Translational Repressor of NRF1

NRF3 increases 20S proteasome activity through *POMP* expression [31], whereas NRF1 maintains 26S proteasome activity by inducing the expression of almost all proteasome-related genes under proteasome inhibition [7], implying the biological relevance of NRF1 and NRF3 for proteasome activity. In fact, the double knockdown of NRF1 and NRF3 impairs basal proteasome activity in living cells [32]. Compared to the single knockdown of NRF1 or NRF3, double knockdown of NRF1 and NRF3 reduces several proteasome-related genes, including *PSMB3*, *PSMB7*, *PSMC2*, *PSMD3*, *PSMG2*, *PSMG3*, and *POMP* [32]. ChIP experiments showed an ARE sequence near the transcription start site of each gene. These results indicated that NRF1 and NRF3 complementarily induce the expression of several proteasome-related genes to maintain the proteasome activity [32].

Interestingly, NRF3 represses the translation of NRF1 proteins by decreasing the amount of *NRF1* mRNA in polysomes, although NRF3 does not only affect the levels of *NRF1* mRNA but also the degradation of NRF1 proteins [32]. Gene expression analysis identified *CPEB3* as the candidate NRF3 target gene for this NRF1 translation repression [32] (Figure 2A). CPEB family proteins recognize a CPEB recognition motif (5′-UUUUA-3′, CPE) in the 3′-untranslated region (UTR) of a target gene for translation regulation [45]. CPEB3 interacts with the *NRF1*–3′-UTR which contains five CPEs, decreasing NRF1 protein levels and the amount of *NRF1* mRNA in polysomes [32] (Figure 2A). Meanwhile, NRF3 deficiency or CPE mutation of the *NRF1*–3′-UTR increases NRF1 translation [32] (Figure 2B). These results indicate that NRF3 directly induces *CPEB3* gene expression, and then CPEB3 inhibits ribosome recruitment to *NRF1* mRNA, resulting in the repression of NRF1 translation.

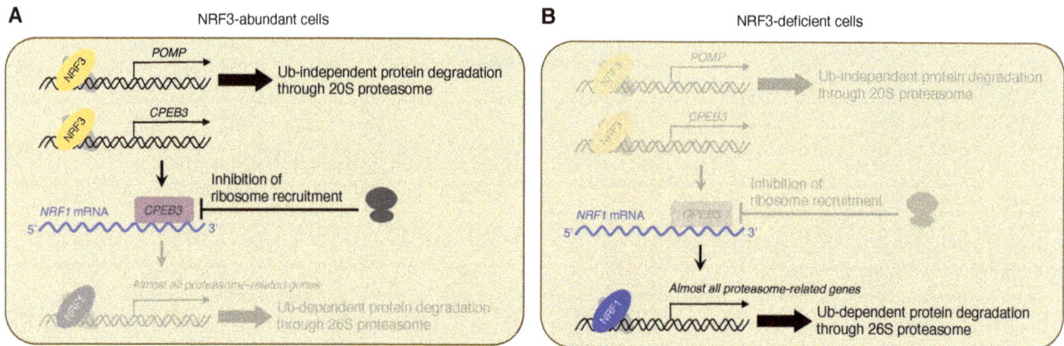

Figure 2. NRF3-CPEB3-NRF1 translational repression axis. (**A**) In NRF3-abundant cells, NRF3 directly induces *CPEB3* expression, resulting in the repression of NRF1 translation. In parallel, NRF3 confers ubiquitin (Ub)-independent protein degradation through the POMP-20S proteasome axis. (**B**) In NRF3-deficient cells, NRF1 escapes from CPEB3-mediated translational repression and confers ubiquitin (Ub)-dependent protein degradation.

3.2. Clinical Significance of the NRF3-CPEB3-NRF1 Translational Repression Axis

In NRF3-deficienct cells, CPEB3 represses NRF1 translation and then reduces the expression levels of *PSMB3*, *PSMB7*, *PSMC2*, *PSMG2*, and *POMP* genes, resulting in the suppression of 26S proteasome activity [32]. CPEB3 also confers resistance to BTZ in NRF3-deficient cells [32]. Furthermore, colorectal cancer patients with higher *CPEB3/NRF3*-expressing tumors exhibited shorter overall survival rates, but higher *CPEB3/NRF1* expression was not associated with poor prognosis [32]. These results suggested that the NRF3-CPEB3-NRF1 translational repression axis is involved in cancer development by shunting ubiquitin-dependent protein degradation through the NRF1–26S proteasome regulatory axis to ubiquitin-independent protein degradation through the POMP-20S proteasome axis (Figure 2).

4. Reprogramming of Lipid Metabolism

4.1. NRF3-SREBP2-HMGCR Axis for Mevalonate Biosynthesis

Lipids, such as cholesterol and fatty acids, influence cell signaling, energy storage, and membrane formation. SREBPs are membrane-bound transcription factors crucial for lipid metabolism [46]. In response to cholesterol depletion, SREBP1 and SREBP2 proteins are cleaved in the Golgi apparatus, resulting in the translocation to the nucleus. SREBP1 induces the gene expression of enzymes required for fatty acid biosynthesis and adipocyte differentiation, whereas SREBP2 induces the gene expression of enzymes required for mevalonate/cholesterol biosynthesis.

NRF3 induces the expression of several SREBP2 target genes, such as *hydroxymethylglutaryl-CoA synthase 1* (*HMGCS1*) and *HMGCR*, encoding a rate-limiting enzyme in mevalonate/cholesterol biosynthesis [47] (Figure 3(1)). ChIP experiments with previously published ChIP sequencing data [48] indicated that NRF3 binds to both AREs in *SREBP2* and *HMGCR* promoters, and that SREBP2 binds to the site nearby *HMGCR*-ARE [33]. Moreover, NRF3 interacts with the active form of SREBP2 [33] (Figure 3(1)), implying that NRF3 and SREBP2 form a transcriptional complex for *HMGCR* gene expression. Luciferase reporter assays containing both ARE and SREBP2 binding sites showed a synergistic transcriptional activity of NRF3 and SREBP2 through the *HMGCR* promoter [33]. Taken together, NRF3 promotes mevalonate biosynthesis by upregulating the SREBP2-HMGCR axis (Figure 3(1)).

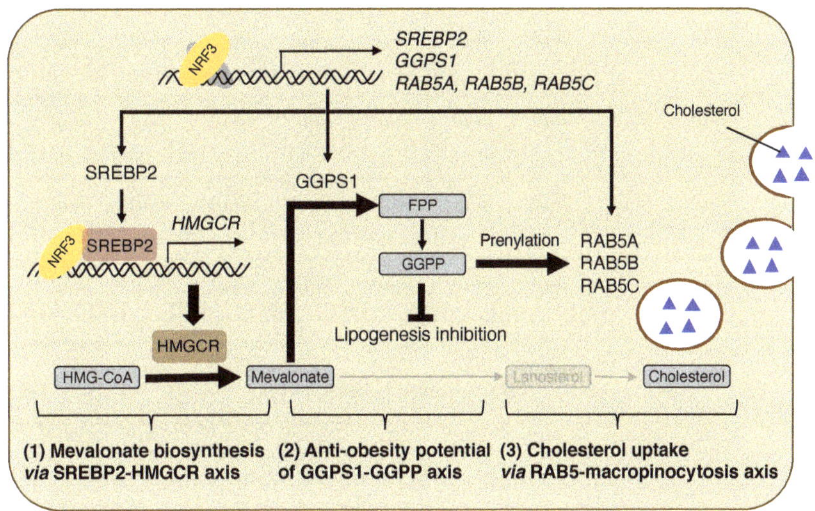

Figure 3. NRF3-regulated lipid metabolism through three axes. (**1**) NRF3 activates *SREBP2* by directly inducing gene expression. NRF3 and SREBP2 synergistically induce *HMGCR* gene expression, promoting mevalonate biosynthesis. (**2**) In parallel, NRF3 induces *GGPS1* expression and then reprograms cholesterol biosynthesis to GGPP production, resulting in lipogenesis inhibition. (**3**) NRF3 also induces the gene expression of three *RAB5* isoforms, resulting in cholesterol uptake through macropinocytosis.

4.2. NRF3-GGPS1-GGPP Production Axis for Lipogenesis Inhibition

Interestingly, NRF3 does not affect the intracellular levels of cholesterol, even if NRF3 increases the expression levels and enzymatic activity of HMGCR. Meanwhile, NRF3 reduces that of lanosterol [33]. Lanosterol is not only a precursor of cholesterol but also a downstream metabolite of farnesyl pyrophosphate, which is also metabolized to GGPP in a reaction catalyzed by GGPS1 (Figure 3(2)). Furthermore, NRF3 directly induces *GGPS1* expression [33], implying that NRF3 reprograms cholesterol biogenesis to the production of GGPP rather than lanosterol. GGPP suppresses SREBP1-dependent fatty acid biosynthesis and intracellular lipid accumulation [49,50]. In fact, DNA microarray analysis showed a negative correlation between the expression levels of *NRF3* and genes related to fatty acid metabolism [33]. More directly, intracellular levels of neutral lipids are increased by NRF3 knockdown and reduced by GGPP treatment [33] (Figure 3(2)). Consistently, a few body mass index-associated genomic loci near the *NRF3* gene have been identified previously [51,52]. These results indicated the potential role of the NRF3-GGPS1-GGPP production axis (Figure 3(2)).

4.3. NRF3-RAB5-Macropincytosis Induction Axis for Cholesterol Uptake

Intracellular cholesterol is derived from not only de novo biosynthesis but also endocytic uptake [53]. NRF3 decreases lanosterol levels, but it does not change cholesterol levels in cells [33] (Figure 3(3)), implying that NRF3 enhances endocytosis for cholesterol uptake to compensate for the potential depletion in cholesterol levels following lanosterol reduction. Low-density lipoprotein receptor (LDLR) is a key endocytosis regulator of LDL [54]. However, NRF3 does not induce *LDLR* gene expression. Meanwhile, NRF3 induces the gene expression of three isoforms of *RAB5A, RAB5B,* and *RAB5C* [33] (Figure 3(3)). These RAB5 proteins act as early endocytosis regulators [55] and are involved in macropinocytosis, a bulk and fluid-phase endocytosis process [56]. NRF3 further increases posttranslational prenylation RAB5 proteins [33] essential for proper localization and function in membranes [57]. The previous section (Figure 3(2)) described that NRF3 induces the production of GGPP, which functions as a required substrate for protein prenylation [58]. Altogether, NRF3 enhances RAB5-mediated endocytosis rather than LDLR-mediated endocytosis for cholesterol uptake through GGPP production (Figure 3, (3)). NRF3 enhances the uptake of fluorogenic LDL in a RAB5-dependent manner. Moreover, NRF3-enhanced uptake of other fluorogenic macropinocytosis indicators based on 70 kDa dextran and bovine serum albumin is abolished by treatment with 5-(*N*-ethyl-*N*-isopropyl)amiloride [33], also known as an inhibitor of macropinocytosis and a selective blocker of Na^+/H^+ exchanger [59]. Similarly, NRF3 enhances the uptake of two fluorogenic cholesterols through macropinocytosis [33]. These results indicated the crucial function of the NRF3-RAB5-macropinocysosis induction (NRF3-RAB5-macropinocysosis) axis on cholesterol uptake (Figure 3(3)). The next section discusses the pathophysiological potential of this axis.

5. Concluding Remarks

Increased ubiquitin-independent proteasomal activity causes tumor growth, metastasis, and resistance to the proteasome inhibitor BTZ. NRF3 induces *POMP* expression, leading to ubiquitin-independent protein degradation of tumor suppressors Rb and p53 (Figure 1). Upregulation of the POMP-20S proteasome axis further results in poor prognosis of colorectal cancer patients [31]. NRF3 also induces the expression of proteasome-related genes in parallel with NRF1 translational repression by inducing *CPEB3* expression [32] (Figure 2A). If the *NRF3* gene is deficient, NRF1 escapes from CPEB3-mediated translational repression, and complementarily plays a transcriptional role for the robust maintenance of basal proteasome activity in cancer cells (Figure 2B). Although NRF3 shares several target genes with NRF1 and NRF2 on ARE [30,60], this review showed the translation-mediated crosstalk between NRF3 and NRF1.

Nrf1 is ubiquitously expressed in normal tissues, and *Nrf1* knockout mice suffer from embryonic lethality [8]. Meanwhile, *Nrf3* expression levels are low, except in several mouse tissues, such as the placenta [5,6], and *Nrf3* knockout mice do not exhibit any obvious abnormalities under normal physiological conditions [18–20]. However, *NRF3* is highly expressed in many cancer cells [31], implying that the proteasome in cancer or normal cells is maintained through the CPEB3-NRF1 axis or the negative feedback regulation of NRF1. Higher *CPEB3/NRF3* expression, but not higher *CPEB3/NRF1* expression, is associated with poor prognosis of cancer patients [32]. Therefore, NRF1 maintains a 26S proteasome activity for normal development, whereas NRF3 alternatively maintains 20S proteasome activity for cancer development through both POMP-20S proteasome and CPEB3-NRF1 axes.

Furthermore, NRF3 is involved in lipid metabolism through three regulatory axes [33] (Figure 3): (1) NRF3 induces the gene expression of *SREBP2* required for cholesterol biosynthesis through the mevalonate pathway. NRF3 also leads to SREBP2 activation through direct induction of gene expression. NRF3 and SREBP2 synergistically induce *HMGCR* expression and the following mevalonate biosynthesis. (2) NRF3 then upregulates GGPS1-mediated GGPP production for lipogenesis inhibition. (3) In parallel, NRF3 confers RAB5-mediated induction of macropinocytosis for cholesterol uptake. This gene expression is induced in colon and/or rectal tissue of newly generated NRF3-transgenic mice [33]. Dietary cholesterol in the blood is absorbed in the intestine [61], and its dysregulation is associated with obesity, resulting in an increased risk of cardiovascular diseases (CVD) and colorectal cancer [62,63]. The gut microbiota has been identified as a CVD risk factor and regulates host cholesterol homeostasis [64–66], suggesting the pathophysiological potential of the NRF3-regulated host lipid metabolism to the gut–heart connection through the gut microbiota.

Epidemiological studies have associated obesity with a range of cancers [67,68]. Furthermore, many efforts have been made to identify the key factor for obesity-induced cancer, including insulin resistance, increased steroid hormones and adipokine, and aberrant inflammation [69]. However, these findings implied possible opposite roles for NRF3 in obesity-induced cancer development (Figure 4): NRF3-SREBP2-HMGCR and the following NRF3-GGPS1-GGPS axes regulate lipid homeostasis and confer resistance to obesity through lipogenesis inhibition, while the NRF3-POMP-20S proteasome and NRF3-CPEB3-NRF1 axes regulate protein homeostasis and confer ubiquitin-independent protein degradation for continuous cancer cell growth. This review further showed that NRF3 maintains cholesterol homeostasis through the RAB5-macropinocytosis axis (Figure 3, (3)). Interestingly, macropinocytosis is associated with obesity-related disorders, such as increased diabetic mouse macrophages and chronic inflammation [70,71]. Furthermore, NRF2 induces macropinocytosis and contributes to the escape of autophagy-deficient cancer cells from metabolic decline and anticancer drugs, such as gemcitabine and doxorubicin, which target the anabolic dependencies of cancer cells [72,73]. These insights implied the possibility that the NRF3-RAB5-macropinocytosis axis paradoxically interferes with obesity-induced cancer development through attenuation of obesity-induced inflammation and resistance to therapy targeting cancer anabolism (Figure 4).

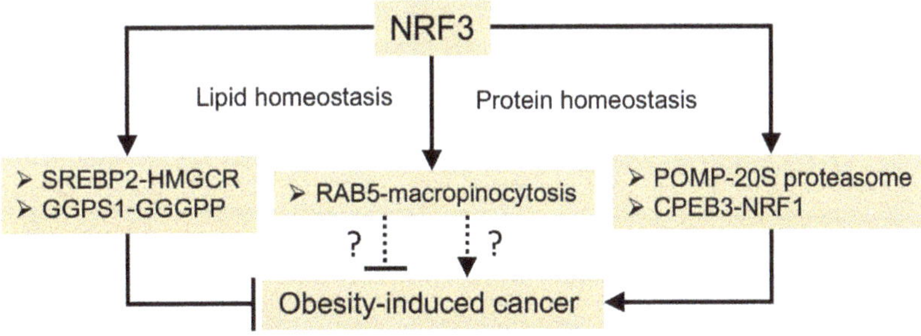

Figure 4. Possible roles for NRF3 in obesity-induced cancer development.

A big issue of the NRF3 study is identifying the endogenous cue of NRF3 activation, although NRF3 is experimentally activated by treatment with a proteasome inhibitor. Recently, NRF1 senses cholesterol levels in the ER membrane through the cholesterol recognition amino acid consensus motif domain (CRAC), and it is activated in response to cholesterol depletion [74]. Because the CRAC domain is conserved in NRF3 proteins [27], NRF3 acts as a cholesterol sensor in the ER membrane similarly to NRF1.

Author Contributions: T.W. and A.K. were involved in the literature survey and wrote the paper. All authors have read and agreed to the published version of the manuscript.

Funding: This work was supported by the JSPS (grant nos. 17K18234 (T.W.), 19K07650 (T.W.), 16H03265 (A.K.), 19K22826 (A.K.), 20H04135 (A.K.), and 21K19743 (A.K.)), the Uehara Memorial Foundation (T.W.), the Harris Research Institute of Doshisha University (T.W.), and the Mitsubishi Foundation (A.K.).

Institutional Review Board Statement: Not applicable.

Informed Consent Statement: Not applicable.

Data Availability Statement: Not applicable.

Acknowledgments: The authors thank all members of our laboratory and the collaborators for their contributions to the research.

Conflicts of Interest: The authors declare no conflict of interest.

References

1. Guang, M.H.Z.; Kavanagh, E.; Dunne, L.; Dowling, P.; Zhang, L.; Lindsay, S.; Bazou, D.; Goh, C.; Hanley, C.; Bianchi, G.; et al. Targeting Proteotoxic Stress in Cancer: A Review of the Role that Protein Quality Control Pathways Play in Oncogenesis. *Cancers* **2019**, *11*, 66. [CrossRef]
2. Su, K.-H.; Dai, C. Metabolic control of the proteotoxic stress response: Implications in diabetes mellitus and neurodegenerative disorders. *Cell. Mol. Life Sci.* **2016**, *73*, 4231–4248. [CrossRef]
3. Munir, R.; Lisec, J.; Swinnen, J.V.; Zaidi, N. Lipid metabolism in cancer cells under metabolic stress. *Br. J. Cancer* **2019**, *120*, 1090–1098. [CrossRef]
4. Yadav, R.S.; Tiwari, N.K. Lipid Integration in Neurodegeneration: An Overview of Alzheimer's Disease. *Mol. Neurobiol.* **2014**, *50*, 168–176. [CrossRef]
5. Kobayashi, A.; Ito, E.; Toki, T.; Kogame, K.; Takahashi, S.; Igarashi, K.; Hayashi, N.; Yamamoto, M. Molecular Cloning and Functional Characterization of a New Cap'n' Collar Family Transcription Factor Nrf3. *J. Biol. Chem.* **1999**, *274*, 6443–6452. [CrossRef]
6. Chénais, B.; Derjuga, A.; Massrieh, W.; Red-Horse, K.; Bellingard, V.; Fisher, S.J.; Blank, V. Functional and Placental Expression Analysis of the Human NRF3 Transcription Factor. *Mol. Endocrinol.* **2005**, *19*, 125–137. [CrossRef]
7. Radhakrishnan, S.K.; Lee, C.S.; Young, P.; Beskow, A.; Chan, J.Y.; Deshaies, R.J. Transcription Factor Nrf1 Mediates the Proteasome Recovery Pathway after Proteasome Inhibition in Mammalian Cells. *Mol. Cell* **2010**, *38*, 17–28. [CrossRef]
8. Chan, J.Y.; Kwong, M.; Lu, R.; Chang, J.; Wang, B.; Yen, T.; Kan, Y.W. Targeted disruption of the ubiquitous CNC-bZIP transcription factor, Nrf-1, results in anemia and embryonic lethality in mice. *EMBO J.* **1998**, *17*, 1779–1787. [CrossRef]
9. Kobayashi, A.; Tsukide, T.; Miyasaka, T.; Morita, T.; Mizoroki, T.; Saito, Y.; Ihara, Y.; Takashima, A.; Noguchi, N.; Fukamizu, A.; et al. Central nervous system-specific deletion of transcription factor Nrf1 causes progressive motor neuronal dysfunction. *Genes Cells* **2011**, *16*, 692–703. [CrossRef]
10. Kim, J.; Xing, W.; Wergedal, J.; Chan, J.Y.; Mohan, S. Targeted disruption of nuclear factor erythroid-derived 2-like 1 in osteoblasts reduces bone size and bone formation in mice. *Physiol. Genom.* **2010**, *40*, 100–110. [CrossRef]
11. Lee, C.S.; Hu, T.; Nguyen, J.M.; Zhang, J.; Martin, M.V.; Vawter, M.P.; Huang, E.J.; Chan, J.Y. Loss of nuclear factor E2-related factor 1 in the brain leads to dysregulation of proteasome gene expression and neurodegeneration. *Proc. Natl. Acad. Sci. USA* **2011**, *108*, 8408–8413. [CrossRef]
12. Xu, Z.; Chen, L.; Leung, L.; Yen, T.S.B.; Lee, C.; Chan, J.Y. Liver-specific inactivation of the Nrf1 gene in adult mouse leads to nonalcoholic steatohepatitis and hepatic neoplasia. *Proc. Natl. Acad. Sci. USA* **2005**, *102*, 4120–4125. [CrossRef]
13. Yamamoto, T.; Suzuki, T.; Kobayashi, A.; Wakabayashi, J.; Maher, J.; Motohashi, H.; Yamamoto, M. Physiological Significance of Reactive Cysteine Residues of Keap1 in Determining Nrf2 Activity. *Mol. Cell. Biol.* **2008**, *28*, 2758–2770. [CrossRef]
14. Romero, R.; Sayin, V.I.; Davidson, S.M.; Bauer, M.R.; Singh, S.X.; Leboeuf, S.E.; Karakousi, T.R.; Ellis, D.C.; Bhutkar, A.; Sánchez-Rivera, F.J.; et al. Keap1 loss promotes Kras-driven lung cancer and results in dependence on glutaminolysis. *Nat. Med.* **2017**, *23*, 1362–1368. [CrossRef]
15. Mitsuishi, Y.; Taguchi, K.; Kawatani, Y.; Shibata, T.; Nukiwa, T.; Aburatani, H.; Yamamoto, M.; Motohashi, H. Nrf2 Redirects Glucose and Glutamine into Anabolic Pathways in Metabolic Reprogramming. *Cancer Cell* **2012**, *22*, 66–79. [CrossRef]
16. Itoh, K.; Chiba, T.; Takahashi, S.; Ishii, T.; Igarashi, K.; Katoh, Y.; Oyake, T.; Hayashi, N.; Satoh, K.; Hatayama, I.; et al. An Nrf2/Small Maf Heterodimer Mediates the Induction of Phase II Detoxifying Enzyme Genes through Antioxidant Response Elements. *Biochem. Biophys. Res. Commun.* **1997**, *236*, 313–322. [CrossRef]
17. Wakabayashi, N.; Itoh, K.; Wakabayashi, J.; Motohashi, H.; Noda, S.; Takahashi, S.; Imakado, S.; Kotsuji, T.; Otsuka, F.; Roop, D.R.; et al. Keap1-null mutation leads to postnatal lethality due to constitutive Nrf2 activation. *Nat. Genet.* **2003**, *35*, 238–245. [CrossRef]
18. Derjuga, A.; Gourley, T.S.; Holm, T.M.; Heng, H.H.Q.; Shivdasani, R.A.; Ahmed, R.; Andrews, N.; Blank, V. Complexity of CNC Transcription Factors As Revealed by Gene Targeting of the Nrf3 Locus. *Mol. Cell. Biol.* **2004**, *24*, 3286–3294. [CrossRef]
19. Kobayashi, A.; Ohta, T.; Yamamoto, M. Unique Function of the Nrf2–Keap1 Pathway in the Inducible Expression of Antioxidant and Detoxifying Enzymes. *Methods Enzymol.* **2004**, *378*, 273–286. [CrossRef]
20. Chevillard, G.; Blank, V. NFE2L3 (NRF3): The Cinderella of the Cap'n'Collar transcription factors. *Cell. Mol. Life Sci.* **2011**, *68*, 3337–3348. [CrossRef]
21. Bury, M.; Le Calvé, B.; Lessard, F.; Maso, T.D.; Saliba, J.; Michiels, C.; Ferbeyre, G.; Blank, V. NFE2L3 Controls Colon Cancer Cell Growth through Regulation of DUX4, a CDK1 Inhibitor. *Cell Rep.* **2019**, *29*, 1469–1481.e9. [CrossRef]
22. Wang, H.; Zhan, M.; Yang, R.; Shi, Y.; Liu, Q.; Wang, J. Elevated expression of NFE2L3 predicts the poor prognosis of pancreatic cancer patients. *Cell Cycle* **2018**, *17*, 2164–2174. [CrossRef]

23. Siegenthaler, B.; Defila, C.; Muzumdar, S.; Beer, H.-D.; Meyer, M.; Tanner, S.; Bloch, W.; Blank, V.; Schäfer, M.; Werner, S. Nrf3 promotes UV-induced keratinocyte apoptosis through suppression of cell adhesion. *Cell Death Differ.* **2018**, *25*, 1749–1765. [CrossRef]

24. Aono, S.; Hatanaka, A.; Hatanaka, A.; Gao, Y.; Hippo, Y.; Taketo, M.M.; Waku, T.; Kobayashi, A. β-Catenin/TCF4 Complex-Mediated Induction of the NRF3 (NFE2L3) Gene in Cancer Cells. *Int. J. Mol. Sci.* **2019**, *20*, 3344. [CrossRef]

25. Chowdhury, A.M.M.A.; Katoh, H.; Hatanaka, A.; Iwanari, H.; Nakamura, N.; Hamakubo, T.; Natsume, T.; Waku, T.; Kobayashi, A. Multiple regulatory mechanisms of the biological function of NRF3 (NFE2L3) control cancer cell proliferation. *Sci. Rep.* **2017**, *7*, 1–14. [CrossRef]

26. Koizumi, S.; Irie, T.; Hirayama, S.; Sakurai, Y.; Yashiroda, H.; Naguro, I.; Ichijo, H.; Hamazaki, J.; Murata, S. The aspartyl protease DDI2 activates Nrf1 to compensate for proteasome dysfunction. *eLife* **2016**, *5*, 1–10. [CrossRef]

27. Zhang, Y.; Kobayashi, A.; Yamamoto, M.; Hayes, J. The Nrf3 Transcription Factor Is a Membrane-bound Glycoprotein Targeted to the Endoplasmic Reticulum through Its N-terminal Homology Box 1 Sequence. *J. Biol. Chem.* **2009**, *284*, 3195–3210. [CrossRef]

28. Yamamoto, M.; Kensler, T.W.; Motohashi, H. The KEAP1-NRF2 System: A Thiol-Based Sensor-Effector Apparatus for Maintaining Redox Homeostasis. *Physiol. Rev.* **2018**, *98*, 1169–1203. [CrossRef]

29. Motohashi, H.; O'Connor, T.; Katsuoka, F.; Engel, J.D.; Yamamoto, M. Integration and diversity of the regulatory network composed of Maf and CNC families of transcription factors. *Gene* **2002**, *294*, 1–12. [CrossRef]

30. Liu, P.; Kerins, M.J.; Tian, W.; Neupane, D.; Zhang, D.D.; Ooi, A. Differential and overlapping targets of the transcriptional regulators NRF1, NRF2, and NRF3 in human cells. *J. Biol. Chem.* **2019**, *294*, 18131–18149. [CrossRef]

31. Waku, T.; Nakamura, N.; Koji, M.; Watanabe, H.; Katoh, H.; Tatsumi, C.; Tamura, N.; Hatanaka, A.; Hirose, S.; Katayama, H.; et al. NRF3-POMP-20S Proteasome Assembly Axis Promotes Cancer Development via Ubiquitin-Independent Proteolysis of p53 and Retinoblastoma Protein. *Mol. Cell. Biol.* **2020**, *40*. [CrossRef]

32. Waku, T.; Katayama, H.; Hiraoka, M.; Hatanaka, A.; Nakamura, N.; Tanaka, Y.; Tamura, N.; Watanabe, A.; Kobayashi, A. NFE2L1 and NFE2L3 Complementarily Maintain Basal Proteasome Activity in Cancer Cells through CPEB3-Mediated Translational Repression. *Mol. Cell. Biol.* **2020**, *40*. [CrossRef]

33. Waku, T.; Hagiwara, T.; Tamura, N.; Atsumi, Y.; Urano, Y.; Suzuki, M.; Iwami, T.; Sato, K.; Yamamoto, M.; Noguchi, N.; et al. NRF3 upregulates gene expression in SREBP2-dependent mevalonate pathway with cholesterol uptake and lipogenesis inhibition. *iScience* **2021**, *24*, 103180. [CrossRef]

34. Coux, O.; Tanaka, K.; Goldberg, A.L. Structure and functions of the 20s and 26s proteasomes. *Annu. Rev. Biochem.* **1996**, *65*, 801–847. [CrossRef]

35. Murata, S.; Yashiroda, H.; Tanaka, K. Molecular mechanisms of proteasome assembly. *Nat. Rev. Mol. Cell Biol.* **2009**, *10*, 104–115. [CrossRef]

36. Gallastegui, N.; Groll, M. The 26S proteasome: Assembly and function of a destructive machine. *Trends Biochem. Sci.* **2010**, *35*, 634–642. [CrossRef]

37. Witt, E.; Zantopf, D.; Schmidt, M.; Kraft, R.; Kloetzel, P.-M.; Krüger, E. Characterisation of the newly identified human Ump1 homologue POMP and analysis of LMP7(β5i) incorporation into 20 S proteasomes. *J. Mol. Biol.* **2000**, *301*, 1–9. [CrossRef]

38. Ben-Nissan, G.; Sharon, M. Regulating the 20S Proteasome Ubiquitin-Independent Degradation Pathway. *Biomolecules* **2014**, *4*, 862–884. [CrossRef]

39. Hyer, M.L.; A Milhollen, M.; Ciavarri, J.; Fleming, P.; Traore, T.; Sappal, D.; Huck, J.; Shi, J.; Gavin, J.; Brownell, J.; et al. A small-molecule inhibitor of the ubiquitin activating enzyme for cancer treatment. *Nat. Med.* **2018**, *24*, 186–193. [CrossRef]

40. Sherr, C.J.; McCormick, F. The RB and p53 pathways in cancer. *Cancer Cell* **2002**, *2*, 103–112. [CrossRef]

41. Waldman, T.; Kinzler, K.W.; Vogelstein, B. p21 is necessary for the p53-mediated G1 arrest in human cancer cells. *Cancer Res.* **1995**, *55*.

42. Nakano, K.; Vousden, K.H. PUMA, a Novel Proapoptotic Gene, Is Induced by p53. *Mol. Cell* **2001**, *7*, 683–694. [CrossRef]

43. Groll, M.; Berkers, C.R.; Ploegh, H.L.; Ovaa, H. Crystal Structure of the Boronic Acid-Based Proteasome Inhibitor Bortezomib in Complex with the Yeast 20S Proteasome. *Structure* **2006**, *14*, 451–456. [CrossRef]

44. Kobayashi, A. Roles of NRF3 in the Hallmarks of Cancer: Proteasomal Inactivation of Tumor Suppressors. *Cancers* **2020**, *12*, 2681. [CrossRef]

45. Fernández-Miranda, G.; Méndez, R. The CPEB-family of proteins, translational control in senescence and cancer. *Ageing Res. Rev.* **2012**, *11*, 460–472. [CrossRef]

46. Horton, J.D.; Goldstein, J.L.; Brown, M.S. SREBPs: Activators of the complete program of cholesterol and fatty acid synthesis in the liver. *J. Clin. Investig.* **2002**, *109*, 1125–1131. [CrossRef]

47. Xue, L.; Qi, H.; Zhang, H.; Ding, L.; Huang, Q.; Zhao, D.; Wu, B.J.; Li, X. Targeting SREBP-2-Regulated Mevalonate Metabolism for Cancer Therapy. *Front. Oncol.* **2020**, *10*, 1510. [CrossRef]

48. Landt, S.G.; Marinov, G.; Kundaje, A.; Kheradpour, P.; Pauli, F.; Batzoglou, S.; Bernstein, B.E.; Bickel, P.; Brown, J.B.; Cayting, P.; et al. ChIP-seq guidelines and practices of the ENCODE and modENCODE consortia. *Genome Res.* **2012**, *22*, 1813–1831. [CrossRef]

49. Bertolio, R.; Napoletano, F.; Mano, M.; Maurer-Stroh, S.; Fantuz, M.; Zannini, A.; Bicciato, S.; Sorrentino, G.; Del Sal, G. Sterol regulatory element binding protein 1 couples mechanical cues and lipid metabolism. *Nat. Commun.* **2019**, *10*, 1–11. [CrossRef]

50. Yeh, Y.-S.; Goto, T.; Takahashi, N.; Egawa, K.; Takahashi, H.; Jheng, H.-F.; Kim, Y.-I.; Kawada, T. Geranylgeranyl pyrophosphate performs as an endogenous regulator of adipocyte function via suppressing the LXR pathway. *Biochem. Biophys. Res. Commun.* **2016**, *478*, 1317–1322. [CrossRef]

51. Lamiquiz-Moneo, I.; Mateo-Gallego, R.; Bea, A.M.; Dehesa-García, B.; Pérez-Calahorra, S.; Marco-Benedí, V.; Baila-Rueda, L.; Laclaustra, M.; Civeira, F.; Cenarro, A. Genetic predictors of weight loss in overweight and obese subjects. *Sci. Rep.* **2019**, *9*, 1 9. [CrossRef]

52. Monda, K.L.; Chen, G.K.; Taylor, K.C.; Palmer, C.; Edwards, T.L.; A Lange, L.; Ng, M.C.Y.; A Adeyemo, A.; A Allison, M.; et al.; NABEC Consortium. A meta-analysis identifies new loci associated with body mass index in individuals of African ancestry. *Nat. Genet.* **2013**, *45*, 690–696. [CrossRef]

53. Steck, T.L.; Lange, Y. Cell cholesterol homeostasis: Mediation by active cholesterol. *Trends Cell Biol.* **2010**, *20*, 680–687. [CrossRef]

54. Goldstein, J.L.; Brown, M.S. The LDL Receptor. *Arter. Thromb. Vasc. Biol.* **2009**, *29*, 431–438. [CrossRef]

55. Zeigerer, A.; Gilleron, J.; Bogorad, R.L.; Marsico, G.; Nonaka, H.; Seifert, S.; Epstein-Barash, H.; Kuchimanchi, S.; Peng, C.G.; Ruda, V.M.; et al. Rab5 is necessary for the biogenesis of the endolysosomal system in vivo. *Nat. Cell Biol.* **2012**, *485*, 465–470. [CrossRef]

56. Kruth, H.S.; Jones, N.L.; Huang, W.; Zhao, B.; Ishii, I.; Chang, J.; Combs, C.A.; Malide, D.; Zhang, W.-Y. Macropinocytosis Is the Endocytic Pathway That Mediates Macrophage Foam Cell Formation with Native Low Density Lipoprotein. *J. Biol. Chem.* **2005**, *280*, 2352–2360. [CrossRef]

57. Barbieri, M.A.; Hoffenberg, S.; Roberts, R.; Mukhopadhyay, A.; Pomrehn, A.; Dickey, B.F.; Stahl, P.D. Evidence for a Symmetrical Requirement for Rab5-GTP in in Vitro Endosome-Endosome Fusion. *J. Biol. Chem.* **1998**, *273*, 25850–25855. [CrossRef]

58. Wang, M.; Casey, P. Protein prenylation: Unique fats make their mark on biology. *Nat. Rev. Mol. Cell Biol.* **2016**, *17*, 110–122. [CrossRef]

59. Commisso, C.; Flinn, R.J.; Bar-Sagi, D. Determining the macropinocytic index of cells through a quantitative image-based assay. *Nat. Protoc.* **2014**, *9*, 182–192. [CrossRef]

60. Sankaranarayanan, K.; Jaiswal, A.K. Nrf3 Negatively Regulates Antioxidant-response Element-mediated Expression and Antioxidant Induction of NAD(P)H:Quinone Oxidoreductase1 Gene. *J. Biol. Chem.* **2004**, *279*, 50810–50817. [CrossRef]

61. Lecerf, J.-M.; de Lorgeril, M. Dietary cholesterol: From physiology to cardiovascular risk. *Br. J. Nutr.* **2011**, *106*, 6–14. [CrossRef]

62. Nelson, R.H. Hyperlipidemia as a Risk Factor for Cardiovascular Disease. *Prim. Care Clin. Off. Pr.* **2013**, *40*, 195–211. [CrossRef]

63. Yao, X.; Tian, Z. Dyslipidemia and colorectal cancer risk: A meta-analysis of prospective studies. *Cancer Causes Control.* **2015**, *26*, 257–268. [CrossRef]

64. Vourakis, M.; Mayer, G.; Rousseau, G. The Role of Gut Microbiota on Cholesterol Metabolism in Atherosclerosis. *Int. J. Mol. Sci.* **2021**, *22*, 8074. [CrossRef]

65. Kenny, D.J.; Plichta, D.R.; Shungin, D.; Koppel, N.; Hall, A.B.; Fu, B.; Vasan, R.S.; Shaw, S.Y.; Vlamakis, H.; Balskus, E.P.; et al. Cholesterol Metabolism by Uncultured Human Gut Bacteria Influences Host Cholesterol Level. *Cell Host Microbe* **2020**, *28*, 245–257.e6. [CrossRef]

66. Le Roy, T.; Lécuyer, E.; Chassaing, B.; Rhimi, M.; Lhomme, M.; Boudebbouze, S.; Ichou, F.; Haro Barceló, J.; Huby, T.; Guerin, M.; et al. The Intestinal Microbiota Regulates Host Cholesterol Homeostasis. *BMC Biol.* **2019**, *17*, 94. [CrossRef]

67. Bianchini, F.; Kaaks, R.; Vainio, H. Overweight, obesity, and cancer risk. *Lancet Oncol.* **2002**, *3*, 565–574. [CrossRef]

68. Calle, E.E.; Rodriguez, C.; Walker-Thurmond, K.; Thun, M.J. Overweight, Obesity, and Mortality from Cancer in a Prospectively Studied Cohort of U.S. Adults. *N. Engl. J. Med.* **2003**, *348*, 1625–1638. [CrossRef]

69. Khandekar, M.J.; Cohen, P.; Spiegelman, B.M. Molecular mechanisms of cancer development in obesity. *Nat. Rev. Cancer* **2011**, *11*, 886–895. [CrossRef]

70. Guest, C.B.; Chakour, K.S.; Freund, G.G. Macropinocytosis is decreased in diabetic mouse macrophages and is regulated by AMPK. *BMC Immunol.* **2008**, *9*, 42–48. [CrossRef]

71. Wang, Y.; Tang, B.; Long, L.; Luo, P.; Xiang, W.; Li, X.; Wang, H.; Jiang, Q.; Tan, X.; Luo, S.; et al. Improvement of obesity-associated disorders by a small-molecule drug targeting mitochondria of adipose tissue macrophages. *Nat. Commun.* **2021**, *12*, 1–16. [CrossRef]

72. Jayashankar, V.; Edinger, A.L. Macropinocytosis confers resistance to therapies targeting cancer anabolism. *Nat. Commun.* **2020**, *11*, 1–15. [CrossRef] [PubMed]

73. Su, H.; Yang, F.; Fu, R.; Li, X.; French, R.; Mose, E.; Pu, X.; Trinh, B.; Kumar, A.; Liu, J.; et al. Cancer cells escape autophagy inhibition via NRF2-induced macropinocytosis. *Cancer Cell* **2021**, *39*, 678–693.e11. [CrossRef]

74. Widenmaier, S.; Snyder, N.A.; Nguyen, T.B.; Arduini, A.; Lee, G.Y.; Arruda, A.P.; Saksi, J.; Bartelt, A.; Hotamisligil, G.S. NRF1 Is an ER Membrane Sensor that Is Central to Cholesterol Homeostasis. *Cell* **2017**, *171*, 1094–1109.e15. [CrossRef]

International Journal of
Molecular Sciences

Review

Molecular Pathways Modulating Sensory Hair Cell Regeneration in Adult Mammalian Cochleae: Progress and Perspectives

Vikrant Rai [†], Shu Tu [‡,§], Joseph R. Frank [‡,‖] and Jian Zuo *

Department of Biomedical Sciences, Creighton University School of Medicine, Omaha, NE 68178, USA;
vrai@westernu.edu (V.R.); sxt735@case.edu (S.T.); joseph.frank@duke.edu (J.R.F.)
* Correspondence: jianzuo@creighton.edu; Tel.: +1-(402)-280-2916
† Present Address: Department of Translational Research, Western University of Health Sciences,
  Pomona, CA 91766, USA.
‡ These authors contributed equally to this work.
§ Present Address: School of Dental Medicine, Case Western Reserve University, Cleveland, OH 44106-7342, USA.
‖ Present Address: Developmental and Stem Cell Biology Program, Duke University, Durham, NC 27708, USA.

Abstract: Noise-induced, drug-related, and age-related disabling hearing loss is a major public health problem and affect approximately 466 million people worldwide. In non-mammalian vertebrates, the death of sensory hair cells (HCs) induces the proliferation and transdifferentiation of adjacent supporting cells into new HCs; however, this capacity is lost in juvenile and adult mammalian cochleae leading to permanent hearing loss. At present, cochlear implants and hearing devices are the only available treatments and can help patients to a certain extent; however, no biological approach or FDA-approved drug is effective to treat disabling hearing loss and restore hearing. Recently, regeneration of mammalian cochlear HCs by modulating molecular pathways or transcription factors has offered some promising results, although the immaturity of the regenerated HCs remains the biggest concern. Furthermore, most of the research done is in neonates and not in adults. This review focuses on critically summarizing the studies done in adult mammalian cochleae and discusses various strategies to elucidate novel transcription factors for better therapeutics.

Keywords: hair cells; adult cochlea; regeneration; transcription factor; bioinformatics

Citation: Rai, V.; Tu, S.; Frank, J.R.; Zuo, J. Molecular Pathways Modulating Sensory Hair Cell Regeneration in Adult Mammalian Cochleae: Progress and Perspectives. *Int. J. Mol. Sci.* **2022**, 23, 66. https://doi.org/10.3390/ijms23010066

Academic Editors:
Amelia Casamassimi,
Alfredo Ciccodicola and
Monica Rienzo

Received: 30 November 2021
Accepted: 20 December 2021
Published: 22 December 2021

Publisher's Note: MDPI stays neutral with regard to jurisdictional claims in published maps and institutional affiliations.

1. Introduction

Noise-induced, drug-related, and age-related disabling hearing loss is a major public health problem. Per a World Health Organization report, they affect nearly 5% of the world population [1]. In non-mammalian vertebrates, sensory hair cell (HC) death induces the proliferation and trans-differentiation of adjacent supporting cells (SCs) into new HCs; however, this capacity is lost in juvenile and adult mammalian cochleae, leading to permanent hearing loss [2,3]. Currently, no biological approach or FDA-approved drug is available to treat disabling hearing loss or to regenerate the sensory HCs in mammalian cochleae. Thus, it is crucial to develop strategies or drugs to either prevent HC loss or promote regeneration in adult mammalian cochleae in vivo. HC regeneration can enhance the number of HCs in the cochlea via two processes: (1) mitotic regeneration, where a SC divides and then the daughter cells (one or both) transdifferentiate into HCs, or (2) direct transdifferentiation where HCs are regenerated via direct phenotypic conversion of SCs without undergoing mitosis [4]. Most of the studies on HC regeneration [3,5–12] have been done either in neonatal cochlear explants (ex vivo) or neonatal mice (in vivo) and only a few studies [5,8–11] have reported regeneration in juvenile and adult mice. Hearing matures around three weeks postnatally in mice. In comparison, during human fetal development hearing becomes mature by the late 2nd trimester and the fetus can hear during the 3rd trimester [13]. It is therefore critical to study HC regeneration in juvenile and adult mice to better understand which approaches are likely to restore hearing in humans. Further, the regenerated HCs in juvenile and adult mice are functionally immature and very few. Improved strategies are needed to increase the number of regenerated HCs and also to promote their maturation.

Recent studies suggest key roles for transcription factors (TFs) Atoh1 and Pou4f3 in HC regeneration and that Atoh1, in particular, is a master regulator of HC differentiation [14] and regeneration [3,8,10]. Further, enhanced regeneration of HCs via modulating the expression of Gata3, Pou4f3, p27^{Kip1} [8], Islet 1 (Isl1) [10], and Gfi1/Pou4f3 [15] with Atoh1 suggests that modulating multiple TFs in combination with Atoh1 is a good strategy to promote regeneration and increase the number of regenerated HCs. Targeting the Notch and Wnt signaling pathways, which are involved in HC development, can also lead to the regeneration of HCs from SCs [6,7,9,16,17]. These studies suggest that SCs have a limited regenerative capacity and that regeneration via transdifferentiation of SCs to HCs is possible, but a small yield of regenerated HCs and functional immaturity remain a major concern. Thus, there is a need to develop strategies to regenerate an increased number of HCs that are also functionally mature. This review focuses on recent literature in sensory HC regeneration in adult mammalian cochleae and briefly discusses molecular pathways, the role of TFs in regeneration, and the challenges and future perspectives of HC regeneration.

2. Targeting Signaling Pathways for Hair Cell Regeneration

The Notch, Wingless-related integration site (Wnt), fibroblast growth factor (FGF), and sonic hedgehog (Shh) pathways are involved in the development and differentiation of HCs and are conserved between various species including zebrafish, birds, and mammals. The crucial role of these pathways in HC development and their conservation among the species are well characterized [18]. Briefly, Notch signaling regulates various cellular processes such as proliferation, differentiation, and cell death in a context-dependent manner. During HC development, Notch signaling is necessary and sufficient for regulating prosensory specification via lateral inhibition, and this process is mediated by its ligands which include jagged 1 (Jag1), Notch intracellular domain (NICD; which interacts with DNA-binding protein and core effector of the canonical Notch pathway, RBPjk), jagged 2 (Jag2), and delta-like 1 (Dll1). Wnt signaling (canonical and noncanonical) is involved in the maintenance of the progenitor cells, cell proliferation, cell fate determination/cell differentiation, and cellular polarization. The FGF signaling pathway plays a crucial role in the induction of the otic placode, development of the otic vesicle, regulation of inner ear morphogenesis, later stages of inner ear development, and HC formation. FGF signaling also helps in regulating the specification of prosensory cells and their differentiation into HCs and SCs during cochlear development [18]. The Shh signaling pathway is involved with regulating prosensory domain formation and auditory function [19], HC formation and differentiation [20], and the spatiotemporal pattern of HC differentiation via regulating the expression of Hey1 and Hey2 [21]. Involvement of the Notch, Wnt, bone morphogenetic protein (BMP), Shh, and fibroblast growth factor (FGF) pathways in the development, differentiation, maturation, and proliferation of HCs in zebrafish, birds, and mice provides strong evidence for significant conservation across species (Table 1). Wnt and Notch signaling play a crucial role in HC regeneration (Figure 1); however, the role of Shh and FGF signaling in regeneration remain unclear. The downstream signaling of these pathways regulates the expression of Atoh1, the master regulator of HC differentiation and regeneration [22]. Since ectopic overexpression of Atoh1 potentiates the regeneration of HCs, it is imperative to hypothesize that targeting these pathways and the genes and transcription factors regulating Atoh1 expression will be effective for the regeneration of HCs (Figure 1). Additionally, if regeneration of HCs from SCs follows development [23], targeting these pathways will be favorable for HC regeneration.

Table 1. Comparative summary of signaling pathways, genes, and transcription factors involved in the development, differentiation, proliferation, and regeneration of HCs among species. ATOH1 (atonal BHLH Transcription Factor 1); Shh (sonic hedgehog); HC (hair cell); SC (supporting cell); OHC (outer hair cells); BP (basilar papilla); FGF (fibroblast growth factor); Fgfr (fibroblast growth factor receptor).

Target	Zebrafish	Birds	Mice	Inference
Notch Atoh1	atoh1a is expressed in all differentiating hair cells [24] Addition of exogenous atoh1a mRNA results in HC overproduction [25]	Atoh1 is immediately upregulated in Sox2+ SCs of the avian BP following HC loss [26]; Atoh1 protein expression is not detectable in mature HCs or SCs in the absence of damage [26]	Math1-null mice fail to develop cochlear and vestibular HCs [27] Atoh1 overexpression can convert SCs into HC-like cells in neonatal mouse cochleae [12]	Atoh1 modulation promotes regeneration in juvenile and adult mice, hence being a potential therapeutic target
Notch Hes5	Does not affect atoh1a expression [28] Hes5 morphants do not generate supernumerary HCs [28]	Notch signaling activates Hes5 expression which inhibits hair cell fate [29] Hes5 is downregulated in SCs following HC damage and loss [30,31]	Beginning at postnatal ages, Hes5 is restricted to supporting cells [32], Hes5 deletion results primarily in supernumerary OHC formation, but also some supernumerary IHC formation [32]	Hes5 inhibition might be a therapeutic target in HC regeneration.
Notch Hey1/Hey 2	Hey1 is downregulated in hair cells [24]	Hey1 and Hey2 are activated by Notch signaling in the basilar papilla and inhibit HC fate [33]	Hey1 and Hey2 negatively regulate Atoh1 to prevent premature HC differentiation [21] Exogenous Shh increases Hey1 and Hey2 mRNA levels in cochlear explants [21]	Hey1/2 inhibition may help regenerate HC-like cells to adopt a more HC-like phenotype.
Wnt β-catenin	Wnt/β-catenin inhibition in embryonic zebrafish reduces proliferation of sox2+ SCs in the developing neuromast [34]; Wnt/β-catenin is upregulated in SCs following HC loss [35] but is not sufficient for regeneration	Increases the proliferation of SCs following HC damage and regulates the number of HCs that form in the embryonic basilar papilla [36], Forced expression of β-catenin and Wnt3a results in the formation of ectopic sensory patches within the embryonic basilar papilla [37]	Activation of Wnt/β-catenin results in proliferation of Sox2+ SCs [17]; Lgr5+ SCs exhibit increased proliferation and differentiation into HCs in vivo in mice which overexpress β-catenin and Atoh1 [16]	β-catenin is a key therapeutic target for expansion of the HC progenitor pool. Wnt/β-catenin is conserved between species and plays a role in HC development and proliferation
Shh	Modifying hedgehog signaling interferes with axial patterning of the zebrafish otic vesicle [38]	Ectopic Shh signaling induces apical hair cell identities in the basal and middle regions of the avian basilar papilla [39]	Constitutive activation of Shh signaling hinders HC differentiation in developing murine cochleae [20] Inhibition of hedgehog signaling in cochlear explants results in an expanded sensory domain and formation of ectopic hair cells [19]	Modifying Shh signaling does not seem to be an effective strategy to promote regeneration

Table 1. *Cont.*

Target	Zebrafish	Birds	Mice	Inference
FGF	Fgf signaling is required for Atoh1 expression and hair cell development [25] During development, weak Fgf inhibition expands the sox2+ prosensory domain while strong Fgf inhibition reduces the sox2+ prosensory domain [40] While Fgf inhibition hinders HC differentiation, by expanding the sox2+ prosensory domain, it ultimately results in the overproduction of hair cells Fgfr3 knockout results in supernumerary HC formation [40] Fgf8 knockout results in reduced HC formation [40] Fgf3 overexpression results in reduced HC number [40]	Inhibition of FGF signaling in E5-E9 chicks results in overproduction of HCs through non-proliferative mechanisms. FGF inhibition increases the number of Sox2+ HCs in the embryonic basilar papilla, suggesting that the formation of extra hair cells is due to transdifferentiation [41] Fgfr3 is restricted to supporting cells in the mature basilar papilla [42] Fgfr3 expression is downregulated in the mature basilar papilla following damage to hair cells [42]	Fgfr1 hypomorphs lack 3rd-row OHCs [43] Fgfr1 plays a role in prosensory specification [44], In the embryo, Fgfr3 is expressed in the area of the cochlear duct that gives rise to pillar cells, OHCs, and Deiter's cells, but Fgfr3 is confined to pillar cells by birth [45], Activation of Fgfr3 with Fgf17 inhibits OHC differentiation without affecting IHCs [46], Pan Fgf inhibition decreases expression of Atoh1 in murine cochlear explants [47], Fgfr3-/- mice lack a row of pillar cells, but have an ectopic additional row of Deiters cells and an additional row of OHCs which appear to have normal bundle morphology [48]	FGF signaling seems to be important signaling to modulate to promote HC regeneration, however, the results seem to be receptor-specific and different receptors have different effects of modulating FGF signaling

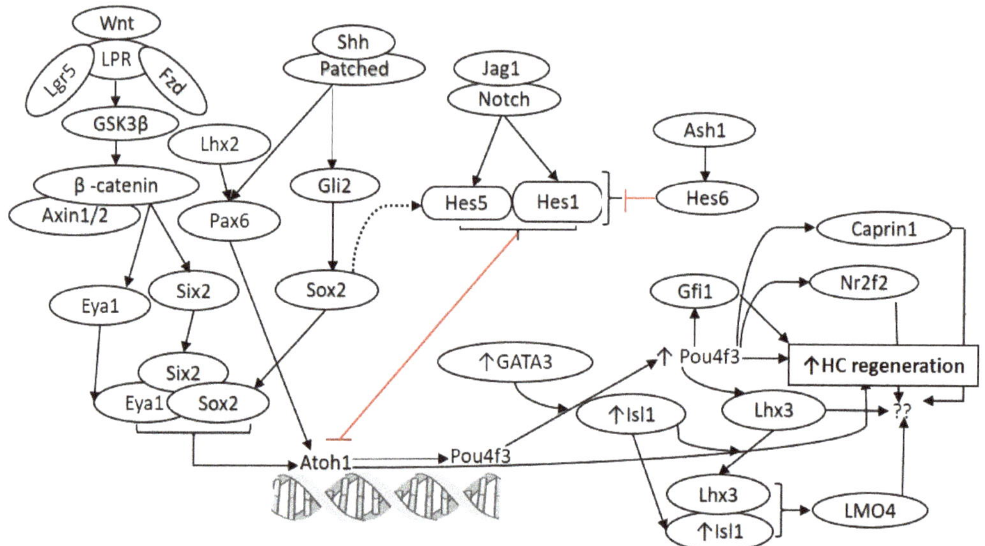

Figure 1. Molecular pathways involved in the development and regeneration of hair cells. Notch and Wnt signaling play a crucial role in the development and differentiation of HCs. Various studies have demonstrated that these pathways can be targeted for HC regeneration. Similarly, targeting Hes1, Gfi1, Pax6, Isl1, Pou4f3, Atoh1, and GATA3 to promote HC regeneration has been reported (as discussed in the text). However, there is a need to find additional candidate genes and transcription factors to promote HC regeneration. The network analysis on the published single-cell RNA-seq data

(Yamashita et al. 2018) predicted other potential targets, including Lhx2, Hes6, Caprin1, Nr2f2, and Lhx3, which may be targeted alone or in combination to promote regeneration of HCs. Atonal BHLH Transcription Factor 1 (Atoh1), frizzled (Fzd), islet 1 (Isl1), jagged 1 (Jag1), lipoprotein receptor-related protein (LPR), POU Class 4 Homeobox 3 (Pou4f3), sonic hedgehog (Shh), Wingless-related integration site (Wnt). Black arrows show stimulatory while red arrows show inhibitory effect.

3. Targeting Notch Signaling for Hair Cell Regeneration

Both fate determination of prosensory epithelial cells into HCs and SCs through lateral inhibition and the prevention of SC to HC conversion during HC development are regulated by active Notch signaling stimulated by ligands on adjacent HCs [49]. Thus, inhibiting Notch signaling might lead to the transdifferentiation of SCs to HCs. Increased number of myosin-VII-positive outer hair cells (OHCs) in vitro with γ-secretase inhibitor, LY411575 suggests that Notch inhibition promotes regeneration via transdifferentiation of SCs. Treatment with LY411575 depleted the supporting cell population, but the number of inner hair cells (IHCs) remained unchanged [9]. In vivo studies with systemic injection of LY411575 (50 mg/kg body weight) for 5 days in noise-deafened mice (4 weeks) showed decreased noise-induced threshold shifts and an increased number of OHCs with apparently innervated stereociliary bundles [9]. A decreased expression of Hes5 and increased expression of Atoh1 associated with SC to HC transdifferentiation suggests an association of Notch inhibition with HC regeneration (apical to mid-apical turn). The regenerated HCs were lineage traced using Sox2-CreER; mT/mG mice with tamoxifen injection at postnatal day 21, confirming their SC origin [9]. The systemic injection of LY411575 was associated with toxicity and a lower dose was not therapeutically potent, however, local injection of LY411575 through the round window membrane showed significant transdifferentiation of SCs to HCs. Note that 3-million-fold higher concentrations of LY411575 (4 mM) than its IC50 (0.14 nM) were used [9]. Notch signaling in the cochlea becomes nonresponsive after the first postnatal week [50]. Additionally, >85% Sox2-CreER activity in SCs when induced at P21 compared to >50% when induced at P1 [51] makes Sox2 a good lineage marker when induction is performed in adult mice but not useful when induction is performed at birth. Additionally, a 92% reduction in the number of fate-mapped regenerated HCs in ROSA-NICD neonatal (P0-P1) mice with NICD (Notch) overexpression compared to the controls [52] supports the need for Notch inhibition in SC to HC transdifferentiation. However, enhanced proliferation of sensory HCs with transient coactivation of cell cycle activator Myc and Notch1 genes by injecting adenovirus (ad)-Myc/ad-Cre into the cochleae of 6-week-old Rosa-NICD transgenic mice seemed contradictory [11]. However, ad-Myc/ad-Cre injection could enable the SCs to proliferate and respond to Atoh1 and transdifferentiate to HC-like cells, which might be due to the differences between direct transdifferentiation vs. induced proliferative regeneration. Regeneration of HCs with sustained release of Hes1 siRNA nanoparticles (siHes1 NPs) in the cochleae of noise-injured adult guinea pigs supports Notch inhibition as a target for HC regeneration. The study reported limited recovery of auditory function over a nine-week follow-up period as well as HC regeneration, evident by the presence of both ectopic and immature HCs across a broad tonotopic range with siHes1 NPs. One of the major limitations of this guinea pig study is that no lineage tracing was performed to prove that newly regenerated HCs are derived from SCs. The advantage of using poly-lactic-co-glycolic acid (PLGA)-mediated siHes1 NPs delivery was its reversible modulation of Hes1 [5].

4. Targeting Wnt Signaling for Hair Cell Regeneration

Wnt signaling plays an important role in cochlear development and its role is context-dependent. Active canonical Wnt/β-catenin signaling is needed for the initial differentiation of HCs but not for maturation and maintenance. Overactive Wnt signaling results in HC proliferation and the formation of ectopic HCs during early embryonic development. This suggests that activating Wnt signaling will favor new HC formation through transdifferentiation or mitotic division [18,53]. The conserved nature of Wnt signaling among species (Table 1) and increased Wnt expression following HC loss appends the notion that increasing Wnt expression might promote SC-to-HC transdifferentiation. The association of activated Wnt/β-catenin signaling with SC proliferation, a transient proliferation of Lgr5+ SCs [54], and HC regeneration [55] support Wnt-mediated transdifferentiation. However, in these reports, it remains elusive whether the regeneration of HCs was due to either activated Wnt signaling or Sox2 haploinsufficiency (loss of one allele of Sox2 due to knockin of CreER in the Sox2 locus results in a haploinsufficient phenotype that produces extra inner hair cells during development and enhances regeneration). Later, Atkinson et al. showed that both β-catenin^{GOF} (gain of function) and Sox2 haploinsufficiency enhance mitotic regeneration in the apical turn whereas Sox2 haploinsufficiency-mediated mitotic regeneration extends into the middle and basal turns [34]. Jan et al. [56] using lineage tracing in P0-P3 Axin2^{lacZ} Wnt reporter mice showed that Wnt responsive Axin-2-positive tympanic border cells proliferate with Wnt activation and generate new HC- and SC-like cells both in vitro and in vivo and can act as a precursor to sensory epithelial HCs. These studies suggest that activation of Wnt signaling in neonatal mice potentiates regeneration of HCs, however, no study has shown increased regeneration by activating Wnt signaling in adults.

5. Combinational Approaches for Hair Cell Regeneration

The role of Wnt activation in SC proliferation and Notch inhibition in transdifferentiation of SCs to HCs is evident by the above studies. Additionally, Wnt activation alone fails to regenerate significant amounts of new HCs in adult mammals, and Notch inhibition alone regenerates HCs at the cost of SCs, resulting in the death of regenerated HCs. Thus, maintaining the population of SCs via proliferation along with SC-to-HC transdifferentiation might enhance the regeneration process by sufficing the SC population for differentiation. Ni et al. reported that Wnt activation with 6-Bromoindirubin-3′-oxime (BIO), a glycogen synthase kinase 3 β (GSK3β) inhibitor, followed by Notch inhibition with DAPT, a γ-secretase inhibitor, preserves the Lgr5+ SC number and strongly promotes the mitotic regeneration of new HCs in both normal and neomycin-damaged cochlear explants (P1; C57/BL6 mice) [17]. Similar findings were reported by Wu et al. [57] by simultaneously inhibiting Notch signaling with DAPT and activating Wnt signaling with Wnt agonist QS11. The first study used the explants from P1 mice while the second study showed it in the utricle of neonatal mice, which itself has some regenerative capacity. Since Notch and Wnt signaling have a reciprocal relationship during HC development, a combined modulation of Notch and Wnt signaling might be a better approach for regeneration. Increased HC regeneration using Notch inhibition followed by Wnt activation in adult and neonatal mouse cochleae has been reported [17,58,59]. Romero-Carvajal et al. highlighted the role of interactions between Notch and Wnt signaling for the regeneration of HCs in zebrafish and demonstrated that inhibition of Notch signaling mimics the expression changes observed during endogenous regeneration [60].

Targeting multiple pathways and factors involved in HC development and insight from their involvement in other regenerative systems could be a promising approach to enhance HC regeneration in the cochlea. Clonal expansion of Lgr5+ SCs isolated from a neonatal cochlea showed that in a matrigel-based 3D culture system, a mixture of growth factors (including epidermal growth factor, basic fibroblast growth factor, and insulin-like growth factor 1), GSK3β inhibitor, histone deacetylase (HDAC) inhibitor, and Notch inhibition led to transcriptional activation, proliferation, and differentiation of SCs [6]. The addition of a stable form of vitamin C and transforming growth factor β (TGF-β) receptor (Alk5) inhibitor individually resulted in increased SC expansion by 2- to 3-fold, and the addition of small molecules in combination with growth factors increased the expansion of Lgr5+ SC numbers by >2000-fold compared to growth factors alone. The addition of these small molecules and γ-secretase inhibitor resulted in the expansion and differentiation of Lgr5+ SCs from a single mouse cochlea to nearly 11,500 HCs in culture organoid. The newly generated HCs in the organoid were myosin VIIa+ cells containing CtBP2+ ribbon synapse-like puncta in the basal region and actin-rich protrusions within the inner lumen. The colonies were both prestin-positive and negative; prestin-negative cells were vesicular glutamate transporter 3 (vGlut3)-positive, reflecting the gene expression of terminally differentiated OHCs and IHCs, respectively. The combination of these small molecules generated a higher number of new HCs in culture and in neonatal cochlear explants compared to the adult mouse Lgr5+ cells. There was no significant difference across the ages. Similarly, the clonal expansion and differentiation of adult human inner ear tissue was also limited [6]. The results of this study suggest that targeting a single gene, TF, or pathway may not be sufficient for HC regeneration and that there is a need for multidimensional approaches to promote transdifferentiation and regeneration. A recent study demonstrated the conversion of mouse embryonic fibroblasts, adult tail-tip fibroblasts, and postnatal supporting cells into induced hair cell-like cells (iHCs) showing HC-like morphology, transcriptomic and epigenetic profiles, electrophysiological properties, and mechanosensory channel expression using a combination of four transcription factors, Six1, Atoh1, Pou4f3, and Gfi1 [61]. Similar results of an increased SC to HC conversion by modulating the expression of p27^{kip1}, GATA3, and Pou4f3 in combination with Atoh1 were reported in adult mouse cochleae by Walters et al. [8]. These studies support the notion of using a combinational approach to promote HC regeneration via direct reprogramming. Notch, Wnt, and other signaling pathways play a crucial role in the development and proliferation of HCs and are conserved among species including zebrafish, birds, and mice (Table 1); targeting these pathways concomitantly might enhance HC regeneration in adult mammals (Figure 1) [34].

6. Modulating Transcription Factors for Hair Cell Regeneration

Transcription factors (TFs) are the proteins initiating and regulating transcription of target genes by binding to their specific regulatory DNA sequences. TFs play a crucial role in the proliferation, differentiation, and survival of HCs. The role of TFs such as Pax2, Sox9, Nor-1, Gbx-2, Neurod1, Neurog1, Fkh10, Tbx1, Brn4, Gata3, Sox2, Atoh1, Six1, Isl1, Pou4f3, Gfi1, and their interactions with cellular and molecular signaling pathways in prosensory cell specification, development, and fate determination of HCs in the inner ear and vestibular apparatus have been described by other groups [62,63]. Since TFs play an important role in HC development and fate determination, investigating their role will help not only in understanding HC development but also in modifying regeneration strategies for improved outcomes. TFs, individually or in combination, play a crucial role in the regeneration of other organ systems [64–66]. Thus, it is necessary to investigate the TFs which might be capable of potentiating the regeneration process in the cochlea and to understand the underlying mechanisms. Costa et al. studied the role of three TFs, namely, Gfi1, Atoh1, and Pou4f3 (GAP) in cell fate determination, and reported that GAP (Figure 1) can induce direct genetic reprogramming of progenitors towards an HC fate, both in vitro and in vivo in the chicken embryo [15]. Another study reported that overexpression of Prox1 suppresses Atoh1 and Gfi1 expression and antagonizes the differentiated HC phenotype; thus, Prox1 inhibition with Atoh1 upregulation might result in a more complete phenotypic conversion [67]. These studies were performed in the embryonic stage and whether SCs can be transdifferentiated to HCs, postnatally or in adults, cannot be determined. These studies allude to Atoh1 as a common denominator target for HC regeneration.

Atoh1 is a master regulator of HC differentiation that is conserved among fish (ortholog atoh1a), birds, and mice (Table 1). Liu et al. [12] investigated the effect of ectopic expression of Atoh1 on regeneration using EGFP reporter mice and reported that ectopic expression of Atoh1 induces the conversion of mouse cochlear SCs (pillar and Deiters' cells; PCs and DCs) to immature HCs, and that this conversion is age-dependent. Ectopic Atoh1 expression was effective in converting PCs and DCs to HCs at neonatal and juvenile ages, but it was insufficient for adult mice. It was found that newly formed HCs reside in the OHC region and survive for 2 months, and that heterogeneity in the reprogramming efficiency among individual Atoh1+ PCs and DCs exists during the conversion process. These studies suggest that transcriptional reprogramming affects the HC phenotype and thus might favor regeneration, but the limitation is that these results were shown at neonatal and juvenile ages and Atoh1 overexpression alone was not effective in adult mice. To address this issue, Walters et al. [8] investigated the role of various TFs in adult mice and reported that ectopic Atoh1 overexpression with p27^{Kip1} deletion circumvents this age-related decline in Atoh1 responsiveness and leads to transdifferentiation of SCs to HCs in mature mouse cochleae after noise damage. Further, upregulation of an Atoh1 cofactor, GATA3, which is lost from SCs with aging, was associated with p27^{Kip1} deletion. Overexpression of POU4F3 alone promoted the conversion of SCs to HCs to a greater degree than Atoh1 alone, and overexpression of Atoh1 combined with POU4F3 or GATA3 resulted in increased conversion of SCs to HCs compared to Atoh1 alone in adult mice [8]. The study concluded that the mature PCs and DCs, which are typically nonresponsive to Atoh1, can be made to respond to ectopic Atoh1 via modulation of additional TFs such as p27^{Kip1}, GATA3, or POU4F3 (Figures 1 and 2). However, the converted HCs were examined only at 3 and 12 weeks following tamoxifen injection, and the long-term survival of these cells was not extensively evaluated [68].

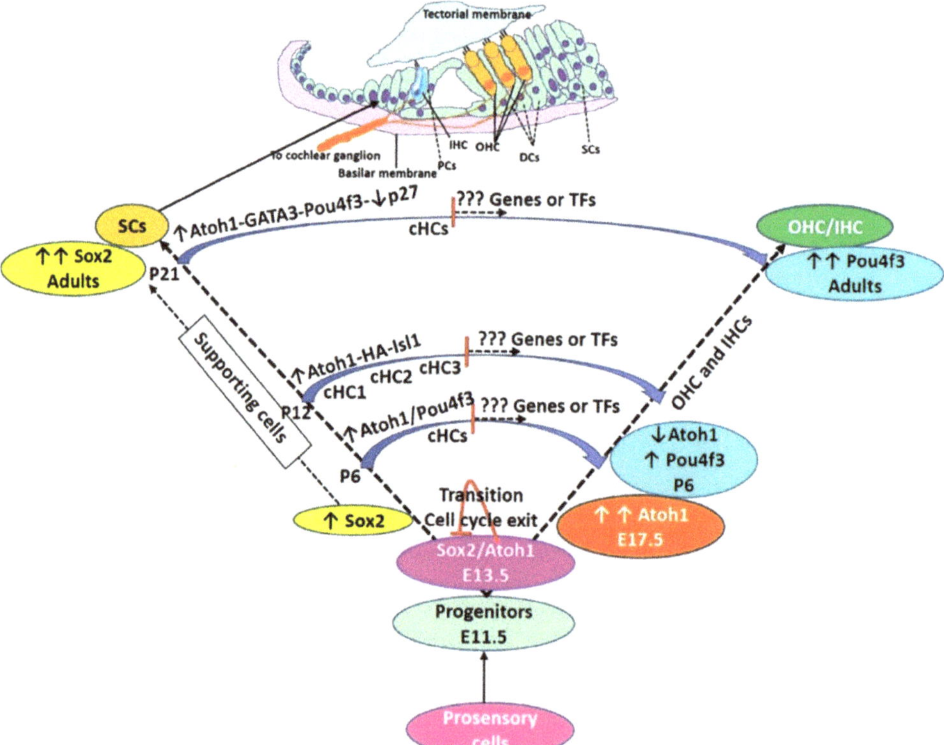

Figure 2. Hair cell development and regeneration: During the embryonic stage of HC development, Atoh1 expression increases, reaches a maximum (E17.5), and then declines (P6). The decline in Atoh1 is associated with increasing levels of Pou4f3 which remain high in adult HCs. During the embryonic stage, autoregulation of Atoh1, Sox2, and cell cycle exit. Some cells have high levels of Sox2 without Atoh1 expression and these cells are deemed to be SCs. Transdifferentiation of supporting cells (SCs) to HCs is mediated by overexpression of transcription factors (TFs) as discussed in the text and shown

here. However, TF-mediated transdifferentiation is not complete and converted HCs (cHCs) remain immature. cHCs show only some features of mature HCs (shown in the middle of the trajectory of conversion), and thus, there is a need to investigate novel targets to push these partially converted cells to mature HCs. Pillar cells (PCs), Deiters' cells (DCs), inner hair cells (IHC), outer hair cells (OHC), postnatal day (P). cHCs are the cells evaluated for their transcriptome by single cells RNA sequencing by Yamashita et al., 2018.

Increased conversion of SCs to HCs by activating Atoh1 conditionally with tamoxifen and 51 differentially expressed TFs between endogenous OHCs, SCs, and converted HCs (cHCs) in the adult cochlea with bulk-RNA sequencing, single-cell RNA sequencing, and single-cell RT-PCR reported by Yamashita et al. [10] supports targeting Atoh1 for regeneration and warrants research for the role of these TFs in HC regeneration. Additionally, a greater number of cHCs with combined overexpression of Atoh1 and Isl1 (one of the differentially expressed TFs in RNA-seq analysis) compared to overexpression of Atoh1 alone both ex vivo and in vivo supports the hypothesis that the conversion process can be pushed further to get a greater number of cHCs by targeting multiple TFs. However, the study only reported the cHCs at two-time points of conversion and the cHCs were not functionally mature (Figure 2).

Prolonged constitutive ectopic Atoh1 expression with tamoxifen using a Cre-inducible mouse might be a cause of immaturity in cHCs because continued Atoh1 expression does not correlate with endogenous HC development [69]. Controlled activation of Atoh1 using tetracycline (e.g., dox)-inducible systems is difficult to achieve in juvenile or adult mice due to the long-term residual activity of tetracycline in the cochlea [70,71]. However, these studies give hope that a greater number of regenerated HCs can be achieved at an adult age. Further, the role of TFs Insm1 [72] and Ikzf2 [73] in the fate determination and functional maturation of OHCs suggests the possibility of other TFs having critical regulatory roles in the regeneration of HCs. Recently, the role of TUB and ZNF532 in promoting Atoh1-mediated hair cell regeneration in mouse cochleae was reported by Xu et al. [74]. Thus, investigating novel TFs and targets upstream or downstream of Atoh1 is warranted, and modulating their expression alone or in combination could provide better results (Figure 1).

7. SC Subpopulations and Hair Cell Regeneration

Transdifferentiation of SCs to HCs has been postulated as the main strategy to regenerate HCs, and most of the studies discussed above have targeted various signaling pathways and TFs in order to do so. However, the debate on which subtype of SCs is more prone to transdifferentiation still exists in the field. Walters et al. [8] reported the unresponsiveness of mature PCs and DCs to Atoh1 and, together with other evidence, concluded that responsiveness to Atoh1 varies across SC subtypes. Thus, it is important to investigate the differential responsiveness of SC subtypes to TFs to achieve a greater number of functionally matured HCs through regeneration. Recently, Hoa et al. [75] reported that adult cochlear SCs are transcriptionally different from perinatal SCs by conducting single-cell RNA-Seq on FACS-sorted GFP expressing adult cochlear SCs from Lfng^{EGFP} adult mice. The study found two different subpopulations of SCs (SC1 and SC2). The SC2 subpopulation expresses transcripts associated with S phase (Mcm4) and G2/M phase (Birc5, Cdk1, Mki67). Cheng et al. [76] also reported differential expression of various cell cycle and signaling pathway genes and TFs in Sox2+ SCs at four different postnatal ages suggesting the existence of age-related transcriptomic landscape changes. The different transcriptomic landscape of the perinatal and postnatal SCs found in this study might be the reason for the differential responsiveness of adult SCs in the study by Walters et al. [8]. Further, the findings of strong expression of the SC genes involved in pathways regulating the cell cycle [75] suggest that these pathways may be targeted to potentiate the transdifferentiation of SCs to HCs by forcing the SCs out of quiescence. This notion is supported by DCs and PCs contributing more to the spontaneously regenerated HCs but inner phalangeal (IPhs) and inner border (IBs) cells having similar regenerative capacity in neonatal mice [77]. This differential response may be because PCs and DCs lose the cell cycle inhibitor p27^{Kip1}

during postnatal development and are capable of mitotic HC regeneration. These findings are supported by a previous study in juvenile mice where ectopic expression of Atoh1 induces SC-to-HC conversion and the newly regenerated HCs are mainly from PCs and DCs [12]. Later, however, a higher, faster, and more complete conversion rate of IBs and IPhs compared to DCs or PCs to IHC-like cells was observed in vivo, as evidenced by straight line-shaped stereociliary bundles, expression of Fgf8 and otoferlin, and by ectopic Atoh1 expression [78]. The study also reported that the conversion rate gradually increases from neonate to adult ages in mice. Differential regenerative capacity of SCs might be due to changing Sox2 expression over time. Changing Sox2 expression was reported by Kempfle et al. [79] suggest that Sox2 is expressed in prosensory cells of the cochlea at E13, in the developing sensory epithelium at E15 and E18, in newly formed IHCs at E15, and its expression continues in newly formed IHCs and OHCs at E18 until P0 and becomes undetectable at P2. Sox2 is strongly expressed in SCs at E18 and continues to be expressed in SCs at P2. Sox2 is necessary for differentiation as deletion of Sox2 at E16 led to no further differentiation of HCs.

The Lgr5+ subtype of SCs has been an attractive target for HC regeneration. Kuo et al. reported an increased number of regenerated HCs via transdifferentiation of Lgr5+ SCs by ectopically co-expressing a constitutively active form of β-catenin and Atoh1 in Lgr5+ cells of the neonatal cochlea. This study suggests that combining proliferation and differentiation of Lgr5+ SCs by coactivating β-catenin and Atoh1 acts synergistically to enhance the process of regeneration, yielding an increased number of regenerated HCs [16]. Although the tamoxifen induction was done at a neonatal age, the study reported the HCs had an adult phenotype. Recently, Zhang et al. reported that activating Frizzled-9 (Fzd9)-positive cells in neonatal mouse cochleae leads to regeneration of a similar number of HCs. Lineage tracing of the tamoxifen-induced cells showed that inner phalangeal cells (IPhCs), inner border cells (IBCs), and third-row Deiters' cells (DCs) were both Fzd9+ and Lgr5+, while pillar cells are Lgr5+ only [7]. The study concluded that the Fzd9+ cells have a similar capacity for HC regeneration, proliferation, and differentiation compared to Lgr5+ cells. These results demonstrate the potential of targeting Notch and Wnt signaling for HC regeneration; however, there is a need to translate these findings to pre-clinical trials and future studies are warranted. Collectively, these studies suggest there is heterogeneity and a changing transcriptomic landscape of SCs over time. Additionally, there are no known differences among different mammalian species in HC regeneration relative to the timing of HC development. Thus, the strategies and timing of manipulating SCs for regeneration are of the utmost importance and warrants further investigation.

8. Finding Additional Transcription Factors as Novel Targets for Hair Cell Regeneration

Atoh1 regulates HC development and differentiation, and overexpression of Atoh1 regenerates HCs from SCs; however, the newly regenerated HCs are fewer, short-lived, and not functionally mature, as evidenced by the absence of prestin, the marker for OHC maturation [8]. Thus, the consensus is to find novel targets upstream or downstream of Atoh1 whose modulation can potentiate regeneration so that increased numbers of functionally mature HCs can be achieved. This notion is supported by the fact that co-activation of Atoh1 with Pou4f3 [8], with Isl1 [10], and with both Pou4f3 and Gfi1 combined [15] yielded a greater number of HCs compared to activation of Atoh1 alone. This suggests that either post-transcriptional modification of Atoh1 targets, Atoh1 itself, or epigenetic regulation of Atoh1 and its targets regulate the expression of various target genes and TFs, and thus HC regeneration. To investigate the direct targets of Atoh1, Cai et al. [80] carried out RNA-seq profiling of purified Atoh1 expressing HCs from neonatal mouse cochleae and identified >600 enriched transcripts with 233 HC genes directly regulated by Atoh1. Atoh1 regulation was verified by the presence of Atoh1 binding sites in the regulatory regions of these genes and by the cerebellum and small intestine Atoh1 ChIP-seq analysis. Anxa4, Rasd2, Rbm24, Srrm4, Chrna10, Mgat5b, Mreg, Pcp4, Scn11a, and Atoh1 were found to be direct targets of Atoh1. The expression of Anxa4, Rasd2, Rbm24, and Srrm4 was completely downregulated within 24 h after knocking out Atoh1, but the expression of Chrna10, Mgat5b, Mreg, Pcp4, and Scn11a were not affected. In the context of epigenetic regulation of Atoh1, Jen et al. reported that the mouse vestibular apparatus has greater Atoh1-mediated regeneration compared to the cochlea due to greater chromatin accessibility [81]. These findings suggest that differential efficiency of Atoh1-mediated regeneration is due to the non-availability of open chromatin in the cochlea and warrants further research using ATAC-seq (Assay for Transposase-Accessible Chromatin using sequencing) and ChIP-seq to unravel the epigenetic regulation and to identify additional targets for regeneration.

9. Investigating Epigenetic Regulation of Hair Cell Development and Regeneration

Coordinated and structured gene expression is a must for cellular development, differentiation, and survival, and epigenetics plays a crucial role in regulating gene transcription and expression. Post-translational histone (basic proteins in the cell nucleus) modification mechanisms include methylation (addition of a methyl group), acetylation (addition of an acetyl group), phosphorylation, and ubiquitination, which regulate chromatin architecture and gene expression. Methylation reduces gene expression by impairing the binding of transcriptional activators whereas acetylation increases gene expression by transcription activation. Histone acetylation is regulated by histone acetyltransferases (HATs) and histone deacetylases (HDACs); methylation and demethylation are regulated by histone methyltransferases (HMTs), DNA methyltransferase (DNMTs), and histone demethylases. Epigenetics play a role in hereditary or syndromic hearing loss by regulating gene expression and HC development [82–84]. Stojanova et al. [85] investigated the epigenetic regulation of Atoh1 and found that progression of Atoh1 expression from poised, to active, to repressive marks is controlled by dynamic changes in histone modifications via methylation and acetylation (H3K4me3/H3K27me3, H3K9ac, and H3K9me3) and correlates with the onset and subsequent silencing of Atoh1 expression in HCs during the perinatal period. The study reported that during HC differentiation, increased Atoh1 expression correlates with increased levels of H3k9ac (H3K9 histone acetylation) and that during HC maturation decreased levels of Atoh1 correlate with decreased levels of H3K9ac and increased levels of H3K9me3. Further, increased expression of HC-related genes and proteins in mouse utricle sensory epithelia-derived progenitor cells with DNMT inhibitor 5-azacytidine suggests an important role for epigenetics in HC differentiation [86]. This notion is also supported by the recent report by McLean et al. [6] where an HDAC inhibitor was used for the regeneration of HCs. However, Layman et al. [87] reported that suberoylanilide hydroxamic acid (SAHA, an HDAC inhibitor) does not affect regeneration in adult cochleae but instead activates pro-survival pathways via regulating the acetylation status of transcription factors and controls the transcriptional activation of pro-survival pathways in response to ototoxic insults. These surprising results suggest that HDAC inhibitors cannot effectively modulate the already fixed epigenetic landscape of adult cochlear SCs and are thus ineffective in reprogramming. HC fate determination and development are highly regulated processes under the influence of various TFs and gene expression, and expression of this transcriptomic landscape changes over time [83,84], with a dramatic change in the transcriptomic landscape between post-natal day (P)5–P7. Thus, investigating the epigenetic regulation of TFs and genes involved in HC development and rescripting the genetic landscape may provide insights to promote HC regeneration.

10. In Silico Approaches to Finding Novel Gene Targets

In silico analysis and the use of the wealth of bioinformatics applications for the acquisition of biological data and data mining have changed the paradigm of research in the field of basic and applied science. In the auditory field, regeneration of HCs deals with the modulation of genes and TFs, thus we can analyze the available databases to uncover better targets to modulate and potentiate the process of regeneration. The binding of TFs to their corresponding TF binding sites (TFBSs) is key to transcriptional regulation. Because information on experimentally validated functional TFBSs is limited, there is a need for the prediction of TFBSs for gene annotation. TFBSs are generally recognized by scanning a position weight matrix (PWM) against DNA sequences using one of several available computer programs. There are also several curated databases of PWMs, applicable to a wide range of species, including the commercial TRANSFAC database [88] and the open-access JASPAR database [89]. Other recent databases include the HOMER motif (http://homer.salk.edu/homer/motif (accessed on 6 June 2020)) HOCOMOCO [90], and CIS-BP [91]. There is a particularly useful program, the Cytoscape plugin iRegulon [92], which can discover master regulators from co-expressed gene sets. Additionally, the methods of inferring co-expression networks from single-cell RNA-seq data and workflow, such as single-cell regulatory network inference and clustering (SCENIC) [34], have been developed to exploit the genomic regulatory code (regulon), guiding the identification of master TFs and revealing different cell states. Such predictions on the master regulators of different cell types/states would be valuable to improve the conversion efficiency from SCs to HCs. Further, network analysis using these tools might predict the master regulators whose modulation, either alone or in combination with other TFs, may promote regeneration.

The network analysis done on an scRNA-seq data of cHCs [10] predicted Lhx3, Six2, Hes2, Irf6, Hes6, and Ikzf2 along with Atoh1 as candidate targets to modulate. Ikzf2 has recently been shown to be crucial for OHC fate and maturation, as prestin and oncomodulin expression is lost in Ikzf2-mutant mice [73]; contrarily, overexpression of Ikzf2 in IHCs leads to downregulation of IHC genes and upregulation of OHC genes. Transformation of adult cochlear SCs into prestin-positive OHCs with concurrent stimulation if Athoh1 and Ikzf2 supports the role of Ikzf2 in transdifferentiation [93]. Hes6 has also been implicated in the differentiation of mammalian HCs [94]. Interestingly, identification of the TFs such as Hes2, Hes6, Irf6, and Atoh1, which have roles in neural development and differentiation [95–98], by our network analysis suggests the feasibility and promising role of using bioinformatics to identify novel targets. Another TF identified in our network analysis, Six2, appears to play a role in regeneration in the mammalian kidney, as it is expressed in self-renewing progenitor cells within this organ [99]. These results suggest that the TFs identified via bioinformatics analysis of cochlear scRNA-seq data play a role in the regeneration and development of other organ systems and hence should be investigated for cochlear HC regeneration, and that further bioinformatics analysis of the existing cochlear scRNA-seq or ATAC-seq data is warranted.

11. Conclusions

Modulating the expression of signaling pathways and genes involved in sensory HC development, as discussed above, has given promising results in adult cochlear HC regeneration; however, the small number and functional immaturity of regenerated HCs remain a challenge. Targeting multiple factors has improved the outcome, but there is still a need to investigate additional targets and to form novel strategies to promote HC regeneration in adult mammals and then to translate these to clinics. The downstream targets of Atoh1 and Pou4f3 might be viable targets for HC regeneration. If regeneration follows development, unraveling the sequential targets for regeneration is of the utmost importance. Similarly, the role of many TFs such as Lhx3, caprin1, Nr2f2, Lmo4, and others in the regeneration process has not been investigated. Analyzing the existing cochlear data using bioinformatics tools investigating endogenous regeneration in zebrafish and birds might give the hearing field an overview and insight into what factors remain to be modified to regenerate HCs that are greater in number and functionally mature. Taken together, investigating the genes and TFs which either alone or in combination can potentiate the transdifferentiation of SCs to HCs should be the focus of current research for better therapeutics.

Author Contributions: V.R. wrote the initial draft, S.T. wrote Section 10, J.R.F. prepared Table 1, S.T., J.R.F., and J.Z. revised and edited manuscript, V.R., S.T., J.R.F., and J.Z. finalized the manuscript. All authors have read and agreed to the published version of the manuscript.

Funding: This work is supported in part by NIHR01DC015010, NIHR01DC015444, ONR-N00014-18-1-2507, USAMRMC-RH170030, and LB692/Creighton to Jian Zuo.

Data Availability Statement: Not applicable.

Acknowledgments: We thank members of the Zuo lab for critical comments and discussions.

Conflicts of Interest: J.Z. is a Co-Founder of Ting Therapeutics L.L.C. Other authors declare no conflict of interest.

References

1. Davis, A.C. Hearing disorders in the population: First phase findings of the MRC National Study of Hearing. In *Hearing Science and Hearing Disorders*; Elsevier: Amsterdam, The Netherlands, 1983; pp. 35–60.
2. Groves, A.K. The challenge of hair cell regeneration. *Exp. Biol. Med.* **2010**, *235*, 434–446. [CrossRef]
3. Cox, B.C.; Chai, R.; Lenoir, A.; Liu, Z.; Zhang, L.; Nguyen, D.H.; Chalasani, K.; Steigelman, K.A.; Fang, J.; Rubel, E.W.; et al. Spontaneous hair cell regeneration in the neonatal mouse cochlea in vivo. *Development* **2014**, *141*, 816–829. [CrossRef] [PubMed]
4. Stone, J.S.; Cotanche, D.A. Hair cell regeneration in the avian auditory epithelium. *Int. J. Dev. Biol.* **2007**, *51*, 633–647. [CrossRef]
5. Du, X.; Cai, Q.; West, M.B.; Youm, I.; Huang, X.; Li, W.; Cheng, W.; Nakmali, D.; Ewert, D.L.; Kopke, R.D. Regeneration of Cochlear Hair Cells and Hearing Recovery through Hes1 Modulation with siRNA Nanoparticles in Adult Guinea Pigs. *Mol. Ther. J. Am. Soc. Gene Ther.* **2018**, *26*, 1313–1326. [CrossRef] [PubMed]
6. McLean, W.J.; Yin, X.; Lu, L.; Lenz, D.R.; McLean, D.; Langer, R.; Karp, J.M.; Edge, A.S. Clonal expansion of Lgr5-positive cells from mammalian cochlea and high-purity generation of sensory hair cells. *Cell Rep.* **2017**, *18*, 1917–1929. [CrossRef]
7. Zhang, S.; Liu, D.; Dong, Y.; Zhang, Z.; Zhang, Y.; Zhou, H.; Guo, L.; Qi, J.; Qiang, R.; Tang, M.; et al. Frizzled-9+ Supporting Cells Are Progenitors for the Generation of Hair Cells in the Postnatal Mouse Cochlea. *Front. Mol. Neurosci.* **2019**, *12*, 184. [CrossRef] [PubMed]
8. Walters, B.J.; Coak, E.; Dearman, J.; Bailey, G.; Yamashita, T.; Kuo, B.; Zuo, J. In Vivo Interplay between p27(Kip1), GATA3, ATOH1, and POU4F3 Converts Non-sensory Cells to Hair Cells in Adult Mice. *Cell Rep.* **2017**, *19*, 307–320. [CrossRef]
9. Mizutari, K.; Fujioka, M.; Hosoya, M.; Bramhall, N.; Okano, H.J.; Okano, H.; Edge, A.S. Notch inhibition induces cochlear hair cell regeneration and recovery of hearing after acoustic trauma. *Neuron* **2013**, *77*, 58–69. [CrossRef] [PubMed]
10. Yamashita, T.; Zheng, F.; Finkelstein, D.; Kellard, Z.; Carter, R.; Rosencrance, C.D.; Sugino, K.; Easton, J.; Gawad, C.; Zuo, J. High-resolution transcriptional dissection of in vivo Atoh1-mediated hair cell conversion in mature cochleae identifies Isl1 as a co-reprogramming factor. *PLoS Genet.* **2018**, *14*, e1007552. [CrossRef]
11. Shu, Y.; Li, W.; Huang, M.; Quan, Y.-Z.; Scheffer, D.; Tian, C.; Tao, Y.; Liu, X.; Hochedlinger, K.; Indzhykulian, A.A. Renewed proliferation in adult mouse cochlea and regeneration of hair cells. *Nat. Commun.* **2019**, *10*, 1–15. [CrossRef]
12. Liu, Z.; Dearman, J.A.; Cox, B.C.; Walters, B.J.; Zhang, L.; Ayrault, O.; Zindy, F.; Gan, L.; Roussel, M.F.; Zuo, J. Age-dependent in vivo conversion of mouse cochlear pillar and Deiters' cells to immature hair cells by Atoh1 ectopic expression. *J. Neurosci.* **2012**, *32*, 6600–6610. [CrossRef]
13. Querleu, D.; Renard, X.; Versyp, F.; Paris-Delrue, L.; Crèpin, G. Fetal hearing. *Eur. J. Obstet. Gynecol. Reprod. Biol.* **1988**, *28*, 191–212. [CrossRef]
14. Edge, A.S.; Chen, Z.-Y. Hair cell regeneration. *Curr. Opin. Neurobiol.* **2008**, *18*, 377–382. [CrossRef] [PubMed]
15. Costa, A.; Sanchez-Guardado, L.; Juniat, S.; Gale, J.E.; Daudet, N.; Henrique, D. Generation of sensory hair cells by genetic programming with a combination of transcription factors. *Development* **2015**, *142*, 1948–1959. [CrossRef] [PubMed]
16. Kuo, B.R.; Baldwin, E.M.; Layman, W.S.; Taketo, M.M.; Zuo, J. In vivo cochlear hair cell generation and survival by coactivation of β-catenin and Atoh1. *J. Neurosci.* **2015**, *35*, 10786–10798. [CrossRef]
17. Ni, W.; Zeng, S.; Li, W.; Chen, Y.; Zhang, S.; Tang, M.; Sun, S.; Chai, R.; Li, H. Wnt activation followed by Notch inhibition promotes mitotic hair cell regeneration in the postnatal mouse cochlea. *Oncotarget* **2016**, *7*, 66754. [CrossRef] [PubMed]
18. Atkinson, P.J.; Najarro, E.H.; Sayyid, Z.N.; Cheng, A.G. Sensory hair cell development and regeneration: Similarities and differences. *Development* **2015**, *142*, 1561–1571. [CrossRef]
19. Driver, E.C.; Pryor, S.P.; Hill, P.; Turner, J.; Rüther, U.; Biesecker, L.G.; Griffith, A.J.; Kelley, M.W. Hedgehog signaling regulates sensory cell formation and auditory function in mice and humans. *J. Neurosci.* **2008**, *28*, 7350–7358. [CrossRef]
20. Tateya, T.; Imayoshi, I.; Tateya, I.; Hamaguchi, K.; Torii, H.; Ito, J.; Kageyama, R. Hedgehog signaling regulates prosensory cell properties during the basal-to-apical wave of hair cell differentiation in the mammalian cochlea. *Development* **2013**, *140*, 3848–3857. [CrossRef]
21. Benito-Gonzalez, A.; Doetzlhofer, A. Hey1 and Hey2 control the spatial and temporal pattern of mammalian auditory hair cell differentiation downstream of Hedgehog signaling. *J. Neurosci.* **2014**, *34*, 12865–12876. [CrossRef]
22. Su, Y.-X.; Hou, C.-C.; Yang, W.-X. Control of hair cell development by molecular pathways involving Atoh1, Hes1 and Hes5. *Gene* **2015**, *558*, 6–24. [CrossRef] [PubMed]

23. Zheng, F.; Zuo, J. Cochlear hair cell regeneration after noise-induced hearing loss: Does regeneration follow development? *Hear. Res.* **2017**, *319*, 182–196. [CrossRef]

24. Lush, M.E.; Diaz, D.C.; Koenecke, N.; Baek, S.; Boldt, H.; St Peter, M.K.; Gaitan-Escudero, T.; Romero-Carvajal, A.; Busch-Nentwich, E.M.; Perera, A.G. scRNA-Seq reveals distinct stem cell populations that drive hair cell regeneration after loss of Fgf and Notch signaling. *Elife* **2019**, *8*, e44431. [CrossRef] [PubMed]

25. Millimaki, B.B.; Sweet, E.M.; Dhason, M.S.; Riley, B.B. Zebrafish atoh1 genes: Classic proneural activity in the inner ear and regulation by Fgf and Notch. *Development* **2007**, *134*, 295–305. [CrossRef] [PubMed]

26. Cafaro, J.; Lee, G.S.; Stone, J.S. Atoh1 expression defines activated progenitors and differentiating hair cells during avian hair cell regeneration. *Dev. Dyn. Off. Publ. Am. Assoc. Anat.* **2007**, *236*, 156–170. [CrossRef]

27. Bermingham, N.A.; Hassan, B.A.; Price, S.D.; Vollrath, M.A.; Ben-Arie, N.; Eatock, R.A.; Bellen, H.J.; Lysakowski, A.; Zoghbi, H.Y. Math1: An essential gene for the generation of inner ear hair cells. *Science* **1999**, *284*, 1837–1841. [CrossRef] [PubMed]

28. Radosevic, M.; Fargas, L.; Alsina, B. The role of her4 in inner ear development and its relationship with proneural genes and Notch signalling. *PLoS ONE* **2014**, *9*, e109860. [CrossRef]

29. Chrysostomou, E.; Gale, J.E.; Daudet, N. Delta-like 1 and lateral inhibition during hair cell formation in the chicken inner ear: Evidence against cis-inhibition. *Development* **2012**, *139*, 3764–3774. [CrossRef] [PubMed]

30. Jiang, L.; Xu, J.; Jin, R.; Bai, H.; Zhang, M.; Yang, S.; Zhang, X.; Zhang, X.; Han, Z.; Zeng, S. Transcriptomic analysis of chicken cochleae after gentamicin damage and the involvement of four signaling pathways (Notch, FGF, Wnt and BMP) in hair cell regeneration. *Hear. Res.* **2018**, *361*, 66–79. [CrossRef]

31. Ku, Y.-C.; Renaud, N.A.; Veile, R.A.; Helms, C.; Voelker, C.C.; Warchol, M.E.; Lovett, M. The transcriptome of utricle hair cell regeneration in the avian inner ear. *J. Neurosci.* **2014**, *34*, 3523–3535. [CrossRef]

32. Zine, A.; Aubert, A.; Qiu, J.; Therianos, S.; Guillemot, F.; Kageyama, R.; de Ribaupierre, F. Hes1 and Hes5 activities are required for the normal development of the hair cells in the mammalian inner ear. *J. Neurosci.* **2001**, *21*, 4712–4720. [CrossRef]

33. Neves, J.; Parada, C.; Chamizo, M.; Giráldez, F. Jagged 1 regulates the restriction of Sox2 expression in the developing chicken inner ear: A mechanism for sensory organ specification. *Development* **2011**, *138*, 735–744. [CrossRef] [PubMed]

34. Aibar, S.; González-Blas, C.B.; Moerman, T.; Imrichova, H.; Hulselmans, G.; Rambow, F.; Marine, J.-C.; Geurts, P.; Aerts, J.; van den Oord, J. SCENIC: Single-cell regulatory network inference and clustering. *Nat. Methods* **2017**, *14*, 1083–1086. [CrossRef] [PubMed]

35. Mi, X.-X.; Yan, J.; Li, Y.; Shi, J.-P. Wnt/β-catenin signaling was activated in supporting cells during exposure of the zebrafish lateral line to cisplatin. *Ann. Anat.-Anat. Anz.* **2019**, *226*, 48–56. [CrossRef]

36. Jacques, B.E.; Montgomery IV, W.H.; Uribe, P.M.; Yatteau, A.; Asuncion, J.D.; Resendiz, G.; Matsui, J.I.; Dabdoub, A. The role of Wnt/β-catenin signaling in proliferation and regeneration of the developing basilar papilla and lateral line. *Dev. Neurobiol.* **2014**, *74*, 438–456. [CrossRef]

37. Stevens, C.B.; Davies, A.L.; Battista, S.; Lewis, J.H.; Fekete, D.M. Forced activation of Wnt signaling alters morphogenesis and sensory organ identity in the chicken inner ear. *Dev. Biol.* **2003**, *261*, 149–164. [CrossRef]

38. Hammond, K.L.; van Eeden, F.J.; Whitfield, T.T. Repression of Hedgehog signalling is required for the acquisition of dorsolateral cell fates in the zebrafish otic vesicle. *Development* **2010**, *137*, 1361–1371. [CrossRef]

39. Son, E.J.; Ma, J.-H.; Ankamreddy, H.; Shin, J.-O.; Choi, J.Y.; Wu, D.K.; Bok, J. Conserved role of Sonic Hedgehog in tonotopic organization of the avian basilar papilla and mammalian cochlea. *Proc. Natl. Acad. Sci. USA* **2015**, *112*, 3746–3751. [CrossRef]

40. Maier, E.C.; Whitfield, T.T. RA and FGF signalling are required in the zebrafish otic vesicle to pattern and maintain ventral otic identities. *PLoS Genet.* **2014**, *10*, e1004858. [CrossRef] [PubMed]

41. Jacques, B.E.; Dabdoub, A.; Kelley, M.W. Fgf signaling regulates development and transdifferentiation of hair cells and supporting cells in the basilar papilla. *Hear. Res.* **2012**, *289*, 27–39. [CrossRef]

42. Bermingham-McDonogh, O.; Stone, J.S.; Reh, T.A.; Rubel, E.W. FGFR3 expression during development and regeneration of the chick inner ear sensory epithelia. *Dev. Biol.* **2001**, *238*, 247–259. [CrossRef] [PubMed]

43. Pirvola, U.; Ylikoski, J.; Trokovic, R.; Hébert, J.M.; McConnell, S.K.; Partanen, J. FGFR1 is required for the development of the auditory sensory epithelium. *Neuron* **2002**, *35*, 671–680. [CrossRef]

44. Hayashi, T.; Ray, C.A.; Younkins, C.; Bermingham-McDonogh, O. Expression patterns of FGF receptors in the developing mammalian cochlea. *Dev. Dyn. Off. Publ. Am. Assoc. Anat.* **2010**, *239*, 1019–1026. [CrossRef]

45. Mueller, K.L.; Jacques, B.E.; Kelley, M.W. Fibroblast growth factor signaling regulates pillar cell development in the organ of corti. *J. Neurosci.* **2002**, *22*, 9368–9377. [CrossRef]

46. Jacques, B.E.; Montcouquiol, M.E.; Layman, E.M.; Lewandoski, M.; Kelley, M.W. Fgf8 induces pillar cell fate and regulates cellular patterning in the mammalian cochlea. *Development* **2007**, *134*, 3021–3029. [CrossRef]

47. Hayashi, T.; Ray, C.A.; Bermingham-McDonogh, O. Fgf20 is required for sensory epithelial specification in the developing cochlea. *J. Neurosci.* **2008**, *28*, 5991–5999. [CrossRef] [PubMed]

48. Hayashi, T.; Cunningham, D.; Bermingham-McDonogh, O. Loss of Fgfr3 leads to excess hair cell development in the mouse organ of Corti. *Dev. Dyn. Off. Publ. Am. Assoc. Anat.* **2007**, *236*, 525–533. [CrossRef]

49. Kelley, M.W. Regulation of cell fate in the sensory epithelia of the inner ear. Nature reviews. *Neuroscience* **2006**, *7*, 837–849. [CrossRef] [PubMed]

50. Maass, J.C.; Gu, R.; Basch, M.L.; Waldhaus, J.; Lopez, E.M.; Xia, A.; Oghalai, J.S.; Heller, S.; Groves, A.K. Changes in the regulation of the Notch signaling pathway are temporally correlated with regenerative failure in the mouse cochlea. *Front. Cell. Neurosci.* **2015**, *9*, 110. [CrossRef] [PubMed]
51. Walters, B.J.; Yamashita, T.; Zuo, J. Sox2-CreER mice are useful for fate mapping of mature, but not neonatal, cochlear supporting cells in hair cell regeneration studies. *Sci. Rep.* **2015**, *5*, 11621. [CrossRef]
52. McGovern, M.M.; Zhou, L.; Randle, M.R.; Cox, B.C. Spontaneous hair cell regeneration is prevented by increased notch signaling in supporting cells. *Front. Cell. Neurosci.* **2018**, *12*, 120. [CrossRef]
53. Jansson, L.; Kim, G.S.; Cheng, A.G. Making sense of Wnt signaling—linking hair cell regeneration to development. *Front. Cell. Neurosci.* **2015**, *9*, 66. [CrossRef]
54. Chai, R.; Kuo, B.; Wang, T.; Liaw, E.J.; Xia, A.; Jan, T.A.; Liu, Z.; Taketo, M.M.; Oghalai, J.S.; Nusse, R. Wnt signaling induces proliferation of sensory precursors in the postnatal mouse cochlea. *Proc. Natl. Acad. Sci. USA* **2012**, *109*, 8167–8172. [CrossRef] [PubMed]
55. Shi, F.; Hu, L.; Edge, A.S. Generation of hair cells in neonatal mice by β-catenin overexpression in Lgr5-positive cochlear progenitors. *Proc. Natl. Acad. Sci. USA* **2013**, *110*, 13851–13856. [CrossRef]
56. Jan, T.A.; Chai, R.; Sayyid, Z.N.; van Amerongen, R.; Xia, A.; Wang, T.; Sinkkonen, S.T.; Zeng, Y.A.; Levin, J.R.; Heller, S. Tympanic border cells are Wnt-responsive and can act as progenitors for postnatal mouse cochlear cells. *Development* **2013**, *140*, 1196–1206. [CrossRef]
57. Wu, J.; Li, W.; Lin, C.; Chen, Y.; Cheng, C.; Sun, S.; Tang, M.; Chai, R.; Li, H. Co-regulation of the Notch and Wnt signaling pathways promotes supporting cell proliferation and hair cell regeneration in mouse utricles. *Sci. Rep.* **2016**, *6*, 1–16. [CrossRef]
58. Li, W.; Wu, J.; Yang, J.; Sun, S.; Chai, R.; Chen, Z.-Y.; Li, H. Notch inhibition induces mitotically generated hair cells in mammalian cochleae via activating the Wnt pathway. *Proc. Natl. Acad. Sci. USA* **2015**, *112*, 166–171. [CrossRef] [PubMed]
59. Ni, W.; Lin, C.; Guo, L.; Wu, J.; Chen, Y.; Chai, R.; Li, W.; Li, H. Extensive supporting cell proliferation and mitotic hair cell generation by in vivo genetic reprogramming in the neonatal mouse cochlea. *J. Neurosci.* **2016**, *36*, 8734–8745. [CrossRef]
60. Romero-Carvajal, A.; Acedo, J.N.; Jiang, L.; Kozlovskaja-Gumbrienė, A.; Alexander, R.; Li, H.; Piotrowski, T. Regeneration of sensory hair cells requires localized interactions between the Notch and Wnt pathways. *Dev. Cell* **2015**, *34*, 267–282. [CrossRef]
61. Menendez, L.; Trecek, T.; Gopalakrishnan, S.; Tao, L.; Markowitz, A.L.; Haoze, V.Y.; Wang, X.; Llamas, J.; Huang, C.; Lee, J. Generation of inner ear hair cells by direct lineage conversion of primary somatic cells. *eLife* **2020**, *9*, e55249. [CrossRef] [PubMed]
62. Li, S.; Qian, W.; Jiang, G.; Ma, Y. Transcription factors in the development of inner ear hair cells. *Front. Biosci.* **2016**, *21*, 1118–1125. [CrossRef]
63. Gálvez, H.; Abelló, G.; Giraldez, F. Signaling and transcription factors during inner ear development: The generation of hair cells and otic neurons. *Front. Cell Dev. Biol.* **2017**, *5*, 21. [CrossRef] [PubMed]
64. Venkatesh, I.; Blackmore, M.G. Selecting optimal combinations of transcription factors to promote axon regeneration: Why mechanisms matter. *Neurosci. Lett.* **2017**, *652*, 64–73. [CrossRef]
65. Dhara, S.P.; Rau, A.; Flister, M.J.; Recka, N.M.; Laiosa, M.D.; Auer, P.L.; Udvadia, A.J. Cellular reprogramming for successful CNS axon regeneration is driven by a temporally changing cast of transcription factors. *Sci. Rep.* **2019**, *9*, 1–12. [CrossRef] [PubMed]
66. Begeman, I.J.; Kang, J. Transcriptional Programs and Regeneration Enhancers Underlying Heart Regeneration. *J. Cardiovasc. Dev. Dis.* **2019**, *6*, 2. [CrossRef]
67. Kirjavainen, A.; Sulg, M.; Heyd, F.; Alitalo, K.; Ylä-Herttuala, S.; Möröy, T.; Petrova, T.V.; Pirvola, U. Prox1 interacts with Atoh1 and Gfi1, and regulates cellular differentiation in the inner ear sensory epithelia. *Dev. Biol.* **2008**, *322*, 33–45. [CrossRef]
68. Sayyid, Z.N.; Wang, T.; Chen, L.; Jones, S.M.; Cheng, A.G. Atoh1 directs regeneration and functional recovery of the mature mouse vestibular system. *Cell Rep.* **2019**, *28*, 312–324.e4. [CrossRef]
69. Chonko, K.T.; Jahan, I.; Stone, J.; Wright, M.C.; Fujiyama, T.; Hoshino, M.; Fritzsch, B.; Maricich, S.M. Atoh1 directs hair cell differentiation and survival in the late embryonic mouse inner ear. *Dev. Biol.* **2013**, *381*, 401–410. [CrossRef]
70. Walters, B.J.; Zuo, J. A Sox10 rtTA/+ Mouse Line Allows for Inducible Gene Expression in the Auditory and Balance Organs of the Inner Ear. *J. Assoc. Res. Otolaryngol.* **2015**, *16*, 331–345. [CrossRef] [PubMed]
71. Cox, B.C.; Dearman, J.A.; Brancheck, J.; Zindy, F.; Roussel, M.F.; Zuo, J. Generation of Atoh1-rtTA transgenic mice: A tool for inducible gene expression in hair cells of the inner ear. *Sci. Rep.* **2014**, *4*, 6885. [CrossRef]
72. Wiwatpanit, T.; Lorenzen, S.M.; Cantu, J.A.; Foo, C.Z.; Hogan, A.K.; Marquez, F.; Clancy, J.C.; Schipma, M.J.; Cheatham, M.A.; Duggan, A. Trans-differentiation of outer hair cells into inner hair cells in the absence of INSM1. *Nature* **2018**, *563*, 691–695. [CrossRef]
73. Chessum, L.; Matern, M.S.; Kelly, M.C.; Johnson, S.L.; Ogawa, Y.; Milon, B.; McMurray, M.; Driver, E.C.; Parker, A.; Song, Y. Helios is a key transcriptional regulator of outer hair cell maturation. *Nature* **2018**, *563*, 696–700. [CrossRef]
74. Xu, Z.; Rai, V.; Zuo, J. TUB and ZNF532 Promote the Atoh1-Mediated Hair Cell Regeneration in Mouse Cochleae. *Front. Cell. Neurosci.* **2021**, *15*, 759223. [CrossRef] [PubMed]
75. Hoa, M.; Olszewski, R.; Li, X.; Taukulis, I.; Gu, S.; DeTorres, A.; Lopez, I.A.; Linthicum Jr, F.H.; Ishiyama, A.; Martin, D. Characterizing Adult cochlear supporting cell transcriptional diversity using single-cell RNA-Seq: Validation in the adult mouse and translational implications for the adult human cochlea. *Front. Mol. Neurosci.* **2020**, *13*, 13. [CrossRef]
76. Cheng, C.; Wang, Y.; Guo, L.; Lu, X.; Zhu, W.; Muhammad, W.; Zhang, L.; Lu, L.; Gao, J.; Tang, M. Age-related transcriptome changes in Sox2+ supporting cells in the mouse cochlea. *Stem Cell Res. Ther.* **2019**, *10*, 1–18. [CrossRef] [PubMed]

77. McGovern, M.M.; Randle, M.R.; Cuppini, C.L.; Graves, K.A.; Cox, B.C. Multiple supporting cell subtypes are capable of spontaneous hair cell regeneration in the neonatal mouse cochlea. *Development* **2019**, *146*, dev171009. [CrossRef]
78. Liu, Z.; Fang, J.; Dearman, J.; Zhang, L.; Zuo, J. In vivo generation of immature inner hair cells in neonatal mouse cochleae by ectopic Atoh1 expression. *PLoS ONE* **2014**, *9*, e89377. [CrossRef]
79. Kempfle, J.S.; Turban, J.L.; Edge, A.S. Sox2 in the differentiation of cochlear progenitor cells. *Sci. Rep.* **2016**, *6*, 23293. [CrossRef] [PubMed]
80. Cai, T.; Jen, H.-I.; Kang, H.; Klisch, T.J.; Zoghbi, H.Y.; Groves, A.K. Characterization of the transcriptome of nascent hair cells and identification of direct targets of the Atoh1 transcription factor. *J. Neurosci.* **2015**, *35*, 5870–5883. [CrossRef]
81. Jen, H.-I.; Hill, M.C.; Tao, L.; Sheng, K.; Cao, W.; Zhang, H.; Haoze, V.Y.; Llamas, J.; Zong, C.; Martin, J.F. Transcriptomic and epigenetic regulation of hair cell regeneration in the mouse utricle and its potentiation by Atoh1. *Elife* **2019**, *8*, e44328. [CrossRef]
82. Li, H. *Epigenetic and Signaling Pathway Regulation in Hair Cell Regeneration*; iMedPub LTD: London, UK, 2015.
83. Layman, W.S.; Zuo, J. Epigenetic regulation in the inner ear and its potential roles in development, protection, and regeneration. *Front. Cell. Neurosci.* **2015**, *8*, 446. [CrossRef]
84. Doetzlhofer, A.; Avraham, K.B. Insights into inner ear-specific gene regulation: Epigenetics and non-coding RNAs in inner ear development and regeneration. *Semin. Cell Dev. Biol.* **2017**, *65*, 69–79. [CrossRef]
85. Stojanova, Z.P.; Kwan, T.; Segil, N. Epigenetic regulation of Atoh1 guides hair cell development in the mammalian cochlea. *Development* **2016**, *143*, 1632. [CrossRef]
86. Zhou, Y.; Hu, Z. Epigenetic DNA demethylation causes inner ear stem cell differentiation into hair cell-like cells. *Front. Cell. Neurosci.* **2016**, *10*, 185. [CrossRef]
87. Layman, W.; Williams, D.; Dearman, J.; Sauceda, M.; Zuo, J. Histone deacetylase inhibition protects hearing against acute ototoxicity by activating the Nf-κ B pathway. *Cell Death Discov.* **2015**, *1*, 1–7. [CrossRef]
88. Matys, V.; Kel-Margoulis, O.V.; Fricke, E.; Liebich, I.; Land, S.; Barre-Dirrie, A.; Reuter, I.; Chekmenev, D.; Krull, M.; Hornischer, K. TRANSFAC® and its module TRANSCompel®: Transcriptional gene regulation in eukaryotes. *Nucleic Acids Res.* **2006**, *34*, D108–D110. [CrossRef]
89. Portales-Casamar, E.; Thongjuea, S.; Kwon, A.T.; Arenillas, D.; Zhao, X.; Valen, E.; Yusuf, D.; Lenhard, B.; Wasserman, W.W.; Sandelin, A. JASPAR 2010: The greatly expanded open-access database of transcription factor binding profiles. *Nucleic Acids Res.* **2010**, *38*, D105–D110. [CrossRef] [PubMed]
90. Kulakovskiy, I.V.; Medvedeva, Y.A.; Schaefer, U.; Kasianov, A.S.; Vorontsov, I.E.; Bajic, V.B.; Makeev, V.J. HOCOMOCO: A comprehensive collection of human transcription factor binding sites models. *Nucleic Acids Res.* **2013**, *41*, D195–D202. [CrossRef] [PubMed]
91. Weirauch, M.T.; Yang, A.; Albu, M.; Cote, A.G.; Montenegro-Montero, A.; Drewe, P.; Najafabadi, H.S.; Lambert, S.A.; Mann, I.; Cook, K. Determination and inference of eukaryotic transcription factor sequence specificity. *Cell* **2014**, *158*, 1431–1443. [CrossRef]
92. Janky, R.; Verfaillie, A.; Imrichova, H.; Van de Sande, B.; Standaert, L.; Christiaens, V.; Hulselmans, G.; Herten, K.; Naval Sanchez, M.; Potier, D.; et al. iRegulon: From a gene list to a gene regulatory network using large motif and track collections. *PLoS Comput. Biol.* **2014**, *10*, e1003731. [CrossRef] [PubMed]
93. Sun, S.; Li, S.; Luo, Z.; Ren, M.; He, S.; Wang, G.; Liu, Z. Dual expression of Atoh1 and Ikzf2 promotes transformation of adult cochlear supporting cells into outer hair cells. *Elife* **2021**, *10*, e66547. [CrossRef]
94. Qian, D.; Radde-Gallwitz, K.; Kelly, M.; Tyrberg, B.; Kim, J.; Gao, W.Q.; Chen, P. Basic helix–loop–helix gene Hes6 delineates the sensory hair cell lineage in the inner ear. *Dev. Dyn. Off. Publ. Am. Assoc. Anat.* **2006**, *235*, 1689–1700. [CrossRef]
95. Sölter, M.; Locker, M.; Boy, S.; Taelman, V.; Bellefroid, E.J.; Perron, M.; Pieler, T. Characterization and function of the bHLH-O protein XHes2: Insight into the mechanisms controlling retinal cell fate decision. *Development* **2006**, *133*, 4097–4108. [CrossRef] [PubMed]
96. Bae, S.; Bessho, Y.; Hojo, M.; Kageyama, R. The bHLH gene Hes6, an inhibitor of Hes1, promotes neuronal differentiation. *Development* **2000**, *127*, 2933–2943. [CrossRef]
97. Kousa, Y.A.; Zhu, H.; Fakhouri, W.D.; Lei, Y.; Kinoshita, A.; Roushangar, R.R.; Patel, N.K.; Agopian, A.; Yang, W.; Leslie, E.J. The TFAP2A–IRF6–GRHL3 genetic pathway is conserved in neurulation. *Hum. Mol. Genet.* **2019**, *28*, 1726–1737. [CrossRef] [PubMed]
98. Klisch, T.J.; Xi, Y.; Flora, A.; Wang, L.; Li, W.; Zoghbi, H.Y. In vivo Atoh1 targetome reveals how a proneural transcription factor regulates cerebellar development. *Proc. Natl. Acad. Sci. USA* **2011**, *108*, 3288–3293. [CrossRef] [PubMed]
99. Kobayashi, A.; Valerius, M.T.; Mugford, J.W.; Carroll, T.J.; Self, M.; Oliver, G.; McMahon, A.P. Six2 defines and regulates a multipotent self-renewing nephron progenitor population throughout mammalian kidney development. *Cell Stem Cell* **2008**, *3*, 169–181. [CrossRef] [PubMed]

International Journal of
Molecular Sciences

Article

Identification of Novel Micropeptides Derived from Hepatocellular Carcinoma-Specific Long Noncoding RNA

Mareike Polenkowski [1,†], Sebastian Burbano de Lara [1,2,†], Aldrige Bernardus Allister [1], Thi Nhu Quynh Nguyen [1], Teruko Tamura [1,‡] and Doan Duy Hai Tran [1,*]

1   Institut fuer Zellbiochemie, OE4310, Medizinische Hochschule Hannover, Carl-Neuberg-Str. 1,
    30623 Hannover, Germany; Polenkowski.Mareike@mh-hannover.de (M.P.);
    sebastian.burbanodelaracarrillo@dkfz-heidelberg.de (S.B.d.L.); Allister.Bernardus@mh-hannover.de (A.B.A.);
    Nguyen.quynh@t-online.de (T.N.Q.N.)
2   Systems Biology of Signal Transduction B200, German Cancer Research Center (DKFZ),
    Im Neuenheimer Feld 280, 69120 Heidelberg, Germany
*   Correspondence: Tran.Doan@MH-Hannover.de; Tel.: +49-511-532-2857
†   These authors contributed equally to this work.
‡   This paper is dedicated in loving memory to Teruko Tamura.

Abstract: Identification of cancer-specific target molecules and biomarkers may be useful in the development of novel treatment and immunotherapeutic strategies. We have recently demonstrated that the expression of long noncoding (lnc) RNAs can be cancer-type specific due to abnormal chromatin remodeling and alternative splicing. Furthermore, we identified and determined that the functional small protein C20orf204-189AA encoded by long intergenic noncoding RNA Linc00176 that is expressed predominantly in hepatocellular carcinoma (HCC), enhances transcription of ribosomal RNAs and supports growth of HCC. In this study we combined RNA-sequencing and polysome profiling to identify novel micropeptides that originate from HCC-specific lncRNAs. We identified nine lncRNAs that are expressed exclusively in HCC cells but not in the liver or other normal tissues. Here, DNase-sequencing data revealed that the altered chromatin structure plays a key role in the HCC-specific expression of lncRNAs. Three out of nine HCC-specific lncRNAs contain at least one open reading frame (ORF) longer than 50 amino acid (aa) and enriched in the polysome fraction, suggesting that they are translated. We generated a peptide specific antibody to characterize one candidate, NONHSAT013026.2/Linc013026. We show that Linc013026 encodes a 68 amino acid micropeptide that is mainly localized at the perinuclear region. Linc013026-68AA is expressed in a subset of HCC cells and plays a role in cell proliferation, suggesting that Linc013026-68AA may be used as a HCC-specific target molecule. Our finding also sheds light on the role of the previously ignored 'dark proteome', that originates from noncoding regions in the maintenance of cancer.

Keywords: HCC-specific small functional protein; NONHSAT013026.2/Linc013026-68AA; fine tuner of cancer formation; dark proteome; hepatocellular carcinoma

Citation: Polenkowski, M.; Burbano de Lara, S.; Allister, A.B.; Nguyen, T.N.Q.; Tamura, T.; Tran, D.D.H. Identification of Novel Micropeptides Derived from Hepatocellular Carcinoma-Specific Long Noncoding RNA. *Int. J. Mol. Sci.* **2022**, *23*, 58. https://doi.org/10.3390/ijms23010058

Academic Editors: Amelia Casamassimi, Alfredo Ciccodicola and Monica Rienzo

Received: 25 November 2021
Accepted: 20 December 2021
Published: 21 December 2021

Publisher's Note: MDPI stays neutral with regard to jurisdictional claims in published maps and institutional affiliations.

1. Introduction

Hepatocellular carcinoma (HCC) is one of the most prevalent tumor types worldwide [1]; however, current treatment options are limited, and precise and effective medical strategies for therapy do not exist [2]. HCC typically occurs on a background of chronic liver disease, with risk factors including viral or autoimmune hepatitis, chronic alcohol abuse, and nonalcoholic fatty liver disease [3]. These risk factors trigger aberrant liver regeneration, which initiates the formation of HCC. However, the underlying molecular mechanism is still largely unknown. It has been recently shown by exome sequencing of HCC that 161 putative driver genes are associated with 11 recurrently altered pathways in HCC development, suggesting that many signaling pathways are altered to a modest degree, and act together [4–6]. Notably, 28% of altered gene products are involved in a chromatin-remodeling complex, suggesting that HCC expresses unique genes that are not expressed in normal hepatocytes. In this context, we have previously shown that a subset of lncRNAs that are predominantly expressed in HCC plays a role as fine tuners in cancer formation and/or maintenance [7,8]. Thus, a potential strategy for cancer therapy may be to target multiple cancer type-specific fine tuners including noncoding RNA.

Traditional annotation of protein-encoding genes relied on assumptions, such as one open reading frame (ORF) encodes one protein and minimal lengths for translated proteins [9]. However, recent data from our laboratory and from others have revealed that RNAs previously considered noncoding, such as long noncoding RNAs (lncRNAs) and circular RNAs are translated into functional small proteins [10–13], suggesting that the proteome is more complex than previously anticipated.

In the present study we utilized RNA sequencing (RNA-seq) and polysome profiling to identify novel micropeptides that originate from HCC-specific lncRNAs. We identified two HCC-specific lncRNAs that are translated into small ORFs. Applying a peptide specific antibody we characterized one lncRNA candidate, NONHSAT013026.2/Linc013026-68AA. Linc013026-68AA is translated into a 68 amino acid micropeptide that is mainly localized at the perinuclear region. Notably, Linc013026-68AA is predominantly expressed in moderately- but not well-differentiated HCC cells and plays a role in cell proliferation, suggesting that Linc013026-68AA may be used as a HCC-specific target molecule. Our data also uncover the important role of previously ignored small ORFs originating from noncoding regions in the maintenance of cancer.

2. Results

2.1. Identification of Hepatocellular Carcinoma-Specific lncRNAs

To identify lncRNAs that are expressed in HCC cells but not in normal hepatocytes, we used publicly available RNA-sequencing (RNA-seq) data generated by the ENCODE Consortium [14] to extract the expression level of lncRNAs. Firstly, we mapped RNA-seq data from normal liver (ENCFF184YUO) and from the HCC cell line HepG2 (ENCFF337WTM) to long intergenic non-coding RNAs (lincRNA) annotated by NONCODE v5.0, an integrated knowledge database of non-coding RNAs [15] (Figure 1A). We limited our study to lincRNAs, because RNA-seq based on second generation sequencing limits the accurate allocation of reads if lncRNAs overlap with coding genes. RNA-seq datasets were normalized using cuffnorm [16]. A total of 906 lincRNAs that expressed only in HepG2 cells but not in the liver were selected. To identify HCC-specific lincRNAs we further examined the expression of our lincRNA candidates in normal tissues including adipose, adrenal, brain, breast, colon, foreskin, heart, kidney, lung, ovary, placenta, prostate, skeletal muscle, testis and thyroid tissues and leukocytes using RNA-seq data generated by the Human Body Map 2.0 [17]. Twelve out of 906 lincRNAs were expressed exclusively in HepG2 cells (Figure 1A, Table S1). We then confirmed the expression of these lincRNAs in HepG2 cells using our previously published RNA-seq data [7]. The expression of nine out of twelve HCC-specific lincRNAs was confirmed (Figure 1B, HepG2 (GSE115139)).

We next asked why these lincRNAs are expressed exclusively in HepG2 cells but not in the liver. We have previously demonstrated that altered chromatin structure in cancer results in the cancer-specific expression of a subset of genes [7,8]. Thus, we examined the chromatin structure at the putative promotor region of nine HCC-specific lincRNAs using DNase-sequencing data (DNase-seq) generated by the ENCODE Consortium [18]. DNase-seq data (Figure 1B, DNase-seq) obtained from human hepatocyte (ENCFF851CVH) and HepG2 (ENCFF474LSZ) revealed that HepG2 contains DNase I hypersensitive sites at the proximal promoter region of seven out of nine lincRNAs (except for NONHSAT204527.1 and NONHSAT223630.1 (Figure 1C)), while normal human liver does not contain these sites at these positions (Figure 1B, blue arrow), suggesting that the chromatin structure in this region is remodeled in HCC cells. To examine whether these open chromatin regions are associated with cis-regulatory elements, we utilized candidate cis-Regulatory Elements (cCREs) database generated by ENCODE consortium which contain 1,063,878 human cCREs [19]. Notably, putative promoter regions of five out of these seven lncRNA candidates contains at least one cCRE (Figure 1B, ENCODE cCRE, blue mark). In addition, ChIP-seq of H3K4 trimethylation and H3K27 acetylation, chromatin marks of active transcription revealed that these putative promoter regions of seven lncRNA candidates are transcriptionally active in HepG2 cells (Figure S2, H3K4me3 and H3K27Ac). These data suggest that transcription may be initiated from these regions. Thus, we utilized the cap analysis of gene expression (CAGE) data in HepG2 cells that mapped the transcription start sites (Figure 1B, CAGE). In agreement with RNA-seq data, the cap site is located at the open promoter regions determined by DNase-seq data (Figure 1B, Transcription Start Site (TSS-black arrow)), suggesting that altered chromatin structure plays a key role in the HCC-specific expression of lincRNAs.

In addition to chromatin structure, tissue-specific transcription factors (TFs) are well known to activate tissue-specific expression program [20]. Thus, we next examined which transcription factor potentially activates the transcription of HCC-specific lncRNA genes by utilizing ChIP-seq datasets of 340 factors generated by ENCODE consortium in HepG2 cells. All HCC-specific lncRNA genes are potentially activated by three to seven TFs (Figure S2). Notably, these TFs are expressed in both normal liver and primary HCC (Figure S3), suggesting that open chromatin structure at the promoter region rather than a transcription factor may play a key role in the HCC-specific expression of lncRNAs.

2.2. Identification of Micropeptide Candidates Derived from HCC-Specific lncRNAs

Six out of nine lincRNAs contain at least one open reading frame (ORF) that is longer than 50 amino acids (AA) (Table 1). To be translated into micropeptides, lincRNAs have to be exported to the cytoplasm. Thus, we examined the mRNA export of these 6 lincRNA candidates using nuclear- and cytoplasmic RNA-seq generated by the ENCODE Consortium. Except for NONHSAT142412.2 the other five lincRNAs were clearly detected in cytoplasmic RNA-seq (Figure 2A), suggesting that they are exported to the cytoplasm. We also confirmed the mRNA export using RT-PCR (Figure 2B). To examine whether these five lincRNA candidates are endogenously translated in HepG2 cells, we isolated the polysome fraction of HepG2 cells using sucrose gradient centrifugation [11] and performed qRT-PCR and RT-PCR for five lincRNAs. Actin mRNA was used as a positive control. Three out of five lincRNA candidates were detected in translated fractions of HepG2 cells (Figures 2C and S1, NONHSAT013026.2, NONHSAT168790.1 and NONHSAT250607.1), suggesting that they are translated.

2.3. NONHSAT013026.2/Linc013026-68AA Is Translated into a 68 Amino Acid Long Micropeptide

Among three lincRNA candidates that were translated, NONHSAT013026.2 had the highest degree of enrichment in a translated fraction, thus we further focused on the characterization of this lincRNA which we renamed Linc013026. Linc013026 potentially encodes two ORFs of 52AA and 68AA. Since ORF-68AA has a Kozak consensus sequence, we further focused on this ORF. First, we examined whether Linc013026-68AA is translated into a stable micropeptide using an in vitro transcription/translation assay and overexpression in cells. Linc013026-68AA is predicted to encode a micropeptide of 8 kDa [21]. The in vitro transcription/translation assay with Linc013026-68AA revealed a single band at 8–10 kDa (Figure 3A, arrow). Furthermore, to examine whether Linc013026-68AA protein is stable in cells, we transfected HeLa cells with C-terminal-GFP- and Myc-tagged Linc013026-68AA. GFP-specific immunoblot revealed a band at ~38 kDa for GFP-tagged Linc013026-68AA that corresponds to a molecular mass of 10 kDa for Linc013026-68AA (Figure 3B, 68AA-GFP), while a band of ~15 kDa was observed for Myc-tagged Linc013026-68AA (Figure 3C). Since some proteins can form a dimer in SDS-PAGE (0.1% SDS) [22], we utilized GST pull down assay to examine the interaction of N-terminal GST -tagged 68AA with C-terminal Myc-tagged 68AA. As shown in Figure 1D, no interaction between GST-68AA and 68AA-Myc was detected. In addition, we also observed a ~15 kDa band for 68AA-Myc using cell lysates pre-treated with 1% and 2% SDS under reduced condition (Figure 3E). These data suggested that Myc-tagged Linc013026-68AA did not form a dimer.

We then examined the peptide sequence of Linc013026-68AA. Linc013026-68AA contains five potential serine, two threonine and one tyrosine phosphorylation sites (Table S3) and one lysine acetylation site (G-AcK) [23]. To clarify whether phosphorylation affects the migration of Linc013026-68AA in SDS PAGE, we treated cell lysates with Lambda Protein Phosphatase (Lambda PP) that dephosphorylates phospho-tyrosine, serine and threonine residues. Upon Lambda PP treatment, we observed two additional bands of ~14 and ~10 kDa (Figure 3F (*)), suggesting that phosphorylation contributes to the slower migration of Linc013026-68AA in SDS PAGE.

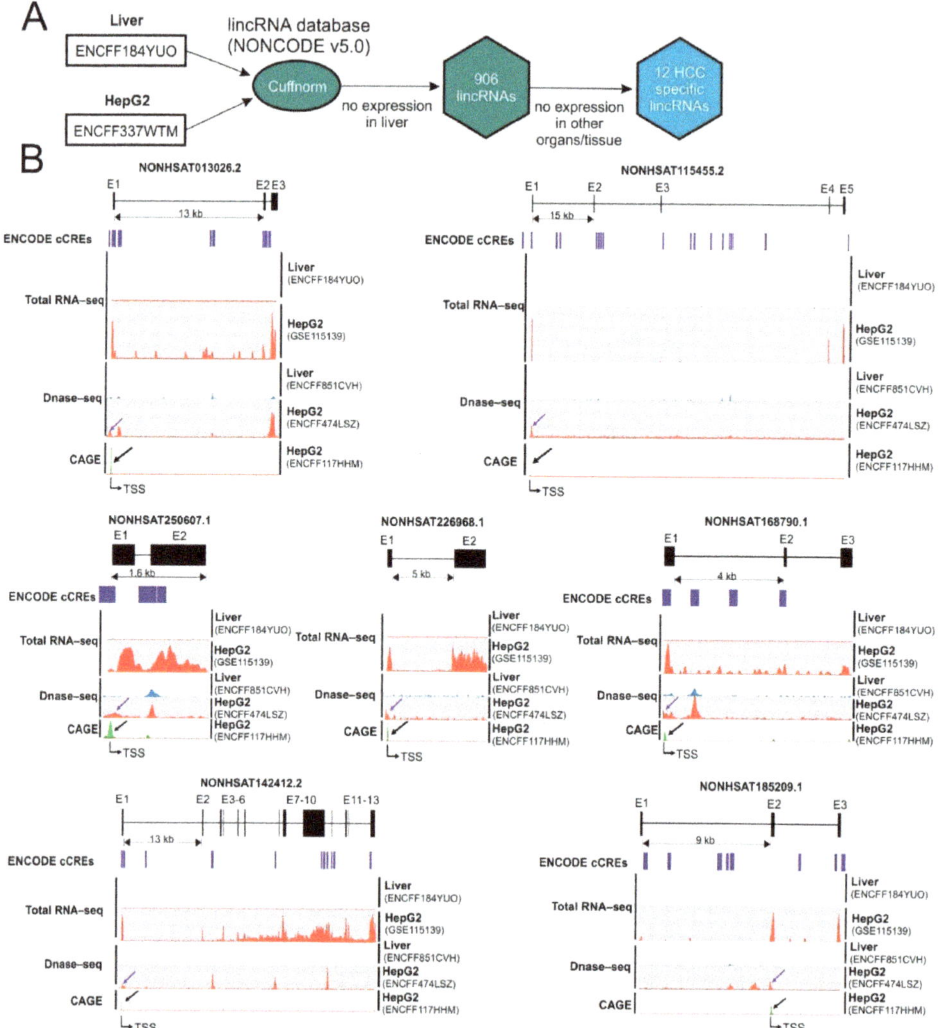

Figure 1. *Cont.*

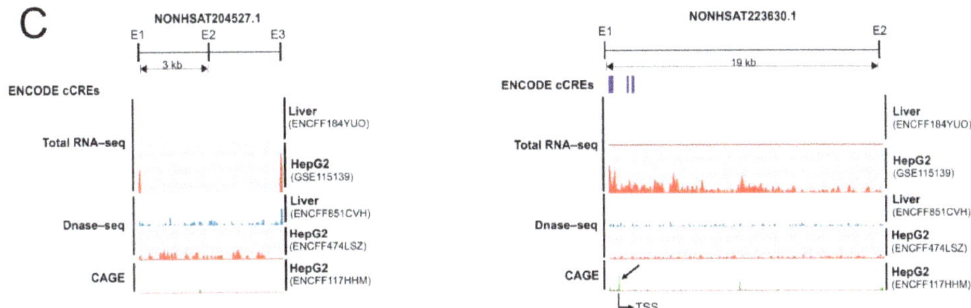

Figure 1. Identification of hepatocellular carcinoma-specific lncRNAs. (**A**) RNA-sequencing (RNA-seq) data from normal liver (ENCFF184YUO) and from the HCC cell line HepG2 (ENCFF337WTM) were aligned to the human reference genome (GRCh38) using Bowtie2. The gene expression values (Fragments per Kilobase Million (FPKM)) were calculated by Cuffnorm using the human NONCODE v5.0 transcript reference. A total of 906 long intergenic noncoding RNAs (lincRNAs) that were expressed in HepG2 cells but not in the normal liver were selected. Among these 906 lincRNAs twelve lincRNAs were expressed exclusively in HepG2 cells but not in the liver and other organs/tissues. (**B,C**) Abnormal chromatin structure induced expression of HCC-specific lincRNAs: Total RNA-seq from liver (ENCFF184YUO) and HepG2 cells (GSE115139), DNase-sequencing (DNase-seq) from normal liver (ENCFF851CVH) and HepG2 (ENCFF474LSZ) and candidate cis-Regulatory Elements (cCREs) generated by ENCODE consortium (ENCODE cCREs, blue mark) and cap analysis of gene expression (CAGE) data in HepG2 cells (ENCFF177HHM) were aligned to the human reference genome (GRCh38). SeqMonk was used to quantitate and visualize the data. Peaks in the wiggle plot represent the normalized RNA-seq, DNase-seq and CAGE read coverage on HCC-specific lincRNAs. E: exon; blue arrow: Open chromatin region at the putative promoter; black arrow: transcription start site (TSS) detected by CAGE.

Table 1. List of six HCC-specific lincRNAs that contain at least one ORF longer than 50AA.

Transcript ID	Expression in the Liver (FPKM) (ENCFF184YUO)	Expression in HepG2 (FPKM) (ENCFF337WTM)	Number of ORFs Longer Than 50AA
NONHSAT226968.1	0	104.9	2
NONHSAT013026.2/Linc013026	0	61.3	2
NONHSAT250607.1	0	39.3	2
NONHSAT142412.2	0	30.5	12
NONHSAT115455.2	0	15.0	2
NONHSAT168790.1	0	13.7	3

To examine the endogenous expression of Linc013026-68AA we generated a rabbit antibody against two mixed synthetic peptides corresponding to amino acid positions 4–17 (peptide I) and 54–68 (peptide II) of Linc013026-68AA (Kaneka Eurogentec S.A. Belgium) (amino acid sequences are shown in Figure 3G). First, we tested the specificity of our antibodies. By immunoblot using anti-peptide I and peptide II antibodies, a 38 kDa band for GFP-tagged Linc013026-68AA was specifically detected (Figure 3H). This band was not detected by peptide absorbed antibody (Figure 3H, anti-peptide II + peptide II). We then examined the subcellular localization of exogenous and endogenous Linc013026-68AA using immunofluorescent (IF) and immunohistochemical (IHC) staining. HeLa cells were transfected with Myc-tagged Linc013026-68AA and stained using the immunofluorescent technique with anti-Linc013026-68AA and Myc-specific antibodies. Myc-tagged Linc013026-68AA was detected mainly at the perinuclear region by a Myc-specific staining (Figure 3I). Anti-peptide II but not peptide I antibody gave a strong IF staining signal that completely overlapped with the Myc-specific signal (Figure 3I, Merged). Next, we tested Linc013026-68AA antibodies for IHC staining. In agreement with IF staining, anti-peptide II but not peptide I antibody gave a strong signal for Myc-tagged Linc013026-68AA at the perinuclear region (Figure 3I).

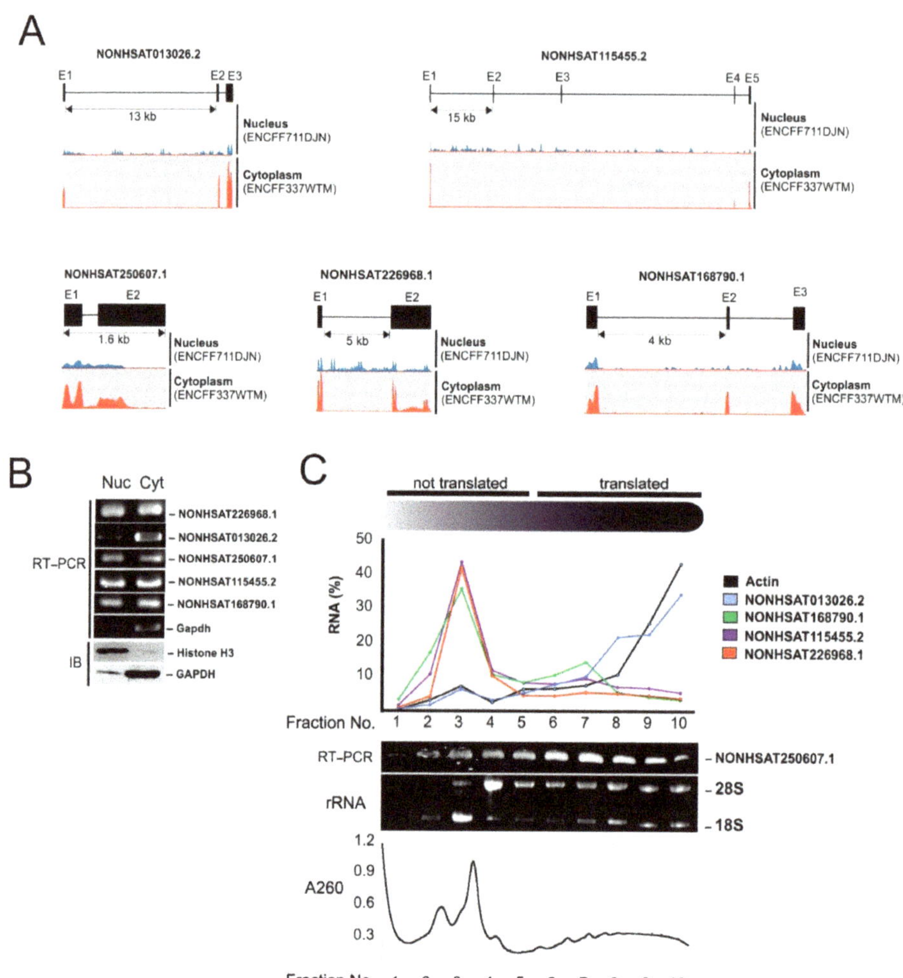

Figure 2. Identification of micropeptide candidates derived from HCC-specific lincRNAs. (**A**) Nuclear- (ENCFF711DJN) and cytoplasmic (ENCFF337WTM) RNA-seq of HepG2 cells generated by the ENCODE Consortium were aligned to the reference human genome (GRCh38). SeqMonk was used to quantitate and visualize the data. Peaks in the wiggle plot represent the normalized RNA-seq read coverage on HCC-specific lincRNAs. E: exon. (**B**) RNA was isolated from the nuclear (Nuc) and cytoplasmic (Cyt) fractions of HepG2 cells and analyzed by RT-PCR. Fractionation quality was measured by immunoblot analysis of THOC5, GAPDH and Histone H3 (Blot). Three independent experiments were performed. (**C**) HepG2 cytoplasmic lysate was prepared and fractionated on sucrose gradients. The distribution of RNA was calculated using the CT values obtained by qRT-PCR. Isolated RNA was supplied in a gel to determine translated fractions. mRNAs were prepared from the indicated fractions and were applied for Actin, NONHSAT013026.2, NONHSAT168790.1, NONHSAT115455.2, and NONHSAT226968.1 qRT-PCR or NONHSAT250607.1 RT-PCR. A representative absorbance profile at 260 nm was obtained during fractionation of gradients. A replicate is shown in Figure S1.

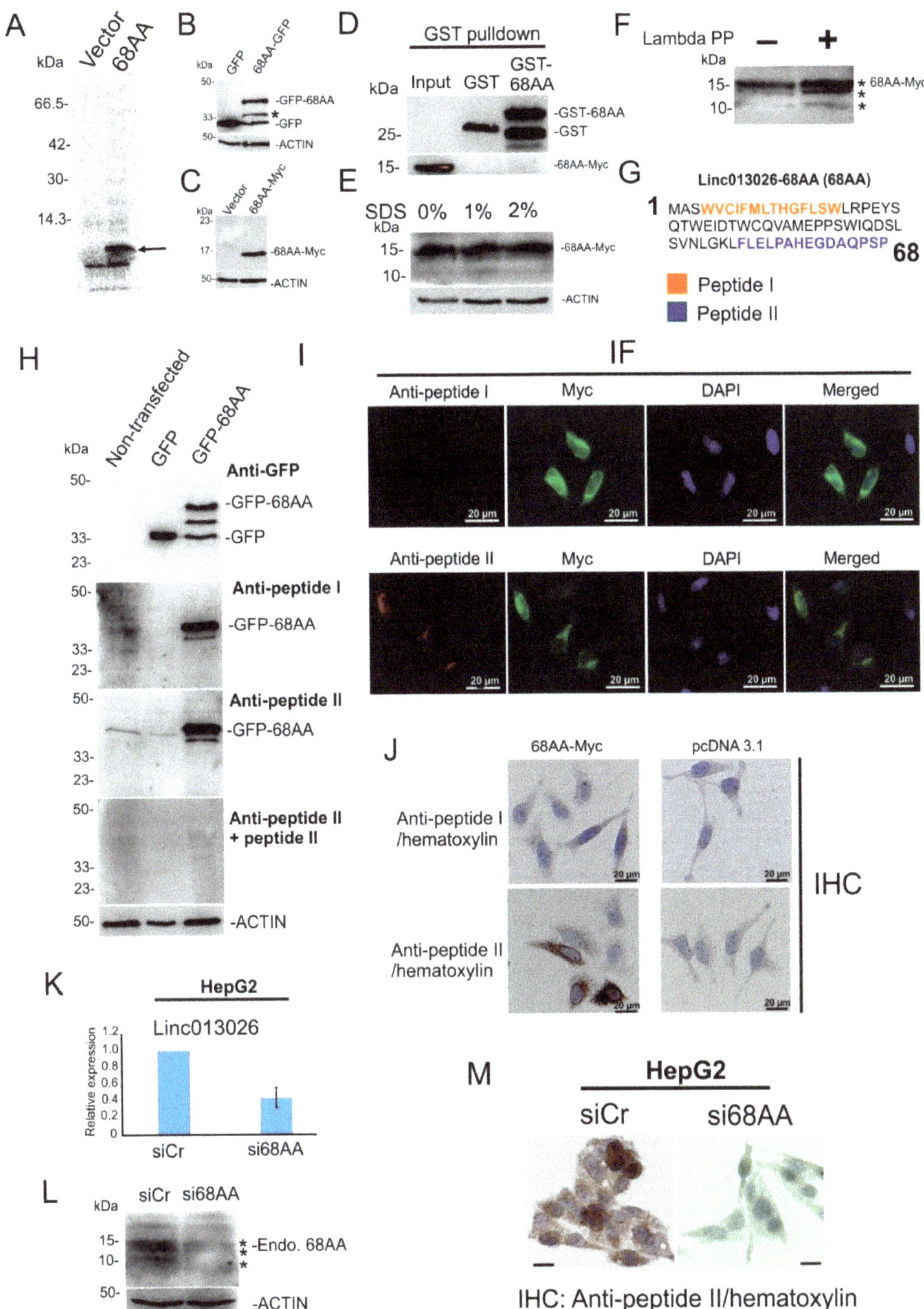

Figure 3. NONHSAT013026.2/Linc013026-68AA is translated into a 68 amino acid long micropeptide. (**A**) In vitro transcription/translation assay of Linc013026-68AA, proteins are labeled with

[^{35}S]methionine. Arrow indicated the translated peptide. (**B,C**) GFP- and Myc-tagged Linc013026-68AA (68AA) and pEGFP-N1 (GFP) or pcDNA3.1 MycHis (Vector) vector was transfected in HeLa cells, and GFP- and Myc-specific immunoblot were performed. (**D**) 68AA-Myc was overexpressed in HeLa cells. Cell extracts were incubated with GST and GST-68AA and then Myc and GST specific immunoblots were performed. (**E**) 68AA-Myc overexpressing cell extracts were pre-treated with 1% and 2% SDS and then Myc and ACTIN specific immunoblots were performed. (**F**) 68AA-Myc was overexpressed in HeLa cells. Forty microliters of cell extracts were treated with 2000 U Lambda Protein Phosphatase (Lambda PP) for 45 min at 30 °C and subsequently supplied for Linc013026-68AA specific immunoblot. (**G**) Amino acid sequence of Linc013026-68AA was depicted. Numbers represent amino acid number. Amino acid sequences of peptide I (orange) and peptide II (blue) were used to generate rabbit antibodies. (**H**) Non-transfected-, pEGFP-N1 (GFP) and GFP-tagged Linc013026-68AA (68AA-GFP) HeLa cell lysates were applied for GFP-, anti-peptide I and anti-peptide II or anti-peptide II + peptide II immunoblot. ACTIN was used as loading control. (**I**) Myc-tagged Linc013026-68AA was expressed in HeLa and stained with anti-Myc, anti-peptide I and anti-peptide II specific antibodies and visualized by the immunofluorescent (IF) technique. (**J**) Myc-tagged Linc013026-68AA (68AA-Myc) and pcDNA3.1 MycHis (Vector) vector were transfected in HeLa cells and immunohistochemically stained with rabbit antibody against Linc013026-68AA-peptide I and peptide II and hematoxylin. (**K**) HepG2 cells were transfected with siCr and siLinc013026-68AA for three days. Total RNAs were isolated and supplied for Linc013026 or Gapdh-specific qRT-PCR. The expression of Linc013026 was normalized by Gapdh. Three independent experiments were performed. Numbers are mean ± standard deviation (SD). (**L**) A sister culture of (**K**) was subjected to anti-peptide II and ACTIN specific immunoblot. (**M**) A sister culture of (**K**) was immunohistochemically stained with anti-peptide II and hematoxylin. All bars represent 20 μm. (*) changes of LINC013026-68AA migration in SDS-PAGE induced by phosphorylation.

Thus, we used anti-peptide II antibody to examine the endogenous expression of Linc013026-68AA in HepG2 cells. We first depleted Linc013026-68AA using siRNA in HepG2 cells (Figure 3K) and performed anti-peptide II specific immunoblot. In control cells, three major bands ranging from 10–15 kDa were detected (Figure 3L, (*)). Upon Linc013026-68AA depletion, the intensity of these bands was reduced. These data suggested that endogenous Linc013026-68AA may also be phosphorylated as observed for exogenous Linc013026-68AA (Figure 3F). We also tested the endogenous expression of Linc013026-68AA using immunohistochemical staining with anti-peptide II antibody. In control cells, endogenous Linc013026-68AA was detected mainly at the perinuclear region (Figure 3M, siCr) which agreed with the subcellular localization of exogenous Linc013026-68AA. Upon Linc013026-68AA depletion, staining signals of Linc013026-68AA were clearly reduced (Figure 3M, si68AA). In sum, immunoblotting and IHC staining suggested that Linc013026-68AA is endogenously translated into 68AA micropeptide.

2.4. Linc013026-68AA Enhances Cell Proliferation

Recent data from our lab suggested that protein derived from lncRNA associates with a biological function [11]. Since HeLa cells do not express Linc013026-68AA (Figure 3I), Myc-tagged Linc013026-68AA in HeLa cells was expressed (Figure 4A). We next examined whether Linc013026-68AA influences cell growth by crystal violet staining assay and Wst-1 assay. Here, within 2 days, growth of Linc013026-68AA-overexpressing HeLa cells was approximately 1.7-fold by crystal violet assay and 1.6-fold by WST assay greater than control vector transfected HeLa cells (Figure 4B,C). Furthermore, depletion of Linc013026 RNA in HepG2 cells reduced cell proliferation approximately two-fold within 3 days measured by crystal violet- (Figure 4D) and Wst-1 assay (Figure 4E), suggesting that Linc013026-68AA promotes cell proliferation. We next examined the Linc013026-68AA transcript in several HCC cell lines, such as HepG2, Hep3B, C3A, Huh7 and HLE. HeLa cells were used as negative control. Linc013026-68AA is expressed in two out of five HCC cell lines (Figure 4F). Thus, we overexpressed Linc013026-68AA in Huh7 and HLE cells, two HCC cell lines that express Linc013026-68AA at low level. These cells also showed an increase in proliferation (1.8-fold in Huh7 cells and 1.6-fold in HLE cells) (Figure 4G), suggesting again that Linc013026-68AA promotes cell proliferation.

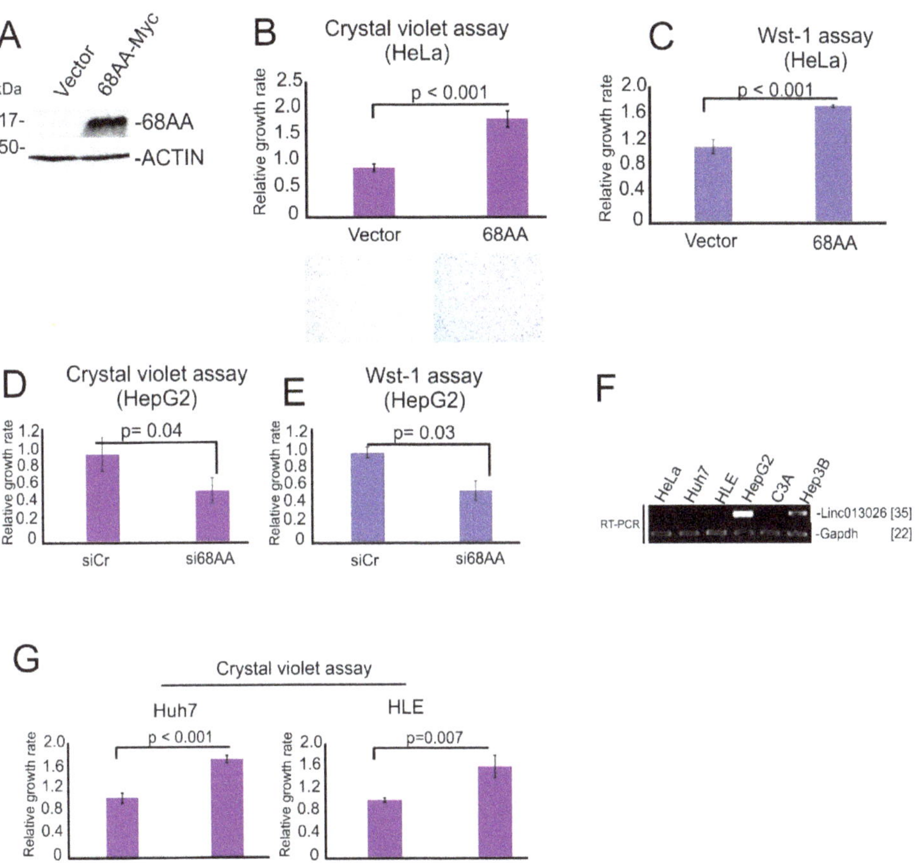

Figure 4. Linc013026-68AA enhances cell proliferation. (**A**) HeLa cells were transfected with Myc-tagged Linc013026-68AA (68AA-Myc) and pcDNA3.1 MycHis (Vector) vector for two days and anti-Linc013026-68AA- and Actin-specific immunoblot were performed. ACTIN was used as loading control. (**B,C**) Sister cultures of (**A**) were supplied for crystal violet- (**B**) and Wst-1 assay (**C**) to examine the effect of Linc013026-68AA on cell proliferation. (**D,E**) HepG2 cells were transfected with siRNA control (siCr) and siLinc013026 (si68AA) for three days and supplied for crystal violet- (**D**) and Wst-1 assay (**E**). (**F**) Total RNAs were isolated from HeLa, Huh7, HLE, HepG2, C3A and Hep3B cells. The expression of Linc013026 was examined using RT-PCR. Gapdh mRNA was used as loading control. [] indicates number of PCR cycles. (**G**) Huh7- and HepG2 cells were transfected with Myc-tagged Linc013026-68AA (68AA-Myc) and pcDNA3.1 MycHis (Vector) vector for two days and supplied for crystal violet- and Wst-1 assay. Three independent experiments were performed for crystal violet- and Wst-1 assay. Numbers are mean ± standard deviation (SD). p: *p*-value.

3. Discussion

In most human cancers, a large number of proteins with driver mutations are involved in tumor development, implying that multiple fine tuners are involved in cancer formation and/or maintenance. A useful strategy for cancer therapy may therefore be to target multiple cancer-specific fine tuners. In this study, using hepatocellular carcinoma as a system we utilized RNA-seq and polysome profiling to identify novel micropeptides derived from cancer-specific lncRNAs. We identified nine lincRNAs that are exclusively expressed in HCC cells but not in normal liver and other tissues (Table S1). Three out of nine lincRNAs encode small ORFs longer than 50 amino acids and are enriched in polysome fractions (Figure 2C), suggesting that they are translated in a cancer-specific manner. Using a peptide specific antibody we characterized NONHSAT013026.2/Linc013026-68AA, one of our candidates. We show that Linc013026-68AA encodes a 68 amino acid micropeptide that is mainly localized at the perinuclear region (Figure 3I,J,M). Linc013026-68AA is expressed in a subset of HCC cells and plays a role in cell proliferation (Figure 4). We are currently performing interactome analysis of Linc013026-68AA to gain insights into molecular mechanism(s) of Linc013026-68AA. It has been shown that a micropeptide is involved in muscle performance [12] and growth [13]. In addition, SPAR polypeptides encoded by the Linc00961 regulate mTORC1 and muscle regeneration [24], and another micropeptide, mitoregulin, is involved in protein complex assembly in mitochondria [25]. Recently, we demonstrated that C20orf204-189AA encoded by a lincRNA, Linc00176 stabilizes nucleolin and promotes ribosomal RNA transcription [11]. These findings shed light on the role of the previously ignored 'dark proteome' in the maintenance of cancer. Thus, further characterization of the coding potency of other cancer-specific lincRNAs (Table 1) may provide clues for identification of novel cancer-specific fine tuners. Furthermore, micropeptides encoded by cancer-specific lncRNAs may also be useful biomarkers for cancer diagnosis.

Why is the expression of a subset of lncRNAs cancer-specific? Recent data identified 161 putative driver genes that are associated with 11 recurrently altered pathways in HCC development [4], and these mutations were not observed in chronic hepatitis or cirrhosis (preneoplastic stages). Interestingly, 28% of the altered gene products play a role in chromatin remodeling, suggesting that abnormal chromatin remodeling results in a cancer-specific expression of a subset of genes [7,8,26]. Indeed, DNase-seq data which map the chromatin accessibility revealed that chromatin at the putative promotor region of seven out of nine HCC-specific lincRNAs is opened in HCC but not in normal liver (Figure 1B). Accessible promoters then enable the recruitment of transcription factors which subsequently activate the transcription in these genes. These data also suggest that cancer cells exhibit remarkable transcriptome alterations, partly by adopting cancer-specific chromatin remodeling events.

One of limitations of this study is the lack of clinical data of HCC-specific lncRNA candidates. Examining the expression of these lncRNAs in RNA-seq data of primary HCC generated by The Cancer Genome Atlas (TCGA) or The International Cancer Genome Consortium (ICGC) will provide clues whether they could be a potentially suitable HCC-specific biomarker. However, retrieving expression from open-access data resource requires the gene annotation by GENCODE [27], while many NONCODE lncRNA genes including lncRNA candidates identified in this study are not yet annotated by GENCODE [28]. Thus it is currently not possible to retrieve expression of our lncRNA candidates from open-access data resource. We are currently examining the protein expression of Linc013026-68AA in primary HCC samples and tumor adjacent normal liver tissues to determine whether it can be a potential HCC biomarker. Furthermore, the role of Linc013026-68AA in in vivo tumor growth should also be examined to clarify whether it may be suitable as a HCC-specific target molecule.

Our study offers novel target molecules as well as biomarkers originating from non-coding RNAs to develop a novel strategy for cancer treatment that targets multiple cancer type-specific fine tuners.

4. Materials and Methods

4.1. Cell Culture, siRNA, and Transfection

HepG2, Huh7, HLE, C3A and HeLa cells were purchased from the American Type Culture Collection (ATCC, Manassas, VA, USA) or the DMSZ-German collection of microorganisms and cell culture (DMSZ, Braunschweig, Germany). They were grown in DMEM supplemented with 10% FCS. All cell lines are free of mycoplasma contamination.

Control siRNA (5′-UAAGGCUAUGAAGAGAUAC-3′), siLinc013026 (5′-AUGGUGU CAGCAUGUGGAU-3′) were purchased from Microsynth AG (Microsynth AG, Balgach, Switzerland). Fifty picomoles of each siRNA were transfected using Lipofectamin 3000 (Thermo Fisher Scientific, Waltham, MA, USA). For ectopic expression of Linc013026-68AA experiments, Linc013026-68AA cDNA was isolated from HepG2 RNA by RT-PCR. The PCR-product was then cloned into pcDNA3.1 MycHis or pEGFP-N1 vector.

4.2. Peptide-Specific Antibodies

Antibodies against the mixture of two synthetic peptides corresponding to amino acid positions 4–17 (peptide I) and 54–68 (peptide II) of Linc013026-68AA were generated in rabbits by Kaneka Eurogentec S.A. (Kaneka Eurogentec S.A., Seraing, Belgium). Two peptide columns were applied for further purification of 4–17 (peptide I) and 54–68 (peptide II) specific antibodies.

4.3. Wst-1 Assay

HeLa cells (500–2000 cells/well) were seeded in duplicate on a 96-well plate and then transfected with vector control and Linc013026-68AA and incubated for 2 days. A Wst-1 proliferation assay kit (Roche Diagnostics, Basel, Switzerland) was employed according to the manufacturer's instructions.

4.4. Crystal Violet Assay

HeLa, HepG2, Huh7 and HLE cells (500–2000 cells/well) were seeded in duplicate on a 96-well plate and then transfected with vector control, Linc013026-68AA or siRNAs and incubated for 2 days. Cells were then washed with phosphate-buffered saline (PBS) and fixated with methanol. Crystal violet dye was applied for 10 min. After air drying the plate, the dye was solubilized in methanol and absorbance was measured at 595 nm.

4.5. Immunohistochemistry/Immunofluorescence

Immunohistochemical and immunofluorescent studies were performed as detailed previously [5,29]. Rabbit monoclonal anti-Myc antibody was from Cell Signaling Technology (cs-2278S, Cambridge, UK).

4.6. In Vitro Transcription/Translation

Radiolabeled substrates were generated by in vitro transcription/translation using the plasmid pcDNA3.1-Linc013026-MYC, the SP6/T7-coupled TNT reticulocyte lysate system (Promega, Madison, WI, USA), and [^{35}S]methionine (370 kBq/µL, >37 TBq/mmol, Hartmann Analytic, Braunschweig, Germany) according to the manufacturer's instructions.

4.7. mRNA Export Assay

Isolation of nuclear- and cytoplasmic RNA was performed as previously described [30,31]. Briefly, cells were washed with ice-cold PBS three times and incubated in cytoplasmic buffer (100 mm Tris-HCl pH 8.0, 150 mm NaCl, 0.5% (v/v) NP-40, protease inhibitor cocktail [Sigma-Aldrich, St. Louis, MO, USA]) and RNase inhibitor (NEB, Ipswich, MA, USA) for 5 min on ice. Cells were then harvested. Nucleus were pelleted by centrifugation. Nuclear- and cytoplasmic RNAs were isolated using the ReliaPrep^{TM} miRNA cell and tissue miniprep system (Promega, Madison, WI, USA) according to the manufacturer's instructions.

4.8. Polysome Profiling

Polysome fractions were prepared using sucrose gradient fractionation as previously described [32]. To prepare polysomes, 1.25×10^7 HepG2 cells were rinsed and scraped in ice-cold PBS containing cycloheximide (0.1 mg/mL). Subsequent steps were carried out in the cold. After pelleting by centrifugation at $500 \times g$ for 7 min, the cells were resuspended in extraction buffer (20 mm Tris-HCl, pH 8.0, 140 mm KCl, 0.5 mm DTT, 5 mm MgCl$_2$, 0.5% Nonidet-P40, 0.1 mg/mL cycloheximide, and 0.5 mg/mL heparin) and incubated for 5 min on ice. Extracts were centrifuged for 10 min at 12,000× *g*. Approximately 0.5 mL of supernatant was layered onto a 12-mL linear sucrose gradient (10–50% sucrose (*w/v*) in 20 mm Tris-HCl, pH 8.0, 140 mm KCl, 0.5 mm DTT, 5 mm MgCl$_2$, 0.1 mg/mL cycloheximide, and 0.5 mg/mL heparin) and centrifuged at 4 °C in an SW40Ti rotor (Beckman, Palo Alto, CA, USA) at 35,000 rpm without brake for 80 min (120 min for experiments examining the distribution of β-globin reporter mRNAs). The gradients were collected into 10–12 1-mL fractions, and absorbance profiles at 260 nm were recorded (ISCO, UA-6 detector). An amount of 0.1 volume of 3 m sodium acetate (pH 5.2) and 1 volume of isopropyl alcohol were added to the probes for overnight precipitation at −20 °C. RNA was purified using the ReliaPrepTM miRNA cell and tissue miniprep system (Promega, Madison, WI, USA).

4.9. Immunoblotting Procedures

Details of immunoblotting have been described previously [31]. Corresponding proteins were visualized by incubation with peroxidase-conjugated anti-mouse, anti-rabbit or anti-goat immunoglobulin, followed by incubation with SuperSignal West FemtoMaximum Sensitivity Substrate (Thermo Fisher Scientific, Waltham, MA, USA). Results were documented on a LAS4000 imaging system (GE Healthcare BioSciences, Uppsala, Sweden). Mouse monoclonal anti-Myc (9E10), anti-GAPDH (sc-32233), anti-GFP (sc-9996) and polyclonal anti-Actin (sc-1616) were purchased from Santa Cruz Biotechnology (Santa Cruz, CA, USA). Polyclonal anti-Histone H3 was from Cell Signaling.

4.10. Semi-Quantitative RT-PCR and qRT-PCR Analysis

RNA was isolated from cells with the ReliaPrep^{TM} miRNA cell and tissue miniprep system (Promega, Madison, WI, USA) according to the manufacturer's instructions. One microgram of RNA was reverse-transcribed using oligo dT primer or random primer and the ProtoScript$^®$ II Reverse Transcriptase (NEB, Ipswich, MA, USA) following the instructions provided. One-twentieth of the cDNA mix was used for real-time PCR with 10 pmol of forward and reverse primer and ORA^{TM} qPCR Green Rox kit (HighQu, Kraichtal, Germany) in a Qiagen Rotorgene machine. The levels of mRNA expression were standardized to the glyceraldehyde-3 phosphate dehydrogenase (GAPDH) mRNA level. Primer sequences are shown in Table S2.

4.11. Statistical Analysis

Cell experiments were performed in triplicate and a minimum of three independent experiments were evaluated. Data were reported as the mean value with standard deviation. The statistical significance of the difference between groups was determined by Student's t-test (two sided).

4.12. RNA Sequencing Data Analysis

Raw sequencing data (FASTQ files) were downloaded from the ENCODE portal or Gene Expression Omnibus (GEO). Galaxy workflow for RNA-Seq (www.usegalaxy.org) (accessed on 20 November 2020) was used for subsequent data analysis. Reads were mapped to the human reference genome (GRCh38) using Bowtie2 (Galaxy Version 2.3.4.1). The gene expression values (Fragments per Kilobase Million (FPKM)) were calculated by Cuffnorm (Galaxy Version 2.2.1.5) using the human NONCODEv5 transcript reference.

4.13. GST Pull-Down Assay

HeLa cells were transfected with pcDNA3.1-Linc013026-MYC for one day and lysed with lysis buffer (10 mm Tris, 150 mm NaCl, 1 mm PMSF, 0.4% NP40, protease inhibitor cocktail (Sigma-Aldrich, Munich, Germany). After centrifugation, supernatants were incubated with GST and GST-Linc013026-68AA fusion protein. Bound proteins were analyzed by Myc- and GST-specific immunoblot.

Supplementary Materials: The following are available online at https://www.mdpi.com/article/10.3390/ijms23010058/s1.

Author Contributions: Conceptualization, T.T. and D.D.H.T.; methodology, S.B.d.L. and D.D.H.T.; data curation, M.P., S.B.d.L. and D.D.H.T.; investigation, M.P., S.B.d.L., A.B.A. and T.N.Q.N.; visualization, M.P., S.B.d.L. and D.D.H.T.; validation, M.P., S.B.d.L. and D.D.H.T.; writing, S.B.d.L. and D.D.H.T.; writing—review and editing, M.P. and D.D.H.T.; project administration, D.D.H.T.; supervision, D.D.H.T.; funding acquisition, D.D.H.T. All authors have read and agreed to the published version of the manuscript.

Funding: This research was supported by DFG Ta-111/16-1, Niedersächsische Krebsgesellschaft and Junge Akademie (MHH) to D.D.H.T. and Ph.D. program Molecular Medicine in HBRS to S.B.d.L. and A.B.A.

Institutional Review Board Statement: Not applicable.

Informed Consent Statement: Not applicable.

Data Availability Statement: All data generated or analyzed during this study are included in this published article and its additional files.

Acknowledgments: We thank C. Bruce Boschek for critically reading the manuscript.

Conflicts of Interest: The authors declare no conflict of interest.

Abbreviations

LincRNA: long intergenic noncoding RNA; HCC: hepatocellular carcinoma; FPKM: fragments per kilobase per million mapped fragments; DAPI: 4′,6-Diamidin-2-phenylindole; IF: immunofluorescent staining; IHC: immunohistochemical staining; CAGE: Cap Analysis of Gene Expression; DNase-seq: DNase I hypersensitive sites sequencing.

References

1. Sung, H.; Ferlay, J.; Siegel, R.L.; Laversanne, M.; Soerjomataram, I.; Jemal, A.; Bray, F. Global Cancer Statistics 2020: GLOBOCAN Estimates of Incidence and Mortality Worldwide for 36 Cancers in 185 Countries. *CA A Cancer J. Clin.* **2021**, *71*, 209–249. [CrossRef] [PubMed]
2. Whittaker, S.; Marais, R.; Zhu, A.X. The role of signaling pathways in the development and treatment of hepatocellular carcinoma. *Oncogene* **2010**, *29*, 4989–5005. [CrossRef]
3. Allaire, M.; Nault, J.C. Type 2 diabetes-associated hepatocellular carcinoma: A molecular profile. *Clin. Liver Dis.* **2016**, *8*, 53–58. [CrossRef] [PubMed]
4. Schulze, K.; Imbeaud, S.; Letouze, E.; Alexandrov, L.B.; Calderaro, J.; Rebouissou, S.; Couchy, G.; Meiller, C.; Shinde, J.; Soysouvanh, F.; et al. Exome sequencing of hepatocellular carcinomas identifies new mutational signatures and potential therapeutic targets. *Nat. Genet.* **2015**, *47*, 505–511. [CrossRef] [PubMed]
5. Saran, S.; Tran, D.D.; Ewald, F.; Koch, A.; Hoffmann, A.; Koch, M.; Nashan, B.; Tamura, T. Depletion of three combined THOC5 mRNA export protein target genes synergistically induces human hepatocellular carcinoma cell death. *Oncogene* **2016**, *35*, 3872–3879. [CrossRef] [PubMed]
6. Tran, D.D.; Saran, S.; Koch, A.; Tamura, T. mRNA export protein THOC5 as a tool for identification of target genes for cancer therapy. *Cancer Lett.* **2016**, *373*, 222–226. [CrossRef]
7. Niehus, S.E.; Allister, A.B.; Hoffmann, A.; Wiehlmann, L.; Tamura, T.; Tran, D.D.H. Myc/Max dependent intronic long antisense noncoding RNA, EVA1A-AS, suppresses the expression of Myc/Max dependent anti-proliferating gene EVA1A in a U2 dependent manner. *Sci. Rep.* **2019**, *9*, 17319. [CrossRef]
8. Tran, D.D.H.; Kessler, C.; Niehus, S.E.; Mahnkopf, M.; Koch, A.; Tamura, T. Myc target gene, long intergenic noncoding RNA, Linc00176 in hepatocellular carcinoma regulates cell cycle and cell survival by titrating tumor suppressor microRNAs. *Oncogene* **2018**, *37*, 75–85. [CrossRef]
9. Orr, M.W.; Mao, Y.; Storz, G.; Qian, S.B. Alternative ORFs and small ORFs: Shedding light on the dark proteome. *Nucleic Acids Res.* **2020**, *48*, 1029–1042. [CrossRef]
10. Ragan, C.; Goodall, G.J.; Shirokikh, N.E.; Preiss, T. Insights into the biogenesis and potential functions of exonic circular RNA. *Sci. Rep.* **2019**, *9*, 2048. [CrossRef]
11. Burbano De Lara, S.; Tran, D.D.H.; Allister, A.B.; Polenkowski, M.; Nashan, B.; Koch, M.; Tamura, T. C20orf204, a hepatocellular carcinoma-specific protein interacts with nucleolin and promotes cell proliferation. *Oncogenesis* **2021**, *10*, 31. [CrossRef] [PubMed]

12. Anderson, D.M.; Anderson, K.M.; Chang, C.L.; Makarewich, C.A.; Nelson, B.R.; McAnally, J.R.; Kasaragod, P.; Shelton, J.M.; Liou, J.; Bassel-Duby, R.; et al. A micropeptide encoded by a putative long noncoding RNA regulates muscle performance. *Cell* **2015**, *160*, 595–606. [CrossRef] [PubMed]
13. Cai, B.; Li, Z.; Ma, M.; Wang, Z.; Han, P.; Abdalla, B.A.; Nie, Q.; Zhang, X. LncRNA-Six1 Encodes a Micropeptide to Activate Six1 in Cis and Is Involved in Cell Proliferation and Muscle Growth. *Front. Physiol.* **2017**, *8*, 230. [CrossRef] [PubMed]
14. Davis, C.A.; Hitz, B.C.; Sloan, C.A.; Chan, E.T.; Davidson, J.M.; Gabdank, I.; Hilton, J.A.; Jain, K.; Baymuradov, U.K.; Narayanan, A.K.; et al. The Encyclopedia of DNA elements (ENCODE): Data portal update. *Nucleic Acids Res.* **2018**, *46*, D794–D801. [CrossRef]
15. Liu, C.; Bai, B.; Skogerbo, G.; Cai, L.; Deng, W.; Zhang, Y.; Bu, D.; Zhao, Y.; Chen, R. NONCODE: An integrated knowledge database of non-coding RNAs. *Nucleic Acids Res.* **2005**, *33*, D112–D115. [CrossRef]
16. Trapnell, C.; Williams, B.A.; Pertea, G.; Mortazavi, A.; Kwan, G.; van Baren, M.J.; Salzberg, S.L.; Wold, B.J.; Pachter, L. Transcript assembly and quantification by RNA-Seq reveals unannotated transcripts and isoform switching during cell differentiation. *Nat. Biotechnol.* **2010**, *28*, 511–515. [CrossRef]
17. Hishiki, T.; Kawamoto, S.; Morishita, S.; Okubo, K. BodyMap: A human and mouse gene expression database. *Nucleic Acids Res.* **2000**, *28*, 136–138. [CrossRef]
18. Consortium, E.P.; Birney, E.; Stamatoyannopoulos, J.A.; Dutta, A.; Guigo, R.; Gingeras, T.R.; Margulies, E.H.; Weng, Z.; Snyder, M.; Dermitzakis, E.T.; et al. Identification and analysis of functional elements in 1% of the human genome by the ENCODE pilot project. *Nature* **2007**, *447*, 799–816. [CrossRef]
19. Consortium, E.P.; Moore, J.E.; Purcaro, M.J.; Pratt, H.E.; Epstein, C.B.; Shoresh, N.; Adrian, J.; Kawli, T.; Davis, C.A.; Dobin, A.; et al. Expanded encyclopaedias of DNA elements in the human and mouse genomes. *Nature* **2020**, *583*, 699–710. [CrossRef]
20. Sonawane, A.R.; Platig, J.; Fagny, M.; Chen, C.Y.; Paulson, J.N.; Lopes-Ramos, C.M.; DeMeo, D.L.; Quackenbush, J.; Glass, K.; Kuijjer, M.L. Understanding Tissue-Specific Gene Regulation. *Cell Rep.* **2017**, *21*, 1077–1088. [CrossRef]
21. Gasteiger, E.; Gattiker, A.; Hoogland, C.; Ivanyi, I.; Appel, R.D.; Bairoch, A. ExPASy: The proteomics server for in-depth protein knowledge and analysis. *Nucleic Acids Res.* **2003**, *31*, 3784–3788. [CrossRef]
22. Gentile, F.; Amodeo, P.; Febbraio, F.; Picaro, F.; Motta, A.; Formisano, S.; Nucci, R. SDS-resistant active and thermostable dimers are obtained from the dissociation of homotetrameric beta-glycosidase from hyperthermophilic Sulfolobus solfataricus in SDS. Stabilizing role of the A-C intermonomeric interface. *J. Biol. Chem.* **2002**, *277*, 44050–44060. [CrossRef] [PubMed]
23. Lundby, A.; Lage, K.; Weinert, B.T.; Bekker-Jensen, D.B.; Secher, A.; Skovgaard, T.; Kelstrup, C.D.; Dmytriyev, A.; Choudhary, C.; Lundby, C.; et al. Proteomic analysis of lysine acetylation sites in rat tissues reveals organ specificity and subcellular patterns. *Cell Rep.* **2012**, *2*, 419–431. [CrossRef]
24. Matsumoto, A.; Pasut, A.; Matsumoto, M.; Yamashita, R.; Fung, J.; Monteleone, E.; Saghatelian, A.; Nakayama, K.I.; Clohessy, J.G.; Pandolfi, P.P. mTORC1 and muscle regeneration are regulated by the LINC00961-encoded SPAR polypeptide. *Nature* **2017**, *541*, 228–232. [CrossRef] [PubMed]
25. Stein, C.S.; Jadiya, P.; Zhang, X.; McLendon, J.M.; Abouassaly, G.M.; Witmer, N.H.; Anderson, E.J.; Elrod, J.W.; Boudreau, R.L. Mitoregulin: A lncRNA-Encoded Microprotein that Supports Mitochondrial Supercomplexes and Respiratory Efficiency. *Cell Rep.* **2018**, *23*, 3710–3720.e8. [CrossRef]
26. Davenport, C.F.; Scheithauer, T.; Dunst, A.; Bahr, F.S.; Dorda, M.; Wiehlmann, L.; Tran, D.D.H. Genome-Wide Methylation Mapping Using Nanopore Sequencing Technology Identifies Novel Tumor Suppressor Genes in Hepatocellular Carcinoma. *Int. J. Mol. Sci.* **2021**, *22*, 3937. [CrossRef] [PubMed]
27. Frankish, A.; Diekhans, M.; Jungreis, I.; Lagarde, J.; Loveland, J.E.; Mudge, J.M.; Sisu, C.; Wright, J.C.; Armstrong, J.; Barnes, I.; et al. Gencode 2021. *Nucleic Acids Res.* **2021**, *49*, D916–D923. [CrossRef]
28. Uszczynska-Ratajczak, B.; Lagarde, J.; Frankish, A.; Guigo, R.; Johnson, R. Towards a complete map of the human long non-coding RNA transcriptome. *Nat. Rev. Genet.* **2018**, *19*, 535–548. [CrossRef]
29. Tran, D.D.; Saran, S.; Dittrich-Breiholz, O.; Williamson, A.J.; Klebba-Farber, S.; Koch, A.; Kracht, M.; Whetton, A.D.; Tamura, T. Transcriptional regulation of immediate-early gene response by THOC5, a member of mRNA export complex, contributes to the M-CSF-induced macrophage differentiation. *Cell Death Dis.* **2013**, *4*, e879. [CrossRef]
30. Guria, A.; Tran, D.D.; Ramachandran, S.; Koch, A.; El Bounkari, O.; Dutta, P.; Hauser, H.; Tamura, T. Identification of mRNAs that are spliced but not exported to the cytoplasm in the absence of THOC5 in mouse embryo fibroblasts. *RNA* **2011**, *17*, 1048–1056. [CrossRef]
31. Tran, D.D.; Saran, S.; Williamson, A.J.; Pierce, A.; Dittrich-Breiholz, O.; Wiehlmann, L.; Koch, A.; Whetton, A.D.; Tamura, T. THOC5 controls 3'end-processing of immediate early genes via interaction with polyadenylation specific factor 100 (CPSF100). *Nucleic Acids Res.* **2014**, *42*, 12249–12260. [CrossRef] [PubMed]
32. Dhamija, S.; Doerrie, A.; Winzen, R.; Dittrich-Breiholz, O.; Taghipour, A.; Kuehne, N.; Kracht, M.; Holtmann, H. IL-1-induced post-transcriptional mechanisms target overlapping translational silencing and destabilizing elements in IkappaBzeta mRNA. *J. Biol. Chem.* **2010**, *285*, 29165–29178. [CrossRef] [PubMed]

International Journal of
Molecular Sciences

Article

The Fusion of *CLEC12A* and *MIR223HG* Arises from a *trans*-Splicing Event in Normal and Transformed Human Cells

Bijay P. Dhungel [1,2,†], Geoffray Monteuuis [3,†], Caroline Giardina [1], Mehdi S. Tabar [1], Yue Feng [1], Cynthia Metierre [1], Sarah Ho [1], Rajini Nagarajah [1], Angela R. M. Fontaine [4], Jaynish S. Shah [1], Divya Gokal [1,2], Charles G. Bailey [1,2,5], Ulf Schmitz [6,*] and John E. J. Rasko [1,2,7,*]

[1] Gene and Stem Cell Therapy Program Centenary Institute, The University of Sydney, Camperdown, NSW 2050, Australia; b.dhungel@centenary.org.au (B.P.D.); c.giardina@centenary.org.au (C.G.); m.tabar@centenary.org.au (M.S.T.); j.feng@centenary.org.au (Y.F.); C.Metierre@centenary.org.au (C.M.); s.ho@centenary.org.au (S.H.); r.nagarajah@centenary.org.au (R.N.); j.shah@centenary.org.au (J.S.S.); d.gokal@centenary.org.au (D.G.); c.bailey@centenary.org.au (C.G.B.)
[2] Faculty of Medicine & Health, The University of Sydney, Camperdown, NSW 2006, Australia
[3] Department of Biochemistry and Developmental Biology, University of Helsinki, 00014 Helsinki, Finland; geoffray.monteuuis@helsinki.fi
[4] Centenary Imaging and Sydney Cytometry Centenary Institute, The University of Sydney, Camperdown, NSW 2050, Australia; a.fontaine@centenary.org.au
[5] Cancer & Gene Regulation Laboratory Centenary Institute, The University of Sydney, Camperdown, NSW 2050, Australia
[6] Department of Molecular & Cell Biology, College of Public Health, Medical & Vet Sciences, James Cook University, Townsville, QLD 4811, Australia
[7] Cell and Molecular Therapies, Royal Prince Alfred Hospital, Camperdown, NSW 2050, Australia
[*] Correspondence: ulf.schmitz@jcu.edu.au (U.S.); j.rasko@centenary.org.au (J.E.J.R.); Tel.: +6-12-9565-6160 (J.E.J.R.); Fax: +6-12-9565-6101 (J.E.J.R.)
[†] Contributed equally.

Citation: Dhungel, B.P.; Monteuuis, G.; Giardina, C.; Tabar, M.S.; Feng, Y.; Metierre, C.; Ho, S.; Nagarajah, R.; Fontaine, A.R.M.; Shah, J.S.; et al. The Fusion of *CLEC12A* and *MIR223HG* Arises from a *trans*-Splicing Event in Normal and Transformed Human Cells. *Int. J. Mol. Sci.* **2021**, 22, 12178. https://doi.org/10.3390/ijms222212178

Academic Editors: Amelia Casamassimi, Alfredo Ciccodicola and Monica Rienzo

Received: 24 September 2021
Accepted: 3 November 2021
Published: 10 November 2021

Publisher's Note: MDPI stays neutral with regard to jurisdictional claims in published maps and institutional affiliations.

Abstract: Chimeric RNAs are often associated with chromosomal rearrangements in cancer. In addition, they are also widely detected in normal tissues, contributing to transcriptomic complexity. Despite their prevalence, little is known about the characteristics and functions of chimeric RNAs. Here, we examine the genetic structure and biological roles of *CLEC12A-MIR223HG*, a novel chimeric transcript produced by the fusion of the cell surface receptor *CLEC12A* and the *miRNA-223* host gene (*MIR223HG*), first identified in chronic myeloid leukemia (CML) patients. Surprisingly, we observed that *CLEC12A-MIR223HG* is not just expressed in CML, but also in a variety of normal tissues and cell lines. *CLEC12A-MIR223HG* expression is elevated in pro-monocytic cells resistant to chemotherapy and during monocyte-to-macrophage differentiation. We observed that *CLEC12A-MIR223HG* is a product of *trans*-splicing rather than a chromosomal rearrangement and that transcriptional activation of *CLEC12A* with the CRISPR/Cas9 Synergistic Activation Mediator (SAM) system increases *CLEC12A-MIR223HG* expression. *CLEC12A-MIR223HG* translates into a chimeric protein, which largely resembles CLEC12A but harbours an altered C-type lectin domain altering key disulphide bonds. These alterations result in differences in post-translational modifications, cellular localization, and protein–protein interactions. Taken together, our observations support a possible involvement of *CLEC12A-MIR223HG* in the regulation of *CLEC12A* function. Our workflow also serves as a template to study other uncharacterized chimeric RNAs.

Keywords: chimeric RNAs; Fusion RNAs encoding protein; fusion transcript; linc-223; miR-223 host gene; trans-splicing; alternative splicing; CCL1; myeloid cell differentiation; C-type lectin; chronic myeloid leukemia

1. Introduction

Chimeric RNAs are transcripts that consist of exons from different parental genes. They can be produced by several mechanisms which may or may not involve chromosomal translocations at the genomic level. Transcription of fusion genes resulting from chromosomal deletion, inversion, or translocation are considered to be a hallmark of cancer [1,2]. These fusion genes can give rise to proteins with important roles in cancer development and progression and may also serve as therapeutic targets. Some examples include the BCR-ABL1 fusion protein in chronic myeloid leukaemia (CML) [3,4] and EML4-ALK in lung cancer [5].

Apart from chimeric RNAs produced by DNA-level gene fusions, studies have uncovered additional mechanisms that are known to produce chimeric transcripts at the RNA level [1,6–8]. These include intergenic splicing by either *cis*-splicing, which involves same-strand neighbouring genes or by *trans*-splicing, whereby exons from two separate RNA transcripts are spliced together [1,9,10]. Although *cis*- and *trans*- splicing events were originally thought to be rare in mammals, high-throughput analyses of transcriptomes have revealed that significant portions of chimeric RNAs are derived from intergenic splicing [11–13].

The expression of chimeric RNAs is significantly higher in cancer compared to normal tissues [1,14,15]. As such, they may serve as potential biomarkers for diagnosis, prognosis or even therapeutic targets. However, the detection of chimeric RNAs in non-malignant tissues has led to investigations of their roles in normal physiology and development [6,16]. Chimeric RNAs could be fundamental to basic cellular processes and provide a means for expansion of the functional genome, i.e., transcriptomic complexity, without an increase in the number of genes [16,17]. For example, *SLC45A3-ELK4* is a recurrent *cis*-spliced chimeric transcript expressed in normal prostate as well as prostate cancer to modulate cell proliferation [18]. *JAZF1-SUZ12* is a protein-coding chimeric transcript produced by a physiologically-regulated *trans*-splicing mechanism, which impairs chromatin repression [19,20]. During normal myogenesis, *PAX3-FOXO1*, a chimeric RNA, is detected transiently without any evidence of chromosomal rearrangement [21] and interferes with the TGF-β pathway [22].

Technological and analytical advancements made in the past decade offer unique opportunities to appreciate the prevalence and explore the relevance of chimeric RNAs in normal physiology and cancer [10,14,16]. While they are now widely recognised as being ubiquitous, chimeric RNAs are largely under-investigated, and their understanding is limited. Currently, most studies use bioinformatics pipelines to search for chimeric RNAs in RNA sequencing data from different biological sources [23–26]. These studies have identified thousands of chimeric transcripts in both malignant and normal tissues [27]. However, very few have been experimentally characterized in detail.

In a recent study, we used RNA sequencing to study the changing transcriptomes in CML patients [28]. In addition to striking changes in alternatively spliced transcripts, we identified known and novel fusion transcripts. In the present study, we analysed the most frequently recurring uncharacterized fusion transcript in CML: namely *CLEC12A-MIR223HG*. This chimeric transcript is produced by the fusion of *CLEC12A*, a myeloid-inhibitory receptor and the host gene of myeloid regulatory miRNA-223 (*MIR223HG*). We examined the genetic architecture and the potential fusion protein encoded by the *CLEC12A-MIR223HG* chimeric transcript. We have also conducted a range of cell biological assays to investigate the function of *CLEC12A-MIR223HG* and its putative role in cancer biology. Our results suggest a possible role of *CLEC12A-MIR223HG* in the regulation of *CLEC12A* function. These results also provide a template for further investigations into the biological significance of *CLEC12A-MIR223HG* and other chimeric RNAs.

2. Results

2.1. CLEC12A-MIR223HG Is Expressed in CML Patients and Healthy Controls

We have previously reported next generation sequencing-based gene fusion analysis on poly(A)-enriched RNA from peripheral blood mononuclear cells (PBMCs) of 16 Philadelphia-positive (Ph+) CML patients at diagnosis and ten matched remission samples from the same individuals, as well as six healthy control subjects [28]. All the CML patients were treated with tyrosine kinase inhibitors (TKIs). Sequencing reads mapping the *BCR-ABL1* locus served as a control. Most frequently recurring, apart from the characteristic *BCR-ABL1* fusion, was *CLEC12A-MIR223HG*, which is a fusion transcript between the transmembrane glycoprotein *CLEC12A* (a.k.a. *CCL1*) and the *miR223* host gene *MIR223HG*, located at chromosomal loci 12p13.31 and Xq12, respectively. Interestingly, we observed nine distinct fusion events involving *CLEC12A*, which encodes a myeloid-inhibitory C-type lectin receptor known to play an important role in haematopoiesis [29]. Seven out of the nine breakpoints were identical (Supplementary Table S1). *CLEC12A* is also a cancer stem cell marker in myelodysplastic syndrome [30] and a therapeutic target in acute myeloid leukemia [31]. Based on its high recurrence frequency and interesting biology associated with both the parental genes, *CLEC12A-MIR223HG* was chosen for further examination (Figure 1A).

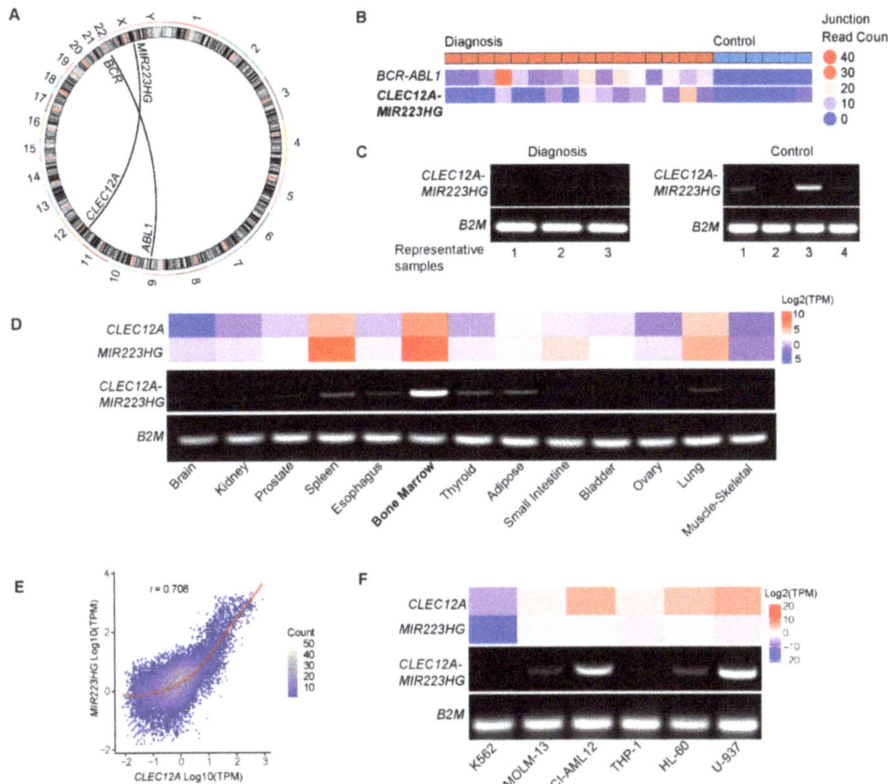

Figure 1. *CLEC12A-MIR223HG* detection in CML patients, controls, normal tissues, and cell lines. (**A**) Reads mapping across the *CLEC12A/MIR223HG* breakpoint. (**B**) Read counts that span across the fusion breakpoint in representative CML diagnostic and control samples. (**C**) RT-PCR-based detection of *CLEC12A-MIR223HG* in CML diagnostic and control samples with primers flanking the breakpoint. Amplicon sizes expected: *B2M*: 96 bp, *CLEC12A-MIR223HG*: 207 bp. (**D**) RT-PCR-based detection of *CLEC12A-MIR223HG* in diverse normal tissues and RNA sequencing-based quantification of *CLEC12A* and *MIR223HG* transcript levels. (**E**) Contour plot illustrating correlation between *CLEC12A* and *MIR223HG* across different tissues. Correlation coefficient (r) was determined using Pearson correlation and the red line indicates the line of best fit. (**F**) RT-PCR-based detection of *CLEC12A-MIR223HG* in leukemic cell lines. TPM (transcripts per million) values for tissues were obtained from the Genotype-Tissue Expression (GTEx) Portal V8 (https://gtexportal.org) (accessed on 25 October 2021). TPM values for cell lines were obtained from the Cancer Cell Line Encyclopedia project portal (https://depmap.org/portal/ccle/) (accessed on 25 October 2021).

CLEC12A-MIR223HG was detected in 10/16 diagnosis samples, 0/10 of the remission samples, and in 1/6 of healthy control samples. While the fusion breakpoints were identical among *CLEC12A-MIR223HG*-positive samples (left: chr12:9982129; right: chrX: 66020127), read counts across the breakpoints differed, indicating varying expression levels of the chimeric transcript (Figure 1B). Surprisingly, using RT-PCR, the *CLEC12A-MIR223HG* fusion could also be detected at low levels in normal control samples (Figure 1C). These results indicate that despite being expressed at CML diagnosis, *CLEC12A-MIR223HG* cannot serve as a prognostic marker for TKI-based therapies.

2.2. CLEC12A-MIR223HG Is Expressed in Normal Human Tissues and Leukemic Cell Lines

As *CLEC12A-MIR223HG* is not specific to CML pathology, we sought to examine its expression in different tissues. As indicated in Figure 1D, the expression of *CLEC12A-MIR223HG* is highest in the bone marrow, where both *CLEC12A* and *MIR223HG* are also individually expressed at high levels. Using Pearson correlation, a strong correlation between the levels of *CLEC12A* and *MIR223HG* (r = 0.708) across 30 different tissues was observed (Figure 1E). We also observed tissue-specific differences in the correlation strength (Figure S1). Next, we studied the expression of *CLEC12A-MIR223HG* in different leukemia cell lines. As depicted in Figure 1F, *CLEC12A-MIR223HG* is expressed at varying levels in several leukemic cell lines. Due to a high expression of *CLEC12A-MIR223HG*, U-937 pro-monocytic cells were used for further characterization.

2.3. CLEC12A-MIR223HG Exhibits a Typical Fusion Transcript Architecture

To determine the architecture of the *CLEC12A-MIR223HG* transcript, we performed RT-PCR with primers specific to different isoforms of *CLEC12A* (Figure 2A,B). As *MIR223HG* is polyadenylated, we sought to determine if *CLEC12A-MIR223HG* also possesses a poly(A)-tail by designing primers around the canonical poly(A) signal of *MIR223HG*. RT-PCR analysis indicates that *CLEC12A-MIR223HG* harbours the poly(A) signal from *MIR223HG* (Figure 2B). Furthermore, as we used oligo(dT) as primer for cDNA synthesis, it is likely that *CLEC12A-MIR223HG* is polyadenylated. Next, we investigated the transcription start site of *CLEC12A-MIR223HG* by identifying the transcript isoform of *CLEC12A* that engages in the fusion event. As indicated by the RT-PCR results (Figure 2B), the primary transcript isoform of *CLEC12A* (*CLEC12A-201*) participates in the fusion.

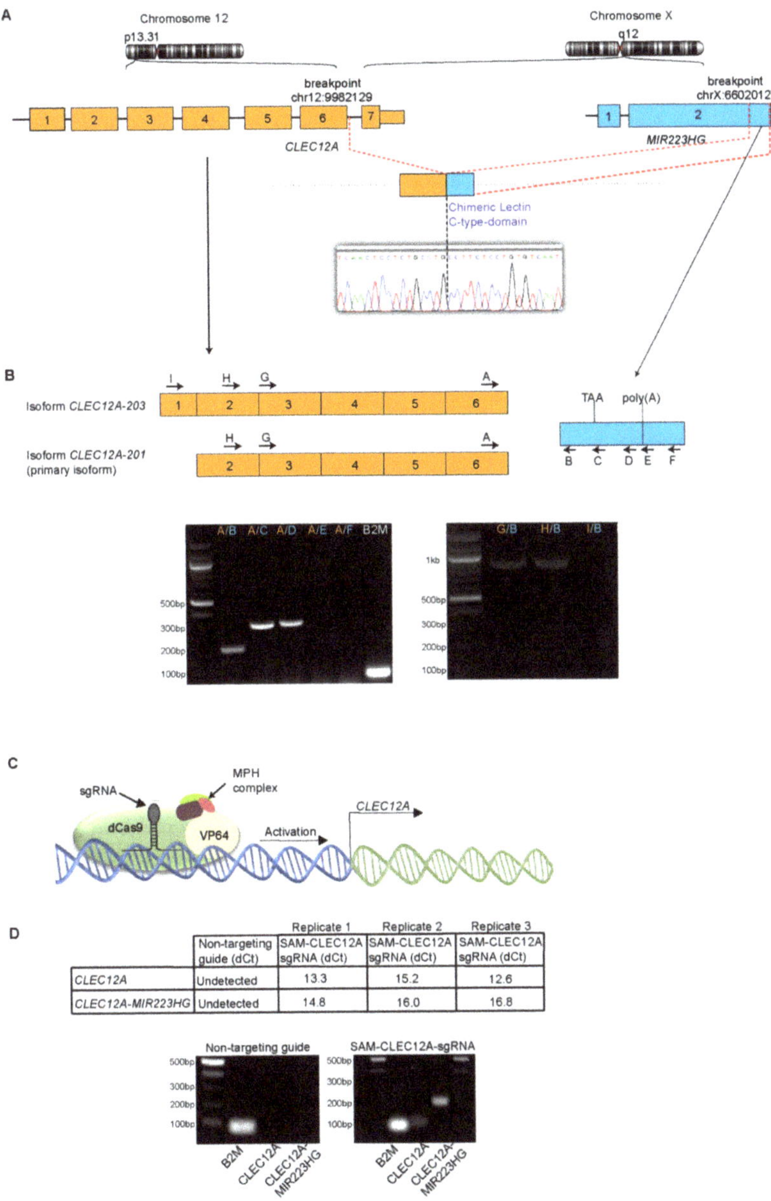

Figure 2. Transcript structure of *CLEC12A-MIR223HG*. (**A**) Illustration of the breakpoint location and the resulting fusion transcript. (**B**) RT-PCR on U-937 mRNA to identify the full length *CLEC12A-MIR223HG* transcript. Expected amplicon sizes are as follows: A/B: 207 bp, A/C: 361 bp, A/D: 371 bp, A/E: 564 bp, A/F: 579 bp, H/B: 829 bp, G/B: 962 bp. The chimeric transcript is produced by the substitution of the final exon of *CLEC12A-201* isoform with a part of *MIR223HG* including an in-frame stop codon followed by a poly(A) sequence. (**C**) Illustration of the CRISPR/SAM-based transcriptional activation system used to increase the endogenous expression of *CLEC12A*. (**D**) Three independent replicates of qRT-PCR accessing expression levels of *CLEC12A* and *CLEC12A-MIR223HG* after CRISPR/SAM-mediated transcriptional activation of *CLEC12A* in HEK293 cells. Expected amplicon sizes are as follows: *B2M*: 91 bp, *CLEC12A*: 105 bp, *CLEC12A-MIR223HG*: 206 bp. B2M was used to determine the dCt values.

2.4. CRISPR-Mediated Transcriptional Activation of CLEC12A Increases CLEC12A-MIR223HG Expression

As *CLEC12A* and *MIR223HG* are located on different chromosomes, *CLEC12A-MIR223HG* could theoretically be produced either by chromosomal rearrangement or *trans*-splicing. To distinguish these alternatives, we used HEK293 cells, which do not express detectable levels of either *CLEC12A* or *CLEC12A-MIR223HG* by RT-qPCR (data not shown). We searched for all the detectable chromosomal translocations in HEK293 cells but found no translocations between chromosome 12 and chromosome X (Supplementary Table S4). Next, we exploited the CRISPR/Cas9 Synergistic Activation Mediator (CRISPR/SAM)-based transcriptional activation technology to further examine the mechanisms regulating *CLEC12A-MIR223HG* expression. We engineered HEK293 cells to stably express the components of the CRISPR activation system (dCas9/MPH) and transduced these cells with a lentivirus that expresses a short guide RNA (sgRNA) that targets the *CLEC12A* promoter region (Figure 2C). Only a sgRNA targeting the *CLEC12A* promoter induced the expression of *CLEC12A-MIR223HG* to levels detectable with qRT-PCR (Figure 2D). Furthermore, *CLEC12A-MIR223HG* was only amplified from the cDNA and not the genomic DNA of U-937 cells (Figure S2E). When combined, these results suggest that *CLEC12A-MIR223HG* is likely to be the result of *trans*-splicing rather than a consequence of chromosomal rearrangement.

2.5. CLEC12A-MIR223HG Encodes a Chimeric Protein That Is Distinct from CLEC12A

Next, we investigated whether *CLEC12A-MIR223HG* translates into a chimeric protein. The chimeric RNA transcript sequence determined by RT-PCR and Sanger sequencing enabled us to analyse the potential protein encoded by *CLEC12A-MIR223HG*. Our analysis indicated that CLEC12A-MIR223HG results in a substitution of the last 52 amino acids (aa 214–265) of CLEC12A with 44 new amino acids and a new 3′ UTR from *MIR223HG*. This would result in a chimeric protein containing the cytoplasmic and transmembrane domain of CLEC12A. However, the extracellular domain, which contains the C-type lectin, would be altered so as to disrupt key disulfide bonds (Figure 3A).

Due to the lack of a suitable antibody specific to the chimeric domain of the fusion protein, we cloned the coding sequences of both *CLEC12A* and *CLEC12A-MIR223HG* into a lentiviral expression vector and N-terminally tagged them with a FLAG epitope. After overexpressing *CLEC12A* in different cell lines, we detected FLAG-tagged CLEC12A ranging between 40 and 75 kDa (Figures 3B and S2A). In contrast, ectopic expression of the *CLEC12A-MIR223HG* coding sequence produced a single prominent band between 37 and 50 kDa (higher than the predicted size of 37 kDa) (Figure 3B). To determine whether the observed higher molecular weight of CLEC12A-MIR223HG was due to glycosylation as previously described [32], U-937 cell lysates overexpressing either *CLEC12A* or *CLEC12A-MIR223HG* were treated with PNGase F to remove N-linked glycosylation and examined by Western blotting (Figure 3C). A predominant band was observed at ~37 kDa for both proteins (Figure 3C), suggesting that CLEC12A and CLEC12A-MIR223HG undergo distinct glycosylation programs. Both proteins have three predicted N-glycosylation sites, but the chimeric CLEC12A-MIR223HG could disrupt normal post-translational modifications of CLEC12A (Figure 3A).

To investigate the fusion protein further, we performed immunoprecipitation coupled with mass spectrometry and identified interacting partners of CLEC12A and CLEC12A-MIR223HG. We identified interactors unique to CLEC12A-MIR223HG, which included CALX, RHG04, A2MG, RCN2, RASL3, and LUC7L. PSA5 was detected in pulldowns of both CLEC12A and CLEC12A-MIR223HG (Figure 3D). A unique peptide sequence originating from the MIR223HG side of the fusion protein (HDLGNCPR) confirmed its expression. These results suggest that the chimeric lectin domain in CLEC12A-MIR223HG facilitates novel protein interactions.

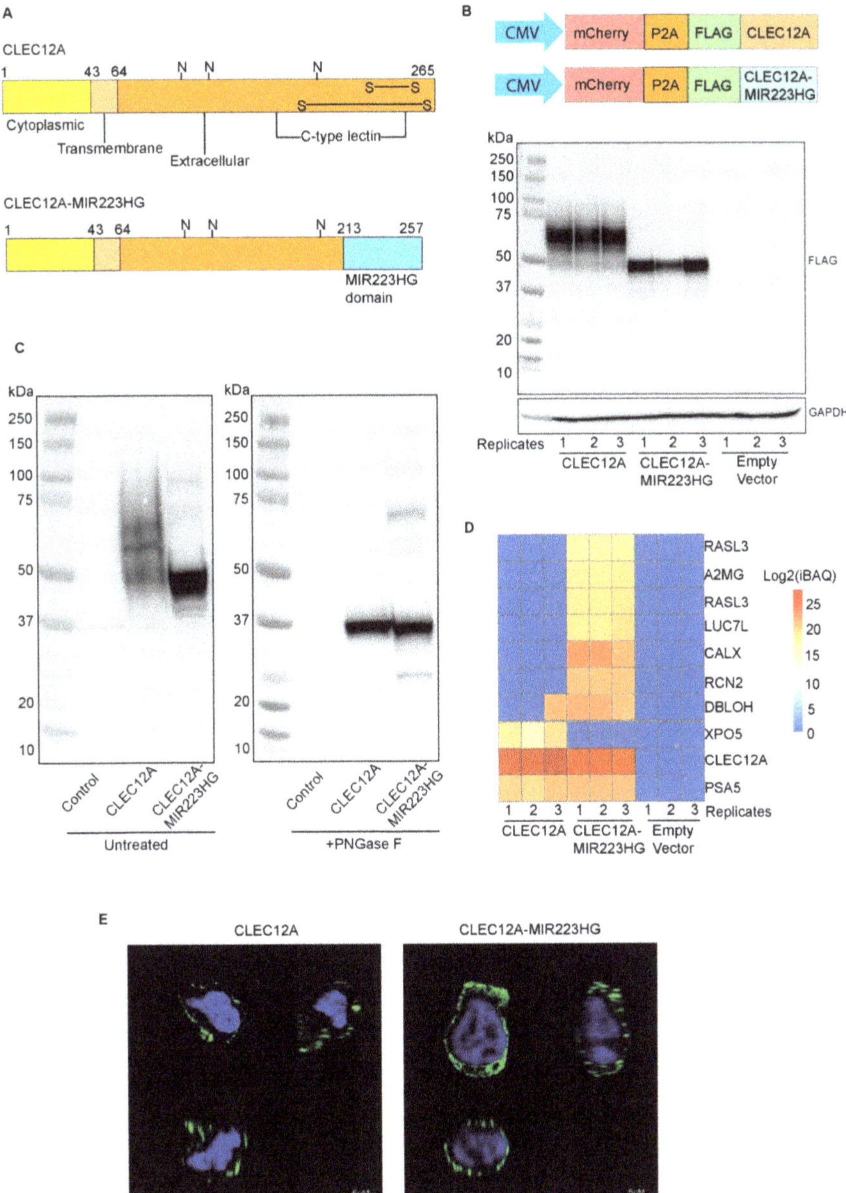

Figure 3. *CLEC12A-MIR223HG* encodes a chimeric protein distinct from CLEC12A. (**A**) Schematic of CLEC12A and CLEC12A-MIR223HG protein architectures. (**B**) Western blot image of cell lysates from HEK293 cells transduced with lentiviral vectors expressing either *CLEC12A* or *CLEC12A-MIR223HG*. (**C**) Western blot image of PNGase F-treated and untreated U-937 cell lysates overexpressing *CLEC12A* or *CLEC12A-MIR223HG*. (**D**) Immunoprecipitation with anti-FLAG antibody coupled with mass spectrometry of U-937 cells expressing either *CLEC12A* or *CLEC12A-MIR223HG*. Total peptide intensity divided by the number of observable peptides for a particular protein (iBAQ) is depicted. Compared to CLEC12A, the chimeric protein has gained interacting partners and lost at least one. Three separate replicates were performed. (**E**) Representative images of immunofluorescence with anti-FLAG antibody (green) and DAPI (blue) of U-937 cells expressing either *CLEC12A* or *CLEC12A-MIR223HG*. Image analysis was performed using Biplane Imaris software. Three different angels of the same cell are depicted.

Next, we performed immunofluorescence on U-937 overexpressing FLAG-tagged-CLEC12A or -CLEC12A-MIR223HG (Figure 3E). Similar to CLEC12A, CLEC12A-MIR223HG localised to the plasma membrane. However, the chimeric protein was also detected in the cytoplasmic compartment like most of its interacting partners (Figure 3E and Supplementary Table S2). When combined, these results suggest that CLEC12A-MIR223HG is a chimeric protein differing from CLEC12A in its patterns of post-translational modifications, interactions with other proteins, and sub-cellular localization.

2.6. Increased CLEC12A-MIR223HG Expression following Cell Differentiation or Chemotherapy

To investigate possible functions of *CLEC12A-MIR223HG* in cancer and normal biology, we performed a range of cell biology assays. We did not observe a significant impact of either *CLEC12A* or *CLEC12A-MIR223HG* overexpression in the proliferation (MTT assay) or survival (clonogenicity assay) of U-937 and THP1 cell lines (Figure S2B). We examined whether *CLEC12A-MIR223HG* could alter chemosensitivity. We treated U-937 cells with cytarabine (AraC; 200 nM) (Figure S2C,D) and observed a 6.5-fold increase in *CLEC12A-MIR223HG* expression when compared to *CLEC12A* ($p < 0.01$, Figure 4A). Similarly, treatment with 400 nM AraC resulted in a 7.2-fold increase ($p < 0.05$, Figure 4A). However, upon overexpression, neither *CLEC12A* nor *CLEC12A-MIR223HG* conferred either resistance or sensitivity to AraC (Figure 4B,C).

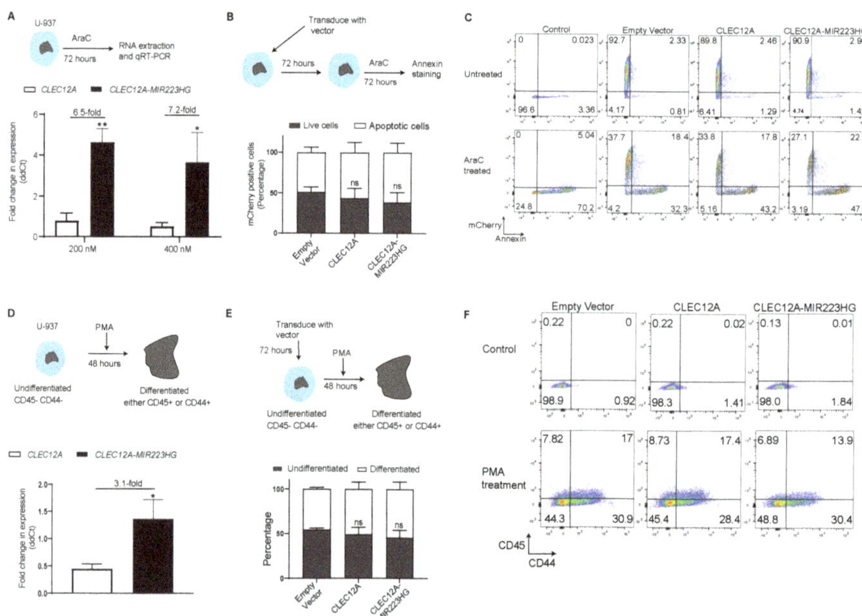

Figure 4. Biological roles of *CLEC12A-MIR223HG*. (**A**) Fold change in the expression of *CLEC12A* and *CLEC12A-MIR223HG* in U-937 cells treated with two different doses of cytarabine (AraC) measured by qRT-PCR. (**B**) Percentage of apoptotic cells (Annexin V-positive cells) after the treatment of U-937 cells overexpressing *CLEC12A* or *CLEC12A-MIR223HG* with cytarabine. (**C**) Representative flow cytometry plots depicting percentage of apoptotic U-937 cells overexpressing either *CLEC12A* or *CLEC12A-MIR223HG* after AraC treatment. X-axis: Annexin V, Y-axis: mCherry (depicting transduction efficiency). (**D**) Fold change in the expression of *CLEC12A* and *CLEC12A-MIR223HG* during the differentiation of U-937 monocytes induced by phorbol 12-myristate 13-acetate (PMA) treatment (measured by qRT-PCR). (**E**) Percentage of either CD45+ or CD44+ cells after the treatment of U-937 cells overexpressing *CLEC12A* or *CLEC12A-MIR223HG* with PMA. (**F**) Representative flow cytometry plots depicting the percentage of either or CD44 (x-axis) or CD45 (y-axis) in PMA-treated U-937 monocytes overexpressing *CLEC12A* or *CLEC12A-MIR223HG* ($n > 3$, * $p < 0.05$, ** $p < 0.01$).

As *CLEC12A* expression decreases upon differentiation of monocytes, we examined whether *CLEC12A-MIR223HG* follows a similar pattern. We treated U-937 pro-monocytic cells with phorbol 12-myristate-13-acetate (PMA) to induce their differentiation. As depicted in Figure 4D, we observed a 3.1-fold increase in *CLEC12A-MIR223HG* expression compared to *CLEC12A* ($p < 0.05$). Next, we tested whether the overexpression of *CLEC12A* or *CLEC12A-MIR223HG* could alter the degree of differentiation of U-937 cells. Interestingly, no significant differences in the differentiation of U-937 were observed after the overexpression of either *CLEC12A* or *CLEC12A-MIR223HG* (Figure 4E,F). These results suggest that despite being upregulated during cell differentiation and chemotherapy, *CLEC12A-MIR223HG* does not drive these processes.

3. Discussion

Gene expression is a finely tuned process, which is tightly regulated by many layers of regulatory mechanisms. Regulation of gene expression can occur at transcriptional, post-transcriptional, translational, and post-translational levels. Chimeric RNAs are produced by the fusion of different parental transcripts and are now widely recognised as an additional layer of transcriptomic complexity. Some fusion transcripts arising from chromosomal rearrangements in cancer like *BCR-ABL1* are well characterized and have significant biological functions [4]. Many other fusion transcripts await further study. In our published investigation, we compared the transcriptomic landscapes of healthy donors and CML patients at diagnosis and remission [28]. We observed a number of fusion transcripts involving *CLEC12A*. In this study, we examined the most frequently recurring fusion transcript, *CLEC12A-MIR223HG*, which results from the fusion between *CLEC12A* and the *miR-223* host gene *MIR223HG*.

CLEC12A is located on chromosome 12 and encodes a type II transmembrane glycoprotein. It is a myeloid cell-inhibitory receptor that can recognize uric acid crystals to alert the immune system of cell death and consequently inhibit an inflammatory response [33]. The *MIR223HG* gene is located on chromosome X [34] and is a key regulator of myeloid cell differentiation [35,36]. We observed a higher expression of *CLEC12A-MIR223HG* in CML diagnostic patient samples compared to remission and control samples. Consistent with similar reports of other chimeric transcripts [12,37], the lower expression of *CLEC12A-MIR223HG* was concurrent with reduced expression of the parental transcripts. We then exploited the CRISPR-based transcriptional activation system to investigate whether *CLEC12A-MIR223HG* results from a *trans*-splicing event. We activated the endogenous expression of *CLEC12A* in HEK293 cells that do not express detectable levels of either *CLEC12A* or *MIR223HG* and do not have a chromosomal translocation between chromosome 12 and chromosome X. Both *CLEC12A* and *CLEC12A-MIR223HG* were detected after transcriptional activation of *CLEC12A*. These observations suggested that a chromosomal translocation is not essential for the production of the chimeric *CLEC12A-MIR223HG* transcript and that it arises from a *trans*-splicing event.

We examined the expression of *CLEC12A-MIR223HG* in a range of normal tissue types and cell lines. As *CLEC12A-MIR223HG* was detected in several healthy controls, it could not serve as a prognostic marker for TKI-based therapies. As expected, the expression of *CLEC12A-MIR223HG* was highest in blood, which also has high expression levels of both the parental genes. Our results mirror other studies that first identified fusion transcripts in cancer but detected their expression in normal cells upon subsequent examination [12,38]. Additionally, the expression of *CLEC12A-MIR223HG* is similar to other chimeric RNAs expressed in normal tissues that are increased in cancer [6,39].

The CLEC12A-MIR223HG fusion protein results in a substitution of 52 amino acids of the C-type lectin domain of CLEC12A with 44 amino acids arising from the DNA sequence of the *MIR223HG* taking part in the fusion. As the C-type lectin domain of CLEC12A is important for its function [33,40,41], it was not surprising that CLEC12A-MIR223HG showed striking differences with CLEC12A in terms of post-translational modifications, cellular localization, and protein–protein interactions. We noted that the molecular weights of both

CLEC12A and CLEC12A-MIR223HG were higher than predicted in Western blots. It was previously reported that this could be a result of N-glycosylation of CLEC12A [32]. While CLEC12A showed multiple bands between 37 and 50kDa, a single band was observed for CLEC12A-MIR223HG. This was surprising because CLEC12A-MIR223HG retains all predicted N-glycosylation sites of CLEC12A. Consistent with a previous report [32], the treatment of cell lysates with PNGase F to remove N-glycosylation, resulted in bands at predicted molecular weights for both CLEC12A and CLEC12A-MIR223HG. We also observed a difference in interacting partners and cellular localizations of CLEC12A and CLEC12A-MIR223HG. Based on previous reports of other receptor proteins, we hypothesize that the change in the extracellular domain of CLEC12A either affects its cell surface trafficking or its stability at the cell surface [42,43]. Furthermore, post-translational modifications like glycosylation can also impact the cell surface expression of a protein [44]. It is also of note that the expression of the native fusion protein in tissues remains to be examined. These results suggest that the substitution of 52 amino acids in CLEC12A with 44 amino acids from MIR223HG results in a fusion protein with an altered glycosylation, sub-cellular localization, protein–protein interactions, possibly leading to different functions.

Consequently, we investigated the biological roles of *CLEC12A-MIR223HG* based on the previously reported functions of *CLEC12A*. Previous studies have reported a higher expression of *CLEC12A* in AML cells resistant to cytarabine (AraC) [31]. In contrast to previous observations, the treatment of U-937 cells with AraC did not result in a significant increase in *CLEC12A* expression. Instead, we observed a higher expression of *CLEC12A-MIR223HG*. These discrepancies could be partly explained by the cell line that was used in our study, which is not derived from AML patients. We then overexpressed either *CLEC12A* or *CLEC12A-MIR223HG* in U-937 cells and treated them with AraC. No significant differences in U-937 cell apoptosis were observed in either group. These observations might suggest that the subset of U-937 cells resistant to AraC may express *CLEC12A-MIR223HG* at a slightly higher level, but *CLEC12A-MIR223HG* by itself does not confer resistance to AraC.

Next, we compared the expression levels of the two genes in monocytes and monocytes differentiated into macrophage-like cells. *CLEC12A* is expressed at high levels in granulocyte-monocyte progenitor cells [29] and downregulated in monocytes treated with PMA, which induces their differentiation into macrophage-like cells [45]. We used CD44 and CD45 as markers of differentiation as the expression of CLEC12A is lower in CD45-positive differentiated monocytes [46]. Interestingly, there was a significant increase in the expression of *CLEC12A-MIR223HG*. However, the rate of differentiation of U-937 did not significantly alter following the overexpression of either *CLEC12A* or *CLEC12A-MIR223HG*. Thus, we concluded that the change in expression of *CLEC12A* and *CLEC12A-MIR223HG* could be a marker, but not a significant driver, of monocyte activation and differentiation.

In conclusion, we have identified and characterized a novel chimeric RNA that differs substantially from its parental genes. Our results invite further studies aimed at understanding the roles of *CLEC12A-MIR223HG* and provide an experimental framework to study other chimeric transcripts in normal physiology and cancer. Our results also highlight the need for caution while discovering and reporting novel and potential diagnostic and prognostic cancer biomarkers.

4. Materials and Methods

4.1. Clinical Samples and Bioinformatics Analysis

Retrieval of patient samples and samples from healthy donors, RNA extraction, library preparation, sequencing, and data analysis has been described previously [28]. In brief, we retrieved 16 diagnostic specimens (total leukocytes from peripheral blood) from treatment naïve CML patients, 10 matched remission samples following successful TKI treatment, and 6 samples from healthy donors. Total RNA was isolated using Trizol and subjected to mRNA sequencing after poly-A-enrichment. Paired-end RNA-sequencing reads (125 nt) were trimmed and mapped to the human reference genome hg38 using

STAR v2.7 [47]. STAR-FUSION v1.4.0 [48] was used for the identification of fusion genes and Fusion Inspector (FusionInspector.github.io) for in silico validation of the predicted gene fusions. In addition, we used Arriba [49] to independently reconfirm the *BCR/ABL1* and *CLEC12A/MIR223HG* fusion predictions. GTEx data of 17,382 samples from 30 different tissues were analysed to access correlation between the expression of *CLEC12A* and *MIR223HG*.

4.2. Cell Lines and Culture

HEK293 cells were cultured in Dulbecco's Modified Eagle Medium (DMEM; Gibco; Waltham, MA, USA, Cat#12430054) supplemented with 10% (v/v) fetal bovine serum (FBS; HyClone; Marlborough, MA, USA, SH30084.03). Cell lines U-937 (ATCC; CRL-1593.2), K-562 (ATCC; CCL-243) and THP-1 (ATCC; TIB-202) were cultured in Roswell Park Memorial Institute 1640 (RPMI-1640; Gibco; Cat#22400089) medium with 10% FBS. THP-1 cells were additionally supplemented with 1X non-essential amino acids (NEAA; Gibco; Cat#11140050) and 1 mM sodium pyruvate (Gibco; Cat#11360070). HL-60 (ATCC; CCL-240) was cultured in Iscove's Modified Dulbecco's Medium (IMDM; Gibco; Cat#12440053) complemented with 20% (v/v) FBS. DMEM containing 20% (v/v) FBS was utilised for cell line MOLM-13 (DSMZ; ACC-554) and MEM Alpha (Gibco; Cat#12571063) containing 10% (v/v) FBS for OCI-AML2 (DSMZ; ACC-99). All growth media contained 100 U/mL penicillin and 100 µg/mL streptomycin (Gibco; Cat#15140122). Cells were maintained in a humidified incubator with 5% CO_2 at 37 °C. All cell lines used in this study were checked regularly for mycoplasma and authenticated using short tandem repeat profiling.

4.3. Expression Vectors, Lentivirus Production, and Transduction

FLAG-CLEC12A and FLAG-CLEC12A-MIR223HG coding sequences were obtained as Geneblocks (IDT Australia) and cloned into a FUW lentiviral plasmid backbone (Addgene #14882) in-frame to an upstream mcherry-P2A sequence. All plasmids were sequence verified with Sanger sequencing (Australian Genome Research Facility). For lentivirus production, HEK293 cells were transfected with packaging plasmids pMD2-VSV-G (Addgene #12259), pRSV-Rev (Addgene #12253), and pMD2-g/pRRE (Addgene #12251) with calcium phosphate. The virus-containing supernatant was collected 48 h post-transfection, filtered through a 0.45 µm filter, and snap-frozen. For transduction, cells (5×10^5) were transferred to FACS tubes (Corning Inc., Corning, NY, USA, Cat#352058). Cells were resuspended in 500 µL of fresh medium and 500 µL of viral supernatant with 4 µg/mL Polybrene (Sigma-Aldrich; St. Louis, MO, USA, Cat#TR-1003). After spinoculation at 1500 rpm for 1.5 h at room temperature, cells were incubated for 4 h at 37 °C, 5% CO_2. Media was then refreshed, and cells were incubated for 72 h. Transduction efficiency was assessed by measuring the percentage of mCherry$^+$ cells by flow cytometry (BD LSRFortessa, Sydney Cytometry) and using FlowJo v10 (BD Biosciences, San Jose, CA, USA) for data analysis.

4.4. RNA Isolation and RT-PCR

Total RNA was extracted using TRIzol™ (Thermo Fisher Scientific, Waltham, MA, USA, Cat#15596026) as instructed by the manufacturer. Complementary DNA was subsequently synthesized from total RNA using SuperScript™ III reverse transcriptase (Thermo Fisher Scientific, Cat#18080093) and PCR was performed using a PCR thermal cycler (Eppendorf, Germany, MasterCycler epgradients). The PCR program used for amplification consisted of 2 min at 92 °C, followed by 35 cycles of 5 s at 92 °C, 10 s at 60 °C, and 20 s at 72 °C and concluded with 10 min at 72 °C. Beta-2-Microglobulin (B2M) gene was used to analyse the relative gene expression with either the dCt or the ddCt method. All the primer sequences are listed in Supplementary Table S3.

4.5. Western Blotting

Total protein extracts were isolated with NP-40 buffer (1% (v/v) NP-40, 0.15 M NaCl, 10 mM EDTA, 10 mM NaN_3, 10 mM Tris-HCl pH 8) containing cOmplete™ Protease Inhibitor Cocktail (Roche, Basel, Switzerland, Cat#116974980001). The protein extracts (20 μg) were separated on SDS PAGE (Bolt™ 4–12% Bis-Tris Plus; Invitrogen; Cat#NW04120BOX) and transferred to Immobilon™-P PVDF membrane (Merck Millipore; Darmstadt, Germany, Cat#IPVH00010). Following the blocking in 5% (*w/v*) bovine serum albumin, the membranes were probed with primary antibodies FLAG-HRP (1:1000; Merck; Cat#F1804) or GAPDH mouse monoclonal antibody (1:5000; Abcam; Cambridge, UK, Cat#ab8245) and secondary antibody (1:5000; Merck Millipore; Cat#AP192P). To remove N-linked glycosylation moieties from cell lysates, equal quantities of protein were incubated with PNGase F (New England Biolabs; Ipswich, MA, USA, Cat#P0704S) at 37 °C for 24 h, prior to running on SDS-PAGE gels.

4.6. Immunofluorescence

U-937 cells were seeded in ibiTreat μ-Slide 8 Well chambers slide (Ibidi; Germany, Cat#80826) (200 000 cells/well) after being subjected to lentiviral transductions. Cells were fixed with 4% (*w/v*) paraformaldehyde (Thermo Fisher Scientific; Cat#28906), permeabilized with Triton X-100 0.2% (Sigma-Aldrich; Cat#X100), and incubated with anti-FLAG mouse monoclonal primary antibody (1:500; Merck; Cat#F1804) and Alexa Fluor 488 rabbit-anti-mouse IgG (1:1000; Invitrogen; Cat# A27023) secondary antibody. Cells were washed with PBS between steps and incubated with DAPI (5 min; 1:5000; Invitrogen; D1306) before mounting with ibidi mounting medium (Ibidi; Cat#50001). Fluorescence images were acquired using a Leica-SP8 confocal microscope with a HC PL APO CS2 40/1.10 water objective and analysed using the Bitplane Imaris software (Imaris, Zurich, Switzerland).

4.7. Liquid Chromatography with Tandem Mass Spectrometry (LC-MS/MS) Analysis

In order to map the protein interaction profile of CLEC12A and CLEC12A-MIR223HG, FLAG-tagged protein pulldown followed by label-free quantification was performed. Whole cell extracts of U-937 cells expressing FLAG-CLEC12A and FLAG-CLEC12A-MIR223HG were incubated with 30 μL anti-FLAG beads (Sigma-Aldrich) for 2 hr. Six washes using a buffer containing 200 mM NaCl, 50 mM Tris-HCl, 0.25% (*v/v*) IGEPAL, pH 7.9 were performed (5 min each; rotating end-to-end). Affinity-purified protein complexes were subjected to on-bead trypsin/LysC digestion in 2 M urea for 1 h at 30 °C. Next, the beads were collected and the supernatant was transferred into low retention tubes. Beads were resuspended in 25 μL of 2 M urea containing 20 mM IAA for 20 min. Tryptic peptides were acidified to a final concentration of 2% (*v/v*) with formic acid (Sigma Aldrich) and desalted using ZipTips (Thermo Fisher Scientific). Liquid Chromatography with tandem mass spectrometry (LC-MS/MS) analysis was performed on a Thermo Scientific Q-Exactive HF-X hybrid quadrupole-Orbitrap mass spectrometer. Raw data were analysed by MaxQuant (version 1.6.6.0) [50] using standard settings. Methionine oxidation (M) and carbamidomethyl cysteine (C) were selected as variable and fixed modifications, respectively. Identified peptides were searched against the reference human proteome. Proteins with less than 2 unique peptides and more than one LFQ missing values were omitted from the analysis. Perseus algorithm was used to impute the missing values [51]. Output files were further processed and keratin, heat shock, and ribosomal proteins were excluded from the analysis. To detect fusion-specific unique peptides, we allowed up to four missed-cleavage and selected match between runs and dependent peptides.

4.8. CRISPR-SAM-Mediated Activation of CLEC12A and Analysis of Chromosomal Translocations in HEK293 Cells

HEK293 cells were transduced with dCas9-VP64_Blast (Addgene #61425) and MS2-p65-HSF1_Hygro (Addgene #61426) lentiviruses. The media was replenished with 2 μg/mL blasticidin (AG Scientific, CA, USA, Cat# B-1247) and 400 μg/mL hygromycin (Roche, Cat#10843555001) 24 hrs post-transduction. The cells were selected for 7 days and the surviving cells were referred to as HEK293-dCas9/MPH. The sgRNA to activate *CLEC12A* was designed as previously described [52]. SgRNA was obtained as single-stranded oligonucleotides (IDT Australia), treated with T4 Polynucleotide Kinase (NEB #M0201), annealed and ligated to lenti sgRNA(MS2)_puro (Addgene #7379) digested with *BsmBI* (NEB, Cat #R0580). A 20 bp non-targeting scrambled sequence in the plasmid backbone was used as a non-targeting control. The plasmids were verified by Sanger Sequencing (Australian Genome Research Facility). For the activation of *CLEC12A*, HEK293-dCas9/MPH cells were transduced with lenti sgRNA(MS2) lentiviruses expressing CLEC12A-targeting sgRNA (SAM-CLEC12A-sgRNA) by spinoculation in a 12-well plate and selected with 2 µg/mL puromycin (Gibco, Cat#A1113803). The expression of *CLEC12A* and *CLEC12A-MIR223HG* was then quantified with qRT-PCR. Whole genome sequencing data for HEK293 were obtained from NCBI SRA (SRR2123657). Raw fastq files were downloaded from ENA and aligned to the hg38 using Speedseq tool that internally uses BWA-MEM for alignment (https://github.com/hall-lab/speedseq#speedseq-realign) (accessed on 26 October 2021) [53].

4.9. U-937 Monocyte Differentiation Assay

In our method, 10^6 U-937 monocytic cells were differentiated with 100 nM PMA (phorbol 12-myristate 13-acetate) (Sigma, Cat#P8139) in 6-well plates. Differentiated cells were collected using the StemPro™ Accutase™ Cell Dissociation Reagent (Life Tech, Cat#A1110501) 48 h post-treatment. Collected cells were washed with PBS and either processed for flow cytometry or RNA extraction. For flow cytometry, cells were washed twice with PBS and stained with CD44-PE (BD, cat# 555479) and CD45-APC (BD Biosciences, cat#559864) for one hour in the dark. Following incubation, the cells were washed twice with PBS and analysed with flow cytometry (BD LSR Fortessa, Sydney Cytometry) and data visualised using FlowJo v10 (BD Biosciences).

4.10. Cell Apoptosis with Annexin V Staining

U-937 cells were seeded in 6-well plates (500,000 cells/well/replicate) and incubated with Cytarabine (AraC, Sigma-Aldrich; Cat#147-94-4) for 72 h. The cells were collected, washed with PBS, and incubated for 15 min in the dark with Annexin V-APC (BD Biosciences; Cat#550474) and DAPI (1:1000). The percentage of apoptotic cells was quantified with flow cytometry and data visualised using FlowJo v10 software.

4.11. Cell Viability Assay

MTT Formazan powder (Sigma-Aldrich; Cat#88417) was dissolved in sterile PBS at a working concentration of 5 mg/mL. The solution was then filtered using a 0.22 µm syringe filter and stored at 4 °C for short-term and −20 °C for long-term use. U-937 cells were seeded in 96-well plates (2000 cells/well/replicate) after lentiviral transductions. Cells were then incubated with AraC for 72 hr. MTT solution was added at 0.5 mg/mL and incubated overnight prior to the addition of 110 µL of the Colour Development Solution (isopropanol with 0.04 N HCl). Absorbance was recorded with the POLARstar microplate reader (BMG LABTECH; VIC, AUS) at 570 and 630 nm wavelength. Calculations were performed using the equation $A_{570nm} - A_{630nm}$. Media only control wells were then subtracted from all the readings. The addition of MTT solution and colour development solution was repeated at 0 and 72 h, with plate absorbance readings occurring at the same time point.

4.12. Colony Formation-Methocult Assay

U-937 cells were diluted in media at a concentration of 100 cells/µL. IMDM + 2% FBS (900 µL) (StemCell Technologies; Vancouver, Canada, Cat#07700) was then added to the cells and mixed thoroughly. Cell mixture (1 mL) was added to 3 mL of Methocult Media H4230 (StemCell Technologies; Cat#04230). The methocult/cell mixture (1 mL) was then aliquoted into gridded 35 mm tissue culture dishes in triplicate (Sarsedt; Germany, Cat#83.3900.002). Dishes were incubated 5% CO_2 at 37 °C for 8 days. Colony numbers were counted and analysed with GraphPad Prism 8.

4.13. Statistical Analysis

All data are reported as mean $\pm$ SD of at least three independent experiments. Data with two groups to compare were analysed using unpaired, two-tailed *t*-test in GraphPad Prism 8. Pearson correlation coefficient was used to access the correlation between the expression of *CLEC12A* and *MIR223HG*. The significance level chosen for the statistical analysis were ** $p < 0.01$, * $p < 0.05$.

Supplementary Materials: The following are available online at https://www.mdpi.com/article/10.3390/ijms222212178/s1, Supplementary Figure S1: Scatterplots illustrating correlation between *CLEC12A* and *MIR223HG* in individual tissues. Figure S2, A: Detection of lentivirus-mediated overexpression of *CLEC12A* and *CLEC12A-MIR223HG* in different cell lines, B: cell growth and proliferation assays after overexpression of *CLEC12A* and *CLEC12A-MIR223HG*, C: Examination of U-937 apoptosis after cytarabine treatment at different doses. D: Percentage of apoptotic U-937 cells after cytarabine treatment. E: PCR on the genomic DNA and cDNA of U-937 cells to detect *CLEC12A-MIR223HG*. Supplementary Table S1: List of fusion transcripts involving *CLEC12A* in CML patients, Supplementary Table S2: Cellular localization of interactors of CLEC12A and CLEC12A-MIR223HG identified with pull down/mass spectrometry, Supplementary Table S3: List of all the primers used in this study, Supplementary Table S4: List of all chromosomal translocations in HEK293 cells.

Author Contributions: Conceptualization, B.P.D., G.M. and U.S.; Methodology, G.M., B.P.D., M.S.T., C.G., J.S.S., S.H., A.R.M.F., D.G., Y.F., R.N. and C.M.; Writing—original draft, G.M. and B.P.D.; Writing—review and editing, G.M., B.P.D., C.G.B., U.S. and J.E.J.R.; Funding acquisition, U.S. and J.E.J.R. All authors have read and agreed to the published version of the manuscript.

Funding: This work was supported by the National Health and Medical Research Council (Investigator Grant 1177305 to J.E.J.R and 1196405 to U.S.; Project Grants #1080530, #1061906, #1128175 and #1129901 to J.E.J.R.; the NSW Genomics Collaborative Grant (J.E.J.R).; Cure the Future (J.E.J.R.); Tour de Cure research grants to J.E.J.R.; and an anonymous foundation (J.E.J.R.). U.S. also held a fellowship from the Cancer Institute of New South Wales. Financial support was also provided by Cancer Council NSW project grants RG11-12 to J.E.J.R. and RG20-12 to U.S.

Institutional Review Board Statement: The study was conducted according to the guidelines of the Declaration of Helsinki, and approved by the Institutional Review Boards: HREC protocol No: 131015 (on 31 October 2013), 081211 (on 15 December 2008), and 101010 (on 18 November 2010)-Royal Adelaide Hospital and 100912—Inserm internal ethical committee, 2012.

Informed Consent Statement: Informed consent was obtained from all subjects involved in the study.

Data Availability Statement: RNA sequencing data can be accessed at Gene Expression Omnibus (GEO, https://www.ncbi.nlm.nih.gov/geo/) accession number: GSE144119.

Acknowledgments: We thank the Sydney Cytometry and Centenary Imaging Facility, Centenary Institute.

Conflicts of Interest: J.E.J.R. has received honoraria or speaker fees (GSK, Miltenyi, Takeda, Gilead, Pfizer, Spark, Novartis, Celgene, bluebird bio); Director of Pathology (Genea); equity ownership (Genea, Rarecyte); consultant (Rarecyte, Imago); chair, Gene Technology Technical Advisory, OGTR, Australian Government. The remaining authors declare no competing financial interests.

References

1. Jia, Y.; Xie, Z.; Li, H. Intergenically Spliced Chimeric RNAs in Cancer. *Trends Cancer* **2016**, *2*, 475–484. [CrossRef] [PubMed]
2. Nambiar, M.; Kari, V.; Raghavan, S.C. Chromosomal translocations in cancer. *Biochim. Biophys. Acta* **2008**, *1786*, 139–152. [CrossRef] [PubMed]
3. Quintás-Cardama, A.; Cortes, J. Molecular biology of bcr-abl1-positive chronic myeloid leukemia. *Blood* **2009**, *113*, 1619–1630. [CrossRef] [PubMed]
4. Apperley, J.F. Part I: Mechanisms of resistance to imatinib in chronic myeloid leukaemia. *Lancet Oncol.* **2007**, *8*, 1018–1029. [CrossRef]
5. Shaw, A.T.; Yeap, B.Y.; Solomon, B.J.; Riely, G.J.; Gainor, J.; Engelman, J.A.; Shapiro, G.I.; Costa, D.B.; Ou, S.H.; Butaney, M.; et al. Effect of crizotinib on overall survival in patients with advanced non-small-cell lung cancer harbouring ALK gene rearrangement: A retrospective analysis. *Lancet Oncol.* **2011**, *12*, 1004–1012. [CrossRef]

6. Babiceanu, M.; Qin, F.; Xie, Z.; Jia, Y.; Lopez, K.; Janus, N.; Facemire, L.; Kumar, S.; Pang, Y.; Qi, Y.; et al. Recurrent chimeric fusion RNAs in non-cancer tissues and cells. *Nucleic Acids Res.* **2016**, *44*, 2859–2872. [CrossRef]
7. Velusamy, T.; Palanisamy, N.; Kalyana-Sundaram, S.; Sahasrabuddhe, A.A.; Maher, C.A.; Robinson, D.R.; Bahler, D.W.; Cornell, T.T.; Wilson, T.E.; Lim, M.S.; et al. Recurrent reciprocal RNA chimera involving YPEL5 and PPP1CB in chronic lymphocytic leukemia. *Proc. Natl. Acad. Sci. USA* **2013**, *110*, 3035–3040. [CrossRef]
8. Li, H.; Wang, J.; Ma, X.; Sklar, J. Gene fusions and RNA trans-splicing in normal and neoplastic human cells. *Cell Cycle* **2009**, *8*, 218–222. [CrossRef]
9. Lei, Q.; Li, C.; Zuo, Z.; Huang, C.; Cheng, H.; Zhou, R. Evolutionary Insights into RNA trans-Splicing in Vertebrates. *Genome Biol. Evol.* **2016**, *8*, 562–577. [CrossRef]
10. Chwalenia, K.; Facemire, L.; Li, H. Chimeric RNAs in cancer and normal physiology. *Wiley Interdiscip. Rev. RNA* **2017**, *8*, e1427. [CrossRef]
11. Nacu, S.; Yuan, W.; Kan, Z.; Bhatt, D.; Rivers, C.S.; Stinson, J.; Peters, B.A.; Modrusan, Z.; Jung, K.; Seshagiri, S.; et al. Deep RNA sequencing analysis of readthrough gene fusions in human prostate adenocarcinoma and reference samples. *BMC Med. Genom.* **2011**, *4*, 11. [CrossRef]
12. Kannan, K.; Wang, L.; Wang, J.; Ittmann, M.M.; Li, W.; Yen, L. Recurrent chimeric RNAs enriched in human prostate cancer identified by deep sequencing. *Proc. Natl. Acad. Sci. USA* **2011**, *108*, 9172–9177. [CrossRef]
13. Annala, M.J.; Parker, B.C.; Zhang, W.; Nykter, M. Fusion genes and their discovery using high throughput sequencing. *Cancer Lett.* **2013**, *340*, 192–200. [CrossRef]
14. Li, Z.; Qin, F.; Li, H. Chimeric RNAs and their implications in cancer. *Curr. Opin. Genet. Dev.* **2018**, *48*, 36–43. [CrossRef]
15. Elfman, J.; Li, H. Chimeric RNA in Cancer and Stem Cell Differentiation. *Stem Cells Int.* **2018**, *2018*, 3178789. [CrossRef]
16. Singh, S.; Qin, F.; Kumar, S.; Elfman, J.; Lin, E.; Pham, L.P.; Yang, A.; Li, H. The landscape of chimeric RNAs in non-diseased tissues and cells. *Nucleic Acids Res.* **2020**, *48*, 1764–1778. [CrossRef]
17. Wu, H.; Singh, S.; Shi, X.; Xie, Z.; Lin, E.; Li, X.; Li, H. Functional heritage: The evolution of chimeric RNA into a gene. *RNA Biol.* **2020**, *17*, 125–134. [CrossRef]
18. Qin, F.; Song, Z.; Babiceanu, M.; Song, Y.; Facemire, L.; Singh, R.; Adli, M.; Li, H. Discovery of CTCF-sensitive Cis-spliced fusion RNAs between adjacent genes in human prostate cells. *PLoS Genet.* **2015**, *11*, e1005001. [CrossRef]
19. Li, H.; Wang, J.; Mor, G.; Sklar, J. A neoplastic gene fusion mimics trans-splicing of RNAs in normal human cells. *Science* **2008**, *321*, 1357–1361. [CrossRef]
20. Ma, X.; Wang, J.; Ma, C.X.; Gao, X.; Patriub, V.; Sklar, J.L. The JAZF1-SUZ12 fusion protein disrupts PRC2 complexes and impairs chromatin repression during human endometrial stromal tumorogenesis. *Oncotarget* **2017**, *8*, 4062–4078. [CrossRef]
21. Yuan, H.; Qin, F.; Movassagh, M.; Park, H.; Golden, W.; Xie, Z.; Zhang, P.; Sklar, J.; Li, H. A chimeric RNA characteristic of rhabdomyosarcoma in normal myogenesis process. *Cancer Discov.* **2013**, *3*, 1394–1403. [CrossRef]
22. Schmitt-Ney, M.; Camussi, G. The PAX3-FOXO1 fusion protein present in rhabdomyosarcoma interferes with normal FOXO activity and the TGF-β pathway. *PLoS ONE* **2015**, *10*, e0121474. [CrossRef]
23. Rodríguez-Martín, B.; Palumbo, E.; Marco-Sola, S.; Griebel, T.; Ribeca, P.; Alonso, G.; Rastrojo, A.; Aguado, B.; Guigó, R.; Djebali, S. ChimPipe: Accurate detection of fusion genes and transcription-induced chimeras from RNA-seq data. *BMC Genom.* **2017**, *18*, 7. [CrossRef]
24. Benelli, M.; Pescucci, C.; Marseglia, G.; Severgnini, M.; Torricelli, F.; Magi, A. Discovering chimeric transcripts in paired-end RNA-seq data by using EricScript. *Bioinformatics* **2012**, *28*, 3232–3239. [CrossRef]
25. Davidson, N.M.; Majewski, I.J.; Oshlack, A. JAFFA: High sensitivity transcriptome-focused fusion gene detection. *Genome Med.* **2015**, *7*, 43. [CrossRef]
26. Singh, S.; Li, H. Prediction, Characterization, and In Silico Validation of Chimeric RNAs. *Methods Mol. Biol.* **2020**, *2079*, 3–12. [CrossRef]
27. Balamurali, D.; Gorohovski, A.; Detroja, R.; Palande, V.; Raviv-Shay, D.; Frenkel-Morgenstern, M. ChiTaRS 5.0: The comprehensive database of chimeric transcripts matched with druggable fusions and 3D chromatin maps. *Nucleic Acids Res.* **2020**, *48*, D825–D834. [CrossRef]
28. Schmitz, U.; Shah, J.S.; Dhungel, B.P.; Monteuuis, G.; Luu, P.L.; Petrova, V.; Metierre, C.; Nair, S.S.; Bailey, C.G.; Saunders, V.A.; et al. Widespread Aberrant Alternative Splicing despite Molecular Remission in Chronic Myeloid Leukaemia Patients. *Cancers* **2020**, *12*, 3738. [CrossRef]
29. Bill, M.; B van Kooten Niekerk, P.; S Woll, P.; Laine Herborg, L.; Stidsholt Roug, A.; Hokland, P.; Nederby, L. Mapping the CLEC12A expression on myeloid progenitors in normal bone marrow; implications for understanding CLEC12A-related cancer stem cell biology. *J. Cell. Mol. Med.* **2018**, *22*, 2311–2318. [CrossRef]
30. Toft-Petersen, M.; Nederby, L.; Kjeldsen, E.; Kerndrup, G.B.; Brown, G.D.; Hokland, P.; Stidsholt Roug, A. Unravelling the relevance of CLEC12A as a cancer stem cell marker in myelodysplastic syndrome. *Br. J. Haematol.* **2016**, *175*, 393–401. [CrossRef]
31. Ma, H.; Padmanabhan, I.S.; Parmar, S.; Gong, Y. Targeting CLL-1 for acute myeloid leukemia therapy. *J. Hematol. Oncol.* **2019**, *12*, 41. [CrossRef] [PubMed]
32. Marshall, A.S.; Willment, J.A.; Pyz, E.; Dennehy, K.M.; Reid, D.M.; Dri, P.; Gordon, S.; Wong, S.Y.; Brown, G.D. Human MICL (CLEC12A) is differentially glycosylated and is down-regulated following cellular activation. *Eur. J. Immunol.* **2006**, *36*, 2159–2169. [CrossRef] [PubMed]

33. Neumann, K.; Castiñeiras-Vilariño, M.; Höckendorf, U.; Hannesschläger, N.; Lemeer, S.; Kupka, D.; Meyermann, S.; Lech, M.; Anders, H.J.; Kuster, B.; et al. Clec12a is an inhibitory receptor for uric acid crystals that regulates inflammation in response to cell death. *Immunity* **2014**, *40*, 389–399. [CrossRef] [PubMed]

34. Fukao, T.; Fukuda, Y.; Kiga, K.; Sharif, J.; Hino, K.; Enomoto, Y.; Kawamura, A.; Nakamura, K.; Takeuchi, T.; Tanabe, M. An evolutionarily conserved mechanism for microRNA-223 expression revealed by microRNA gene profiling. *Cell* **2007**, *129*, 617–631. [CrossRef]

35. Fazi, F.; Rosa, A.; Fatica, A.; Gelmetti, V.; De Marchis, M.L.; Nervi, C.; Bozzoni, I. A minicircuitry comprised of microRNA-223 and transcription factors NFI-A and C/EBPalpha regulates human granulopoiesis. *Cell* **2005**, *123*, 819–831. [CrossRef]

36. Fatica, A.; Rosa, A.; Fazi, F.; Ballarino, M.; Morlando, M.; De Angelis, F.G.; Caffarelli, E.; Nervi, C.; Bozzoni, I. MicroRNAs and hematopoietic differentiation. *Cold Spring Harb. Symp. Quant. Biol.* **2006**, *71*, 205–210. [CrossRef]

37. Qin, F.; Song, Y.; Zhang, Y.; Facemire, L.; Frierson, H.; Li, H. Role of CTCF in Regulating SLC45A3-ELK4 Chimeric RNA. *PLoS ONE* **2016**, *11*, e0150382. [CrossRef]

38. Fang, W.; Wei, Y.; Kang, Y.; Landweber, L.F. Detection of a common chimeric transcript between human chromosomes 7 and 16. *Biol. Direct* **2012**, *7*, 49. [CrossRef]

39. Rickman, D.S.; Pflueger, D.; Moss, B.; VanDoren, V.E.; Chen, C.X.; de la Taille, A.; Kuefer, R.; Tewari, A.K.; Setlur, S.R.; Demichelis, F.; et al. SLC45A3-ELK4 is a novel and frequent erythroblast transformation-specific fusion transcript in prostate cancer. *Cancer Res.* **2009**, *69*, 2734–2738. [CrossRef]

40. Morsink, L.M.; Walter, R.B.; Ossenkoppele, G.J. Prognostic and therapeutic role of CLEC12A in acute myeloid leukemia. *Blood Rev.* **2019**, *34*, 26–33. [CrossRef]

41. Li, K.; Neumann, K.; Duhan, V.; Namineni, S.; Hansen, A.L.; Wartewig, T.; Kurgyis, Z.; Holm, C.K.; Heikenwalder, M.; Lang, K.S.; et al. The uric acid crystal receptor Clec12A potentiates type I interferon responses. *Proc. Natl. Acad. Sci. USA* **2019**, *116*, 18544–18549. [CrossRef]

42. Dintzis, S.M.; Velculescu, V.E.; Pfeffer, S.R. Receptor extracellular domains may contain trafficking information. Studies of the 300-kDa mannose 6-phosphate receptor. *J. Biol. Chem.* **1994**, *269*, 12159–12166. [CrossRef]

43. Drake, M.T.; Shenoy, S.K.; Lefkowitz, R.J. Trafficking of G protein-coupled receptors. *Circ. Res.* **2006**, *99*, 570–582. [CrossRef]

44. Luo, H.; Zhao, L.; Ji, X.; Zhang, X.; Jin, Y.; Liu, W. Glycosylation affects the stability and subcellular distribution of human PAT1 protein. *FEBS Lett.* **2017**, *591*, 613–623. [CrossRef]

45. Marshall, A.S.; Willment, J.A.; Lin, H.H.; Williams, D.L.; Gordon, S.; Brown, G.D. Identification and characterization of a novel human myeloid inhibitory C-type lectin-like receptor (MICL) that is predominantly expressed on granulocytes and monocytes. *J. Biol. Chem.* **2004**, *279*, 14792–14802. [CrossRef]

46. Wang, J.; Chen, S.; Xiao, W.; Li, W.; Wang, L.; Yang, S.; Wang, W.; Xu, L.; Liao, S.; Liu, W.; et al. CAR-T cells targeting CLL-1 as an approach to treat acute myeloid leukemia. *J. Hematol. Oncol.* **2018**, *11*, 7. [CrossRef]

47. Dobin, A.; Davis, C.A.; Schlesinger, F.; Drenkow, J.; Zaleski, C.; Jha, S.; Batut, P.; Chaisson, M.; Gingeras, T.R. STAR: Ultrafast universal RNA-seq aligner. *Bioinformatics* **2013**, *29*, 15–21. [CrossRef]

48. Haas, B.J.; Dobin, A.; Li, B.; Stransky, N.; Pochet, N.; Regev, A. Accuracy assessment of fusion transcript detection via read-mapping and de novo fusion transcript assembly-based methods. *Genome Biol.* **2019**, *20*, 213. [CrossRef]

49. Uhrig, S.; Ellermann, J.; Walther, T.; Burkhardt, P.; Fröhlich, M.; Hutter, B.; Toprak, U.H.; Neumann, O.; Stenzinger, A.; Scholl, C.; et al. Accurate and efficient detection of gene fusions from RNA sequencing data. *Genome Res.* **2021**, *31*, 448–460. [CrossRef]

50. Cox, J.; Mann, M. MaxQuant enables high peptide identification rates, individualized p.p.b.-range mass accuracies and proteome-wide protein quantification. *Nat. Biotechnol.* **2008**, *26*, 1367–1372. [CrossRef]

51. Tyanova, S.; Temu, T.; Sinitcyn, P.; Carlson, A.; Hein, M.Y.; Geiger, T.; Mann, M.; Cox, J. The Perseus computational platform for comprehensive analysis of (prote)omics data. *Nat. Methods* **2016**, *13*, 731–740. [CrossRef]

52. Konermann, S.; Brigham, M.D.; Trevino, A.E.; Joung, J.; Abudayyeh, O.O.; Barcena, C.; Hsu, P.D.; Habib, N.; Gootenberg, J.S.; Nishimasu, H.; et al. Genome-scale transcriptional activation by an engineered CRISPR-Cas9 complex. *Nature* **2015**, *517*, 583–588. [CrossRef]

53. Chiang, C.; Layer, R.M.; Faust, G.G.; Lindberg, M.R.; Rose, D.B.; Garrison, E.P.; Marth, G.T.; Quinlan, A.R.; Hall, I.M. SpeedSeq: Ultra-fast personal genome analysis and interpretation. *Nat. Methods* **2015**, *12*, 966–968. [CrossRef]

International Journal of
Molecular Sciences

Article

Parallel Reporter Assays Identify Altered Regulatory Role of rs684232 in Leading to Prostate Cancer Predisposition

Naixia Ren, Qingqing Liu, Lingjie Yan and Qilai Huang *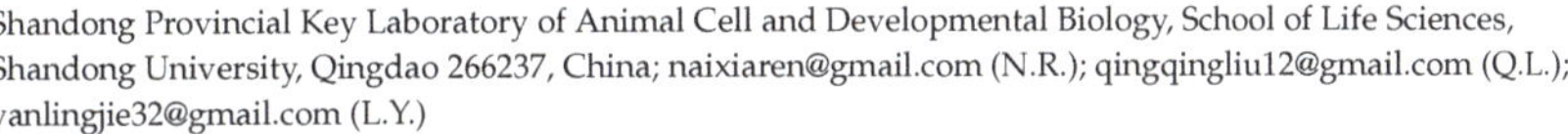

Shandong Provincial Key Laboratory of Animal Cell and Developmental Biology, School of Life Sciences, Shandong University, Qingdao 266237, China; naixiaren@gmail.com (N.R.); qingqingliu12@gmail.com (Q.L.); yanlingjie32@gmail.com (L.Y.)
* Correspondence: qlhuang@sdu.edu.cn

Abstract: Functional characterization of cancer risk-associated single nucleotide polymorphism (SNP) identified by genome-wide association studies (GWAS) has become a big challenge. To identify the regulatory risk SNPs that can lead to transcriptional misregulation, we performed parallel reporter gene assays with both alleles of 213 prostate cancer risk-associated GWAS SNPs in 22Rv1 cells. We disclosed 32 regulatory SNPs that exhibited different regulatory activities with two alleles. For one of the regulatory SNPs, rs684232, we found that the variation altered chromatin binding of transcription factor FOXA1 on the DNA region and led to aberrant gene expression of *VPS53*, *FAM57A*, and *GEMIN4*, which play vital roles in prostate cancer malignancy. Our findings reveal the roles and underlying mechanism of rs684232 in prostate cancer progression and hold great promise in benefiting prostate cancer patients with prognostic prediction and target therapies.

Keywords: parallel reporter assay; rs684232; FOXA1; *VPS53*; FAM57A; *GEMIN4*

Citation: Ren, N.; Liu, Q.; Yan, L.; Huang, Q. Parallel Reporter Assays Identify Altered Regulatory Role of rs684232 in Leading to Prostate Cancer Predisposition. *Int. J. Mol. Sci.* **2021**, *22*, 8792. https://doi.org/10.3390/ijms22168792

Academic Editor: Amelia Casamassimi

Received: 9 July 2021
Accepted: 13 August 2021
Published: 16 August 2021

Publisher's Note: MDPI stays neutral with regard to jurisdictional claims in published maps and institutional affiliations.

1. Introduction

More than 2000 genome-wide association (GWASs) studies have been published, identifying many loci associated with susceptibility to over 1000 unique traits and common diseases since 2005 [1,2]. Prostate cancer (MIM:176807) is the second most common cancer in males and the fifth leading cause of cancer death in men worldwide [3,4]. As with other complex diseases, the genetic heritability of prostate cancer is caused by both rarely occurring but higher penetrant genetic variants and moderate to commonly occurring variants conferring lower risks. So far, GWAS has identified over 170 low-penetrance prostate cancer susceptibility loci, including more than 1000 SNPs, predominantly in populations of mixed European ancestry [5–9]. Current researches on prostate cancer susceptibility variants can explain 34.4% of the familial risk of prostate cancer, with approximately 6% accounted for by rarely occurring variants and 28.4% attributed to more commonly occurring [minor allele frequency (MAF) > 1%] SNPs as well as some rarer single nucleotide variants [10]. Importantly, a significant number of susceptibility variants have been elucidated for their roles and underlying mechanism in leading to disease susceptibility [11–23]. Nevertheless, there is still a substantial knowledge gap between SNP-disease associations derived from GWASs and an understanding of how these risk SNPs contribute to the biology of human diseases [24].

A significant challenge remains to identify the functional SNPs from a large number of risk variants. These causal SNPs often locate in gene regulatory elements and can lead to transcriptional misregulation of cancer-related genes [14,15,25]. A parallel reporter gene assay method is urgently needed to evaluate the potential regulatory function of these SNPs. So far, several DNA barcode-based parallel reporter methods have been applied to the screening of regulatory risk sites [26–34]. Among them, the dinucleotide reporter system (DiR) was developed to realize parallel reporter assay with minimized tag composition bias, which made it suitable for investigating the subtle regulatory effect from the causal SNPs [34].

In this study, we applied the DiR-seq method to evaluate the prostate cancer risk-related SNPs in 22Rv1 cells. From 213 SNPs, we disclosed 32 regulatory SNPs with their two alleles conferring different regulatory activities. The rs684232 site is one of the regulatory sites and has been widely reported for association with prostate cancer susceptibility in European ancestry men [7,8,23,35,36]. However, the function and mechanism leading to cancer progression still remain unknown. We discovered that the rs684232 T allele increased forkhead box A1 (FOXA1 [MIM:602294]) binding and led to elevated gene expression of *VPS53* subunit of GARP complex (*VPS53* [MIM:615850]), family with sequence similarity 57 member A (*FAM57A* [MIM:611627]), and gem nuclear organelle associated protein 4 (*GEMIN4* [MIM:606969]). The upregulation of the three genes often occurred in prostate cancer tissues and was associated with low disease-specific survival probability for prostate cancer patients. Our findings reveal the roles and underlying mechanism of risk SNP in prostate cancer progression and contribute to defining it as a biomarker for prostate cancer susceptibility or therapeutic responses.

2. Results

2.1. DiR Assay Discovers Regulatory SNPs in Prostate Cancer Cells

The majority of the SNPs that have associations with increased cancer risk function as potential gene regulatory elements [14,15,25]. Notably, up to 57.1% of GWAS SNPs are located in the DHSs (DNase hypersensitive sites), which indicates that most GWAS SNPs have potential regulatory functions themselves [25]. To identify functional prostate cancer risk-associated variants displaying transcription regulatory function, we used the DiR-seq approach to evaluate the 213 prostate cancer risk SNPs with both risk and protective alleles from the previous GWAS catalog (Table S1). We cloned the 55 bp fragments bearing individual alleles in the middle and inserted them into the DiR vectors right upstream of the basic SV40 promoter (Figure 1A). After transfection of the plasmid pool in the cells, the RNA was extracted, reverse transcribed, prepared next-generation sequencing (NGS) library, and subjected to sequencing (Figure 1A). For all variants, a blanked DiR construct without any insertion was taken as a control. Considering that the typical transcription factors only occupy 6–12 bp DNA sequences [37], and short DNA stretches bearing transcription factor binding sites were widely used in the reporter gene assay [38], the 55 bp fragment will generally be enough to assess the effect of SNPs, even though we might not observe the full activity of the potential herein enhancer element. We used 150-bp paired-end sequencing to ensure the detection of the variants at the DiR barcode sequences (450 bp). Using this method, we could derive the allelic regulatory activities in DNA sequencing compared with those in RNA sequencing.

Our DiR-seq analysis in 22Rv1 cells showed that the tag expression levels had high consistency between individual replicates (Figure 1B), and some SNP sites exhibited elevated tag expression levels compared to the template (Figure 1C). Since the two alleles of the functional SNPs are supposed to drive gene expression differentially, we picked SNPs based on the ratio of reporter expression level from the risk and protective alleles (Figure 1D). In the 22Rv1 prostate cancer cell line, 14 SNPs exhibited decreased expression levels for the risk alleles (risk/protective < 0.8, $p < 0.05$), and 18 SNPs showed increased expression to the contrary (risk/protective > 1.2, $p < 0.05$) (Figure 1D). All of the SNPs picked out in the DiR-seq analysis are listed in Table S6.

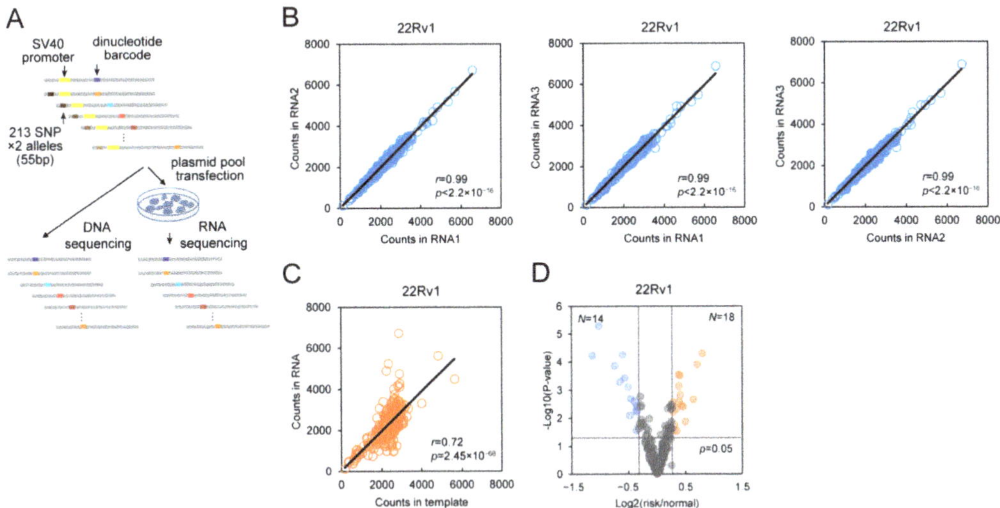

Figure 1. DiR-seq assays discovered regulatory SNPs in prostate cancer cells. (**A**) Flowchart of the DiR system for reporter gene assay in prostate cancer cells. The DiR plasmid library was developed based on the pGL3-promoter vector by optimizing the luciferase coding sequence. The 55 bp SNPs fragment was inserted upstream of the SV40 promoter, 450 bp dinucleotide-barcoded sequences were used as the reporter gene. (**B**) Consistency evaluation between individual biological replicates of DiR-seq assay of 213 prostate cancer risk SNPs in 22Rv1 cells. The correlation coefficient values and p values were calculated with Pearson correlation analysis (**C**) Scatter plot of DiR-seq tag counts in RNA and DNA template in 22Rv1 cells. The correlation coefficient values and p values were calculated with Pearson correlation analysis. (**D**) Volcano plot of the allelic ratio of reporter expression levels in DiR-seq analysis in the 22Rv1 cell line. p values came from a two-tailed Student's t-test of the reporter expression of individual alleles. The blue dots represent SNP sites satisfying the criteria of fold change < 0.8, $p < 0.05$, and the orange dots represent SNP sites satisfying the criteria of fold change > 1.2, $p < 0.05$.

2.2. Chromatin Status of 32 Functional SNPs

In eukaryotes, transcription activating elements usually locate in chromatin regions having high open status. We evaluated the chromatin open status of the 32 regulatory SNPs identified in the 22Rv1 DiR-seq analysis by the FAIRE qPCR method [39] and found more than half significantly enriched in the FAIRE DNA (Figure 2). Among the 32 regulatory SNPs, thirteen heterozygous SNP sites are highlighting in orange in Figure 2. Interestingly, Sanger sequencing chromatography of the FAIRE DNA showed strongly allele-specific openness for rs684232, rs887391, and rs5759167 sites in 22Rv1 cells (Figure S1A,B and S2A,B). We then chose the two most enriched heterozygous sites, rs684232 and rs887391, for further exploration.

2.3. Allele-Specific Activity of rs684232 and rs887391

In 22Rv1 cells, DiR-seq (Figure 3A), DiR-qPCR (Figure 3B), and luciferase reporter assays (Figure 3C) indicated that the rs684232 region could drive reporter expression, and the T allele exhibited significantly higher activity than the C allele. Interestingly, the T allele was also highly preferred in the active chromatin region, as shown in Sanger sequencing chromatography of the FAIRE DNA (Figure 3D), and the allele-specific enrichment was also confirmed by AS-qPCR (Figure 3E). For the site rs887391, the T allele exhibited significantly higher enrichment than the C allele in FAIRE DNA as determined by AS-qPCR (Figure S2C). Besides, the rs11672691 site, one SNP in LD (Linkage disequilibrium) with rs887391, also enriched in the FAIRE DNA in an allele-specific manner (Figure S2D–F), whose biological function and underlying mechanism had been illustrated previously [15,40].

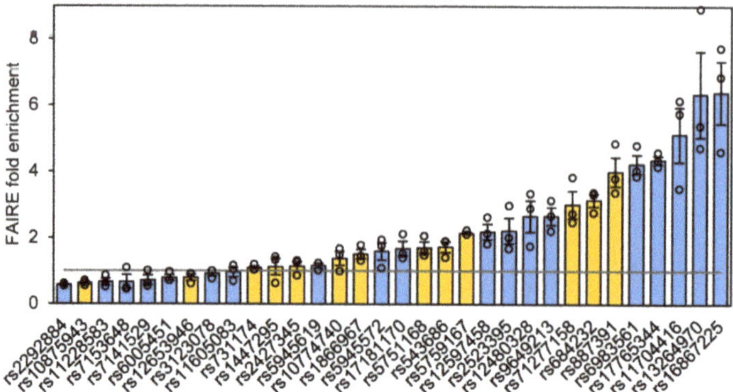

Figure 2. Chromatin status of the 32 functional SNPs in 22Rv1 cells. FAIRE-qPCR analysis of the 32 regulatory SNPs identified by DiR-seq analysis in 22Rv1. Heterozygous SNPs are highlighted in orange. Mean ± SEM of three independent experiments.

The H3K4me3 and H3K27ac histone modifications are usually the markers of active gene regulatory elements [14,41,42]. Our ChIP qPCR analysis showed that rs684232 (Figure 3F), rs887391, and rs11672691 (Figure S2G) were significantly enriched in these two types of histone modifications. Interestingly, the enrichment of the rs684232 site in histone modifications also displayed a strong preference for the T allele (Figure 3G). Further AS-qPCR analysis of the ChIP DNA confirmed the allele-specific enrichment (Figure 3H). Correspondingly, both rs887391 and rs11672691 also exhibited allele-specific enrichment in the H3K4me3 and H3K27ac ChIP DNA (Figure S2H–K). The results suggest that all three risk SNPs are potential gene regulatory variants. Since the rs11672691 and rs887391 have been reported previously for their biological function and underlying mechanism [15,40], we focused on the rs684232 site in the subsequent mechanism study.

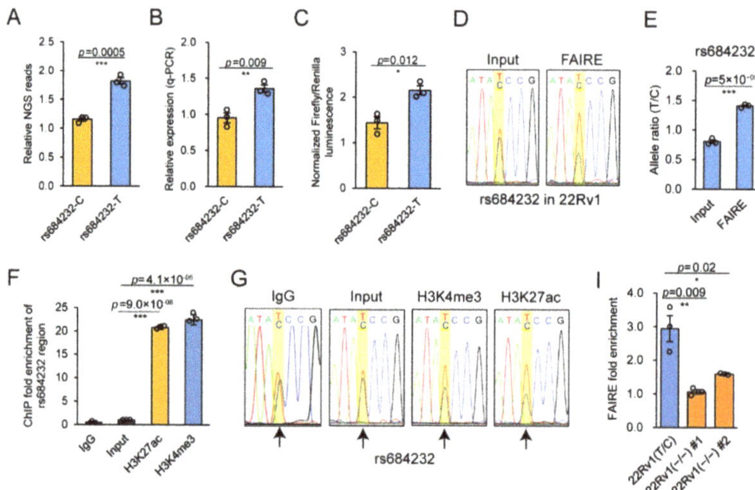

Figure 3. *Cont.*

H

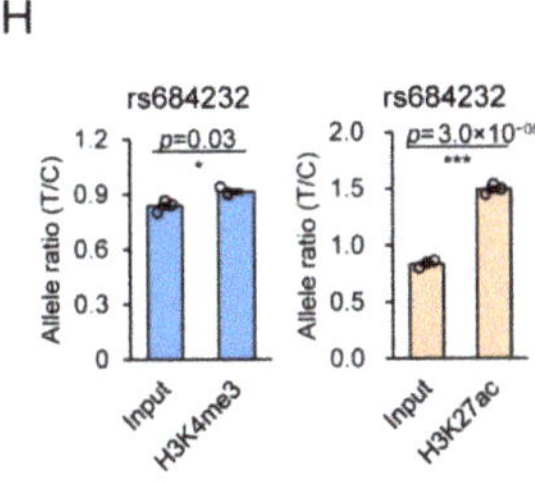

J

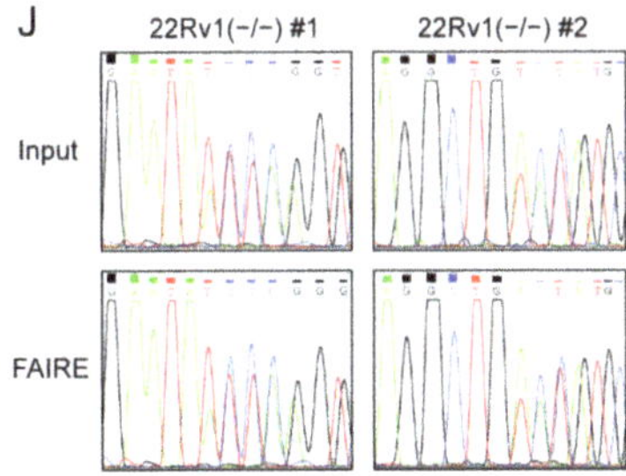

Figure 3. Chromatin status of the functional SNP rs684232 in 22Rv1 cells. (**A**) Reporter gene expression level for rs684232 SNP region in the DiR-seq analysis in 22Rv1 cells. T allele showed increased enhancer activity relative to the C allele. Mean ± SEM of three independent experiments. *** $p < 0.001$, two-tailed Student's *t*-test. (**B**) Reporter gene expression level for rs684232 SNP region in the DiR-qPCR assay in 22Rv1 cells. Mean ± SEM of three independent experiments. ** $p < 0.01$, two-tailed Student's *t*-test. (**C**) Normalized luciferase activity for the rs684232 region in 22Rv1 cells. T allele showed increased enhancer activity relative to the C allele. Mean ± SEM of three independent experiments. * $p < 0.05$, two-tailed Student's *t*-test. (**D**) Sanger sequencing chromatography of FAIRE DNA for the rs684232 site in 22Rv1. The position of rs684232 is highlighted in a yellow square. (**E**) Allele-specific enrichment of rs684232 site in FAIRE DNA determined in 22Rv1 cells by AS-qPCR. Mean ± SEM of three independent experiments. *** $p < 0.001$, two-tailed Student's *t*-test. (**F**) Fold enrichment of the rs684232 region in H3K27ac and H3K4me3 ChIP DNA in 22Rv1. Nonspecific immunoglobulin G (IgG) and Input as the negative control. Mean ± SD of three technical replicates. *** $p < 0.001$, two-tailed Student's *t*-test. (**G**) Sanger sequencing chromatography of rs684232 in H3K27ac and H3K4me3 ChIP DNA in 22Rv1 cells. IgG and Input as the negative control. The position of rs684232 is highlighted in a yellow square. (**H**) Allele-specific enrichment of rs684232 region in ChIP DNA of H3K27ac and H3K4me3 modification determined by AS-qPCR in 22Rv1. Mean ± SD of three technical replicates. * $p < 0.05$, *** $p < 0.001$, two-tailed Student's *t*-test. (**I**) Chromatin open status analysis of the rs684232 region in the two genome-edited cell lines 22Rv1(−/−) #1 and 22Rv1(−/−) #2. Mean ± SEM of three independent experiments. * $p < 0.05$, ** $p < 0.01$, two-tailed Student's *t*-test. (**J**) Sanger sequencing chromatography of the FAIRE DNA around rs684232 sites in the two genome-edited 22Rv1 cell lines.

2.4. The Gene Regulatory Function of SNP rs684232

The rs684232 site has been reported for association with prostate cancer susceptibility [7,8,36] and determined to be an important expression quantitative trait locus (eQTL) [42–45]. To further investigate the potential function of rs684232, we obtained two rs684232-edited single-cell clones, named 22Rv1(−/−) #1 and 22Rv1(−/−) #2 (Figure S3), through the CRISPR/Cas9 technology. Notably, indel mutation of the rs684232 element in the edited 22Rv1 cell clones led to hindered chromatin openness of this SNP region as evaluated by FAIRE qPCR analysis (Figure 3I). The Sanger sequencing chromatography of FAIRE DNA indicated that the allele preference also diminished upon the genome editing (Figure 3J).

The rs684232 site locates in 17p13.3 loci, 2 kb upstream of gene *VPS53*, and its LD SNPs rs2955626 and rs461251 were proposed to be the possible functional variants [7]. In our FAIRE analysis in 22Rv1, even though both SNPs were significantly enriched in the open chromatin regions (Figure S4A), neither exhibited significant allele preference as evaluated with Sanger sequencing and AS-qPCR analysis (Figure S4B–G). Besides, both SNPs were also enriched significantly in the H3K4me3 and H3K27ac histone modification regions in ChIP analysis (Figure S4H,I). Nevertheless, rs2955626, the higher enrichment site, did not exhibit allele specificity in both histone modifications as determined in Sanger sequencing and AS-qPCR analysis (Figure S4J,K), and rs461251 displayed allele specificity only for the H3K27ac modification (Figure S4L,M). In the luciferase reporter assay, the rs2955626 site did not exhibit apparent regulatory activity (Figure S5A). Even though the rs461251 genomic region showed apparent reporter gene activity, the two alleles did not drive gene expression differentially. (Figure S5B). It indicated that rs684232 should be the causal SNP on this locus, and the allele preference of rs461251 in H3K27ac modification might attribute to their closeness to the rs684232 site.

2.5. rs684232 Affects FOXA1 Chromatin Binding

Next, we explored the potential transcription factors that participate in the biological function of the rs684232 site using the HaploReg v4.1 [46]. We found that FOXA1 was the potential transcription factor that bound the rs684232 region. Our ChIP qPCR analysis further showed that the rs684232 region was significantly enriched in the FOXA1 cistrome in 22Rv1 cells (Figure 4A). Notably, the T allele was significantly preferred for the FOXA1 chromatin binding, as shown in the Sanger sequencing (Figure 4B) and AS-qPCR assay (Figure 4C). These results indicate that the rs684232 might affect the chromatin binding of FOXA1.

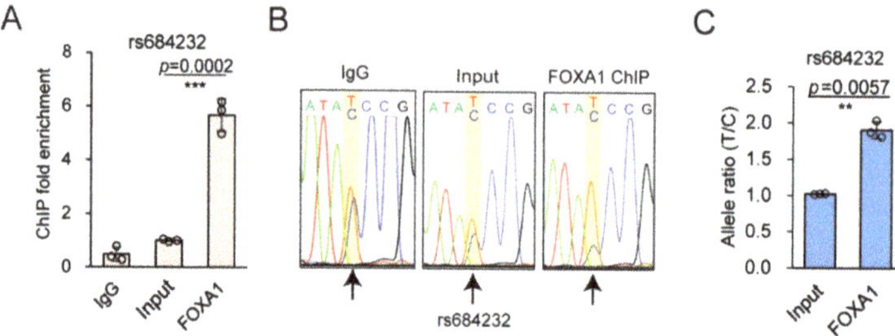

Figure 4. SNP rs684232 alters FOXA1 binding. (**A**) Enrichment of rs684232 region in the FOXA1 ChIP DNA in 22Rv1 cells. IgG and Input as the negative control. Mean ± SD of three technical replicates. *** $p < 0.001$, two−tailed Student's *t*-test. (**B**) Sanger sequencing chromatography of FOXA1 ChIP DNA for the rs684232 site in 22Rv1 cells. IgG and Input as the negative control. The position of rs684232 is highlighted with a yellow square and arrow. (**C**) Allele−specific enrichment of rs684232 site in the FOXA1 ChIP DNA determined by AS−qPCR in 22Rv1 cells. Mean ± SD of three technical replicates. ** $p < 0.01$, two−tailed Student's *t*-test.

2.6. rs684232 Regulates Gene Expression of VPS53, FAM57A, and GEMIN4 through FOXA1

The rs684232 site locates in 17p13.3 loci accompanied by three nearby genes, *VPS53*, *FAM57A*, and *GEMIN4* (Figure 5A). In eQTL analysis using the genotype-tissue expression (GTEx) database, *VPS53*, *FAM57A*, and *GEMIN4* genes all exhibited significant associations with the rs684232 variation by the normalized effect size (NES) of −0.40, −0.27, and −0.26, respectively (Figure 5B–D). Notably, the T allele, which was preferred in the active chromatin and exhibited higher activity in the reporter gene assay, also corresponded to higher expression levels for all three genes. We also found that the expression level of the three target genes significantly decreased upon indel mutation of the rs684232 site in 22Rv1(−/−) #1 and 22Rv1(−/−) #2 cells (Figure 5E). Furthermore, as the causal SNP, the heterozygous rs684232 site should drive allele-specific expression of the three genes in 22Rv1 cells. To investigate the allele imbalance, we picked three heterozygous SNPs, rs11558129, rs113201579, and rs3744741, in the exon of *VPS53*, *FAM57A*, and *GEMIN4*, respectively. Sanger sequencing chromatography of the 22Rv1 cDNA showed that all three genes had allele-specific expression (Figure S6A). Remarkably, when the rs684232 were mutated by indel, the allele preference in all the three genes diminished accordingly (Figure S6A). Moreover, FOXA1 knockdown with shRNA led to significant down-regulation of *VPS53*, *FAM57A*, and *GEMIN4* genes in 22Rv1 cells (Figure 5F). The results indicate that the rs684232 site might regulate gene expression of *VPS53*, *FAM57A*, and *GEMIN4* by affecting FOXA1 binding.

To further explore the relationship between FOXA1 and the three target genes, *VPS53*, *FAM57A*, and *GEMIN4*, we performed a gene express association assay using RNA-seq data from The Cancer Genome Atlas (TCGA, The Cancer Genome Atlas, RRID:SCR_003193) prostate cancer tissues (TCGA-PRAD, dbGaP Study Accession: phs000178). We found that the mRNA level of FOXA1 positively correlated strongly with *VPS53* (Spearman correlation coefficient = 0.62, Figure 5G), weakly with *FAM57A* (Spearman r = 0.37, Figure 5H), and moderately with *GEMIN4* (Spearman r = 0.48, Figure 5I) genes. Interestingly, for the 33 cancer types from the TCGA Pan-Cancer analysis project, up to 19, 16, and 16 cancer types displayed positive correlations for *FOXA1* vs. *VPS53*, *FOXA1* vs. *FAM57A*, and *FOXA1* vs. *GEMIN4*, respectively (Figure 5J–L). These results provide evidence that transcription factor FOXA1 regulates gene expression of *VPS53*, *FAM57A*, and *GEMIN4*.

Surprisingly, we also observed that the three target genes strongly correlated with each other in TCGA prostate cancer tissues with a Pearson correlation coefficient of 0.61, 0.7, and 0.78, respectively (Figure S6B). When explored in the TCGA Pan-Cancer tissues, all the 33 cancer types displayed a positive pairwise correlation between *VPS53*, *FAM57A*, and *GEMIN4* (Figure S6C). The results indicate that the three target genes *VPS53*, *FAM57A*, and *GEMIN4* may have a very important regulatory role in prostate cancer disease, and there may also be a synergistic promoting effect between them.

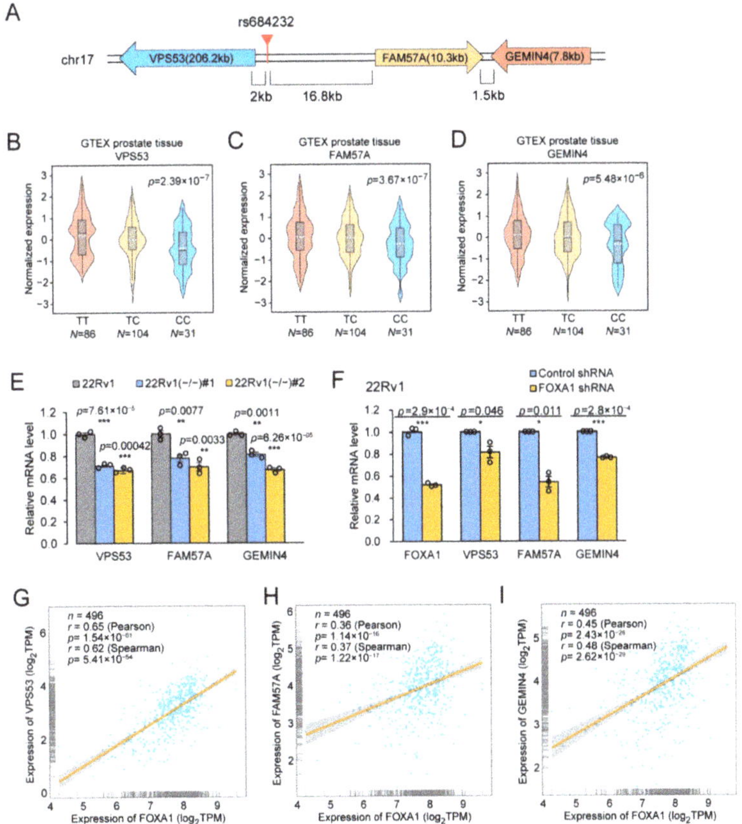

Figure 5. *Cont.*

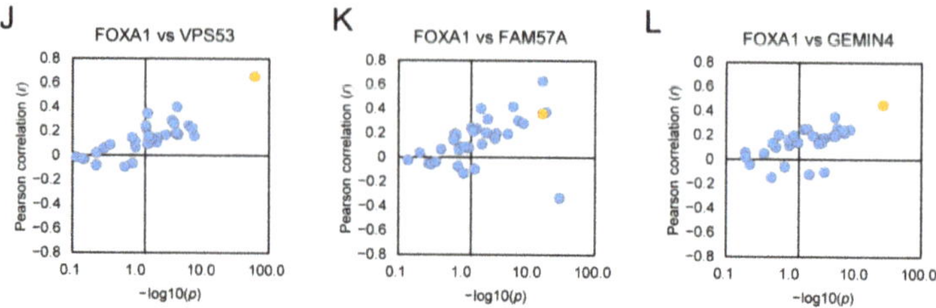

Figure 5. rs684232 regulates gene expression of *VPS53*, *FAM57A*, and *GEMIN4* through FOXA1. (**A**) The location of the rs684232 relative to three target genes. (**B–D**) eQTL analysis in GTEx prostate tissues to reveal the association between alleles of rs684232 and *VPS53* (**B**), *FAM57A* (**C**), and *GEMIN4* (**D**) genes. *p* values are from a linear regression model. (**E**) Gene expression quantification of *VPS53*, *FAM57A*, and *GEMIN4* by RT-qPCR in three 22Rv1 cells, including parental 22Rv1 cells and mutated cells 22Rv1($-/-$) #1/ #2. Mean ± SEM of three biological replicates. ** $p < 0.01$, *** $p < 0.001$, two-tailed Student's *t*-test. (**F**) Gene expression quantification of *VPS53*, *FAM57A*, and *GEMIN4* by RT-qPCR in 22Rv1 cells treated with *FOXA1* shRNA. Mean ± SD of three technical replicates. * $p < 0.05$, *** $p < 0.001$, two-tailed Student's *t*-test. (**G–I**) Gene expression correlation analysis between transcription factor FOXA1 and the three target genes *VPS53* (**G**), *FAN57A* (**H**), and *GEMIN4* (**I**) in prostate tumor tissues from TCGA-PRAD database. Correlation coefficient (*r*) values and *p* values were from Pearson or Spearman correlation analysis, respectively. (**J–L**) Gene expression correlation analysis between FOXA1 and *VPS53* (**J**), *FAM57A* (**K**), or *GEMIN4* (**L**) in 33 kinds of cancer tissues from TCGA. The Pearson correlation coefficient value of each cancer type was plotted vs. the $-\log 10$ of *p*-value. The dot in yellow represents the PRAD. Correlation coefficient (*r*) values and *p* values were from the Pearson correlation analysis.

2.7. VPS53, FAM57A, and GEMIN4 Knockdown Impedes Cancerous Phenotypes

To further understand the biological function of rs684232, we first assessed the effect of the *VPS53*, *FAM57A*, and *GEMIN4* genes on cancerous phenotypes in 22Rv1 cells. We found that the lentiviral shRNA knockdown for the three individual genes all impeded cell proliferation dramatically in a time-course CCK-8 assay (Figure 6A, B). Additionally, their capabilities in forming single-cell colonies were also hindered upon gene downregulation (Figure 6C). The results indicate that all three rs684232 target genes, *VPS53*, *FAM57A*, and *GEMIN4*, play essential roles during cell proliferation and colony formation of prostate cancer cells.

Next, we evaluated the cancerous phenotypes of the two genome-edited 22Rv1 cell lines, 22Rv1($-/-$) #1 and 22Rv1($-/-$) #2, that had rs6842323 site mutated. Remarkably, the genome-edited 22Rv1 cells exhibited decreased capabilities for both cell proliferation (Figure 6D) and colony formation (Figure 6E) dramatically. What is more, the mutation of rs684232 elements also delayed cancer cell migration dramatically, as demonstrated in the wound healing assay (Figure 6F,G). The results indicate that the rs684232 element is vital for cancer malignancy.

2.8. VPS53, FAM57A, and GEMIN4 Affect Cancer Progression

We then investigated the expression level of the three rs684232 target genes, *VPS53*, *FAM57A*, and *GEMIN4*, in cancer tissues and their effect on the clinical prognosis of cancer patients. In the TCGA prostate cohort, cancer tissues displayed significantly higher expression levels for the *VPS53* ($p = 0.014$, Figure 7A), *FAM57A* ($p = 0.013$, Figure 7B), and *GEMIN4* ($p = 1.2 \times 10^{-5}$, Figure 7C) genes in comparison to adjacent normal tissues. Furthermore, we performed the Kaplan–Meier survival analysis for the three genes using the TCGA prostate cancer cohort (Figure 7D–F). We found that patients with higher expression levels of *VPS53* (Log-rank $p = 0.03$, Figure 7D), *FAM57A* (Log-rank $p = 0.018$, Figure 7E), and *GEMIN4* (Log-rank $p = 0.016$, Figure 7F) had worse prostate cancer-specific survival probability. Remarkably, the TCGA Pan-Cancer patients with a higher expression level of *VPS53*, *FAM57A*, and *GEMIN4* also displayed decreased overall survival probability (Figure S7).

We next explored how the three genes affect the disease recurrence. Interestingly we found that patients with lower expression levels of *VPS53* (Log-rank $p = 0.031$) and *GEMIN4* (Log-rank $p = 0.017$) had shorter disease-free intervals (Figure S8). A similar trend was observed with the *FAM57A* gene (Log-rank $p = 0.109$), but without reaching statistical significance.

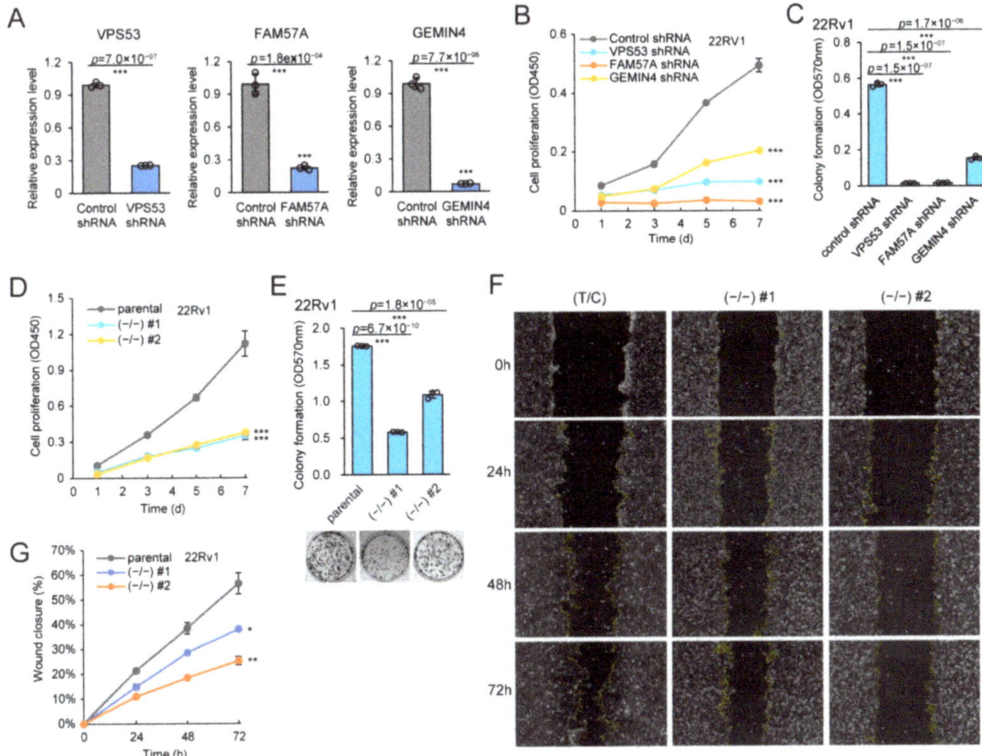

Figure 6. *VPS53*, *FAM57A*, and *GEMIN4* knockdown impede cancerous phenotypes. (**A**) Effect evaluation of lentiviral shRNAs targeting *VPS53*, *FAM57A*, and *GEMIN4* in 22Rv1 cells. Mean ± SD of three technical replicates. *** $p < 0.001$, two-tailed Student's *t*-test. (**B**) Cell proliferation assay of 22Rv1 cells undergoing lentiviral shRNA gene knockdown targeting *VPS53*, *FAM57A*, and *GEMIN4*, respectively. Cell viability was determined using the CCK-8 method at 1–7 d post-seeding. Mean ± SD of three biological replicates. *** $p < 0.001$, two-tailed Student's *t*-test. (**C**) Colony formation assay of 22Rv1 cells undergoing lentiviral shRNA gene knockdown targeting *VPS53*, *FAM57A*, and *GEMIN4*. Cell colonies were quantified through the Crystal Violet staining method. Mean ± SD of three biological replicates. *** $p < 0.001$, two-tailed Student's *t*-test. (**D**) Cell proliferation assay for the two genome-edited 22Rv1 cells with rs684232 site mutated through indels. Cell viability was determined using the CCK-8 method at 1–7 d post-seeding. Mean ± SD of three biological replicates. *** $p < 0.001$, two-tailed Student's *t*-test. (**E**) Colony formation assay for the two genome-edited 22Rv1 cells. Cell colonies were quantified through the Crystal Violet staining method. Representative images from triplicate experiments on the bottom. Mean ± SD of three biological replicates. *** $p < 0.001$, two-tailed Student's *t*-test. (**F**) Wound healing assay of the two rs684232 knockout 22Rv1 cell lines. Representative images from triplicate experiments. (**G**) The wound closure percentages in the wound healing assay experiments were quantified using Image J software. Mean ± SEM of three biological replicates. * $p < 0.05$, ** $p < 0.01$, two-tailed Student's *t*-test.

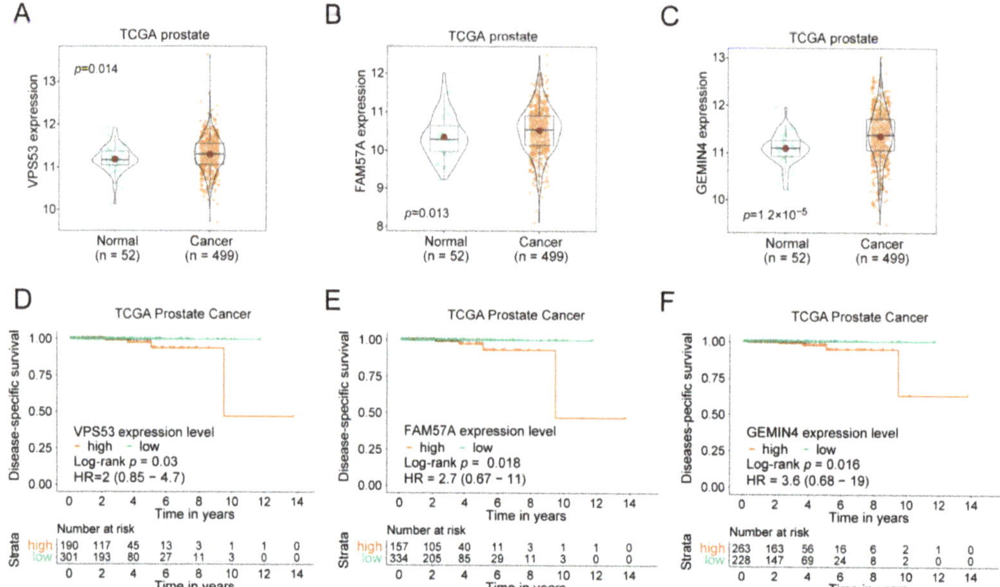

Figure 7. *VPS53*, *FAM57A*, and *GEMIN4* affect cancer progression. (**A–C**) Violin box plot to compare gene expression of *VPS53* (**A**), *FAM57A* (**B**), and *GEMIN4* (**C**) at mRNA level between normal and cancer tissues from TCGA. Gene expression value was showed as log2 value of reads data, with the mean, median, 0.25, and 0.75 quantiles represented. *p* values were examined by Mann-Whitney U tests. (**D–F**) Kaplan–Meier disease-specific survival analysis of prostate cancer patients that were stratified into two groups according to the expression level of *VPS53* (strata point = 2.52) (**D**), *FAM57A* (strata point = 3.92) (**E**), and *GEMIN4* (strata point = 3.06) (**F**). *p* values were calculated by the log-rank test, with a 95% confidence interval.

In brief, the results indicate that the rs684232 site and the target genes *VPS53*, *FAM57A*, and *GEMIN4* are positively associated with prostate cancer cell malignancy, and the high expression for the three genes are associated with poor prognosis for cancer patients.

3. Discussion

Risk SNPs have become a hot spot in the cancer research field with the advent of the post-GWAS era [7,47,48]. The DiR-seq screening system has high accuracy and is suitable for functional screening of the risk SNPs, which usually have a modest impact [33,34]. In this study, we applied the DiR system in prostate cancer cells to screen the causal risk SNPs that possess potential gene regulatory functions. We identified 32 regulatory SNPs based on the ratio of reporter expression level from the risk and normal alleles. Among them, fourteen SNPs exhibited decreased expression levels for the risk alleles, and eighteen SNPs showed increased expression to the contrary. The results provide valuable clues to further mechanism elucidation of the functional prostate cancer risk SNPs. However, since only 55 bp SNP site-centered genomic regions were used in our reporter gene assays, an eQTL will be missed if its function involves a larger genomic region or interactions with other molecules bound on distal enhancer sites.

In addition, we disclosed the regulatory pathway for rs684232 sites, in which the SNP site altered the chromatin binding of FOXA1 and led to the misregulated expression of *VPS53*, *FAM57A*, and *GEMIN4*. Notably, mutating the rs684232 element through genome editing or knockdown the expression of *VPS53*, *FAM57A*, and *GEMIN4* genes led to impeded cancer malignancy of 22Rv1 cells. Patients with higher expression of *VPS53*, *FAM57A*, and *GEMIN4* exhibited worse disease-specific survival probability, as demonstrated in the Kaplan–Meier survival analysis on the TCGA prostate cancer cohort. Remarkably, all three target genes were upregulated significantly in tumor tissues compared to adjacent normal tissues in TCGA prostate clinical samples.

Interestingly, we also found that downregulation of *VPS53*, *GEMIN4*, and *FAM57A* genes led to shorter disease-free intervals for prostate cancer patients to the contrary. Similarly, Ramanand et al. [45] recently reported that the impeded expression of *VPS53*, *FAM57A*, and *GEMIN4* genes might cause the increased biochemical recurrence risk for prostate cancer patients. We think that the rs684232 site and its target genes are supposed to be multifaceted in prostate cancer. In prostate cancer patients, the T allele of rs684232 leads to elevated expression of the target genes and is associated with worse disease-specific survival probability. However, when it turns to cancer susceptibility to cancer incidence and the risk for recurrence, the C allele might lead to higher prostate cancer susceptibility on the contrary. Even though complicated to understand, disclosing the seemingly opposite effect for the risk SNP and the three genes is crucial for understanding their biological functions, especially in translational medicine. Otherwise, cancer patients might receive the wrong suggestions and administrations based on one-sided knowledge, and lead to undesirable consequences. However, it is still unclear why that might be, and further systematic investigation of the functions of the three genes is necessary to address this question. The results potentially highlight the complexity of genetic susceptibility to cancer, and more works involving multiple variants and other factors are demanded to fully understand their contribution to cancer susceptibility.

FOXA1 encodes a pioneer factor that induces open chromatin conformation to allow the binding of other transcription factors. FOXA1 has been proved as a driver of prostate cancer onset and progression [49–52]. The transcription factor FOXA1 has been proven to regulate transcriptional programs in both normal prostate tissue and cancer tissues by directly interacting with AR [53–55]. Therefore, other factors such as AR might also participate in the function of rs684232. So, in the future, more depth researches on transcription factor FOXA1 in prostate cancer are needed.

The *VPS53* gene encodes the VPS53 subunit of the GARP complex that functions in retrograde transport from endosomes to the trans-Golgi network (TGN). The *FAM57A* gene encodes a membrane-associated protein that might involve in amino acid transport and glutathione metabolism. The *GEMIN4* gene product is s part of the Gemini bodies that function in spliceosome snRNP assembly and spliceosome regeneration required for pre-mRNA splicing. However, the roles of all three genes in cancer progression remain entirely unknown. To illustrate their functions and underlying mechanisms in affecting cancer susceptibility will be essential topics in the future and give vital clinical implications and translational value for cancer patients.

In general, we identified regulatory prostate cancer risk SNPs by DiR-seq analysis in prostate cancer cell lines and elucidated the function and mechanism of rs684232 in leading to prostate cancer progression. The results described here should be valuable for accurate prognostic prediction of prostate cancer patients in clinical. Further studies on mouse models and clinical samples might be demanded before applied in translational medicine.

4. Materials and Methods

4.1. Construction of the DiR Reporter Pool for Prostate Cancer Risk SNPs

Prostate cancer risk SNPs list were obtained from the GWAS Catalog (https://www.ebi.ac.uk/gwas/, accessed on 1 August 2016) in 2016, which contained 213 prostate cancer risk SNPs (Table S1) at that time. They are tag SNPs reported in the previous GWAS studies and are significant associated with prostate cancer risk (p-value $< 10^{-5}$). We obtained the 55 bp SNP-centered DNA region sequence for both protective and risk alleles from the UCSC genome browser on GRCh38/hg38. The annealed oligos (Table S2) were inserted into the DiR vectors between SmaI and BglII sites using T4 DNA Ligase (EL0011, Thermo Scientific, Waltham, MA, USA) as described previously [34]. DiR constructs were confirmed correct through Sanger sequencing. The 426 reporter constructs for 213 SNPs were mixed with the DiR-Promoter and the DiR-Control vector and then subjected to reporter assays in prostate cancer cells.

4.2. Cell Culture

The 22Rv1 (ATCC Cat# CRL-2505, RRID:CVCL_1045) cells used in this study were purchased from the American Type Culture Collection (ATCC) and grown in RPMI-1640 (Gibco, New York, NY, USA) supplied with 10% FBS (Gibco, New York, NY, USA) and 1% antibiotics (Penicillin-Streptomycin, Sigma, St. Louis, MO, USA). The Lenti-X 293T cells were purchased from Clontech Laboratories (Clontech, CA, USA) and maintained in DMEM (Gibco, New York, NY, USA) supplied with 10% FBS (Gibco, New York, NY, USA) and 1% Penicillin-Streptomycin. The cells were cultured at 37 °C with 95% air and 5% CO_2 and routinely confirmed to be mycoplasma free using the Myco-Blue Mycoplasma Detector (D101-01, Vazyme, Nanjing, China). 22Rv1 cells used in our study were cultured following the ATCC instructions.

4.3. Cell Transfection

Plasmids were transfected cells using Lipofectamine 2000 Reagent (11668-019, Invitrogen, Carlsbad, CA, USA) or Polyethylenimine (PEI, 408727-sigma, St. Louis, MO, USA) dependent on cell types. Lipofectamine 2000 was used for 22Rv1, and PEI was used for Lenti-X 293T cells following the manufacturer's instructions. Cell transfections were performed at 8–24 h post cell seeding, depending on the cell density and cell growth status. Generally, a 70–90% confluent cell culture was optimal. The DNA/transfection reagent ratio was 1:3 for Lipofectamine 2000 and 1:1.5 for PEI. The DNA was diluted in Opti-MEM and then added to the diluted transfection reagent. After gently mixing and 10–15 min incubation, the DNA complex was added to cells by drops and incubated for 1–2 days at 37 °C.

4.4. RNA Isolation and Reverse Transcription

The 22Rv1 cells were washed twice and harvested in 1 × PBS twenty-four hours post-transfection, and total RNA was extracted from the surviving cells using RNeasy Plus Mini Kit (74136, QIAGEN, Dusseldorf, Germany). We treated the mRNA with RapidOut DNA Removal Kit (K2981, Thermo Scientific, Waltham, MA, USA) to remove the trace amount of genomic DNA residue according to the product manual. The purified RNA was then subjected to reverse transcription with High-Capacity cDNA Reverse Transcription Kits (4374967, Applied Biosystems, Waltham, MA, USA). Briefly, 1.5 µg RNA was added into 10 µL of 2× RT Master Mix and made up to the final 20 µL with nuclease-free water. The reactions were incubated at 25 °C for 10 min, followed by 120 min at 37 °C, then were inactivated Reverse Transcriptase by heating to 85 °C for 5 min. The cDNA products were stored at −20 °C or −80 °C and ready for qPCR analysis and NGS sequencing library preparation. For the DiR analysis, the sequence-specific primer BarP6 (CACGATCTGTC-CGCACTGCTTGG) was used for reverse transcription, and random primer supplied in the reverse transcription kit was used for reverse transcription for other applications.

4.5. Quantitative PCR

We performed RT-qPCR, ChIP-qPCR, and FAIRE-qPCR assays using the AceQ qPCR SYBR Green Master Mix (Q111-03, Vazyme, Nanjing, China) on the thermocyclers Rotor-Gene Q (Qiagen, Dusseldorf, Germany) or LightCycler 96 thermal cycler Instrument (Roche Applied Science, Indianapolis, IN, USA). All the qPCR primer pairs were confirmed to have reasonable specificity and amplification efficiency before qPCR assays, and all the qPCR assays were performed in three technical replications. In the RT-qPCR analysis to analyze the gene expression, the endogenous *ACTB* gene was used for normalization control. For ChIP-qPCR assays, the relative enrichment of the target DNA region was determined by calculating the immunoprecipitation efficiency over input control and then normalized to the control region. In FAIRE qPCR analysis, the enrichment fold of the given region was calculated similarly. Specifically, in the AS-qPCR assay, primers were designed with allele-specific nucleotide placed at the 3′ terminal to enable selective amplification of SNP regions. The DiR-qPCR primers are listed in Table S3, and all the other qPCR primers are listed in Table S4.

4.6. DiR-Seq Library Preparation for Illumina Sequencing

The DiR-seq libraries were prepared with two rounds of PCR amplification with cDNA as templates using 2× Phusion Hot Start II High-Fidelity PCR Master Mix (F565L, Thermo Scientific, Waltham, MA, USA). To adapt the 150 bp paired-end sequencing strategy on the Illumina HiSeq X-TEN platform, we divided the 450 bp barcoding region into two amplicons of 271 bp and 270 bp, respectively, in the first round of PCR. During this step, the binding sites of Illumina sequencing primers were introduced at both ends. In the second-round PCR, adaptors for cluster generation and the index sequences were added. Twenty-four sets of primers tiling the flank sequence of the barcoding region in the first round of PCR, in combination with 12 sequencing indexes introduced in the second-round PCR, will enable up to 288 treatments to be analyzed in parallel in one NGS library. The first-round PCR was performed with 2× Phusion Hot Start II High-Fidelity PCR Master Mix using the program: 98 °C for the 30 s of initial denaturation, then 7 cycles of 98 °C for 10 s, 72 °C for 45 s and followed by a final extension at 72 °C for 5 min. The PCR products were purified using 1× VAHTS DNA Clean Beads (N411, Vazyme, Nanjing, China), eluted in 10 μL water, pooled every twenty-four sets of products equally, and then subjected to the second round PCR, which was performed using 2× Phusion HS II HF Master Mix with 1 ng template DNA (98 °C for 30 s, 10 cycles of 98 °C for 10 s, 68 °C for 15 s, 72 °C for 30 s, followed by 72 °C for 5 min). The products were purified using 1× VAHTS DNA Clean Beads and eluted in 15 μL 1× TE buffer. We also subjected the template plasmid pool to NGS library preparation as input control for calculating the expression level. The purified DiR-seq libraries were subjected to 150 bp paired-end sequencing on the Illumina HiSeq X-TEN platform run by Genewiz(NJ, USA), generating about 1 million reads per library. Primers used for DiR-seq library construction are shown in Table S5.

4.7. NGS Data Processing

Illumina sequencing raw data were performed quality control using the software FastP (https://github.com/OpenGene/fastp, accessed on 3 December 2018). It is important to note that the 5′ terminal 'N' base should not be removed during the cleaning step. The clean Illumina reads were assembled for the paired reads using the software Pandaseq [56], and the sub-libraries were then sorted out using the R package 'ShortRead' [57]. Further, we counted the read number of each dinucleotide barcodes using the R package 'ShortRead' for each sub-library and normalized the barcode counts by making each sub-library 1 M total reads to eliminate the influence of sequencing depth variation. For each dinucleotide barcode, the expression level was counted by dividing the reads number in cDNA by template DNA. The statistical significance of the expression difference between the two SNP alleles was evaluated with the Two-tailed Student's *t*-test. All the SNPs determined to have regulatory functions were listed in Table S6.

4.8. Luciferase Reporter Assays

DNA fragments bearing the rs684232, rs2955626, rs461251, rs887391, and rs11672691 sites were inserted upstream to the SV40 promoter in the pGL3 Promoter vector. The corresponding DNA fragments sizes are 852 bp, 835 bp, 852 bp, 766 bp, and 766 bp, respectively. The internal Renilla control plasmid pGL4.75 [hRluc/CMV] (E6931, Promega, Fitchburg, WI, USA, RRID:Addgene_24348) and each reporter plasmid were co-transfected into 22Rv1 cells in Nunc™ F96 MicroWell White Polystyrene Plate (136101, Thermo Scientific, Waltham, MA, USA) in a reverse transfection manner using Lipofectamine 2000 Transfection Reagent (11668-019, Invitrogen, Carlsbad, CA, USA) according to the protocol provided by the manufacturer. The luciferase activity was measured with Dual-Glo Luciferase Assay System

(E2920, Promega, Fitchburg, WI, USA) at 48 h post transfection, and the luminescence was acquired using the EnSpire Multimode Plate Reader from PerkinElmer (Manchester, UK). All data were obtained from at least three replicate wells, and statistical analyses were performed with the Two-tailed Student's *t*-test.

4.9. Formaldehyde-Assisted Isolation of Regulatory Elements (FAIRE)

FAIRE assays were performed as previously described [39]. Briefly, cells were fixed with 1% formaldehyde (F8775, Sigma-Aldrich, St. Louis, MO, USA) for 10 min at room temperature, and the fix reaction was quenched with 125 mM glycine (0167, Amresco Radnor, PA, USA). After washing twice with cold PBS, the cells were collected and resuspended in hypotonic lysis buffer (20 mM Tris-HCl, pH 8.0, with 10 mM KCl, 10% glycerol, 2 mM DTT supplied with cOmplete EDTA-free Protease Inhibitor Cocktail) followed by rotation at 4 °C for 30 min. The cell nuclei were washed with cold PBS and then resuspended in 2% SDS lysis buffer (50 mM Tris-HCl, pH 8.1, with 2% SDS, 10 mM EDTA supplied with cOmplete EDTA-free Protease Inhibitor Cocktail) and incubated at 4 °C for 30–60 min. The chromatin was sheared to an average size of 200 bp with a Bioruptor (Bioruptor pico), and the lysate was then cleared by 5 min centrifugation at 13,000× *g* at 4 °C. Chromatin lysate containing 0.5 µg DNA was subjected to twice phenol/chloroform/isoamyl alcohol extraction followed by one chloroform/isoamyl alcohol extraction. The top aqueous layers containing DNA were collected and subjected to ethanol precipitation with the presence of 20 µg of glycogen. The DNA was pelleted and resuspended in 10 mM Tris-HCl (pH 7.4). After being treated with RNase A, the FAIRE DNA and Input DNA were subjected to reverse cross-linking overnight at 65 °C in the presence of proteinase K and purified using 1×VAHTS DNA Clean Beads. The FAIRE DNA was then applied to qPCR analysis to determine the enrichment of the given DNA region in open chromatin or to PCR amplification of SNP regions for Sanger sequencing. All primers are shown in Table S4.

4.10. Chromatin Immunoprecipitation (ChIP)

ChIP experiments were performed as described previously [14] with slight modifications. Briefly, chromatin lysate was prepared in the same way as the FAIRE analysis. Immunoprecipitation was performed with antibodies targeting FOXA1 (Santa Cruz Biotechnology, Santa Cruz, CA, USA, Cat# sc-22841, RRID:AB_2104862), H3K27ac (Abcam, Cambridge, UK, Cat# ab4729, RRID:AB_2118291), and H3K4me3 (Abcam, Cambridge, UK, Cat# ab8580, RRID:AB_306649). They were coated onto the Magna ChIP Protein A + G Magnetic Beads (16-663, EMD Millipore, MA, USA) in blocking buffer, which contains 0.5% BSA in IP buffer (20 mM Tris-HCl, pH 8.0, with 2 mM EDTA, 150 mM NaCl, 1% Triton X-100 supplied with cOmplete EDTA-free Protease Inhibitor Cocktail). DNA-protein complex bounded to magnetic beads were washed in turn with wash buffer I (20 mM Tris-HCl, pH 8.0, with 2 mM EDTA, 0.1% SDS, 1% Triton X-100 and 150 mM NaCl), wash buffer II (20 mM Tris-HCl, pH 8.0, with 2 mM EDTA, 0.1% SDS, 1% Triton X-100 and 500 mM NaCl), wash buffer III (10 mM Tris-HCl, pH 8.0, with 1 mM EDTA, 250 mM lithium chloride, 1% deoxycholate and 1% NP-40), and buffer IV (10 mM Tris-HCl, pH 8.0, and 1 mM EDTA) for twice and eluted in extraction buffer (10 mM Tris-HCl, pH 8.0, with 1 mM EDTA and 1% SDS). The complex was incubated with 0.2 mg/mL RNase A (Thermo Scientific, Waltham, MA, USA) for 30 min at 37 °C and subjected to overnight reverse cross-linking at 65 °C with proteinase K (Thermo Scientific, Waltham, MA, USA) and purified using 1×VAHTS DNA Clean Beads. The ChIP DNA was then subjected to qPCR analysis to determine the enrichment of the given DNA region or to PCR amplification of SNP regions to observe the allele selectivity by Sanger sequencing. All primers are shown in Table S4.

4.11. Lentiviral Constructs, Lentivirus Production, and Infection

The shRNA constructs targeting *FOXA1* were the same as previously described [15], and the *VPS53*, *FAM57A*, and *GEMIN4* shRNA in pLKO.1-puro were designed according to the validated shRNA clones in MISSION® shRNA Library (Sigma-Aldrich, St. Louis,

MO, USA). Detailed information on these shRNA constructs is provided in Table S7. Lentivirus expressing given shRNA was produced with the third-generation packaging system in Lenti-X 293T cells (Clontech). Briefly, 70–80% confluent Lenti-X 293T cells in 6-well plates were transfected with 3 µg shRNA plasmid, 1 µg pVSVG (envelope plasmid, RRID:Addgene_85140), 1 µg pMDLg/pRRE (packaging plasmid, RRID:Addgene_12251), and 1µg pRSV-Rev (packaging plasmid, RRID:Addgene_12253) in a 3:1:1:1 ratio using PEI (408727, Sigma, St. Louis, MO, USA) in an FBS and antibiotics free medium [38]. The medium was replaced with fresh complete DMEM containing 10% FBS (Gibco, New York, NY, USA) and 1% Penicillin-Streptomycin (Sigma-Aldrich, St. Louis, MO, USA) after 4–8 h, and the virus supernatant was collected every 12 h for up to six times. The supernatant containing viral particles was cleared by centrifugation at $1000 \times g$ and passed through a 0.45 µm filter unit (Millipore) and then stored at $-80\ ^{\circ}$C in aliquots or used directly for subsequent experiments.

For viral infection, target cells were seeded in a 6-well plate and grown for 16–24 h until they reach 60–70% confluence. The growth medium was replaced with the virus supernatant supplied with 8 µg/mL polybrene (Sigma-Aldrich, St. Louis, MO, USA). Twenty-four hours later, the virus-containing medium was replaced with the complete medium with puromycin (Sigma, St. Louis, MO, USA) at 4 µg/mL for 22Rv1. When control cells without virus infection were all dead, the surviving cells were split and cultured in the same growth medium. After three days, the cells were collected for RNA preparation and RT-qPCR gene expression quantification. The most efficient shRNA for each gene was selected for subsequent analysis, including cell proliferation assays and cell colony formation assays.

4.12. rs684232 Knockout Using CRISPR/Cas9

We designed the gRNA sequences that guide Cas9 cleavage on both alleles precisely to the left of the rs684232 site. The oligos were annealed and jointed with the BbsI (FD1014, Thermo Scientific, Waltham, MA, USA) digested pSpCas9(BB)-2A-Puro (PX459) V2.0 (RRID:Addgene_62988) [58]. For negative control, the sgRNA was designed to target a non-Mammalian sequence. All the oligos sequences for gRNA sequences are listed in Table S7. The CRISPR plasmids were transfected into 22Rv1 cells at 70% confluence in a 12-well plate using Lipofectamine 2000 Transfection Reagent (11668-019, Invitrogen, Carlsbad, CA, USA) according to the manufacturer's instructions. After 24 h, the medium was replaced with fresh medium supplied with puromycin at a final concentration of 4 µg/mL. When non-transfected cells all died, the surviving cells were split and cultured in the complete medium and subjected to editing efficiency evaluation by getPCR analysis. The surviving cells were then trypsinized and seeded into 96-well plates at a dilution to have less than one cell at each well. The single-cell clones were propagated for 1 to 2 months and screened through the getPCR method. The clones that had desired rs684232 mutation were further genotyped through the Sanger sequencing method.

4.13. Genome Editing Efficiency Determination and Single-Cell Clone Screening

To detect genome editing, we performed getPCR assays as previously described [59]. Briefly, for determining the genome editing efficiency, the tested primer was designed with four watching nucleotides. The getPCR was performed using 7.5 µL AceQ qPCR SYBR Green Master Mix (Vazyme, Nanjing, China) on Roche LightCycler96. While screening the single-cell clones that happened anticipated modification, we used the watching primers with their 3′ end located on the rs684232 site. The control amplification was designed 200 bp away from the cutting site, which was used for normalization purposes in calculating the percentage of wild-type DNA in the edited genomic DNA. The primers used in getPCR experiments are listed in Table S4.

4.14. Cell Viability and Proliferation Assays

To investigate the cell proliferation, the 22Rv1 cells that underwent infection with lentiviral particles or single-cell clones with rs684232 site deleted through genome editing were counted and seeded into 96-well cell culture plates at 5×103 per well. Cell viability and proliferation were measured with a CCK-8 kit (MA0218, Meilun, Dalian, China), and the optical density at 450 nm was acquired on an M200 PRO multimode plate reader (Tecan, Sunnyvale, USA) every 24 h. The results were obtained from three independent experiments, and the statistical significance was calculated with the Two-tailed Student's *t*-test.

4.15. Colony-Forming Assay

Cells were trypsinized into single cells and seeded into 6-well plates with 1500 cells per well. The medium was replaced with fresh medium every three days. After 15–20 days, the medium was discarded, and cells were washed twice with 1 mL cold $1 \times$PBS carefully. After fixation with 2 mL 100% methanol for 30 min, the cells were further stained with 2 mL of 0.05% Crystal Violet staining solution (HY-B0324A, MCE, Shanghai, China) for 30 min. The cells were washed twice with deionized water and dried overnight and lysed with 1% SDS in 0.2 N NaOH for 1 h, and the optical density at 570 nm was acquired on an M200 PRO multimode plate reader (Tecan). A blank well without cell was set as a control to minus the background staining. All data came from three replicate wells, and the statistical significance was calculated with the Two-tailed Student's *t*-test.

4.16. Wound Healing Assays

The wound-healing assay was performed as previously described [60]. Briefly, the 22Rv1 cells were seeded in a 6-well plate in the serum-free medium at a density that made 90% confluence 12 h later. Made a scratch wound on the cell monolayer using a 200 µL pipette tip and washed the cells three times with fresh medium to remove the debris and smooth the edge. The cells were grown in a complete medium containing 10% FBS, and the wound healing process was imaged ($10\times$) using an inverted fluorescence microscope (Olympus, Tokyo, Japan) every 24 h. The wound closure area in each well was analyzed using ImageJ software (ImageJ, RRID:SCR_003070).

4.17. Statistical Analysis

For the DiR-qPCR, DiR-seq, RT-qPCR, ChIP-qPCR, and FAIRE-qPCR analysis as well as for the evaluation of cell proliferation, cell migration, and colony formation, we used the Two-tailed Student's *t*-test.

The transcriptome data was downloaded from The Cancer Genome Atlas (TCGA, The Cancer Genome Atlas, RRID:SCR_003193) database using the R package "TCGAbiolinks" (TCGAbiolinks, RRID:SCR_017683) for differential gene expression analysis for prostate cancer tissues and Para-cancerous tissues. The GDC.h38 GENCODE v22 GTF file for gene annotation was used to match the data file to TCGA ID, and the transcriptome counts data were further processed with the R package "Deseq2" (DESeq2, RRID:SCR_015687). The differential expression levels of *VPS53*, *FAM57A*, and *GEMIN4* gene in prostate cancer tissues and Para-cancerous tissues were visualized as violin box plots using "ggplot2" (ggplot2, RRID:SCR_014601). We used the Mann-Whitney U test to evaluate the statistical significance of gene expression differences between normal and tumor tissues.

For the correlation analysis of gene expression in tissues of prostate cancer or pan-cancer of 33 cancer types, we obtained the gene expression RNA-seq data as "TOIL RSEM tpm" file from the TCGA Pan-Cancer (PANCAN) cohort and annotated it with the gtf file "genecode. v23.annotation". The cancerous tissues were then extracted out and subjected to calculating the correlation coefficient and *p*-value using the "ggplot2" package.

For Kaplan-Meier survival analysis in prostate cancer patients or Pan-Cancer of 33 cancer types, we obtained the integrated TCGA Pan-Cancer Clinical Data from Liu's work [61] and merged to the gene expression matrix of Pan-Cancer tissues aforementioned. We then used the R package "survival" (survival, RRID:SCR_021137) and "survminer" (survminer, RRID:SCR_021094) to perform Kaplan–Meier survival analysis and visualization. Patients were sub-grouped based on the optimal cut-off point determined using the "survminer" R package. We used the Cox proportional hazards model to assess the hazard ratio (HR) and log-rank test to assess the statistical significance between the two groups of patients. We used R-4.0.2 for running the R packages.

Supplementary Materials: The following are available online at https://www.mdpi.com/article/10.3390/ijms22168792/s1.

Author Contributions: N.R.: Data curation, formal analysis, investigation, methodology, visualization, writing—original draft. Q.L.: investigation, methodology, validation, visualization, writing—review and editing. L.Y.: methodology, visualization, writing—review and editing. Q.H.: conceptualization, investigation, project administration, supervision, writing—review and editing, Funding acquisition. All authors have read and agreed to the published version of the manuscript.

Funding: This research was funded by the National Natural Science Foundation of China [grant number 31872809], Shandong Provincial Natural Science Foundation, China (ZR2016CM50), and Qilu Young Scholar to Q.H.

Institutional Review Board Statement: Not applicable.

Informed Consent Statement: Not applicable.

Data Availability Statement: The raw sequence data generated using the Illumina Hiseq-PE150 platform for DiR-seq and 10-nucleotide tag reporter assay have been publicly available in the Gene Expression Omnibus (GEO) database under the accession number GSE165765.

Acknowledgments: We thank Caiyun Sun, Xiangmei Ren, and Rui Wang from the State Key Laboratory of Microbial Technology of Shandong University for assistance with the EnSpire multimode plate reader and M200 PRO multimode plate reader.

Conflicts of Interest: The authors declare no conflict of interest.

References

1. Sud, A.; Kinnersley, B.; Houlston, R.S. Genome-wide association studies of cancer: Current insights and future perspectives. *Nat. Rev. Cancer* **2017**, *17*, 692–704. [CrossRef]
2. Welter, D.; MacArthur, J.; Morales, J.; Burdett, T.; Hall, P.; Junkins, H.; Klemm, A.; Flicek, P.; Manolio, T.; Hindorff, L.; et al. The NHGRI GWAS Catalog, a curated resource of SNP-trait associations. *Nucleic Acids Res.* **2014**, *42*, D1001–D1006. [CrossRef]
3. Bray, F.; Ferlay, J.; Soerjomataram, I.; Siegel, R.L.; Torre, L.A.; Jemal, A. Global cancer statistics 2018: GLOBOCAN estimates of incidence and mortality worldwide for 36 cancers in 185 countries. *CA Cancer J. Clin.* **2020**, *68*, 394–424. [CrossRef] [PubMed]
4. Siegel, R.L.; Miller, K.D.; Jemal, A. Cancer statistics, 2019. *CA Cancer J. Clin.* **2019**, *69*, 7–34. [CrossRef] [PubMed]
5. Al Olama, A.A.; Kote-Jarai, Z.; Berndt, S.I.; Conti, D.V.; Schumacher, F.; Han, Y.; Benlloch, S.; Hazelett, D.J.; Wang, Z.M.; Saunders, E.; et al. A meta-analysis of 87,040 individuals identifies 23 new susceptibility loci for prostate cancer. *Nat. Genet.* **2014**, *46*, 1103–1109. [CrossRef]
6. Eeles, R.; Goh, C.; Castro, E.; Bancroft, E.; Guy, M.; Al Olama, A.A.; Easton, D.; Kote-Jarai, Z. The genetic epidemiology of prostate cancer and its clinical implications. *Nat. Rev. Urol.* **2014**, *11*, 18–31. [CrossRef] [PubMed]
7. Eeles, R.A.; Al Olama, A.A.; Benlloch, S.; Saunders, E.J.; Leongamornlert, D.A.; Tymrakiewicz, M.; Ghoussaini, M.; Luccarini, C.; Dennis, J.; Jugurnauth-Little, S.; et al. Identification of 23 new prostate cancer susceptibility loci using the iCOGS custom genotyping array. *Nat. Genet.* **2013**, *45*, 385–391. [CrossRef]
8. Schumacher, F.R.; Al Olama, A.A.; Berndt, S.I.; Benlloch, S.; Ahmed, M.; Saunders, E.J.; Dadaev, T.; Leongamornlert, D.; Anokian, E.; Cieza-Borrella, C.; et al. Association analyses of more than 140,000 men identify 63 new prostate cancer susceptibility loci. *Nat. Genet.* **2019**, *50*, 928–936. [CrossRef] [PubMed]
9. Virlogeux, V.; Graff, R.E.; Hoffmann, T.J.; Witte, J.S. Replication and Heritability of Prostate Cancer Risk Variants: Impact of Population-Specific Factors. *Cancer Epidemiol. Prev. Biomark.* **2015**, *24*, 938–943. [CrossRef]
10. Benafif, S.; Kote-Jarai, Z.; Eeles, R.A.; Consortium, P. A Review of Prostate Cancer Genome-Wide Association Studies (GWAS). *Cancer Epidemiol. Prev. Biomark.* **2018**, *27*, 845–857. [CrossRef]
11. Khurana, E.; Fu, Y.; Chakravarty, D.; Demichelis, F.; Rubin, M.A.; Gerstein, M. Role of non-coding sequence variants in cancer. *Nat. Rev. Genet.* **2016**, *17*, 93–108. [CrossRef]
12. Farashi, S.; Kryza, T.; Clements, J.; Batra, J. Post-GWAS in prostate cancer: From genetic association to biological contribution. *Nat. Rev. Cancer* **2019**, *19*, 46–59. [CrossRef] [PubMed]

13. Ward, L.D.; Kellis, M. Interpreting noncoding genetic variation in complex traits and human disease. *Nat. Biotechnol.* **2012**, *30*, 1095–1106. [CrossRef] [PubMed]
14. Huang, Q.; Whitington, T.; Gao, P.; Lindberg, J.F.; Yang, Y.; Sun, J.; Vaisanen, M.R.; Szulkin, R.; Annala, M.; Yan, J.; et al. A prostate cancer susceptibility allele at 6q22 increases *RFX6* expression by modulating HOXB13 chromatin binding. *Nat. Genet.* **2014**, *46*, 126–135. [CrossRef]
15. Gao, P.; Xia, J.H.; Sipeky, C.; Dong, X.M.; Zhang, Q.; Yang, Y.; Zhang, P.; Cruz, S.P.; Zhang, K.; Zhu, J.; et al. Biology and Clinical Implications of the 19q13 Aggressive Prostate Cancer Susceptibility Locus. *Cell* **2018**, *174*, 576–589.e18. [CrossRef]
16. Takeda, D.Y.; Spisak, S.; Seo, J.H.; Bell, C.; O'Connor, E.; Korthauer, K.; Ribli, D.; Csabai, I.; Solymosi, N.; Szallasi, Z.; et al. A Somatically Acquired Enhancer of the Androgen Receptor Is a Noncoding Driver in Advanced Prostate Cancer. *Cell* **2018**, *174*, 422–432.e13. [CrossRef]
17. Wang, X.; Tian, J.; Zhao, Q.; Yang, N.; Ying, P.; Peng, X.; Zou, D.; Zhu, Y.; Zhong, R.; Gao, Y.; et al. Functional characterization of a low-frequency V1937I variant in *FASN* associated with susceptibility to esophageal squamous cell carcinoma. *Arch. Toxicol.* **2020**, *94*, 2039–2046. [CrossRef]
18. Yu, C.Y.; Han, J.X.; Zhang, J.; Jiang, P.; Shen, C.; Guo, F.; Tang, J.; Yan, T.; Tian, X.; Zhu, X.; et al. A 16q22.1 variant confers susceptibility to colorectal cancer as a distal regulator of ZFP90. *Oncogene* **2020**, *39*, 1347–1360. [CrossRef]
19. Li, Q.; Seo, J.H.; Stranger, B.; McKenna, A.; Pe'er, I.; Laframboise, T.; Brown, M.; Tyekucheva, S.; Freedman, M.L. Integrative eQTL-based analyses reveal the biology of breast cancer risk loci. *Cell* **2013**, *152*, 633–641. [CrossRef] [PubMed]
20. Oldridge, D.A.; Wood, A.C.; Weichert-Leahey, N.; Crimmins, I.; Sussman, R.; Winter, C.; McDaniel, L.D.; Diamond, M.; Hart, L.S.; Zhu, S.; et al. Genetic predisposition to neuroblastoma mediated by a *LMO1* super-enhancer polymorphism. *Nature* **2015**, *528*, 418–421. [CrossRef] [PubMed]
21. Ma, S.; Ren, N.; Huang, Q. rs10514231 Leads to Breast Cancer Predisposition by Altering *ATP6AP1L* Gene Expression. *Cancers* **2021**, *13*, 3752. [CrossRef]
22. Li, B.; Huang, Q.; Wei, G.H. The Role of HOX Transcription Factors in Cancer Predisposition and Progression. *Cancers* **2019**, *11*, 528. [CrossRef]
23. Brandao, A.; Paulo, P.; Teixeira, M.R. Hereditary Predisposition to Prostate Cancer: From Genetics to Clinical Implications. *Int. J. Mol. Sci.* **2020**, *21*, 5036. [CrossRef] [PubMed]
24. Tak, Y.G.; Farnham, P.J. Making sense of GWAS: Using epigenomics and genome engineering to understand the functional relevance of SNPs in non-coding regions of the human genome. *Epigenet. Chromatin* **2015**, *8*, 57. [CrossRef]
25. Maurano, M.T.; Humbert, R.; Rynes, E.; Thurman, R.E.; Haugen, E.; Wang, H.; Reynolds, A.P.; Sandstrom, R.; Qu, H.; Brody, J.; et al. Systematic localization of common disease-associated variation in regulatory DNA. *Science* **2012**, *337*, 1190–1195. [CrossRef]
26. Patwardhan, R.P.; Lee, C.; Litvin, O.; Young, D.L.; Pe'er, D.; Shendure, J. High-resolution analysis of DNA regulatory elements by synthetic saturation mutagenesis. *Nat. Biotechnol.* **2009**, *27*, 1173–1175. [CrossRef] [PubMed]
27. Patwardhan, R.P.; Hiatt, J.B.; Witten, D.M.; Kim, M.J.; Smith, R.P.; May, D.; Lee, C.; Andrie, J.M.; Lee, S.I.; Cooper, G.M.; et al. Massively parallel functional dissection of mammalian enhancers in vivo. *Nat. Biotechnol.* **2012**, *30*, 265–270. [CrossRef]
28. Smith, R.P.; Taher, L.; Patwardhan, R.P.; Kim, M.J.; Inoue, F.; Shendure, J.; Ovcharenko, I.; Ahituv, N. Massively parallel decoding of mammalian regulatory sequences supports a flexible organizational model. *Nat. Genet.* **2013**, *45*, 1021–1028. [CrossRef]
29. Melnikov, A.; Murugan, A.; Zhang, X.; Tesileanu, T.; Wang, L.; Rogov, P.; Feizi, S.; Gnirke, A.; Callan, C.G., Jr.; Kinney, J.B.; et al. Systematic dissection and optimization of inducible enhancers in human cells using a massively parallel reporter assay. *Nat. Biotechnol.* **2012**, *30*, 271–277. [CrossRef]
30. Arnold, C.D.; Gerlach, D.; Stelzer, C.; Boryn, L.M.; Rath, M.; Stark, A. Genome-wide quantitative enhancer activity maps identified by STARR-seq. *Science* **2013**, *339*, 1074–1077. [CrossRef] [PubMed]
31. Zhang, P.; Xia, J.H.; Zhu, J.; Gao, P.; Tian, Y.J.; Du, M.; Guo, Y.C.; Suleman, S.; Zhang, Q.; Kohli, M.; et al. High-throughput screening of prostate cancer risk loci by single nucleotide polymorphisms sequencing. *Nat. Commun.* **2018**, *9*, 2022. [CrossRef] [PubMed]
32. Vanhille, L.; Griffon, A.; Maqbool, M.A.; Zacarias-Cabeza, J.; Dao, L.T.; Fernandez, N.; Ballester, B.; Andrau, J.C.; Spicuglia, S. High-throughput and quantitative assessment of enhancer activity in mammals by CapStarr-seq. *Nat. Commun.* **2015**, *6*, 6905. [CrossRef]
33. Liu, S.; Liu, Y.; Zhang, Q.; Wu, J.; Liang, J.; Yu, S.; Wei, G.H.; White, K.P.; Wang, X. Systematic identification of regulatory variants associated with cancer risk. *Genome Biol.* **2017**, *18*, 194. [CrossRef]
34. Ren, N.; Li, B.; Liu, Q.; Yang, L.; Liu, X.; Huang, Q. A dinucleotide tag-based parallel reporter gene assay method. *bioRxiv* **2021**. [CrossRef]
35. Al Olama, A.A.; Dadaev, T.; Hazelett, D.J.; Li, Q.; Leongamornlert, D.; Saunders, E.J.; Stephens, S.; Cieza-Borrella, C.; Whitmore, I.; Benlloch Garcia, S.; et al. Multiple novel prostate cancer susceptibility signals identified by fine-mapping of known risk loci among Europeans. *Hum. Mol. Genet.* **2015**, *24*, 5589–5602. [CrossRef] [PubMed]
36. Dadaev, T.; Saunders, E.J.; Newcombe, P.J.; Anokian, E.; Leongamornlert, D.A.; Brook, M.N.; Cieza-Borrella, C.; Mijuskovic, M.; Wakerell, S.; Olama, A.A.A.; et al. Fine-mapping of prostate cancer susceptibility loci in a large meta-analysis identifies candidate causal variants. *Nat. Commun.* **2018**, *9*, 2256. [CrossRef]
37. Levo, M.; Segal, E. In pursuit of design principles of regulatory sequences. *Nat. Rev. Genet.* **2014**, *15*, 453–468. [CrossRef]

38. Huang, Q.; Gong, C.; Li, J.; Zhuo, Z.; Chen, Y.; Wang, J.; Hua, Z.C. Distance and helical phase dependence of synergistic transcription activation in cis-regulatory module. *PLoS ONE* **2012**, *7*, e31198. [CrossRef] [PubMed]
39. Simon, J.M.; Giresi, P.G.; Davis, I.J.; Lieb, J.D. Using formaldehyde-assisted isolation of regulatory elements (FAIRE) to isolate active regulatory DNA. *Nat. Protoc.* **2012**, *7*, 256–267. [CrossRef]
40. Hua, J.T.; Ahmed, M.; Guo, H.; Zhang, Y.; Chen, S.; Soares, F.; Lu, J.; Zhou, S.; Wang, M.; Li, H.; et al. Risk SNP-Mediated Promoter-Enhancer Switching Drives Prostate Cancer through lncRNA *PCAT19*. *Cell* **2018**, *174*, 564–575.e18. [CrossRef]
41. Igolkina, A.A.; Zinkevich, A.; Karandasheva, K.O.; Popov, A.A.; Selifanova, M.V.; Nikolaeva, D.; Tkachev, V.; Penzar, D.; Nikitin, D.M.; Buzdin, A. H3K4me3, H3K9ac, H3K27ac, H3K27me3 and H3K9me3 Histone Tags Suggest Distinct Regulatory Evolution of Open and Condensed Chromatin Landmarks. *Cells* **2019**, *8*, 1034. [CrossRef]
42. Whitington, T.; Gao, P.; Song, W.; Ross-Adams, H.; Lamb, A.D.; Yang, Y.; Svezia, I.; Klevebring, D.; Mills, I.G.; Karlsson, R.; et al. Gene regulatory mechanisms underpinning prostate cancer susceptibility. *Nat. Genet.* **2016**, *48*, 387–397. [CrossRef] [PubMed]
43. Larson, N.B.; McDonnell, S.; French, A.J.; Fogarty, Z.; Cheville, J.; Middha, S.; Riska, S.; Baheti, S.; Nair, A.A.; Wang, L.; et al. Comprehensively evaluating cis-regulatory variation in the human prostate transcriptome by using gene-level allele-specific expression. *Am. J. Hum. Genet.* **2015**, *96*, 869–882. [CrossRef] [PubMed]
44. Houlahan, K.E.; Shiah, Y.J.; Gusev, A.; Yuan, J.; Ahmed, M.; Shetty, A.; Ramanand, S.G.; Yao, C.Q.; Bell, C.; O'Connor, E.; et al. Genome-wide germline correlates of the epigenetic landscape of prostate cancer. *Nat. Med.* **2019**, *25*, 1615–1626. [CrossRef] [PubMed]
45. Ramanand, S.G.; Chen, Y.; Yuan, J.; Daescu, K.; Lambros, M.B.; Houlahan, K.E.; Carreira, S.; Yuan, W.; Baek, G.; Sharp, A.; et al. The landscape of RNA polymerase II-associated chromatin interactions in prostate cancer. *J. Clin. Investig.* **2020**, *130*, 3987–4005. [CrossRef]
46. Ward, L.D.; Kellis, M. HaploReg: A resource for exploring chromatin states, conservation, and regulatory motif alterations within sets of genetically linked variants. *Nucleic Acids Res.* **2012**, *40*, D930–D934. [CrossRef] [PubMed]
47. Fachal, L.; Dunning, A.M. From candidate gene studies to GWAS and post-GWAS analyses in breast cancer. *Curr. Opin. Genet. Dev.* **2015**, *30*, 32–41. [CrossRef]
48. Chen, H.; Yu, H.; Wang, J.; Zhang, Z.; Gao, Z.; Chen, Z.; Lu, Y.; Liu, W.; Jiang, D.; Zheng, S.L.; et al. Systematic enrichment analysis of potentially functional regions for 103 prostate cancer risk-associated loci. *Prostate* **2015**, *75*, 1264–1276. [CrossRef] [PubMed]
49. Sahu, B.; Laakso, M.; Ovaska, K.; Mirtti, T.; Lundin, J.; Rannikko, A.; Sankila, A.; Turunen, J.P.; Lundin, M.; Konsti, J.; et al. Dual role of FoxA1 in androgen receptor binding to chromatin, androgen signalling and prostate cancer. *EMBO J.* **2011**, *30*, 3962–3976. [CrossRef]
50. Barbieri, C.E.; Baca, S.C.; Lawrence, M.S.; Demichelis, F.; Blattner, M.; Theurillat, J.P.; White, T.A.; Stojanov, P.; Van Allen, E.; Stransky, N.; et al. Exome sequencing identifies recurrent *SPOP*, *FOXA1* and *MED12* mutations in prostate cancer. *Nat. Genet.* **2012**, *44*, 685–689. [CrossRef] [PubMed]
51. Gerhardt, J.; Montani, M.; Wild, P.; Beer, M.; Huber, F.; Hermanns, T.; Muntener, M.; Kristiansen, G. FOXA1 Promotes Tumor Progression in Prostate Cancer and Represents a Novel Hallmark of Castration-Resistant Prostate Cancer. *Am. J. Pathol.* **2012**, *180*, 848–861. [CrossRef]
52. Grasso, C.S.; Wu, Y.M.; Robinson, D.R.; Cao, X.H.; Dhanasekaran, S.M.; Khan, A.P.; Quist, M.J.; Jing, X.; Lonigro, R.J.; Brenner, J.C.; et al. The mutational landscape of lethal castration-resistant prostate cancer. *Nature* **2012**, *487*, 239–243. [CrossRef]
53. Sahu, B.; Laakso, M.; Pihlajamaa, P.; Ovaska, K.; Sinielnikov, I.; Hautaniemi, S.; Janne, O.A. FoxA1 Specifies Unique Androgen and Glucocorticoid Receptor Binding Events in Prostate Cancer Cells. *Cancer Res.* **2013**, *73*, 1570–1580. [CrossRef] [PubMed]
54. Gao, N.; Zhang, J.F.; Rao, M.A.; Case, T.C.; Mirosevich, J.; Wang, Y.Q.; Jin, R.J.; Gupta, A.; Rennie, P.S.; Matusik, R.J. The role of hepatocyte nuclear factor-3 alpha (forkhead box A1) and androgen receptor in transcriptional regulation of prostatic genes. *Mol. Endocrinol.* **2003**, *17*, 1484–1507. [CrossRef]
55. Pomerantz, M.M.; Li, F.; Takeda, D.Y.; Lenci, R.; Chonkar, A.; Chabot, M.; Cejas, P.; Vazquez, F.; Cook, J.; Shivdasani, R.A.; et al. The androgen receptor cistrome is extensively reprogrammed in human prostate tumorigenesis. *Nat. Genet.* **2015**, *47*, 1346–1351. [CrossRef]
56. Masella, A.P.; Bartram, A.K.; Truszkowski, J.M.; Brown, D.G.; Neufeld, J.D. PANDAseq: Paired-end assembler for illumina sequences. *BMC Bioinform.* **2012**, *13*, 31. [CrossRef] [PubMed]
57. Morgan, M.; Anders, S.; Lawrence, M.; Aboyoun, P.; Pages, H.; Gentleman, R. ShortRead: A bioconductor package for input, quality assessment and exploration of high-throughput sequence data. *Bioinformatics* **2009**, *25*, 2607–2608. [CrossRef]
58. Ran, F.A.; Hsu, P.D.; Wright, J.; Agarwala, V.; Scott, D.A.; Zhang, F. Genome engineering using the CRISPR-Cas9 system. *Nat. Protoc.* **2013**, *8*, 2281–2308. [CrossRef] [PubMed]
59. Li, B.; Ren, N.; Yang, L.; Liu, J.; Huang, Q. A qPCR method for genome editing efficiency determination and single-cell clone screening in human cells. *Sci. Rep.* **2019**, *9*, 18877. [CrossRef]
60. Liang, C.C.; Park, A.Y.; Guan, J.L. In vitro scratch assay: A convenient and inexpensive method for analysis of cell migration in vitro. *Nat. Protoc.* **2007**, *2*, 329–333. [CrossRef]
61. Liu, J.; Lichtenberg, T.; Hoadley, K.A.; Poisson, L.M.; Lazar, A.J.; Cherniack, A.D.; Kovatich, A.J.; Benz, C.C.; Levine, D.A.; Lee, A.V.; et al. An Integrated TCGA Pan-Cancer Clinical Data Resource to Drive High-Quality Survival Outcome Analytics. *Cell* **2018**, *173*, 400–416.e11. [CrossRef] [PubMed]

International Journal of *Molecular Sciences*

Review

Transcriptional Regulation of Endogenous Retroviruses and Their Misregulation in Human Diseases

Qian Zhang, Juan Pan, Yusheng Cong * and Jian Mao *

Key Laboratory of Aging and Cancer Biology of Zhejiang Province, Hangzhou Normal University School of Basic Medical Sciences, Hangzhou 311121, China
* Correspondence: yscong@hznu.edu.cn (Y.C.); maojian@hznu.edu.cn (J.M.)

Abstract: Endogenous retroviruses (ERVs), deriving from exogenous retroviral infections of germ line cells occurred millions of years ago, represent ~8% of human genome. Most ERVs are highly inactivated because of the accumulation of mutations, insertions, deletions, and/or truncations. However, it is becoming increasingly apparent that ERVs influence host biology through genetic and epigenetic mechanisms under particular physiological and pathological conditions, which provide both beneficial and deleterious effects for the host. For instance, certain ERVs expression is essential for human embryonic development. Whereas abnormal activation of ERVs was found to be involved in numbers of human diseases, such as cancer and neurodegenerative diseases. Therefore, understanding the mechanisms of regulation of ERVs would provide insights into the role of ERVs in health and diseases. Here, we provide an overview of mechanisms of transcriptional regulation of ERVs and their dysregulation in human diseases.

Keywords: endogenous retroviruses (ERVs); transcriptional regulation; cancer; neurodegenerative diseases

Citation: Zhang, Q.; Pan, J.; Cong, Y.; Mao, J. Transcriptional Regulation of Endogenous Retroviruses and Their Misregulation in Human Diseases. *Int. J. Mol. Sci.* **2022**, *23*, 10112. https://doi.org/10.3390/ijms231710112

Academic Editors: Alfredo Ciccodicola, Amelia Casamassimi and Monica Rienzo

Received: 11 July 2022
Accepted: 1 September 2022
Published: 4 September 2022

Publisher's Note: MDPI stays neutral with regard to jurisdictional claims in published maps and institutional affiliations.

1. Introduction

Transposable elements (TEs) are repetitive genetic sequences that once had or still have the ability to transpose, that is, to mobilize and insert elsewhere in the genome [1]. Nearly half of the human genome consists of TEs [2,3] (Figure 1A). TEs can be categorized into two classes: elements that can be transposed via a DNA intermediate and a cut-and-paste mechanism (transposons), and those using an RNA and a copy-paste mechanism (retrotransposons) [4]. Retrotransposons are further divided into long terminal repeat (LTR) elements and more primitive and ancient non-LTR elements with an obligate intracellular life cycle [5]. Non-LTR retrotransposons consist of two main groups: long interspersed nuclear elements (LINEs), which encode their own proteins necessary for retrotransposition; and short interspersed nuclear elements (SINEs), which are short, noncoding RNAs that hijack the LINE protein machinery [5] (Figure 1A,B). Retrotransposons flanked by LTRs that have high similarities to exogenous retroviruses are termed endogenous retroviruses (ERVs), which are the remnants of ancient exogenous retroviral infections [6] (Figure 1B). These endogenized forms of viral sequences were derived from exogenous retroviral infections and integrations for germ cells and transmitted vertically through Mendelian inheritance [7]. In human, ERVs account for ~8% of the human genome [2] (Figure 1A). The complete genomic structure of ERVs is composed of *gag*, *pro*, *pol*, and *env*, flanked by two LTRs (Figure 1B). Among them, *gag* encodes for capsid, nucleocapsid, and matrix protein; *pro* encodes for protease; *pol* encodes for reverse transcriptase and integrase; and *env* encodes for envelope protein [7]. LTRs are non-coding regions that contain many regulatory functions (promoter, enhancer, polyA signal, and others) [7]. However, most of ERVs are non-protein coding due to the accumulation of mutations, insertions, deletions, and truncations [8,9]. Based on the sequence similarity of their *pol* regions with reverse transcriptase sequences of exogenous retroviruses, ERVs are divided into three main classes: class I (Gamma- and Epsilonretrovirus-like), class II (Alpha-, Beta-, Deltaretrovirus-, and Lentivirus-like), and class III (Spumaretrovirus-like) [10]. However, no Alpha-, Deltaretrovirus-, or Lentivirus-like elements were detectable in human genome [9]. The major classes of ERVs are shown in Table 1. Human ERVs (HERVs) are further classified into several groups based on the tRNA binding to the viral primer binding site (PBS) to prime reverse transcription. For example, HERV-K implies a group of proviruses using a lysine (K) tRNA as primer [11]. In some cases, the PBS sequence was unclear when novel elements were discovered, resulting in their names based on neighboring genes (e.g., HERV-ADP), clone number (e.g., HERV-S71), or amino acid motifs (e.g., HERV-FRD) [12]. Recently, a unified nomenclature system for ERVs that provides the ERV group, the genomic loci, and species was proposed, which can aid genome annotation and research of ERVs [12].

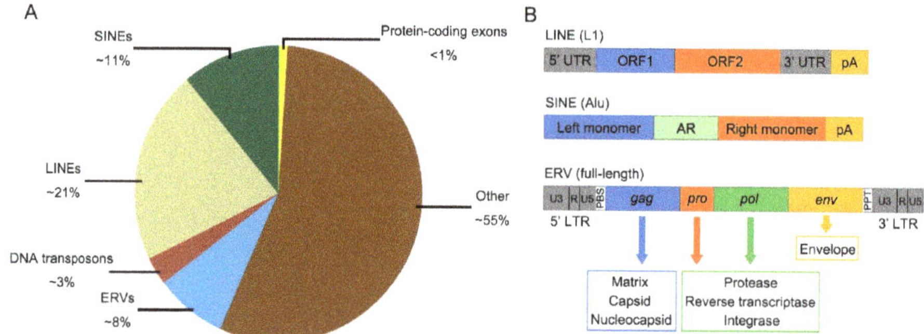

Figure 1. Organization and structure of transposable elements (TEs). (**A**) Pie chart shows the proportion of various selected genomic features within the human genome. (**B**) Genomic structures of LINE, SINE, and ERV. The general structure of a full-length ERV is shown. AR, adenine (A)-rich region. pA, poly (A) tail. PBS, primer binding site. PPT, polypurine tract.

Table 1. Classification of ERVs.

ERVs Classes	Exogenous Counterpart	Representative ERVs
Class I	Gammaretrovirus Epsilonretrovirus	FeLV, GALV, KoRV, McERV, MDEV, MuERV-C, MuRRS, MuRVY, MuLV, GLN, VL30, PERV, **HERV-E, F, H, I, P, R, T, W, HERV-FRD**
Class II	Alpharetrovirus Betaretrovirus Deltaretrovirus Lentivirus	ALV, IAP, MMTV, MPMV, MusD/ETn, MINERVa, RELIK, **HERV-K (HML-1, 2, 3, 4, 5, 6, 7, 8, 9, 10)**
Class III	Spumaretrovirus	MuERV-L, **HERV-L**

Human ERVs are shown in bold. No Alpha-, Deltaretrovirus-, or Lentivirus-like elements are detectable in human genome.

ERVs have been considered as "junk DNA sequences" for a long time [5]. However, it is becoming increasingly apparent that ERVs influence host biology through genetic and epigenetic mechanisms under particular physiological and pathological conditions, which provides both beneficial and deleterious effects for the host. For example, certain ERVs expression is essential for human embryonic development, whereas abnormal activation of ERVs is involved in numbers of human diseases, such as cancer and neurodegenerative diseases. Therefore, ERVs are under strict epigenetic regulation by the host, among which the methylation modifications of histone and DNA play significant roles. In this review, we will summarize recent findings on the mechanisms of transcriptional regulation of ERVs and their transcriptional dysregulation in human diseases.

2. Silencing and Transcriptional Regulation of ERVs

Currently, most discoveries about ERVs silencing and transcriptional regulation have been studied in mice, especially during embryonic development and in germ cells. Mouse embryonic stem cells (mESCs) are usually used as a cellular model to study the transcriptional regulation of ERVs, as the pluripotent state is capable of suppressing both exogenous and endogenous retroviruses [13–15]. Many mechanisms and concepts of transcriptional regulation of mouse ERVs may be applicable to human ERVs, but there may be discrepancies between them.

2.1. KRAB-ZFPs/TRIM28 Pathway Is a Master Regulator for ERVs Silencing

A common target for ERVs silencing is the PBS, an essential sequence used to prime reverse transcription by a host tRNA as primer. The first example for this mechanism is ZFP809, a member of the family of Krüppel-associated box-containing zinc-finger proteins (KRAB-ZFPs), which binds the PBS^{Pro} of provirus [16]. Interestingly, the DNA-binding specificity of ZFP809 is evolutionarily conserved and predates the endogenization of retroviruses presently targeted by ZFP809 in *Mus musculus* [17]. ZFP809 contains two domains, a KRAB box at the N terminus that is responsible for the interaction with TRIM28 and a zinc-finger domain containing seven zinc fingers that provide its sequence-specific DNA-binding activity [18]. Besides ZFP809, additional KRAB-ZFPs have also been identified to bind ERVs sequences, through which mediate provirus silencing, including ZFP708, ZNF91/93, ZFP819, ZFP932, Gm15446, and YY1 [19–23]. The KRAB domain of KRAB-ZFPs mediates the recruitment of TRIM28, the master regulator of ERVs silencing [6,24]. TRIM28 (also known as KAP1, TIF1β, or KRIP-1), which was identified to bind to the KRAB domain of KRAB-ZFPs, functions as a scaffold for other repressive histone-modifying and -binding factors, including the histone methyltransferase SETDB1, the human silencing hub (HUSH) complex, and heterochromatin protein 1 (HP1), which catalyze heterochromatin formation and transcriptional repression [25] (Figure 2). *Trim28* is expressed in a variety of cell types with especially high levels during early embryonic development, in brain and mESCs [24,26,27]. Knockout of *Trim28* results in embryonic lethal at E8.5, highlighting its essential role in early development [26].

SETDB1 (also known as ESET or KMT1E) is a protein lysine methyltransferase methylating histone H3 at lysine 9 (H3K9) [28]. Unlike other H3K9 methyltransferases, SETDB1 and SETDB1-mediated H3K9me3 play critical roles for silencing of ERVs [29]. SETDB1 is mainly localized in cytoplasm [30] while ATF7IP promotes its nuclear import and inhibits its nuclear export [31]. SETDB1 interacts with TRIM28 and is recruited to ERVs by KRAB-ZFPs/TRIM28 pathway, then establishes H3K9me3 in ERVs [29] (Figure 2). Actually, the KRAB-ZFPs/TRIM28 pathway is central for the de novo recruitment of SETDB1 to ERVs [28]. In addition to SETDB1, several other histone methyltransferases have also been described for ERVs silencing, including SUV39H1, SUV39H2, G9a (also known as EHMT2), GLP (also known as EHMT1), and NSD2 [32–35]. Another well-known interaction partner of TRIM28 is the human silencing hub (HUSH) complex, comprising TASOR (also known as FAM208A), MPP8 (MPHOSPH8), and PPHLN1 (Periphilin 1) [36]. The HUSH complex is recruited to genomic loci rich in H3K9me3 and interacts with SETDB1 and MORC2, mainly repressing evolutionarily young retrotransposons, such as young L1 [36]. Furthermore, epigenetic silencing by the HUSH complex also mediates position-effect variegation in human cells [37]. Recently, a study described a functional connection between the mouse-orthologous "nuclear exosome targeting" (NEXT) and HUSH complexes, involved in nuclear RNA decay and the epigenetic silencing of TEs, respectively, suggesting that transcriptional and post-transcriptional machineries synergize to suppress the genotoxic potential of TE RNAs [38]. H3K9me3 reader proteins, such as HP1, can bind to pre-existing H3K9me3 and bridge with SETDB1 through direct interaction [39].

Despite the reported function of HP1 proteins in H3K9me-dependent gene repression and the critical role of H3K9me3 in transcriptional silencing of ERVs, the depletion of all three HP1 isoforms (HP1α, HP1β, and HP1γ) in mESCs is not sufficient for the derepression of selected ERVs [40]. This surprising finding is attributed that H3K9me3 may repress ERVs transcription via inhibiting deposition of covalent histone modifications required for transcription [40]. Regardless, additional studies aimed at characterizing the functional significance of H3K9 readers are clearly warranted.

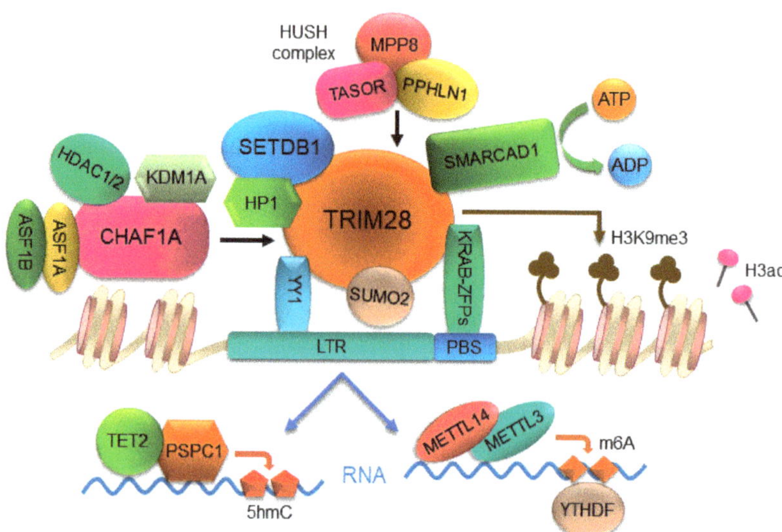

Figure 2. An overview model of ERVs silencing. ERVs are predominantly silenced by H3K9me3 through the canonical KRAB-ZFPs/TRIM28 pathway. KRAB-ZFPs bind to PBS region of ERVs and recruit TRIM28. Sumoylation of TRIM28 by SUMO2 enhances the recruitment of TRIM28 to ERVs. The ATPase activity of chromatin remodeler SMARCAD1 contributes to the occupancy of TRIM28 at ERVs. TRIM28 provides a scaffolding platform allowing for the recruitment of SETDB1, HP1, and HUSH complex, and the formation of macromolecular ensembles, which establish H3K9me3 in ERVs. Histone chaperone CHAF1A interacts with HP1, SETDB1, KDM1A, and HDAC1/2, modifying proviral chromatin with the repressive histone mark H3K9me3 and reducing the acquisition of active H3K4me3 and H3Ac marks. Histone chaperone isoforms ASF1A and ASF1B promote the localization of CHAF1A to ERVs. In addition to transcription-based silencing, RNA-mediated regulation of ERVs, such as epigenetic modifications of ERV RNAs, also play a critical role in silencing of ERVs.

2.2. Chromatin Remodeler and Histone Chaperone Maintain ERVs Silencing through KRAB-ZFPs/TRIM28 Pathway

The chromatin remodelers and the histone chaperones differing from well-known RNA chaperones or Janus chaperones have been considered as two important classes of factors involved in transcriptional regulation of ERVs, which are dependent on the KRAB-ZFPs/TRIM28 pathway. Recently, the SWI/SNF-like remodeler SMARCAD1 was identified as a key factor in the control of ERVs in mESCs [41]. As key regulators of nucleosome positioning, the SWI/SNF family of chromatin-remodeling complexes use energy generated through hydrolysis of ATP to slide or eject nucleosomes and promote chromatin access by moving nucleosomes, by which either activates or represses transcription [42,43]. For the transcriptional regulation of ERVs, SMARCAD1 is enriched at ERVs subfamilies class I and II, particularly at active IAPs, where it preserves repressive histone methylation marks. Importantly, recruitment of SMARCAD1 to ERVs is dependent on TRIM28 and the ATPase function of SMARCAD1 is required for SMARCAD1 and TRIM28 occupancy at ERVs (Figure 2), highlighting a critical role for SWI/SNF-like chromatin-remodeling activities in the establishment of ERVs silencing in mammals [41].

The association of histones with specific chaperone complexes is important for their folding, oligomerization, post-translational modification, nuclear import, stability, assembly, and genomic localization, which affects all chromosomal processes, including gene expression, chromosome segregation, and genome replication and repair [44]. Recently, a systematic genome-wide siRNA screen identified CHAF1A, a histone chaperone that assembles histones H3/H4 during DNA replication and repair [45,46], as a significant factor for silencing of ERVs [39]. It is shown that CHAF1A interacts with HP1, SETDB1, KDM1A, and HDAC1/2 [39,47,48] (Figure 2), modifying proviral chromatin with the repressive histone mark H3K9me3 and reducing the acquisition of active H3K4me3 and H3Ac marks [39]. ASF1 is also a chaperone that forms a complex with histones H3 and H4 [49]. The nucleosome assembly function of the two ASF1 isoforms, ASF1A and ASF1B, is shown to be responsible for localizing CHAF1A to proviral sequences [39].

ATRX is a chromatin remodeler and interacts with DAXX to form a histone chaperone complex, which deposits histone variant H3.3 into repetitive heterochromatin, including regions of retrotransposons, pericentric heterochromatin, and telomeres [50]. A series of studies revealed that ATRX and DAXX play roles for heterochromatin formation on ERVs through deposition of histone H3.3 [51–53]. The histone variant H3.3 belongs to the replication-independent class of variants and associated to both active chromatin states (e.g., H3K4me and H3K27ac) and heterochromatin states (e.g., H3K9me3 and H3K27me3) [54]. In mESCs, a study reported that recruitment of DAXX, H3.3 and TRIM28 to ERVs is codependent and occurs upstream of SETDB1, and H3.3 deletion leads to reduced H3K9me3 at ERVs regions and derepression of IAPs, establishing an important role for H3.3 in control of ERVs transcription in mESCs [53].

2.3. Sumoylation of TRIM28 Contributes to ERVs Silencing

Post-translational modification with small ubiquitin-related modifier (SUMO) proteins is one of the key regulatory protein modifications in eukaryotic cells. Hundreds of proteins involved in processes, such as chromatin organization, transcription, DNA repair, macromolecular assembly, protein homeostasis, trafficking, and signal transduction, are subject to reversible sumoylation [55]. Recent studies have shown that H3K9me3 deposition requires protein sumoylation, suggesting that the SUMO pathway functions as an important module in gene silencing and heterochromatin formation [56]. Importantly, the genome-wide screen for provirus silencing factors further confirmed the significant role of sumoylation for ERVs repression [39]. The SUMO family in mammals consists of four members: SUMO1, SUMO2, SUMO3, and SUMO4 [55]. Among them, SUMO2 orchestrates viral silencing through sumoylation modification of TRIM28 [39]. Sumoylation enhances the recruitment of TRIM28 to the proviral DNA, which in turn results in the modification of proviral chromatin with repressive histone H3K9me3 marks [39] (Figure 2). Nonetheless, further studies are needed to determine the mechanism of sumoylation in transcriptional regulation of ERVs.

2.4. DNA Methylation in ERVs Silencing

In addition to histone-based silencing, ERVs exhibit distinctive DNA methylation patterns [6]. Interestingly, KRAB-ZFPs/TRIM28 and SETDB1 are necessary to target ERV-containing loci for rapid de novo DNA methylation [57]. Three DNA methyltransferases in mammals (DNMT1, DNMT3A, and DNMT3B) have been intensively studied. The roles of DNA methylation in ERVs silencing appears to be cell type dependent [24,29,58,59]. For example, knockout of all three DNA methyltransferases in mESCs showed a complete loss of DNA methylation on ERVs, but only subtle derepression of ERVs was observed [29,60,61]. However, loss of DNA methylation activates ERVs expression in differentiated or somatic cells, such as mouse embryonic fibroblasts (MEFs) [58,59], while deletion of *Trim28* or *Setdb1* in MEFs does not lead to significant activation of ERVs [24,29]. Interestingly, deletion of

Trim28 or *Setdb1* in neural progenitor cells or pro-B cells results in strong ERVs derepression with only a slight reduction in DNA methylation [62–65]. These data indicate that the KRAB-ZFPs/TRIM28 pathway is primarily used for ERVs silencing in cells with stemness, whereas differentiated cells primarily rely on DNA methylation to suppress ERVs. As a member of DNMT3 family, DNMT3L (DNMT3-like) has no DNA methyltransferase activity but is capable of interacting with both DNMT3A and DNMT3B to stimulate their enzymatic activities [66]. Deletion of *Dnmt3l* in mouse testis prevents the de novo methylation of both LTR and non-LTR retrotransposons, leading to the activation of IAPs and L1, as well as meiotic failure [67]. Notably, a recent study reported a correlation between the silencing mechanism and the evolutionary age of ERVs [68]. Young LTRs tend to be CpG rich and are mainly suppressed by DNA methylation, while intermediate age LTRs are associated predominantly with histone modifications, particularly H3K9 methylation [68].

2.5. RNA-Mediated Regulation of ERVs

In addition to DNA-specific binding by co-repressors, histone chaperones, and chromatin remodelers, RNA-mediated targeting of ERVs also play a significant role for the silencing and transcriptional regulation of ERVs. It has been reported that siRNA- or antisense transcripts-based silencing pathways suppress IAPs and non-LTR retrotransposons such as L1 [69,70]. The most representative RNA-dependent gene silencing is *Xist*, the master regulator of X chromosome inactivation in mammals [71]. SPEN is a key factor for establishment of *Xist*-mediated silencing through directly recruited to *Xist* RNA [71]. A recent study showed that SPEN binds to retroviral RNA and performs a surveillance role to recruit chromatin-silencing machinery to these parasitic loci, suggesting that *Xist* may coopt ERVs RNA–protein interactions to repurpose powerful antiviral chromatin-silencing machinery [72]. Another mechanism for RNA-mediated transcriptional regulation of ERVs is the Piwi-interacting RNA (piRNA) pathway. piRNAs are a class of small RNAs that are 24–31 nucleotides in length and associate with PIWI proteins to form effector complexes known as piRNA-induced silencing complexes, which repress retrotransposons via transcriptional or post-transcriptional mechanisms [73]. It is in *Drosophila* that piRNA was first found to induce silencing through H3K9me3 formation [74]. In mice, depletion of Piwi proteins leads to derepression of IAPs and L1 [75,76].

It is worth noting that RNA epigenetic modifications play a significant role in regulation of ERVs. TET2, a member of the Ten-eleven translocation (TET) family, can be recruited to actively transcribed MuERVL RNAs by the RNA-binding protein PSPC1, then catalyzes 5hmC modification of MuERVL RNAs, resulting in their destabilization (Figure 2), which provides evidence for a functional role of transcriptionally active ERVs as specific docking sites for RNA epigenetic modulation [77]. m6A RNA methylation, which is catalyzed by the complex of methyltransferase-like METTL3-METTL14 proteins [78], is shown to reduce the half-life of IAP mRNAs by recruiting the m6A reader proteins YTHDF family (Figure 2), indicating that RNA methylation provides a protective effect in maintaining cellular integrity by clearing reactive ERVs-derived RNA species [79].

2.6. Exogenous Viruses Are Associated with ERVs Activation

Human ERVs activation can be triggered by infections of exogenous viruses such as HIV-1, hepatitis B virus (HBV), hepatitis C virus (HCV), human T-lymphotropic tumor virus-1 (HTLV-1), influenza A virus, and Kaposi's Sarcoma-associated herpesvirus (KSHV) [80,81]. For HIV-1, the recombinant Tat protein upregulates HERV-K (HML-2) *gag* RNA transcripts in lymphocytes and monocytic cells through transcription factors NF-κB or NF-AT, indicating that exogenous viral infection activates transcription factors, which also bind to ERVs LTR regions and induce their activation [82]. An in-depth understanding of how ERVs are activated by exogenous viruses would facilitate the search for novel targets of virus-mediated diseases and therapeutic intervention.

2.7. Additional Factors in ERVs Transcriptional Regulation

Additional factors also contribute to transcriptional regulation of ERVs. TIP60, a lysine acetyltransferase, was found to be involved in silencing of ERVs, through positively regulating the expression of SUV39H1 and SETDB1, and thereby establishing global H3K9me3 levels [83]. KDM1A (also known as LSD1), a lysine-specific demethylase, was shown to be required to silence ERVs through regulating histone methylation and acetylation at LTR sequences. *Kdm1a* mutant mESCs exhibit increased methylation of histone H3K4, increased acetylation of H3K27, and decreased methylation of H3K9, indicating that chromatin modification mediated by KDM1A is part of the host's defense against excessive ERVs activity [84]. Recently, the histone chaperone FACT, which is critical for nucleosome reorganization during replication, transcription, and DNA repair [85], was reported to recruit USP7 to repress MuERVL and MuERVL-fused 2C genes in mESCs by impeding the ubiquitination of H2Bub, providing insights into the regulation of TE-derived cryptic promoters during mammalian development and in diseases [86].

As a DNA-binding protein that is specifically expressed in two cell-stage embryos during mouse development [87], ZSCAN4C is positively associated with H3K27ac, H3K4me1 and H3K14ac deposition on MT2 (MuERVL LTR) and interacts with GBAF chromatin-remodeling complex to activate MT2 enhancer activity, indicating that ZSCAN4C plays a significant role in regulating MuERVL in mESCs [88]. *DUX4*, a eutherian-specific multicopy retrogene, encodes a transcription factor that can activate hundreds of retroviral elements (MuERVL/HERVL family) that define the cleavage-specific transcriptional programs in humans and mice [89]. In addition, it is shown that female sex hormones activate HERV-K through the OCT4 transcription factor in T47D breast cancer cells [90]. Notably, a recent study reported that TERT, the catalytic subunit of telomerase, can activate a subclass of ERVs independent of its telomerase activity to form double-stranded RNAs (dsRNAs), which trigger interferon signaling in cancer cells and promote an immunosuppressive tumor microenvironment [91].

3. Transcriptional Dysregulation of ERVs in Human Diseases

Several studies have suggested that TEs are domesticated for the benefit of the host. This process, in which the host makes use of TEs (including ERVs)-derived functions, are called exaptation, co-option, or repurposing [6]. Either *cis*-regulatory element activities or encoded proteins of ERVs can be beneficial to the host. For example, syncytin-1 and syncytin-2, which are specifically expressed in the placenta, are envelope proteins encoded by HERV-W and HERV-FRD, respectively, and with cell–cell fusogenic activities, contributing to the formation of placenta syncytiotrophoblast layer at the materno–fetal interface [92,93]. Therefore, capture of retroviral envelope genes may play a critical role in the emergence of placental mammals. Another example of exaptation is the *Fv1* gene of mice, which is an endogenous *gag* gene related to ERV-L family [94,95]. Fv1 confers host resistance to MuLV by blocking the incoming viral capsid cores shortly after entry [96,97]. Fv1 orthologues have been identified in a wide range of rodent species [98,99] and some Fv1 homologues restrict non-MuLV retroviruses [100], suggesting that Fv1 does not recognize conserved amino acid motifs but may instead detect structurally conserved spatial patterns in the hexameric lattice typical of retroviral capsid cores [97,101]. Notably, the neuronal Arc protein, which evolved from a Ty3/Gypsy retrotransposon Gag domain and has retained the topology of a retroviral Gag protein [102], is able to self-assemble into virus-like capsids that encapsulate RNA [103]. The Arc protein is released from neurons in extracellular vesicles and transfer the *Arc* mRNA into new target cells, where it can undergo activity-dependent translation, suggesting that Gag retroelements have been repurposed during evolution to mediate intercellular communication in the nervous system [103].

In spite of the exaptations of ERVs by the host, the dysregulation of them is involved in numbers of pathological processes. Although there is no direct evidence for ERVs causing diseases, aberrant expression profiles of the ERVs transcripts and their regulatory activities on proximal host genes have been identified in different diseases, such as cancer and neurodegenerative diseases. Mechanistically, ERVs may participate in pathological processes through several pathways: (i) ERVs act as promoters or enhance cellular gene expression through LTR *cis*-regulatory element activities; (ii) insertion of ERVs sequences induces chromosomal rearrangements and genome instability; (iii) ERVs encode proteins, long non-coding RNAs (lncRNAs), and double-stranded RNAs (dsRNAs) to affect host physiology.

3.1. HERVs in Cancer

The transcriptional activation of HERVs is a common feature in human cancers, suggesting that ERVs are causative elements or cofactors contributing to the onset and progression of human cancer [104]. So far, several studies have strongly suggested that ERVs play roles in various human cancers (Table 2).

Table 2. HERVs and oncogenic mechanisms in human cancers.

HERVs	HERVs Products/Activities	Oncogenic Mechanisms
HERV-K (HML-2)	ENV protein	ENV induces EMT and activates ERK pathway in breast cancer [105]. ENV mediates intercellular fusion in melanoma [106]. ENV maintains CD133+ melanoma cells with stemness features [107]. ENV promotes pancreatic cancer proliferation, tumorigenesis, and metastasis by activating RAS/MEK/ERK and JNK/c-Jun signaling pathways [108].
	Rec protein	Rec activates *c-MYC* by overcoming the transcriptional repression of testicular zinc-finger protein (TZFP) for *c-MYC* promoter [109]. Rec relieves the repression of androgen receptor (AR) activity by forming a trimeric complex with TZFP and AR [109], or binds to the human small glutamine-rich tetratricopeptide repeat protein (hSGT) [110].
	Np9 protein	Np9 interacts with ligand of Numb protein X, affecting tumorigenesis through the LNX/Numb/Notch pathway [111]. Np9 as a critical molecular switch of multiple signaling pathways in leukemia [112]. Rec and Np9 derepressed *c-MYC* through the inhibition of promyelocytic leukemia zinc-finger protein (PLZF) [113].
HERV-H	lncRNA	*UCA1* enhances proliferation, motility, invasion, and drug resistance of bladder cancer [114]. linc-ROR contributes to progression, metastasis, or chemoresistance in breast cancer [115], pancreatic cancer [116], and hepatocellular carcinoma [117].
	cis-regulatory element	LTR acts as alternative promoter for *GSDML* in cervical cancer [118,119].
HERV-E	*cis*-regulatory element	Modulating *PLA2G4A* transcription in urothelial cancer [120].
syncytin-1/ERVW-1	ENV protein	Syncytin-1 mediates cancer–endothelial cell fusions in breast cancer [121].
ERV-9	lncRNA	*PRLH1* plays an important role in the formation of RNA–protein complex that promotes the HR-mediated DSB repair [122].
MaLR	*cis*-regulatory element	LTR acts as alternative promoter for *CSF1R* in Hodgkin's lymphoma [123].
MER52A	lncRNA	lncMER52A promotes invasion and metastasis of hepatocellular carcinoma cells by stabilizing p120-catenin [124].
MER48	lncRNA/*cis*-regulatory element	lncRNA EVADR is associated with adenocarcinomas, and a MER48 ERV element acts as an active promoter for its specific activation [125].
Multiple HERVs	dsRNA	dsRNAs derived from the bi-directional transcription of HERVs induce an immunosuppressive tumor microenvironment [91,126].

Given the potential transposable ability of retrotransposon, it is the belief that the tumorigenicity of HERVs can depend on retroviral movement, thereby destabilizing the host genome [104]. Indeed, new insertions of TEs, especially ERVs, have been reported in several tumors [127]. LTRs can act as alternative promoter or enhancer, leading to the deregulation of proto-oncogenes or tumor suppressor genes [7]. A representative example is in Hodgkin's lymphoma, where *CSF1R* transcription initiates at an LTR element of the MaLR THE1B family, rather than from its own promoter [123]. However, it should be noted that the LTRs activity may have an anti-oncogenic effect by driving the expression of tumor suppressor genes, such as *TP63* and *TNFRSF10B*, which are regulated by upstream LTRs belonging to the ERV9 group of HERVs [128,129].

HERVs can take a direct action via their own proteins in cancer. The envelop proteins of HERVs, such as syncytin-1 and HERV-K (HML-2) ENV, have been reported to contribute to tumorigenesis by inducing cell–cell fusion in melanoma [106], endometrial carcinoma [130], and breast cancer [121]. Furthermore, HERV-K (HML-2) ENV has also been shown to activate Ras/Raf/MEK/ERK and JNK/c-Jun signaling pathways, thereby promoting tumorigenesis and development [105,108,131], suggesting a direct interaction of ENV with cellular signaling pathways. Another mechanism by which ENV supports tumor progression is to promote immune escape by abolishing the anti-oncogenic cytolytic immune responses through its immunosuppressive domain (ISD) [132]. In addition to ENV, HERV proteins Rec and Np9 encoded by HERV-K (HML-2) are also regarded as tumor-specific biomarkers and act oncogenically by activating oncogene *c-MYC* or signaling pathways such as Notch, Wnt/β-catenin, Ras/ERK, and AKT [109–113].

lncRNAs play significant roles in various biological processes, including cancer progression. Strikingly, 75–83% of lncRNAs have been identified to contain TE sequences, especially ERVs [133]. Several HERVs-derived lncRNAs have been characterized in tumorigenesis and development. *UCA1*, a lncRNA consists of LTR7Y and HERV-H, has been shown to enhance proliferation, motility, invasion, and drug resistance of bladder cancer [114]. The HERVs-derived lncRNAs *SAMMSON* and *BANCR* are involved in melanoma progression [134,135], and linc-ROR contributes to progression, metastasis or chemoresistance in breast cancer [115], pancreatic cancer [116], and hepatocellular carcinoma [117]. A recent study identified a novel HCC (hepatocellular carcinoma)-specific lncRNA derived from MER52A, lncMER52A, which promotes invasion and metastasis of HCC cells by stabilizing p120-catenin [124]. Higher lncMER52A is associated with advanced TNM stage, less differentiated tumors, and shorter overall survival, and can serve as biomarker and therapeutic target for patients with HCC [124]. Another HERVs-derived lncRNA, EVADR, is revealed a striking association with adenocarcinomas, which are tumors of glandular origin, including colon, rectal, lung, pancreas, and stomach adenocarcinomas, and EVADR expression correlates with decreased patient survival [125]. Interestingly, a MER48 ERV element provides an active promoter to drive the specific activation of EVADR [125].

It has been reported that HERVs contribute to the modulation of innate immune response in different physiological and pathological conditions [104]. For example, lnc-EPAV, a full-length ERV-derived lncRNA, is a positive regulator of host innate immune responses by regulating expression of RELA, an NF-κB subunit that plays a critical role in antiviral responses [136]. Notably, dsRNAs derived from the bi-directional transcription of HERVs have opposite effects on modulating immune response in tumorigenesis and development. They may be involved in both anti-tumor defense and oncogenic process. On the one hand, dsRNAs from HERVs activated by DNMT inhibitors (DNMTis) in tumor cells can induce a growth-inhibiting immune response, and the high expression of genes associated with anti-viral response potentiates the response to immune checkpoint therapy [137,138]. On the other hand, HERVs-derived dsRNAs can also induce immune-suppressed microenvironment of tumors, similar to a chronic virally infected state [91,126]. These findings suggest significant implications of HERVs in cancer immunotherapy.

HERV deoxyuridine triphosphate nucleotidohydrolase (dUTPase) can trigger innate and adaptive immune responses [139]. In pulmonary arterial hypertension (PAH), the HERV-K dUTPase activates B cells, elevates cytokines in monocytes and pulmonary arterial endothelial cells, and increases pulmonary artery vulnerability to apoptosis, contributing to sustained inflammation and immune dysregulation [140]. Increased production and release of elastase, neutrophil extracellular traps, and vinculin-mediated increased adhesion in PAH are attributed to an increased in HERV-K dUTPase [141]. Another example of pro-inflammatory potential of HERV-K dUTPase is psoriasis, where HERV-K dUTPase proteins induce the activation of NF-κB through TLR2 to trigger the secretion of TH1 and TH17 cytokines involved in the formation of psoriatic plaques, supporting HERV-K dUTPase as a potential contributor to psoriasis pathophysiology [142]. Moreover, expression of dUTPase was identified in colorectal cancer and could be a predictive biomarker for the metastatic potential of colorectal cancer [143,144]. Interestingly, a recent study revealed that the expression of dUTPase determines whether elevation of the ribonucleotide reductase subunit R2 can lead to genome stress and chromosomal instability, and the combination of low dUTPase and high R2 in clinical tumor samples predicts poor survival in patients with colorectal cancer or breast cancers [145].

Considering the activation of HERVs in many human cancers and that HERVs expression has been shown to be associated with proliferation, metastasis, TNM stage, and overall survival, HERVs can be used as biomarkers for tumor diagnosis and/or prognosis [104]. For instance, a study reported that the combination testing of HERV-K (HML-2) with traditional prostate-specific antigen improves the efficacy of prostate cancer detection, specifically for older men and smokers who tend to develop a more aggressive disease [146]. HERVs have the potential to be targets for new cancer therapeutic opportunities as well. In this view, anti-HERV-K (HML-2) ENV antibodies have been shown to inhibit growth and induce apoptosis of breast cancer cells in vitro, and reduce growth of xenograft tumors in mice [147]. Consistently, HERV-K ENV-specific CAR$^+$ T cells are able to lyse melanoma tumor cells in an antigen-specific manner [148]. Moreover, DNMTis activate the viral recognition and interferon response pathway by inducing dsRNAs transcribed by HERVs, which potentiates the response to immune checkpoint therapy [137,138].

3.2. HERVs in Aging and Neurodegenerative Diseases

As mentioned above, ERVs are largely transcriptionally silenced through heterochromatic structures. However, there may be a net loss of heterochromatin with aging, leading to the abnormal activation of TEs, including ERVs, in aging individuals [149,150]. It has been reported that IAPs and MusD are activated in aging mice [151,152]. In humans, HERV-K (HML-2) and HERV-W exhibit distinct expression patterns between young and old individuals [153]. Interestingly, the expression of HERV-H and HERV-W in peripheral blood mononuclear cells (PBMCs) was shown to be significantly positively correlated with age over 30 years [154]. Notably, HERV-W expression has been shown to significantly increase in individuals over 40 years old, and neurodegenerative diseases such as multiple sclerosis (MS) also occur in this age range [154]. Nowadays, ERVs have been implicated in the occurrence and development of neurodegenerative diseases, such as MS, amyotrophic lateral sclerosis (ALS), and autism spectrum disorder (ASD).

MS is an autoimmune-mediated neurodegenerative disease of the central nervous system characterized by inflammatory demyelination with axonal transection [155]. Although the underlying etiology of MS is still not fully understood, the development of MS has been associated with activation of HERVs, especially HERV-W [4]. The presence of retroviral particles was first found in MS patients approximately 30 years ago [156,157] and subsequent studies revealed that these particles originated from HERV elements, originally called MS-associated retrovirus (MSRV), and now named HERV-W because it uses a tryptophan (W) tRNA as a primer for reverse transcription [158,159]. Mechanistically, HERV-W ENV can activate the innate immune system through a TLR4/CD14-dependent pathway and promote the development of a Th1 type of immune response upon DC activation [160]. HERV-W ENV-mediated activation of TLR4 leads to the induction of proinflammatory cytokines and inducible nitric oxide synthase, as well as the formation of nitrotyrosine groups and a subsequent reduction in myelin protein expression, resulting in an overall reduction of the oligodendroglial differentiation capacity and remyelination failure in MS [161]. Moreover, HERV-W ENV is also a potent superantigen associated with demyelination in MS, possibly related to molecular mimicry with myelin oligodendrocyte glycoprotein [162]. A recent study reported that HERV-W ENV induces a degenerative phenotype in microglial cells and drives them toward a close spatial association with myelinated axons, suggesting that HERV-W ENV-mediated microglial polarization contributes to neurodegeneration in MS [163]. Accordingly, treatment with neutralizing antibodies against HERV-W ENV abrogates the oligodendroglial maturation blockade [164]. In this view, the neutralizing antibodies have been used in a recently completed clinical study in MS patients, which showed that the antibody-mediated neutralization exerts neuroprotective effects [163]. In addition to HERV-W, other HERV elements have also been found in MS, such as HERV-H and HERV-K (HML-2) [165–168]. Taken together, these data suggest that activation of multiple HERVs families is linked to MS, among which HERV-W play a significant role.

Amyotrophic lateral sclerosis (ALS), a neurodegenerative disease characterized by progressive loss of cortical and spinal motor neurons, is another neurodegenerative disease associated with HERVs [162]. Activation of retroviral elements in ALS was first found through a study that identified RNA-directed DNA polymerase activity in brain tissue extracts from ALS patients, whereas no virus or transmissible agent was detected [169]. Subsequent studies confirmed the presence of reverse transcriptase in serum of ALS patients [170–172]; however, the attempts to search for exogenous retroviruses in ALS patients were unsuccessful [171,173], leading to the investigation of HERVs in ALS pathogenesis. As expected, a study revealed that HERV-K (HML-2) *pol* transcripts are upregulated in patients with ALS but not detectable in Parkinson disease or in healthy controls [174]. A subsequent study further identified the expression of HERV-K (HML-2) *pol, env*, and *gag* genes in brains of ALS patients [175]. Moreover, HERV-K (HML-2) ENV has also been detected in cortical and spinal neurons of ALS patients, but not in neurons from healthy individuals, which contributes to neurite retraction and beading, and neurodegeneration [175]. Several mechanisms by which HERV-K (HML-2) is activated in ALS have been revealed. For example, the nuclear translocation of IRF1 and NF-κB isoforms p50 and p65 has been revealed to contribute to the neuronal HERV-K (HML-2) activation in ALS brain tissue, implicating the critical role of neuroinflammation [176]. As a multifunctional protein dysregulated in ALS, *TDP-43* expression strongly correlates with HERV-K (HML-2) [174], and has been shown to activate HERV-K (HML-2) through binding to the LTR region of the provirus [175]. Besides HERV-K (HML-2), HERV-W ENV is also detected in muscle cells of ALS patients [177]. Nonetheless, although activation of HERVs is common in ALS, its pathogenic mechanisms require further investigation.

Recent studies have revealed aberrant expression of HERVs in neurodevelopmental disorder ASD. HERV-H is more abundantly expressed while HERV-W shows lower expression levels in PBMCs from ASD patients compared to healthy controls [178]. Furthermore, the expression of HERV-H is significantly upregulated in ASD patients with severe disease development [178]. Notably, HEMO, an ERV envelope protein of MER34 family [179], was reported to be altered in ASD patients and may be useful for the disease diagnosis [180]. Therefore, HERVs expression can be considered as a biomarker that is easily detectable in blood and may be helpful for early diagnosis of ASD. In addition, several studies have revealed that HERV-H, HERV-K (HML-2), HERV-L, and HERV-W are activated in Alzheimer's disease (AD) [181–183]. In schizophrenia, several HERVs families have been shown to be dysregulated, including HERV-K (HML-2), HERV-W, ERV9, HERV-FRD, and HERV-H [184–189]. In addition, a recent study revealed that HERV-W ENV alters the NMDAR-mediated synaptic organization and plasticity through glia- and cytokine-dependent changes, leading to defective glutamate synapse maturation, behavioral impairments, and psychosis [190].

4. Conclusions and Perspective

ERVs are involved in various biological processes by encoding proteins, lncRNAs, dsRNAs, or acting as promoters/enhancers, thereby affecting human health and disease. Recent progress suggests that the implication of ERVs in cancer and neurodegenerative diseases provides an opportunity to develop novel therapeutic strategies. For example, nucleoside reverse transcriptase inhibitors (NRTIs) have shown promise in the treatment of neurodegenerative diseases [5]. DNMTis have been revealed to induce dsRNAs transcribed by ERVs in tumor cells, which activate the viral recognition and interferon response pathway, thereby enhancing the response to immune checkpoint therapy [137,138]. However, although many players in ERVs regulation have been identified, the detailed mechanisms of ERVs silencing and activation, especially the mechanisms of their action in human health and diseases, are not fully understood. Clearly, there are many species-, cell-type-, and disease-specific mechanisms, and unraveling which ERVs are silenced or activated, and how they are sensed in different contexts will be a major undertaking. The application of omics approaches, such as high-throughput sequencing, single-cell RNA-seq, genome editing technology, and proteomics can help to address these issues. Previously, ERVs have not received enough attention due to technical difficulties in analyzing these highly repetitive elements. With the development of technology, the mysteries of ERVs are being revealed step by step. However, the story of ERVs transcriptional regulation and the identification of specific ERV loci associated with specific diseases remains incomplete. Therefore, it is crucial and promising to enrich the knowledge of ERVs, our ancient "roommates" making up ~8% of human genome, in health and diseases.

Author Contributions: Conceptualization, Y.C. and J.M.; writing—original draft preparation, Q.Z. and J.M.; writing—review and editing, Q.Z., J.P., Y.C. and J.M.; supervision, Y.C.; funding acquisition, Q.Z., Y.C. and J.M. All authors have read and agreed to the published version of the manuscript.

Funding: This work was funded by grants from the National Natural Science Foundation of China (31730020, 32000512, 31801155), and the Hangzhou Science and Technology Bureau (20182014B01).

Institutional Review Board Statement: Not applicable.

Informed Consent Statement: Not applicable.

Data Availability Statement: Not applicable.

Acknowledgments: The authors thank all members of our laboratory for critical comments and discussions.

Conflicts of Interest: The authors declare no conflict of interest.

References

1. Senft, A.D.; Macfarlan, T.S. Transposable elements shape the evolution of mammalian development. *Nat. Rev. Genet.* **2021**, *22*, 691–711. [CrossRef] [PubMed]
2. Lander, E.S.; Linton, L.M.; Birren, B.; Nusbaum, C.; Zody, M.C.; Baldwin, J.; Devon, K.; Dewar, K.; Doyle, M.; FitzHugh, W.; et al. Initial sequencing and analysis of the human genome. *Nature* **2001**, *409*, 860–921. [CrossRef] [PubMed]
3. Kazazian, H.H.; Moran, J.V. Mobile DNA in Health and Disease. *N. Engl. J. Med.* **2017**, *377*, 361–370. [CrossRef] [PubMed]
4. Kury, P.; Nath, A.; Creange, A.; Dolei, A.; Marche, P.; Gold, J.; Giovannoni, G.; Hartung, H.P.; Perron, H. Human Endogenous Retroviruses in Neurological Diseases. *Trends Mol. Med.* **2018**, *24*, 379–394. [CrossRef]
5. Gorbunova, V.; Seluanov, A.; Mita, P.; McKerrow, W.; Fenyö, D.; Boeke, J.D.; Linker, S.B.; Gage, F.H.; Kreiling, J.A.; Petrashen, A.P.; et al. The role of retrotransposable elements in ageing and age-associated diseases. *Nature* **2021**, *596*, 43–53. [CrossRef]
6. Geis, F.K.; Goff, S.P. Silencing and Transcriptional Regulation of Endogenous Retroviruses: An Overview. *Viruses* **2020**, *12*, 884. [CrossRef]
7. Mao, J.; Zhang, Q.; Cong, Y.-S. Human endogenous retroviruses in development and disease. *Comput. Struct. Biotechnol. J.* **2021**, *19*, 5978–5986. [CrossRef]
8. De Parseval, N.; Heidmann, T. Human endogenous retroviruses: From infectious elements to human genes. *Cytogenet. Genome Res.* **2005**, *110*, 318–332. [CrossRef]
9. Vargiu, L.; Rodriguez-Tomé, P.; Sperber, G.O.; Cadeddu, M.; Grandi, N.; Blikstad, V.; Tramontano, E.; Blomberg, J. Classification and characterization of human endogenous retroviruses; mosaic forms are common. *Retrovirology* **2016**, *13*, 7. [CrossRef]
10. Johnson, W.E. Endogenous Retroviruses in the Genomics Era. *Annu. Rev. Virol.* **2015**, *2*, 135–159. [CrossRef]
11. Mager, D.L.; Stoye, J.P. Mammalian Endogenous Retroviruses. *Microbiol. Spectr.* **2015**, *3*, MDNA3-0009-2014. [CrossRef]
12. Gifford, R.J.; Blomberg, J.; Coffin, J.M.; Fan, H.; Heidmann, T.; Mayer, J.; Stoye, J.; Tristem, M.; Johnson, W.E. Nomenclature for endogenous retrovirus (ERV) loci. *Retrovirology* **2018**, *15*, 59. [CrossRef]
13. Teich, N.M.; Weiss, R.A.; Martin, G.R.; Lowy, D.R. Virus infection of murine teratocarcinoma stem cell lines. *Cell* **1977**, *12*, 973–982. [CrossRef]
14. Niwa, O.; Yokota, Y.; Ishida, H.; Sugahara, T. Independent mechanisms involved in suppression of the Moloney leukemia virus genome during differentiation of murine teratocarcinoma cells. *Cell* **1983**, *32*, 1105–1113. [CrossRef]
15. Feuer, G.; Taketo, M.; Hanecak, R.C.; Fan, H. Two blocks in Moloney murine leukemia virus expression in undifferentiated F9 embryonal carcinoma cells as determined by transient expression assays. *J. Virol.* **1989**, *63*, 2317–2324. [CrossRef]
16. Wolf, D.; Goff, S.P. Embryonic stem cells use ZFP809 to silence retroviral DNAs. *Nature* **2009**, *458*, 1201–1204. [CrossRef]

17. Wolf, G.; Yang, P.; Füchtbauer, A.C.; Füchtbauer, E.-M.; Silva, A.M.; Park, C.; Wu, W.; Nielsen, A.L.; Pedersen, F.S.; Macfarlan, T.S. The KRAB zinc finger protein ZFP809 is required to initiate epigenetic silencing of endogenous retroviruses. *Genes Dev.* **2015**, *29*, 538–554. [CrossRef]
18. Wang, C.; Goff, S.P. Differential control of retrovirus silencing in embryonic cells by proteasomal regulation of the ZFP809 retroviral repressor. *Proc. Natl. Acad. Sci. USA* **2017**, *114*, E922–E930. [CrossRef]
19. Seah, M.K.Y.; Wang, Y.; Goy, P.-A.; Loh, H.M.; Peh, W.J.; Low, D.H.P.; Han, B.Y.; Wong, E.; Leong, E.L.; Wolf, G.; et al. The KRAB-zinc-finger protein ZFP708 mediates epigenetic repression at RMER19B retrotransposons. *Development* **2019**, *146*, dev170266. [CrossRef]
20. Jacobs, F.M.J.; Greenberg, D.; Nguyen, N.; Haeussler, M.; Ewing, A.D.; Katzman, S.; Paten, B.; Salama, S.R.; Haussler, D. An evolutionary arms race between KRAB zinc-finger genes ZNF91/93 and SVA/L1 retrotransposons. *Nature* **2014**, *516*, 242–245. [CrossRef]
21. Tan, X.; Xu, X.; Elkenani, M.; Smorag, L.; Zechner, U.; Nolte, J.; Engel, W.; Pantakani, D.V.K. Zfp819, a novel KRAB-zinc finger protein, interacts with KAP1 and functions in genomic integrity maintenance of mouse embryonic stem cells. *Stem Cell Res.* **2013**, *11*, 1045–1059. [CrossRef] [PubMed]
22. Ecco, G.; Cassano, M.; Kauzlaric, A.; Duc, J.; Coluccio, A.; Offner, S.; Imbeault, M.; Rowe, H.M.; Turelli, P.; Trono, D. Transposable Elements and Their KRAB-ZFP Controllers Regulate Gene Expression in Adult Tissues. *Dev. Cell* **2016**, *36*, 611–623. [CrossRef] [PubMed]
23. Schlesinger, S.; Lee, A.H.; Wang, G.Z.; Green, L.; Goff, S.P. Proviral silencing in embryonic cells is regulated by Yin Yang 1. *Cell Rep.* **2013**, *4*, 50–58. [CrossRef] [PubMed]
24. Rowe, H.M.; Jakobsson, J.; Mesnard, D.; Rougemont, J.; Reynard, S.; Aktas, T.; Maillard, P.V.; Layard-Liesching, H.; Verp, S.; Marquis, J.; et al. KAP1 controls endogenous retroviruses in embryonic stem cells. *Nature* **2010**, *463*, 237–240. [CrossRef]
25. Yang, P.; Wang, Y.; Macfarlan, T.S. The Role of KRAB-ZFPs in Transposable Element Repression and Mammalian Evolution. *Trends Genet.* **2017**, *33*, 871–881. [CrossRef]
26. Cammas, F.; Mark, M.; Dollé, P.; Dierich, A.; Chambon, P.; Losson, R. Mice lacking the transcriptional corepressor TIF1beta are defective in early postimplantation development. *Development* **2000**, *127*, 2955–2963. [CrossRef]
27. Grassi, D.A.; Jönsson, M.E.; Brattås, P.L.; Jakobsson, J. TRIM28 and the control of transposable elements in the brain. *Brain Res.* **2019**, *1705*, 43–47. [CrossRef]
28. Fukuda, K.; Shinkai, Y. SETDB1-Mediated Silencing of Retroelements. *Viruses* **2020**, *12*, 596. [CrossRef]
29. Matsui, T.; Leung, D.; Miyashita, H.; Maksakova, I.A.; Miyachi, H.; Kimura, H.; Tachibana, M.; Lorincz, M.C.; Shinkai, Y. Proviral silencing in embryonic stem cells requires the histone methyltransferase ESET. *Nature* **2010**, *464*, 927–931. [CrossRef]
30. Tachibana, K.; Gotoh, E.; Kawamata, N.; Ishimoto, K.; Uchihara, Y.; Iwanari, H.; Sugiyama, A.; Kawamura, T.; Mochizuki, Y.; Tanaka, T.; et al. Analysis of the subcellular localization of the human histone methyltransferase SETDB1. *Biochem. Biophys. Res. Commun.* **2015**, *465*, 725–731. [CrossRef]
31. Tsusaka, T.; Shimura, C.; Shinkai, Y. ATF7IP regulates SETDB1 nuclear localization and increases its ubiquitination. *EMBO Rep.* **2019**, *20*, e48297. [CrossRef] [PubMed]
32. Bulut-Karslioglu, A.; De La Rosa-Velázquez, I.A.; Ramirez, F.; Barenboim, M.; Onishi-Seebacher, M.; Arand, J.; Galán, C.; Winter, G.E.; Engist, B.; Gerle, B.; et al. Suv39h-dependent H3K9me3 marks intact retrotransposons and silences LINE elements in mouse embryonic stem cells. *Mol. Cell* **2014**, *55*, 277–290. [CrossRef] [PubMed]
33. Leung, D.C.; Dong, K.B.; Maksakova, I.A.; Goyal, P.; Appanah, R.; Lee, S.; Tachibana, M.; Shinkai, Y.; Lehnertz, B.; Mager, D.L.; et al. Lysine methyltransferase G9a is required for de novo DNA methylation and the establishment, but not the maintenance, of proviral silencing. *Proc. Natl. Acad. Sci. USA* **2011**, *108*, 5718–5723. [CrossRef] [PubMed]
34. Maksakova, I.A.; Thompson, P.J.; Goyal, P.; Jones, S.J.; Singh, P.B.; Karimi, M.M.; Lorincz, M.C. Distinct roles of KAP1, HP1 and G9a/GLP in silencing of the two-cell-specific retrotransposon MERVL in mouse ES cells. *Epigenetics Chromatin* **2013**, *6*, 15. [CrossRef]
35. Gao, T.; Chen, F.; Zhang, W.; Zhao, X.; Hu, X.; Lu, X. Nsd2 Represses Endogenous Retrovirus MERVL in Embryonic Stem Cells. *Stem Cells Int.* **2021**, *2021*, 6663960. [CrossRef] [PubMed]
36. Robbez-Masson, L.; Tie, C.H.C.; Conde, L.; Tunbak, H.; Husovsky, C.; Tchasovnikarova, I.A.; Timms, R.T.; Herrero, J.; Lehner, P.J.; Rowe, H.M. The HUSH complex cooperates with TRIM28 to repress young retrotransposons and new genes. *Genome Res.* **2018**, *28*, 836–845. [CrossRef]
37. Tchasovnikarova, I.A.; Timms, R.T.; Matheson, N.J.; Wals, K.; Antrobus, R.; Göttgens, B.; Dougan, G.; Dawson, M.A.; Lehner, P.J. GENE SILENCING. Epigenetic silencing by the HUSH complex mediates position-effect variegation in human cells. *Science* **2015**, *348*, 1481–1485. [CrossRef]
38. Garland, W.; Müller, I.; Wu, M.; Schmid, M.; Imamura, K.; Rib, L.; Sandelin, A.; Helin, K.; Jensen, T.H. Chromatin modifier HUSH co-operates with RNA decay factor NEXT to restrict transposable element expression. *Mol. Cell* **2022**, *82*, 1691–1707.e8. [CrossRef]
39. Yang, B.X.; El Farran, C.A.; Guo, H.C.; Yu, T.; Fang, H.T.; Wang, H.F.; Schlesinger, S.; Seah, Y.F.S.; Goh, G.Y.L.; Neo, S.P.; et al. Systematic identification of factors for provirus silencing in embryonic stem cells. *Cell* **2015**, *163*, 230–245. [CrossRef]
40. Maksakova, I.A.; Goyal, P.; Bullwinkel, J.; Brown, J.P.; Bilenky, M.; Mager, D.L.; Singh, P.B.; Lorincz, M.C. H3K9me3-binding proteins are dispensable for SETDB1/H3K9me3-dependent retroviral silencing. *Epigenetics Chromatin* **2011**, *4*, 12. [CrossRef]

41. Sachs, P.; Ding, D.; Bergmaier, P.; Lamp, B.; Schlagheck, C.; Finkernagel, F.; Nist, A.; Stiewe, T.; Mermoud, J.E. SMARCAD1 ATPase activity is required to silence endogenous retroviruses in embryonic stem cells. *Nat. Commun.* **2019**, *10*, 1335. [CrossRef] [PubMed]

42. Mittal, P.; Roberts, C.W.M. The SWI/SNF complex in cancer—Biology, biomarkers and therapy. *Nat. Rev. Clin. Oncol.* **2020**, *17*, 435–448. [CrossRef] [PubMed]

43. Clapier, C.R.; Iwasa, J.; Cairns, B.R.; Peterson, C.L. Mechanisms of action and regulation of ATP-dependent chromatin-remodelling complexes. *Nat. Rev. Mol. Cell Biol.* **2017**, *18*, 407–422. [CrossRef]

44. Hammond, C.M.; Strømme, C.B.; Huang, H.; Patel, D.J.; Groth, A. Histone chaperone networks shaping chromatin function. *Nat. Rev. Mol. Cell Biol.* **2017**, *18*, 141–158. [CrossRef]

45. Kaufman, P.D.; Kobayashi, R.; Kessler, N.; Stillman, B. The p150 and p60 subunits of chromatin assembly factor I: A molecular link between newly synthesized histones and DNA replication. *Cell* **1995**, *81*, 1105–1114. [CrossRef]

46. Gaillard, P.-H.L.; Martini, E.M.; Kaufman, P.D.; Stillman, B.; Moustacchi, E.; Almouzni, G. Chromatin assembly coupled to DNA repair: A new role for chromatin assembly factor I. *Cell* **1996**, *86*, 887–896. [CrossRef]

47. Lechner, M.S.; Begg, G.E.; Speicher, D.W.; Rauscher, F.J. Molecular determinants for targeting heterochromatin protein 1-mediated gene silencing: Direct chromoshadow domain-KAP-1 corepressor interaction is essential. *Mol. Cell. Biol.* **2000**, *20*, 6449–6465. [CrossRef]

48. Thiru, A.; Nietlispach, D.; Mott, H.R.; Okuwaki, M.; Lyon, D.; Nielsen, P.R.; Hirshberg, M.; Verreault, A.; Murzina, N.V.; Laue, E.D. Structural basis of HP1/PXVXL motif peptide interactions and HP1 localisation to heterochromatin. *EMBO J.* **2004**, *23*, 489–499. [CrossRef]

49. Simon, B.; Lou, H.J.; Huet-Calderwood, C.; Shi, G.; Boggon, T.J.; Turk, B.E.; Calderwood, D.A. Tousled-like kinase 2 targets ASF1 histone chaperones through client mimicry. *Nat. Commun.* **2022**, *13*, 749. [CrossRef]

50. Dyer, M.A.; Qadeer, Z.A.; Valle-Garcia, D.; Bernstein, E. ATRX and DAXX: Mechanisms and Mutations. *Cold Spring Harb. Perspect. Med.* **2017**, *7*, a026567. [CrossRef]

51. Sadic, D.; Schmidt, K.; Groh, S.; Kondofersky, I.; Ellwart, J.; Fuchs, C.; Theis, F.J.; Schotta, G. Atrx promotes heterochromatin formation at retrotransposons. *EMBO Rep.* **2015**, *16*, 836–850. [CrossRef] [PubMed]

52. He, Q.; Kim, H.; Huang, R.; Lu, W.; Tang, M.; Shi, F.; Yang, D.; Zhang, X.; Huang, J.; Liu, D.; et al. The Daxx/Atrx Complex Protects Tandem Repetitive Elements during DNA Hypomethylation by Promoting H3K9 Trimethylation. *Cell Stem Cell* **2015**, *17*, 273–286. [CrossRef] [PubMed]

53. Elsässer, S.J.; Noh, K.-M.; Diaz, N.; Allis, C.D.; Banaszynski, L.A. Histone H3.3 is required for endogenous retroviral element silencing in embryonic stem cells. *Nature* **2015**, *522*, 240–244. [CrossRef] [PubMed]

54. Trovato, M.; Patil, V.; Gehre, M.; Noh, K.M. Histone Variant H3.3 Mutations in Defining the Chromatin Function in Mammals. *Cells* **2020**, *9*, 2716. [CrossRef]

55. Flotho, A.; Melchior, F. Sumoylation: A regulatory protein modification in health and disease. *Annu. Rev. Biochem.* **2013**, *82*, 357–385. [CrossRef]

56. Ninova, M.; Fejes Tóth, K.; Aravin, A.A. The control of gene expression and cell identity by H3K9 trimethylation. *Development* **2019**, *146*, dev181180. [CrossRef]

57. Rowe, H.M.; Friedli, M.; Offner, S.; Verp, S.; Mesnard, D.; Marquis, J.; Aktas, T.; Trono, D. De novo DNA methylation of endogenous retroviruses is shaped by KRAB-ZFPs/KAP1 and ESET. *Development* **2013**, *140*, 519–529. [CrossRef]

58. Hutnick, L.K.; Huang, X.; Loo, T.-C.; Ma, Z.; Fan, G. Repression of retrotransposal elements in mouse embryonic stem cells is primarily mediated by a DNA methylation-independent mechanism. *J. Biol. Chem.* **2010**, *285*, 21082–21091. [CrossRef]

59. Walsh, C.P.; Chaillet, J.R.; Bestor, T.H. Transcription of IAP endogenous retroviruses is constrained by cytosine methylation. *Nat. Genet.* **1998**, *20*, 116–117. [CrossRef]

60. Karimi, M.M.; Goyal, P.; Maksakova, I.A.; Bilenky, M.; Leung, D.; Tang, J.X.; Shinkai, Y.; Mager, D.L.; Jones, S.; Hirst, M.; et al. DNA methylation and SETDB1/H3K9me3 regulate predominantly distinct sets of genes, retroelements, and chimeric transcripts in mESCs. *Cell Stem Cell* **2011**, *8*, 676–687. [CrossRef] [PubMed]

61. Groh, S.; Schotta, G. Silencing of endogenous retroviruses by heterochromatin. *Cell. Mol. Life Sci.* **2017**, *74*, 2055–2065. [CrossRef] [PubMed]

62. Fasching, L.; Kapopoulou, A.; Sachdeva, R.; Petri, R.; Jönsson, M.E.; Männe, C.; Turelli, P.; Jern, P.; Cammas, F.; Trono, D.; et al. TRIM28 represses transcription of endogenous retroviruses in neural progenitor cells. *Cell Rep.* **2015**, *10*, 20–28. [CrossRef] [PubMed]

63. Tan, S.-L.; Nishi, M.; Ohtsuka, T.; Matsui, T.; Takemoto, K.; Kamio-Miura, A.; Aburatani, H.; Shinkai, Y.; Kageyama, R. Essential roles of the histone methyltransferase ESET in the epigenetic control of neural progenitor cells during development. *Development* **2012**, *139*, 3806–3816. [CrossRef]

64. Collins, P.L.; Kyle, K.E.; Egawa, T.; Shinkai, Y.; Oltz, E.M. The histone methyltransferase SETDB1 represses endogenous and exogenous retroviruses in B lymphocytes. *Proc. Natl. Acad. Sci. USA* **2015**, *112*, 8367–8372. [CrossRef] [PubMed]

65. Pasquarella, A.; Ebert, A.; Pereira de Almeida, G.; Hinterberger, M.; Kazerani, M.; Nuber, A.; Ellwart, J.; Klein, L.; Busslinger, M.; Schotta, G. Retrotransposon derepression leads to activation of the unfolded protein response and apoptosis in pro-B cells. *Development* **2016**, *143*, 1788–1799. [CrossRef]

66. Veland, N.; Lu, Y.; Hardikar, S.; Gaddis, S.; Zeng, Y.; Liu, B.; Estecio, M.R.; Takata, Y.; Lin, K.; Tomida, M.W.; et al. DNMT3L facilitates DNA methylation partly by maintaining DNMT3A stability in mouse embryonic stem cells. *Nucleic Acids Res.* **2019**, *47*, 152–167. [CrossRef]

67. Bourc'his, D.; Bestor, T.H. Meiotic catastrophe and retrotransposon reactivation in male germ cells lacking Dnmt3L. *Nature* **2004**, *431*, 96–99. [CrossRef]

68. Ohtani, H.; Liu, M.; Zhou, W.; Liang, G.; Jones, P.A. Switching roles for DNA and histone methylation depend on evolutionary ages of human endogenous retroviruses. *Genome Res.* **2018**, *28*, 1147–1157. [CrossRef]

69. Stein, P.; Rozhkov, N.V.; Li, F.; Cárdenas, F.L.; Davydenko, O.; Davydenk, O.; Vandivier, L.E.; Gregory, B.D.; Hannon, G.J.; Schultz, R.M. Essential Role for endogenous siRNAs during meiosis in mouse oocytes. *PLoS Genet.* **2015**, *11*, e1005013. [CrossRef]

70. Li, J.; Kannan, M.; Trivett, A.L.; Liao, H.; Wu, X.; Akagi, K.; Symer, D.E. An antisense promoter in mouse L1 retrotransposon open reading frame-1 initiates expression of diverse fusion transcripts and limits retrotransposition. *Nucleic Acids Res.* **2014**, *42*, 4546–4562. [CrossRef]

71. Brockdorff, N.; Bowness, J.S.; Wei, G. Progress toward understanding chromosome silencing by Xist RNA. *Genes Dev.* **2020**, *34*, 733–744. [CrossRef] [PubMed]

72. Carter, A.C.; Xu, J.; Nakamoto, M.Y.; Wei, Y.; Zarnegar, B.J.; Shi, Q.; Broughton, J.P.; Ransom, R.C.; Salhotra, A.; Nagaraja, S.D.; et al. Spen links RNA-mediated endogenous retrovirus silencing and X chromosome inactivation. *eLife* **2020**, *9*, e54508. [CrossRef] [PubMed]

73. Iwasaki, Y.W.; Siomi, M.C.; Siomi, H. PIWI-Interacting RNA: Its Biogenesis and Functions. *Annu. Rev. Biochem.* **2015**, *84*, 405–433. [CrossRef] [PubMed]

74. Pal-Bhadra, M.; Leibovitch, B.A.; Gandhi, S.G.; Chikka, M.R.; Rao, M.; Bhadra, U.; Birchler, J.A.; Elgin, S.C.R. Heterochromatic silencing and HP1 localization in Drosophila are dependent on the RNAi machinery. *Science* **2004**, *303*, 669–672. [CrossRef]

75. Aravin, A.A.; Sachidanandam, R.; Girard, A.; Fejes-Toth, K.; Hannon, G.J. Developmentally regulated piRNA clusters implicate MILI in transposon control. *Science* **2007**, *316*, 744–747. [CrossRef]

76. Aravin, A.A.; Sachidanandam, R.; Bourc'his, D.; Schaefer, C.; Pezic, D.; Toth, K.F.; Bestor, T.; Hannon, G.J. A piRNA pathway primed by individual transposons is linked to de novo DNA methylation in mice. *Mol. Cell* **2008**, *31*, 785–799. [CrossRef]

77. Guallar, D.; Bi, X.; Pardavila, J.A.; Huang, X.; Saenz, C.; Shi, X.; Zhou, H.; Faiola, F.; Ding, J.; Haruehanroengra, P.; et al. RNA-dependent chromatin targeting of TET2 for endogenous retrovirus control in pluripotent stem cells. *Nat. Genet.* **2018**, *50*, 443–451. [CrossRef]

78. Shi, H.; Wei, J.; He, C. Where, When, and How: Context-Dependent Functions of RNA Methylation Writers, Readers, and Erasers. *Mol. Cell* **2019**, *74*, 640–650. [CrossRef]

79. Chelmicki, T.; Roger, E.; Teissandier, A.; Dura, M.; Bonneville, L.; Rucli, S.; Dossin, F.; Fouassier, C.; Lameiras, S.; Bourc'his, D. m^{6}A RNA methylation regulates the fate of endogenous retroviruses. *Nature* **2021**, *591*, 312–316. [CrossRef]

80. Chen, J.; Foroozesh, M.; Qin, Z. Transactivation of human endogenous retroviruses by tumor viruses and their functions in virus-associated malignancies. *Oncogenesis* **2019**, *8*, 6. [CrossRef]

81. Tovo, P.-A.; Garazzino, S.; Daprà, V.; Alliaudi, C.; Silvestro, E.; Calvi, C.; Montanari, P.; Galliano, I.; Bergallo, M. Chronic HCV Infection Is Associated with Overexpression of Human Endogenous Retroviruses that Persists after Drug-Induced Viral Clearance. *Int. J. Mol. Sci.* **2020**, *21*, 3980. [CrossRef] [PubMed]

82. Vincendeau, M.; Göttesdorfer, I.; Schreml, J.M.H.; Wetie, A.G.N.; Mayer, J.; Greenwood, A.D.; Helfer, M.; Kramer, S.; Seifarth, W.; Hadian, K.; et al. Modulation of human endogenous retrovirus (HERV) transcription during persistent and de novo HIV-1 infection. *Retrovirology* **2015**, *12*, 27. [CrossRef] [PubMed]

83. Rajagopalan, D.; Tirado-Magallanes, R.; Bhatia, S.S.; Teo, W.S.; Sian, S.; Hora, S.; Lee, K.K.; Zhang, Y.; Jadhav, S.P.; Wu, Y.; et al. TIP60 represses activation of endogenous retroviral elements. *Nucleic Acids Res.* **2018**, *46*, 9456–9470. [CrossRef]

84. Macfarlan, T.S.; Gifford, W.D.; Agarwal, S.; Driscoll, S.; Lettieri, K.; Wang, J.; Andrews, S.E.; Franco, L.; Rosenfeld, M.G.; Ren, B.; et al. Endogenous retroviruses and neighboring genes are coordinately repressed by LSD1/KDM1A. *Genes Dev.* **2011**, *25*, 594–607. [CrossRef] [PubMed]

85. Winkler, D.D.; Luger, K. The histone chaperone FACT: Structural insights and mechanisms for nucleosome reorganization. *J. Biol. Chem.* **2011**, *286*, 18369–18374. [CrossRef]

86. Chen, F.; Zhang, W.; Xie, D.; Gao, T.; Dong, Z.; Lu, X. Histone chaperone FACT represses retrotransposon MERVL and MERVL-derived cryptic promoters. *Nucleic Acids Res.* **2020**, *48*, 10211–10225. [CrossRef]

87. Falco, G.; Lee, S.-L.; Stanghellini, I.; Bassey, U.C.; Hamatani, T.; Ko, M.S.H. Zscan4: A novel gene expressed exclusively in late 2-cell embryos and embryonic stem cells. *Dev. Biol.* **2007**, *307*, 539–550. [CrossRef]

88. Zhang, W.; Chen, F.; Chen, R.; Xie, D.; Yang, J.; Zhao, X.; Guo, R.; Zhang, Y.; Shen, Y.; Göke, J.; et al. Zscan4c activates endogenous retrovirus MERVL and cleavage embryo genes. *Nucleic Acids Res.* **2019**, *47*, 8485–8501. [CrossRef]

89. Hendrickson, P.G.; Doráis, J.A.; Grow, E.J.; Whiddon, J.L.; Lim, J.-W.; Wike, C.L.; Weaver, B.D.; Pflueger, C.; Emery, B.R.; Wilcox, A.L.; et al. Conserved roles of mouse DUX and human DUX4 in activating cleavage-stage genes and MERVL/HERVL retrotransposons. *Nat. Genet.* **2017**, *49*, 925–934. [CrossRef]

90. Nguyen, T.D.; Davis, J.; Eugenio, R.A.; Liu, Y. Female Sex Hormones Activate Human Endogenous Retrovirus Type K Through the OCT4 Transcription Factor in T47D Breast Cancer Cells. *AIDS Res. Hum. Retrovir.* **2019**, *35*, 348–356. [CrossRef]

91. Mao, J.; Zhang, Q.; Wang, Y.; Zhuang, Y.; Xu, L.; Ma, X.; Guan, D.; Zhou, J.; Liu, J.; Wu, X.; et al. TERT activates endogenous retroviruses to promote an immunosuppressive tumour microenvironment. *EMBO Rep.* **2022**, *23*, e52984. [CrossRef]

92. Mi, S.; Lee, X.; Li, X.; Veldman, G.M.; Finnerty, H.; Racie, L.; LaVallie, E.; Tang, X.Y.; Edouard, P.; Howes, S.; et al. Syncytin is a captive retroviral envelope protein involved in human placental morphogenesis. *Nature* **2000**, *403*, 785–789. [CrossRef] [PubMed]

93. Blaise, S.; de Parseval, N.; Bénit, L.; Heidmann, T. Genomewide screening for fusogenic human endogenous retrovirus envelopes identifies syncytin 2, a gene conserved on primate evolution. *Proc. Natl. Acad. Sci. USA* **2003**, *100*, 13013–13018. [CrossRef] [PubMed]

94. Best, S.; Le Tissier, P.; Towers, G.; Stoye, J.P. Positional cloning of the mouse retrovirus restriction gene Fv1. *Nature* **1996**, *382*, 826–829. [CrossRef]

95. Bénit, L.; De Parseval, N.; Casella, J.F.; Callebaut, I.; Cordonnier, A.; Heidmann, T. Cloning of a new murine endogenous retrovirus, MuERV-L, with strong similarity to the human HERV-L element and with a gag coding sequence closely related to the Fv1 restriction gene. *J. Virol.* **1997**, *71*, 5652–5657. [CrossRef] [PubMed]

96. Pincus, T.; Hartley, J.W.; Rowe, W.P. A major genetic locus affecting resistance to infection with murine leukemia viruses. I. Tissue culture studies of naturally occurring viruses. *J. Exp. Med.* **1971**, *133*, 1219–1233. [CrossRef] [PubMed]

97. Johnson, W.E. Origins and evolutionary consequences of ancient endogenous retroviruses. *Nat. Rev. Microbiol.* **2019**, *17*, 355–370. [CrossRef]

98. Boso, G.; Buckler-White, A.; Kozak, C.A. Ancient Evolutionary Origin and Positive Selection of the Retroviral Restriction Factor in Muroid Rodents. *J. Virol.* **2018**, *92*, e00850-18. [CrossRef]

99. Young, G.R.; Yap, M.W.; Michaux, J.R.; Steppan, S.J.; Stoye, J.P. Evolutionary journey of the retroviral restriction gene. *Proc. Natl. Acad. Sci. USA* **2018**, *115*, 10130–10135. [CrossRef]

100. Yap, M.W.; Colbeck, E.; Ellis, S.A.; Stoye, J.P. Evolution of the retroviral restriction gene Fv1: Inhibition of non-MLV retroviruses. *PLoS Pathog.* **2014**, *10*, e1003968. [CrossRef]

101. Mortuza, G.B.; Haire, L.F.; Stevens, A.; Smerdon, S.J.; Stoye, J.P.; Taylor, I.A. High-resolution structure of a retroviral capsid hexameric amino-terminal domain. *Nature* **2004**, *431*, 481–485. [CrossRef] [PubMed]

102. Myrum, C.; Moreno-Castilla, P.; Rapp, P.R. 'Arc'-hitecture of normal cognitive aging. *Ageing Res. Rev.* **2022**, *80*, 101678. [CrossRef] [PubMed]

103. Pastuzyn, E.D.; Day, C.E.; Kearns, R.B.; Kyrke-Smith, M.; Taibi, A.V.; McCormick, J.; Yoder, N.; Belnap, D.M.; Erlendsson, S.; Morado, D.R.; et al. The Neuronal Gene Arc Encodes a Repurposed Retrotransposon Gag Protein that Mediates Intercellular RNA Transfer. *Cell* **2018**, *172*, 275–288.e18. [CrossRef] [PubMed]

104. Matteucci, C.; Balestrieri, E.; Argaw-Denboba, A.; Sinibaldi-Vallebona, P. Human endogenous retroviruses role in cancer cell stemness. *Semin. Cancer Biol.* **2018**, *53*, 17–30. [CrossRef]

105. Lemaitre, C.; Tsang, J.; Bireau, C.; Heidmann, T.; Dewannieux, M. A human endogenous retrovirus-derived gene that can contribute to oncogenesis by activating the ERK pathway and inducing migration and invasion. *PLoS Pathog.* **2017**, *13*, e1006451. [CrossRef] [PubMed]

106. Huang, G.; Li, Z.; Wan, X.; Wang, Y.; Dong, J. Human endogenous retroviral K element encodes fusogenic activity in melanoma cells. *J. Carcinog.* **2013**, *12*, 5. [CrossRef]

107. Argaw-Denboba, A.; Balestrieri, E.; Serafino, A.; Cipriani, C.; Bucci, I.; Sorrentino, R.; Sciamanna, I.; Gambacurta, A.; Sinibaldi-Vallebona, P.; Matteucci, C. HERV-K activation is strictly required to sustain CD133+ melanoma cells with stemness features. *J. Exp. Clin. Cancer Res.* **2017**, *36*, 20. [CrossRef]

108. Li, M.; Radvanyi, L.; Yin, B.; Rycaj, K.; Li, J.; Chivukula, R.; Lin, K.; Lu, Y.; Shen, J.; Chang, D.Z.; et al. Downregulation of Human Endogenous Retrovirus Type K (HERV-K) Viral RNA in Pancreatic Cancer Cells Decreases Cell Proliferation and Tumor Growth. *Clin. Cancer Res.* **2017**, *23*, 5892–5911. [CrossRef]

109. Kaufmann, S.; Sauter, M.; Schmitt, M.; Baumert, B.; Best, B.; Boese, A.; Roemer, K.; Mueller-Lantzsch, N. Human endogenous retrovirus protein Rec interacts with the testicular zinc-finger protein and androgen receptor. *J. Gen. Virol.* **2010**, *91*, 1494–1502. [CrossRef]

110. Hanke, K.; Chudak, C.; Kurth, R.; Bannert, N. The Rec protein of HERV-K(HML-2) upregulates androgen receptor activity by binding to the human small glutamine-rich tetratricopeptide repeat protein (hSGT). *Int J. Cancer* **2013**, *132*, 556–567. [CrossRef]

111. Armbruester, V.; Sauter, M.; Roemer, K.; Best, B.; Hahn, S.; Nty, A.; Schmid, A.; Philipp, S.; Mueller, A.; Mueller-Lantzsch, N. Np9 protein of human endogenous retrovirus K interacts with ligand of numb protein X. *J. Virol.* **2004**, *78*, 10310–10319. [CrossRef]

112. Chen, T.; Meng, Z.; Gan, Y.; Wang, X.; Xu, F.; Gu, Y.; Xu, X.; Tang, J.; Zhou, H.; Zhang, X.; et al. The viral oncogene Np9 acts as a critical molecular switch for co-activating β-catenin, ERK, Akt and Notch1 and promoting the growth of human leukemia stem/progenitor cells. *Leukemia* **2013**, *27*, 1469–1478. [CrossRef]

113. Denne, M.; Sauter, M.; Armbruester, V.; Licht, J.D.; Roemer, K.; Mueller-Lantzsch, N. Physical and functional interactions of human endogenous retrovirus proteins Np9 and rec with the promyelocytic leukemia zinc finger protein. *J. Virol.* **2007**, *81*, 5607–5616. [CrossRef]

114. Wang, F.; Li, X.; Xie, X.; Zhao, L.; Chen, W. UCA1, a non-protein-coding RNA up-regulated in bladder carcinoma and embryo, influencing cell growth and promoting invasion. *FEBS Lett.* **2008**, *582*, 1919–1927. [CrossRef]

115. Hou, P.; Zhao, Y.; Li, Z.; Yao, R.; Ma, M.; Gao, Y.; Zhao, L.; Zhang, Y.; Huang, B.; Lu, J. LincRNA-ROR induces epithelial-to-mesenchymal transition and contributes to breast cancer tumorigenesis and metastasis. *Cell Death Dis.* **2014**, *5*, e1287. [CrossRef]

116. Zhan, H.X.; Wang, Y.; Li, C.; Xu, J.W.; Zhou, B.; Zhu, J.K.; Han, H.F.; Wang, L.; Wang, Y.S.; Hu, S.Y. LincRNA-ROR promotes invasion, metastasis and tumor growth in pancreatic cancer through activating ZEB1 pathway. *Cancer Lett.* **2016**, *374*, 261–271. [CrossRef]
117. Takahashi, K.; Yan, I.K.; Kogure, T.; Haga, H.; Patel, T. Extracellular vesicle-mediated transfer of long non-coding RNA ROR modulates chemosensitivity in human hepatocellular cancer. *FEBS Open Bio* **2014**, *4*, 458–467. [CrossRef]
118. Sin, H.S.; Huh, J.W.; Kim, D.S.; Kang, D.W.; Min, D.S.; Kim, T.H.; Ha, H.S.; Kim, H.H.; Lee, S.Y.; Kim, H.S. Transcriptional control of the HERV-H LTR element of the GSDML gene in human tissues and cancer cells. *Arch. Virol* **2006**, *151*, 1985–1994. [CrossRef]
119. Sun, Q.; Yang, J.; Xing, G.; Sun, Q.; Zhang, L.; He, F. Expression of GSDML Associates with Tumor Progression in Uterine Cervix Cancer. *Transl. Oncol.* **2008**, *1*, 73–83. [CrossRef]
120. Gosenca, D.; Gabriel, U.; Steidler, A.; Mayer, J.; Diem, O.; Erben, P.; Fabarius, A.; Leib-Mösch, C.; Hofmann, W.-K.; Seifarth, W. HERV-E-mediated modulation of PLA2G4A transcription in urothelial carcinoma. *PLoS ONE* **2012**, *7*, e49341. [CrossRef]
121. Bjerregaard, B.; Holck, S.; Christensen, I.J.; Larsson, L.I. Syncytin is involved in breast cancer-endothelial cell fusions. *Cell. Mol. Life Sci.* **2006**, *63*, 1906–1911. [CrossRef]
122. Deng, B.; Xu, W.; Wang, Z.; Liu, C.; Lin, P.; Li, B.; Huang, Q.; Yang, J.; Zhou, H.; Qu, L. An LTR retrotransposon-derived lncRNA interacts with RNF169 to promote homologous recombination. *EMBO Rep.* **2019**, *20*, e47650. [CrossRef]
123. Lamprecht, B.; Walter, K.; Kreher, S.; Kumar, R.; Hummel, M.; Lenze, D.; Köchert, K.; Bouhlel, M.A.; Richter, J.; Soler, E.; et al. Derepression of an endogenous long terminal repeat activates the CSF1R proto-oncogene in human lymphoma. *Nat. Med.* **2010**, *16*, 571–579. [CrossRef]
124. Wu, Y.; Zhao, Y.; Huan, L.; Zhao, J.; Zhou, Y.; Xu, L.; Hu, Z.; Liu, Y.; Chen, Z.; Wang, L.; et al. An LTR Retrotransposon-Derived Long Noncoding RNA lncMER52A Promotes Hepatocellular Carcinoma Progression by Binding p120-Catenin. *Cancer Res.* **2020**, *80*, 976–987. [CrossRef]
125. Gibb, E.A.; Warren, R.L.; Wilson, G.W.; Brown, S.D.; Robertson, G.A.; Morin, G.B.; Holt, R.A. Activation of an endogenous retrovirus-associated long non-coding RNA in human adenocarcinoma. *Genome Med.* **2015**, *7*, 22. [CrossRef]
126. Cañadas, I.; Thummalapalli, R.; Kim, J.W.; Kitajima, S.; Jenkins, R.W.; Christensen, C.L.; Campisi, M.; Kuang, Y.; Zhang, Y.; Gjini, E.; et al. Tumor innate immunity primed by specific interferon-stimulated endogenous retroviruses. *Nat. Med.* **2018**, *24*, 1143–1150. [CrossRef]
127. Lee, E.; Iskow, R.; Yang, L.; Gokcumen, O.; Haseley, P.; Luquette, L.J., III; Lohr, J.G.; Harris, C.C.; Ding, L.; Wilson, R.K.; et al. Landscape of somatic retrotransposition in human cancers. *Science* **2012**, *337*, 967–971. [CrossRef]
128. Beyer, U.; Moll-Rocek, J.; Moll, U.M.; Dobbelstein, M. Endogenous retrovirus drives hitherto unknown proapoptotic p63 isoforms in the male germ line of humans and great apes. *Proc. Natl. Acad. Sci. USA* **2011**, *108*, 3624–3629. [CrossRef]
129. Beyer, U.; Kronung, S.K.; Leha, A.; Walter, L.; Dobbelstein, M. Comprehensive identification of genes driven by ERV9-LTRs reveals TNFRSF10B as a re-activatable mediator of testicular cancer cell death. *Cell Death Differ.* **2016**, *23*, 64–75. [CrossRef]
130. Strick, R.; Ackermann, S.; Langbein, M.; Swiatek, J.; Schubert, S.W.; Hashemolhosseini, S.; Koscheck, T.; Fasching, P.A.; Schild, R.L.; Beckmann, M.W.; et al. Proliferation and cell-cell fusion of endometrial carcinoma are induced by the human endogenous retroviral Syncytin-1 and regulated by TGF-beta. *J. Mol. Med.* **2007**, *85*, 23–38. [CrossRef]
131. Zhou, F.; Li, M.; Wei, Y.; Lin, K.; Lu, Y.; Shen, J.; Johanning, G.L.; Wang-Johanning, F. Activation of HERV-K Env protein is essential for tumorigenesis and metastasis of breast cancer cells. *Oncotarget* **2016**, *7*, 84093–84117. [CrossRef]
132. Grandi, N.; Tramontano, E. HERV Envelope Proteins: Physiological Role and Pathogenic Potential in Cancer and Autoimmunity. *Front. Microbiol.* **2018**, *9*, 462. [CrossRef]
133. Goke, J.; Ng, H.H. CTRL+INSERT: Retrotransposons and their contribution to regulation and innovation of the transcriptome. *EMBO Rep.* **2016**, *17*, 1131–1144. [CrossRef]
134. Leucci, E.; Vendramin, R.; Spinazzi, M.; Laurette, P.; Fiers, M.; Wouters, J.; Radaelli, E.; Eyckerman, S.; Leonelli, C.; Vanderheyden, K.; et al. Melanoma addiction to the long non-coding RNA SAMMSON. *Nature* **2016**, *531*, 518–522. [CrossRef]
135. Flockhart, R.J.; Webster, D.E.; Qu, K.; Mascarenhas, N.; Kovalski, J.; Kretz, M.; Khavari, P.A. BRAFV600E remodels the melanocyte transcriptome and induces BANCR to regulate melanoma cell migration. *Genome Res.* **2012**, *22*, 1006–1014. [CrossRef]
136. Zhou, B.; Qi, F.; Wu, F.; Nie, H.; Song, Y.; Shao, L.; Han, J.; Wu, Z.; Saiyin, H.; Wei, G.; et al. Endogenous Retrovirus-Derived Long Noncoding RNA Enhances Innate Immune Responses via Derepressing RELA Expression. *mBio* **2019**, *10*, e00937-19. [CrossRef]
137. Chiappinelli, K.B.; Strissel, P.L.; Desrichard, A.; Li, H.; Henke, C.; Akman, B.; Hein, A.; Rote, N.S.; Cope, L.M.; Snyder, A.; et al. Inhibiting DNA Methylation Causes an Interferon Response in Cancer via dsRNA Including Endogenous Retroviruses. *Cell* **2015**, *162*, 974–986. [CrossRef] [PubMed]
138. Roulois, D.; Loo Yau, H.; Singhania, R.; Wang, Y.; Danesh, A.; Shen, S.Y.; Han, H.; Liang, G.; Jones, P.A.; Pugh, T.J.; et al. DNA-Demethylating Agents Target Colorectal Cancer Cells by Inducing Viral Mimicry by Endogenous Transcripts. *Cell* **2015**, *162*, 961–973. [CrossRef]
139. Otsuki, S.; Saito, T.; Taylor, S.; Li, D.; Moonen, J.-R.; Marciano, D.P.; Harper, R.L.; Cao, A.; Wang, L.; Ariza, M.E.; et al. Monocyte-released HERV-K dUTPase engages TLR4 and MCAM causing endothelial mesenchymal transition. *JCI Insight* **2021**, *6*, e146416. [CrossRef]
140. Saito, T.; Miyagawa, K.; Chen, S.-Y.; Tamosiuniene, R.; Wang, L.; Sharpe, O.; Samayoa, E.; Harada, D.; Moonen, J.-R.A.J.; Cao, A.; et al. Upregulation of Human Endogenous Retrovirus-K Is Linked to Immunity and Inflammation in Pulmonary Arterial Hypertension. *Circulation* **2017**, *136*, 1920–1935. [CrossRef]

141. Taylor, S.; Isobe, S.; Cao, A.; Contrepois, K.; Benayoun, B.A.; Jiang, L.; Wang, L.; Melemenidis, S.; Ozen, M.O.; Otsuki, S.; et al. Endogenous Retroviral Elements Generate Pathologic Neutrophils in Pulmonary Arterial Hypertension. *Am. J. Respir. Crit. Care Med.* **2022**. [CrossRef] [PubMed]

142. Ariza, M.-E.; Williams, M.V. A human endogenous retrovirus K dUTPase triggers a TH1, TH17 cytokine response: Does it have a role in psoriasis? *J. Investig. Dermatol.* **2011**, *131*, 2419–2427. [CrossRef] [PubMed]

143. Fleischmann, J.; Kremmer, E.; Müller, S.; Sommer, P.; Kirchner, T.; Niedobitek, G.; Grässer, F.A. Expression of deoxyuridine triphosphatase (dUTPase) in colorectal tumours. *Int. J. Cancer* **1999**, *84*, 614–617. [CrossRef]

144. Kawahara, A.; Akagi, Y.; Hattori, S.; Mizobe, T.; Shirouzu, K.; Ono, M.; Yanagawa, T.; Kuwano, M.; Kage, M. Higher expression of deoxyuridine triphosphatase (dUTPase) may predict the metastasis potential of colorectal cancer. *J. Clin. Pathol.* **2009**, *62*, 364–369. [CrossRef]

145. Chen, C.-W.; Tsao, N.; Huang, L.-Y.; Yen, Y.; Liu, X.; Lehman, C.; Wang, Y.-H.; Tseng, M.-C.; Chen, Y.-J.; Ho, Y.-C.; et al. The Impact of dUTPase on Ribonucleotide Reductase-Induced Genome Instability in Cancer Cells. *Cell Rep.* **2016**, *16*, 1287–1299. [CrossRef]

146. Wallace, T.A.; Downey, R.F.; Seufert, C.J.; Schetter, A.; Dorsey, T.H.; Johnson, C.A.; Goldman, R.; Loffredo, C.A.; Yan, P.; Sullivan, F.J.; et al. Elevated HERV-K mRNA expression in PBMC is associated with a prostate cancer diagnosis particularly in older men and smokers. *Carcinogenesis* **2014**, *35*, 2074–2083. [CrossRef]

147. Wang-Johanning, F.; Rycaj, K.; Plummer, J.B.; Li, M.; Yin, B.; Frerich, K.; Garza, J.G.; Shen, J.; Lin, K.; Yan, P.; et al. Immunotherapeutic potential of anti-human endogenous retrovirus-K envelope protein antibodies in targeting breast tumors. *J. Natl. Cancer Inst.* **2012**, *104*, 189–210. [CrossRef]

148. Krishnamurthy, J.; Rabinovich, B.A.; Mi, T.; Switzer, K.C.; Olivares, S.; Maiti, S.N.; Plummer, J.B.; Singh, H.; Kumaresan, P.R.; Huls, H.M.; et al. Genetic Engineering of T Cells to Target HERV-K, an Ancient Retrovirus on Melanoma. *Clin. Cancer Res.* **2015**, *21*, 3241–3251. [CrossRef]

149. Villeponteau, B. The heterochromatin loss model of aging. *Exp. Gerontol.* **1997**, *32*, 383–394. [CrossRef]

150. Pal, S.; Tyler, J.K. Epigenetics and aging. *Sci. Adv.* **2016**, *2*, e1600584. [CrossRef]

151. Barbot, W.; Dupressoir, A.; Lazar, V.; Heidmann, T. Epigenetic regulation of an IAP retrotransposon in the aging mouse: Progressive demethylation and de-silencing of the element by its repetitive induction. *Nucleic Acids Res.* **2002**, *30*, 2365–2373. [CrossRef] [PubMed]

152. De Cecco, M.; Criscione, S.W.; Peterson, A.L.; Neretti, N.; Sedivy, J.M.; Kreiling, J.A. Transposable elements become active and mobile in the genomes of aging mammalian somatic tissues. *Aging* **2013**, *5*, 867–883. [CrossRef] [PubMed]

153. Nevalainen, T.; Autio, A.; Mishra, B.H.; Marttila, S.; Jylha, M.; Hurme, M. Aging-associated patterns in the expression of human endogenous retroviruses. *PLoS ONE* **2018**, *13*, e0207407. [CrossRef]

154. Balestrieri, E.; Pica, F.; Matteucci, C.; Zenobi, R.; Sorrentino, R.; Argaw-Denboba, A.; Cipriani, C.; Bucci, I.; Sinibaldi-Vallebona, P. Transcriptional activity of human endogenous retroviruses in human peripheral blood mononuclear cells. *Biomed. Res. Int.* **2015**, *2015*, 164529. [CrossRef] [PubMed]

155. McGinley, M.P.; Goldschmidt, C.H.; Rae-Grant, A.D. Diagnosis and Treatment of Multiple Sclerosis: A Review. *JAMA* **2021**, *325*, 765–779. [CrossRef] [PubMed]

156. Perron, H.; Geny, C.; Laurent, A.; Mouriquand, C.; Pellat, J.; Perret, J.; Seigneurin, J.M. Leptomeningeal cell line from multiple sclerosis with reverse transcriptase activity and viral particles. *Res. Virol.* **1989**, *140*, 551–561. [CrossRef]

157. Perron, H.; Lalande, B.; Gratacap, B.; Laurent, A.; Genoulaz, O.; Geny, C.; Mallaret, M.; Schuller, E.; Stoebner, P.; Seigneurin, J.M. Isolation of retrovirus from patients with multiple sclerosis. *Lancet* **1991**, *337*, 862–863. [CrossRef]

158. Komurian-Pradel, F.; Paranhos-Baccala, G.; Bedin, F.; Ounanian-Paraz, A.; Sodoyer, M.; Ott, C.; Rajoharison, A.; Garcia, E.; Mallet, F.; Mandrand, B.; et al. Molecular cloning and characterization of MSRV-related sequences associated with retrovirus-like particles. *Virology* **1999**, *260*, 1–9. [CrossRef]

159. Blond, J.L.; Besème, F.; Duret, L.; Bouton, O.; Bedin, F.; Perron, H.; Mandrand, B.; Mallet, F. Molecular characterization and placental expression of HERV-W, a new human endogenous retrovirus family. *J. Virol.* **1999**, *73*, 1175–1185. [CrossRef]

160. Rolland, A.; Jouvin-Marche, E.; Viret, C.; Faure, M.; Perron, H.; Marche, P.N. The envelope protein of a human endogenous retrovirus-W family activates innate immunity through CD14/TLR4 and promotes Th1-like responses. *J. Immunol.* **2006**, *176*, 7636–7644. [CrossRef]

161. Kremer, D.; Schichel, T.; Förster, M.; Tzekova, N.; Bernard, C.; van der Valk, P.; van Horssen, J.; Hartung, H.-P.; Perron, H.; Küry, P. Human endogenous retrovirus type W envelope protein inhibits oligodendroglial precursor cell differentiation. *Ann. Neurol.* **2013**, *74*, 721–732. [CrossRef] [PubMed]

162. Römer, C. Viruses and Endogenous Retroviruses as Roots for Neuroinflammation and Neurodegenerative Diseases. *Front. Neurosci.* **2021**, *15*, 648629. [CrossRef]

163. Kremer, D.; Gruchot, J.; Weyers, V.; Oldemeier, L.; Göttle, P.; Healy, L.; Ho Jang, J.; Kang T Xu, Y.; Volsko, C.; Dutta, R.; et al. pHERV-W envelope protein fuels microglial cell-dependent damage of myelinated axons in multiple sclerosis. *Proc. Natl. Acad. Sci. USA* **2019**, *116*, 15216–15225. [CrossRef] [PubMed]

164. Kremer, D.; Förster, M.; Schichel, T.; Göttle, P.; Hartung, H.-P.; Perron, H.; Küry, P. The neutralizing antibody GNbAC1 abrogates HERV-W envelope protein-mediated oligodendroglial maturation blockade. *Mult. Scler. J.* **2015**, *21*, 1200–1203. [CrossRef] [PubMed]

165. Hansen, B.; Oturai, A.B.; Harbo, H.F.; Celius, E.G.; Nissen, K.K.; Laska, M.J.; Søndergaard, H.B.; Petersen, T.; Nexø, B.A. Genetic association of multiple sclerosis with the marker rs391745 near the endogenous retroviral locus HERV-Fc1: Analysis of disease subtypes. *PLoS ONE* **2011**, *6*, e26438. [CrossRef]

166. Christensen, T.; Dissing Sørensen, P.; Riemann, H.; Hansen, H.J.; Munch, M.; Haahr, S.; Møller-Larsen, A. Molecular characterization of HERV-H variants associated with multiple sclerosis. *Acta Neurol. Scand.* **2000**, *101*, 229–238. [CrossRef]

167. Brudek, T.; Christensen, T.; Aagaard, L.; Petersen, T.; Hansen, H.J.; Møller-Larsen, A. B cells and monocytes from patients with active multiple sclerosis exhibit increased surface expression of both HERV-H Env and HERV-W Env, accompanied by increased seroreactivity. *Retrovirology* **2009**, *6*, 104. [CrossRef]

168. Muradrasoli, S.; Forsman, A.; Hu, L.; Blikstad, V.; Blomberg, J. Development of real-time PCRs for detection and quantitation of human MMTV-like (HML) sequences HML expression in human tissues. *J. Virol. Methods* **2006**, *136*, 83–92. [CrossRef]

169. Viola, M.V.; Frazier, M.; White, L.; Brody, J.; Spiegelman, S. RNA-instructed DNA polymerase activity in a cytoplasmic particulate fraction in brains from Guamanian patients. *J. Exp. Med.* **1975**, *142*, 483–494. [CrossRef]

170. Andrews, W.D.; Tuke, P.W.; Al-Chalabi, A.; Gaudin, P.; Ijaz, S.; Parton, M.J.; Garson, J.A. Detection of reverse transcriptase activity in the serum of patients with motor neurone disease. *J. Med. Virol.* **2000**, *61*, 527–532. [CrossRef]

171. McCormick, A.L.; Brown, R.H.; Cudkowicz, M.E.; Al-Chalabi, A.; Garson, J.A. Quantification of reverse transcriptase in ALS and elimination of a novel retroviral candidate. *Neurology* **2008**, *70*, 278–283. [CrossRef] [PubMed]

172. Steele, A.J.; Al-Chalabi, A.; Ferrante, K.; Cudkowicz, M.E.; Brown, R.H.; Garson, J.A. Detection of serum reverse transcriptase activity in patients with ALS and unaffected blood relatives. *Neurology* **2005**, *64*, 454–458. [CrossRef] [PubMed]

173. Kim, Y.J.; Fan, Y.; Laurie, P.; Kim, J.M.H.; Ravits, J. No evidence of HIV pol gene in spinal cord tissues in sporadic ALS by real-time RT-PCR. *Amyotroph. Lateral Scler.* **2010**, *11*, 91–96. [CrossRef] [PubMed]

174. Douville, R.; Liu, J.; Rothstein, J.; Nath, A. Identification of active loci of a human endogenous retrovirus in neurons of patients with amyotrophic lateral sclerosis. *Ann. Neurol.* **2011**, *69*, 141–151. [CrossRef]

175. Li, W.; Lee, M.-H.; Henderson, L.; Tyagi, R.; Bachani, M.; Steiner, J.; Campanac, E.; Hoffman, D.A.; von Geldern, G.; Johnson, K.; et al. Human endogenous retrovirus-K contributes to motor neuron disease. *Sci. Transl. Med.* **2015**, *7*, 307ra153. [CrossRef]

176. Manghera, M.; Ferguson-Parry, J.; Lin, R.; Douville, R.N. NF-κB and IRF1 Induce Endogenous Retrovirus K Expression via Interferon-Stimulated Response Elements in Its 5′ Long Terminal Repeat. *J. Virol.* **2016**, *90*, 9338–9349. [CrossRef]

177. Oluwole, S.O.A.; Yao, Y.; Conradi, S.; Kristensson, K.; Karlsson, H. Elevated levels of transcripts encoding a human retroviral envelope protein (syncytin) in muscles from patients with motor neuron disease. *Amyotroph. Lateral Scler.* **2007**, *8*, 67–72. [CrossRef]

178. Balestrieri, E.; Arpino, C.; Matteucci, C.; Sorrentino, R.; Pica, F.; Alessandrelli, R.; Coniglio, A.; Curatolo, P.; Rezza, G.; Macciardi, F.; et al. HERVs expression in Autism Spectrum Disorders. *PLoS ONE* **2012**, *7*, e48831. [CrossRef]

179. Heidmann, O.; Béguin, A.; Paternina, J.; Berthier, R.; Deloger, M.; Bawa, O.; Heidmann, T. HEMO, an ancestral endogenous retroviral envelope protein shed in the blood of pregnant women and expressed in pluripotent stem cells and tumors. *Proc. Natl. Acad. Sci. USA* **2017**, *114*, E6642–E6651. [CrossRef]

180. Balestrieri, E.; Cipriani, C.; Matteucci, C.; Benvenuto, A.; Coniglio, A.; Argaw-Denboba, A.; Toschi, N.; Bucci, I.; Miele, M.T.; Grelli, S.; et al. Children with Autism Spectrum Disorder and Their Mothers Share Abnormal Expression of Selected Endogenous Retroviruses Families and Cytokines. *Front. Immunol* **2019**, *10*, 2244. [CrossRef]

181. Johnston, J.B.; Silva, C.; Holden, J.; Warren, K.G.; Clark, A.W.; Power, C. Monocyte activation and differentiation augment human endogenous retrovirus expression: Implications for inflammatory brain diseases. *Ann. Neurol* **2001**, *50*, 434–442. [CrossRef] [PubMed]

182. Sun, W.; Samimi, H.; Gamez, M.; Zare, H.; Frost, B. Pathogenic tau-induced piRNA depletion promotes neuronal death through transposable element dysregulation in neurodegenerative tauopathies. *Nat. Neurosci.* **2018**, *21*, 1038–1048. [CrossRef]

183. Dembny, P.; Newman, A.G.; Singh, M.; Hinz, M.; Szczepek, M.; Krüger, C.; Adalbert, R.; Dzaye, O.; Trimbuch, T.; Wallach, T.; et al. Human endogenous retrovirus HERV-K(HML-2) RNA causes neurodegeneration through Toll-like receptors. *JCI Insight* **2020**, *5*, e131093. [CrossRef]

184. Mak, M.; Samochowiec, J.; Frydecka, D.; Pełka-Wysiecka, J.; Szmida, E.; Karpiński, P.; Sąsiadek, M.M.; Piotrowski, P.; Samochowiec, A.; Misiak, B. First-episode schizophrenia is associated with a reduction of HERV-K methylation in peripheral blood. *Psychiatry Res.* **2019**, *271*, 459–463. [CrossRef] [PubMed]

185. Perron, H.; Hamdani, N.; Faucard, R.; Lajnef, M.; Jamain, S.; Daban-Huard, C.; Sarrazin, S.; LeGuen, E.; Houenou, J.; Delavest, M.; et al. Molecular characteristics of Human Endogenous Retrovirus type-W in schizophrenia and bipolar disorder. *Transl. Psychiatry* **2012**, *2*, e201. [CrossRef] [PubMed]

186. Li, F.; Sabunciyan, S.; Yolken, R.H.; Lee, D.; Kim, S.; Karlsson, H. Transcription of human endogenous retroviruses in human brain by RNA-seq analysis. *PLoS ONE* **2019**, *14*, e0207353. [CrossRef]

187. Huang, W.; Li, S.; Hu, Y.; Yu, H.; Luo, F.; Zhang, Q.; Zhu, F. Implication of the env gene of the human endogenous retrovirus W family in the expression of BDNF and DRD3 and development of recent-onset schizophrenia. *Schizophr. Bull.* **2011**, *37*, 988–1000. [CrossRef]

188. Huang, W.-J.; Liu, Z.-C.; Wei, W.; Wang, G.-H.; Wu, J.-G.; Zhu, F. Human endogenous retroviral pol RNA and protein detected and identified in the blood of individuals with schizophrenia. *Schizophr. Res.* **2006**, *83*, 193–199. [CrossRef]
189. Karlsson, H.; Bachmann, S.; Schröder, J.; McArthur, J.; Torrey, E.F.; Yolken, R.H. Retroviral RNA identified in the cerebrospinal fluids and brains of individuals with schizophrenia. *Proc. Natl. Acad. Sci. USA* **2001**, *98*, 4634–4639. [CrossRef]
190. Johansson, E.M.; Bouchet, D.; Tamouza, R.; Ellul, P.; Morr, A.S.; Avignone, E.; Germi, R.; Leboyer, M.; Perron, H.; Groc, L. Human endogenous retroviral protein triggers deficit in glutamate synapse maturation and behaviors associated with psychosis. *Sci. Adv.* **2020**, *6*, eabc0708. [CrossRef]

International Journal of
Molecular Sciences

Article

Transcriptional Alterations in X-Linked Dystonia–Parkinsonism Caused by the SVA Retrotransposon

Jelena Pozojevic [1,2], Shela Marie Algodon [1], Joseph Neos Cruz [1], Joanne Trinh [1], Norbert Brüggemann [1,3], Joshua Laß [1], Karen Grütz [1], Susen Schaake [1], Ronnie Tse [1], Veronica Yumiceba [2], Nathalie Kruse [2], Kristin Schulz [2], Varun K. A. Sreenivasan [2], Raymond L. Rosales [4], Roland Dominic G. Jamora [5], Cid Czarina E. Diesta [6], Jakob Matschke [7], Markus Glatzel [7], Philip Seibler [1], Kristian Händler [2], Aleksandar Rakovic [1], Henriette Kirchner [2], Malte Spielmann [2,8,9], Frank J. Kaiser [10,11], Christine Klein [1,*] and Ana Westenberger [1,*]

1   Institute of Neurogenetics, University of Lübeck, 23538 Lübeck, Germany; jelena.pozojevic@uksh.de (J.P.); shela.algodon@neuro.uni-luebeck.de (S.M.A.); josephneos1010@gmail.com (J.N.C.); joanne.trinh@neuro.uni-luebeck.de (J.T.); norbert.brueggemann@neuro.uni-luebeck.de (N.B.); joshua.lass@student.uni-luebeck.de (J.L.); karen.gruetz@neuro.uni-luebeck.de (K.G.); susen.schaake@neuro.uni-luebeck.de (S.S.); ronnie.tse@neuro.uni-luebeck.de (R.T.); philip.seibler@neuro.uni-luebeck.de (P.S.); aleksandar.rakovic@neuro.uni-luebeck.de (A.R.)
2   Institute of Human Genetics, University of Lübeck, 23538 Lübeck, Germany; veronica.yumicebacorral@student.uni-luebeck.de (V.Y.); nathalie.kruse@uksh.de (N.K.); kristin.schultz@uksh.de (K.S.); varun.sreenivasan@uksh.de (V.K.A.S.); kristian.haendler@uksh.de (K.H.); henriette.kirchner@uksh.de (H.K.); malte.spielmann@uksh.de (M.S.)
3   Department of Neurology, University Hospital Schleswig Holstein, 23538 Lübeck, Germany
4   The Hospital Neuroscience Institute, Department of Neurology and Psychiatry and The FMS-Research Center for Health Sciences, University of Santo Tomas, Manila 1008, Philippines; rlrosales@ust.edu.ph
5   Department of Neurosciences, College of Medicine-Philippine General Hospital, University of the Philippines Manila, Manila 1000, Philippines; rgjamora@up.edu.ph
6   Department of Neurosciences, Movement Disorders Clinic, Makati Medical Center, Makati City 1229, Philippines; ciddiesta@gmail.com
7   Institute of Neuropathology, University Medical Center Hamburg-Eppendorf, 20246 Hamburg, Germany; matschke@uke.de (J.M.); m.glatzel@uke.de (M.G.)
8   Human Molecular Genomics Group, Max Planck Institute for Molecular Genetics, 14195 Berlin, Germany
9   DZHK (German Centre for Cardiovascular Research), Partner Site Hamburg/Lübeck/Kiel, 23538 Lübeck, Germany
10 Institut für Humangenetik, Universitätsklinikum Essen, Universität Duisburg-Essen, 45147 Essen, Germany; frank.kaiser@uk-essen.de
11 Essener Zentrum für Seltene Erkrankungen, Universitätsmedizin Essen, 45147 Essen, Germany
*   Correspondence: christine.klein@neuro.uni-luebeck.de (C.K.); ana.westenberger@neuro.uni-luebeck.de (A.W.)

Citation: Pozojevic, J.; Algodon, S.M.; Cruz, J.N.; Trinh, J.; Brüggemann, N.; Laß, J.; Grütz, K.; Schaake, S.; Tse, R.; Yumiceba, V.; et al. Transcriptional Alterations in X-Linked Dystonia–Parkinsonism Caused by the SVA Retrotransposon. *Int. J. Mol. Sci.* **2022**, *23*, 2231. https://doi.org/10.3390/ijms23042231

Academic Editors: Amelia Casamassimi, Alfredo Ciccodicola and Monica Rienzo

Received: 15 December 2021
Accepted: 14 February 2022
Published: 17 February 2022

Publisher's Note: MDPI stays neutral with regard to jurisdictional claims in published maps and institutional affiliations.

Abstract: X-linked dystonia–parkinsonism (XDP) is a severe neurodegenerative disorder that manifests as adult-onset dystonia combined with parkinsonism. A SINE-VNTR-Alu (SVA) retrotransposon inserted in an intron of the *TAF1* gene reduces its expression and alters splicing in XDP patient-derived cells. As a consequence, increased levels of the *TAF1* intron retention transcript *TAF1-32i* can be found in XDP cells as compared to healthy controls. Here, we investigate the sequence of the deep intronic region included in this transcript and show that it is also present in cells from healthy individuals, albeit in lower amounts than in XDP cells, and that it undergoes degradation by nonsense-mediated mRNA decay. Furthermore, we investigate epigenetic marks (e.g., DNA methylation and histone modifications) present in this intronic region and the spanning sequence. Finally, we show that the SVA evinces regulatory potential, as demonstrated by its ability to repress the *TAF1* promoter in vitro. Our results enable a better understanding of the disease mechanisms underlying XDP and transcriptional alterations caused by SVA retrotransposons.

Keywords: XDP; retrotransposon; SVA; splicing; epigenetics; transcription

1. Introduction

X-linked dystonia–parkinsonism (XDP) is an adult-onset neurodegenerative movement disorder endemic to the Philippines, predominantly affecting men due to the X-linked mode of inheritance. It typically presents in the third to fifth decade of life as a focal dystonia that progresses and becomes generalized, severely incapacitating patients. In patients that survive this disease stage, parkinsonism sets in, overlaps with the dystonia, and predominates from the tenth year of illness onward [1,2]. XDP was initially considered as a pure disorder of the basal ganglia, due to considerable neuronal loss and mosaic gliosis described in the striatum [3,4]. However, more recent findings of reduced cortical thickness and cerebellar gray matter pathology implicate these additional regions in the pathogenesis of XDP [5].

All patients identified to date share a common haplotype, including the likely disease-causing variant, the SVA (SINE-VNTR-Alu) retrotransposon insertion in intron 32 of the *TAF1* gene on the X chromosome [6–10]. Consequently, dysfunction of *TAF1* has been postulated to underlie XDP pathogenesis. Consistent with striatal degeneration, a neuron-specific *TAF1* transcript is reduced in the caudate nucleus of XDP patients, as well as all *TAF1* transcripts in various tissues and cell lines [7,9–12]. In addition to the reduced *TAF1* expression, the SVA retrotransposon insertion seems to cause increased levels of an alternative splicing isoform, termed *TAF1-32i* and is composed of canonical exon 32 spliced to a cryptic exon within intron 32, terminating 5′ to the SVA insertion [9]. Expression levels of this transcript were found to be higher in XDP cell lines as compared to healthy controls and excision of the SVA restored levels of this transcript in cellular models [9,13]. Furthermore, an inherent part of the SVA is a $(CCCTCT)_n$ hexamer, where the repeat number n varies among patients (range: 30–55 repeats) and correlates inversely with the age at disease onset, disease severity, and *TAF1* expression [14,15]. Of note, *TAF1* encodes the transcription initiation factor TATA-box binding protein associated factor 1, a subunit of the TFIID complex that mediates transcription by RNA polymerase II, functioning as an important regulator in the expression of a large number of genes [16,17].

Here, we confirm that the alternative *TAF1* splicing isoform, *TAF1-32i*, originally reported only in patient-derived cell lines [9], can be found in cell lines of healthy individuals as well, albeit in significantly lower amounts than in XDP patients. Moreover, we show that this transcript undergoes nonsense-mediated mRNA decay (NMD). By further functional investigations, we observed H3K36me3 within this intronic region, an epigenetic mark present within transcribed regions, while alterations of DNA methylation adjacent to the SVA insertion were not detected. Finally, we found that the SVA alters promoter activity in vitro, suggesting that it may recruit transcription factors or alter chromatin architecture to modulate gene expression.

2. Results

2.1. The Intron Retention Transcript Is Present in Healthy and XDP Cell Lines

Using patient-derived fibroblasts, induced pluripotent stem cells (iPSCs), and blood samples, we reproduced and confirmed the previous observations that the *TAF1-32i* isoform can be found in various cell lines of patients with XDP (Figure 1a,b, Figures S1 and S2) [9,13]. We next investigated the association of *TAF1-32i* expression in the blood of 50 XDP patients with repeat number, age at disease onset (AAO), disease duration, or age at blood collection, and detected no significant correlation (Figure S3). It seems that this alternative transcript variant is also physiologically present in low amounts in non-carriers of the SVA insertion, as we have detected this transcript in various cell lines and tissues, including fibroblasts, iPSCs, and blood samples derived from our healthy controls and SH-SY5Y and HEK293 cells (Figure 1a,b and Figure S4). Our analysis of the levels of *TAF1-32i* in fibroblasts, iPSCs, and blood samples of healthy controls revealed that the expression of this transcript is significantly lower in samples from non-carriers of the SVA insertion when compared to XDP patients (Figure 1a,b). Nevertheless, the ratio between the levels of this transcript in patients and controls was much lower than previously reported [13]. Furthermore, *TAF1-32i* levels were increased in XDP iPSC lines when compared to the iPSC lines in which the SVA was edited out (Figure 1a). Sanger sequencing of the obtained PCR bands revealed that the deep intronic region of this alternative transcript variant, spliced to exon 32, starts at nucleotide position 15,560 from the last nucleotide in exon 32 (Figure 1c). The sequence of the intronic region included in the transcript was identical among different cell types, patients and controls.

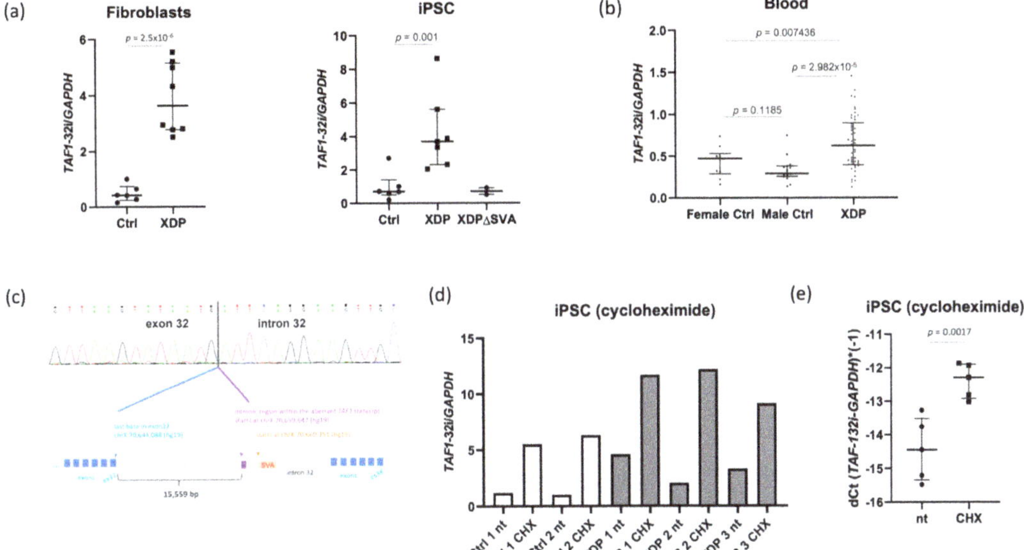

Figure 1. *TAF1-32i* transcript in cells of healthy controls and patients with XDP. (**a**) qPCR results of fibroblasts (n = 6 controls; n = 8 patients) and induced pluripotent stem cells (iPSC; n = 6 controls; n = 7 patients; n = 2 XDPΔSVA cell lines), normalized to *GAPDH* levels. XDPΔSVA refers to the patient-derived cell lines where the SVA was excised by CRISPR/Cas9. *t* test was performed on dCt values. (**b**) qPCR results (n = 10 female controls; n = 15 male controls; n = 50 patients) on blood-derived cDNA samples, relative to *GAPDH*. *p* values result from pair-wise Wilcoxon rank-sum test. To overcome possible batch effects, two independent samples were measured repeatedly in each batch, and all other samples were corrected according to the mean changes in measurement for these two samples. (**c**) Sanger sequencing showing the sequence of the intronic region included in the transcript and the scheme explaining the genomic locations, with genomic coordinates in the hg19/GRCh37 assembly. Blue squares represent canonical *TAF1* exons, the violet square represents the intronic region within the transcript (not drawn to scale). (**d**) qPCR results showing increased levels of the *TAF1-32i* transcript in control (n = 2) and patient-derived cells (n = 3) after cycloheximide (CHX) treatment, as compared to the non-treated (nt) cells. (**e**) *t* test on dCt values, comparing non-treated and CHX-treated samples. Ctrl, control.

2.2. The Intron Retention Transcript Undergoes Nonsense-Mediated mRNA Decay

Given that the *TAF1-32i* transcript is present in low amounts in healthy control iPSCs, we hypothesized that either its synthesis is increased or that its degradation is decreased in XDP. To test whether it undergoes degradation by nonsense-mediated mRNA decay (NMD), we treated the cells with cycloheximide. NMD functions both as an RNA quality control mechanism (via degradation of aberrant transcripts such as those containing disease-causing variants) and a regulator of gene expression (via degradation of normal transcripts, e.g., alternative splicing isoforms), at the interface between transcription and translation (reviewed in [18,19]), and can thus be blocked indirectly by cycloheximide that interferes with protein synthesis. This experiment was performed on healthy control- and patient-derived iPSCs, since the intron retention transcript is the most abundant in this cell type in XDP [9]. We observed increased amounts of this transcript in comparison to untreated cells (2–7-fold changes) upon cycloheximide treatment, both in healthy controls and XDP cells, suggesting that the *TAF1-32i* transcript undergoes NMD (Figure 1d). In addition, our results indicate increased synthesis of *TAF1-32i* in XDP, given that even upon cycloheximide treatment, levels of this transcript are higher in cells from XDP patients as compared to non-carriers of the SVA insertion.

2.3. The Intronic Region Included in the Transcript Is Associated with H3K36me3 in Control and XDP Cells

To further understand the molecular processes underlying the inclusion of the deep intronic region in the transcript, we aimed to investigate epigenetic marks within the region. Visualization of histone modifications from the ENCODE (Encyclopedia Of DNA Elements) project [20] revealed that H3K36me3 (trimethylation of lysine 36 on histone H3) is present in this region, with a particularly strong signal in NT2-D1 and U2OS cells (Figure 2a). Furthermore, chromatin state segmentation, a computational prediction of chromatin states based on chromatin immunoprecipitation-sequencing (ChIP-seq) data, indicated weak transcription in six out of nine cell types. In contrast, this region was annotated as heterochromatic in human embryonic stem cells (H1-hESC), which is generally considered as the cell type most similar to iPSCs (Figure 2a). H3K36me3 is present in gene bodies, marking both exons that undergo active transcription and alternative exons, consistent with its role in alternative splicing [21,22]. Furthermore, it marks constitutive and facultative heterochromatin and plays a role in the DNA damage response by recruiting the DNA repair machinery [23,24]. To experimentally test for the presence of this histone mark in the region, we performed chromatin immunoprecipitation followed by next-generation sequencing (NGS) and qPCR. We did not observe a difference in qPCR results between two healthy controls and three patient-derived iPSC lines, with primers targeting the region predicted to be enriched in this histone mark and included in the transcript (Figure 2b). To cover a wider region of the intron, we performed NGS and confirmed that there was no difference in H3K36me3 levels between a patient and a control line, with a weak signal for this histone mark in intron 32 (Figure 2c).

2.4. DNA Methylation Is Not Altered in the 5′ Region Adjacent to the SVA

Given the significance of DNA methylation in alternative splicing regulation and suppression of transposable elements (reviewed in [25,26]), we aimed to investigate CpG methylation of the SVA and the adjacent regions in intron 32. Since the human genome contains >2700 SVA elements [27], we applied long-read nanopore sequencing to precisely target the XDP-specific SVA and bisulfite pyrosequencing to verify the results at selected CpGs in intron 32. We have previously shown that the SVA is heavily methylated in various XDP tissues and cell lines [28]. When comparing the methylation frequency of the regions proximal to the SVA insertion in patients and controls (which also includes the intronic region within the *TAF1-32i* transcript), a CpG site at genomic position chrX:70,659,134 (hg19) strikingly deviated from others in brain samples (Figure 3a). Thus, we selected this CpG along with the neighboring one (chrX:70,659,225; hg19) for quantification by pyrosequencing in multiple samples. Our results were consistent while using the two methods, but there was no significant difference in methylation levels between patients and controls across the investigated tissues (Figure 3b). Specifically, although the intron retention transcript is prominent in patient-derived iPSCs, we did not see any difference in methylation levels between healthy controls and XDP patients. The methylation frequency remained high and unchanged even in the "ΔSVA" cell line, an XDP-derived cell line where the SVA was excised by CRISPR/Cas9. These results suggest that increased *TAF1-32i* transcript levels in XDP iPSCs likely cannot be attributed to alterations in DNA methylation.

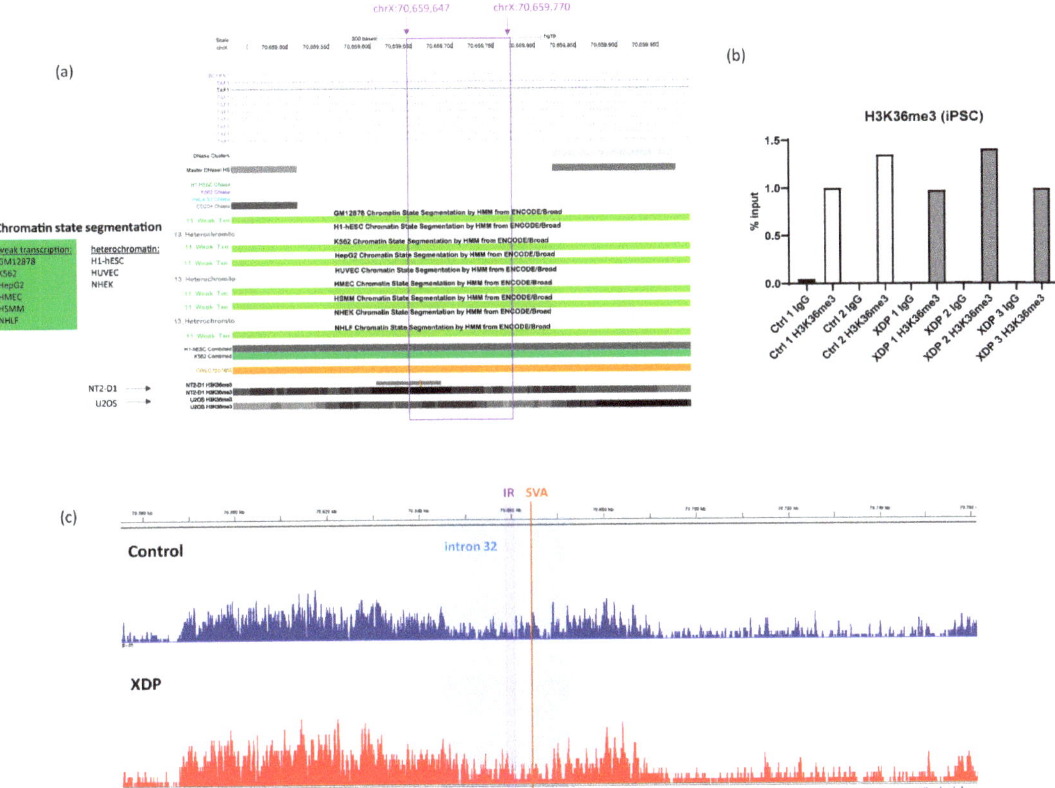

Figure 2. H3K36me3 levels at the intronic region included in the transcript. (**a**) USCS browser screenshot, showing chromatin state segmentation and H3K36me3 histone mark in different cell lines. Note that the strongest signal is present in NT2-D1 cells. The region marked in violet is included in the transcript, and its genomic coordinates are shown above. (**b**) ChIP-qPCR results from control (*n* = 2) and XDP (*n* = 3) iPSC lines. Results are calculated relative to the corresponding input sample and shown as % input. (**c**) ChIP-seq results showing the *TAF1* locus in control (blue) and XDP-derived (red) iPSCs. The region highlighted in blue depicts *TAF1* intron 32; the narrower region marked as IR indicates the intronic region retained in the *TAF1-32i* transcript, and the orange line marks the position in which the SVA is inserted.

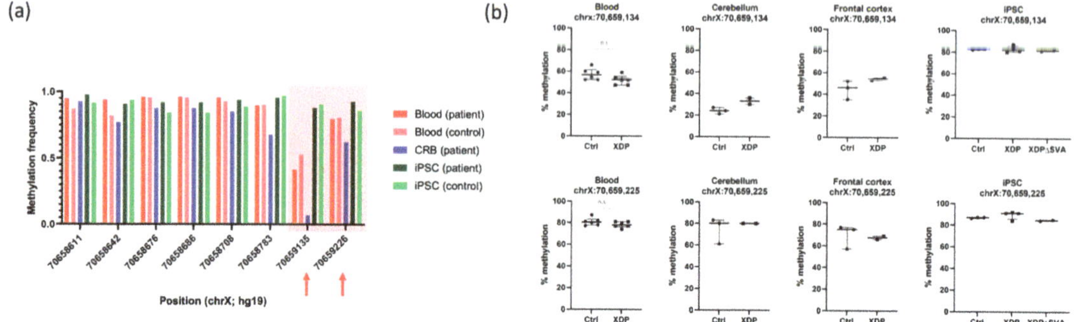

Figure 3. DNA methylation in intron 32 of *TAF1*, proximal to the SVA insertion. (**a**) Long-read nanopore sequencing on DNA from different tissues and cell lines from a healthy control and a patient with XDP (i.e., blood, iPSCs, cerebellum). Arrows indicate the two CpGs selected for analysis by pyrosequencing. (**b**) Pyrosequencing results showing methylation levels at the two selected CpGs: chrX:70,659,134 and chrX:70,659,225 (hg19), in blood (n = 6 controls, n = 6 patients), cerebellum (n = 3 controls; n = 2 patients), frontal cortex (n = 3 controls; n = 2 patients), and iPSCs (n = 3 controls; n = 4 patients; n = 2 XDPΔSVA cell lines). Unpaired *t* test was performed on blood samples after testing for normality with Kolmogorov–Smirnov test. n.s., not significant.

2.5. The SVA Represses TAF1 Promoter Activity In Vitro

An emerging body of evidence indicates that transposable elements function to regulate gene expression by affecting gene transcription, chromatin structure, pre-mRNA processing, and various aspects of mRNA metabolism (reviewed in [29]). SINE retrotransposons can cause epigenetic reprogramming of adjacent gene promoters and can serve as transcriptional enhancers by recruiting various transcription factors [30,31]. Thus, we aimed to test whether the XDP-specific SVA exerts transcriptional activity in a classical enhancer–promoter experiment, using a luciferase assay. First, we characterized the *TAF1* promoter region by inserting either the full-length region (chrX:70,585,177-70,586,242; hg19) or one of its fragments into a promoterless firefly luciferase reporter vector, pGL4.10 (Figure 4a). Promoter fragments were created based on histone marks and DNase sensitivity in order to define the most active region. In this experimental setup, the *TAF1* promoter controls expression of the luciferase gene, and any change in the promoter activity will be detected as a change in the luciferase signal. Measurement of relative luciferase activity upon transfecting HEK293 cells with these constructs narrowed down the most active *TAF1* promoter region to a fragment of approximately 400 bp (chrX:70,585,696-70,586,107; hg19) (Figure 4b). Next, we inserted the full-length SVA containing the hexanucleotide repeat with the minimum reported number of units (($CCCTCT$)$_{30}$) in either sense or antisense orientation into the pGL4.10 vector containing this 400 bp *TAF1* promoter region. Our results show that the SVA (inserted in either direction) strongly suppresses *TAF1* promoter activity in comparison to a size-matched control, suggesting its regulatory potential and possible recruitment of transcription factors (Figure 4c).

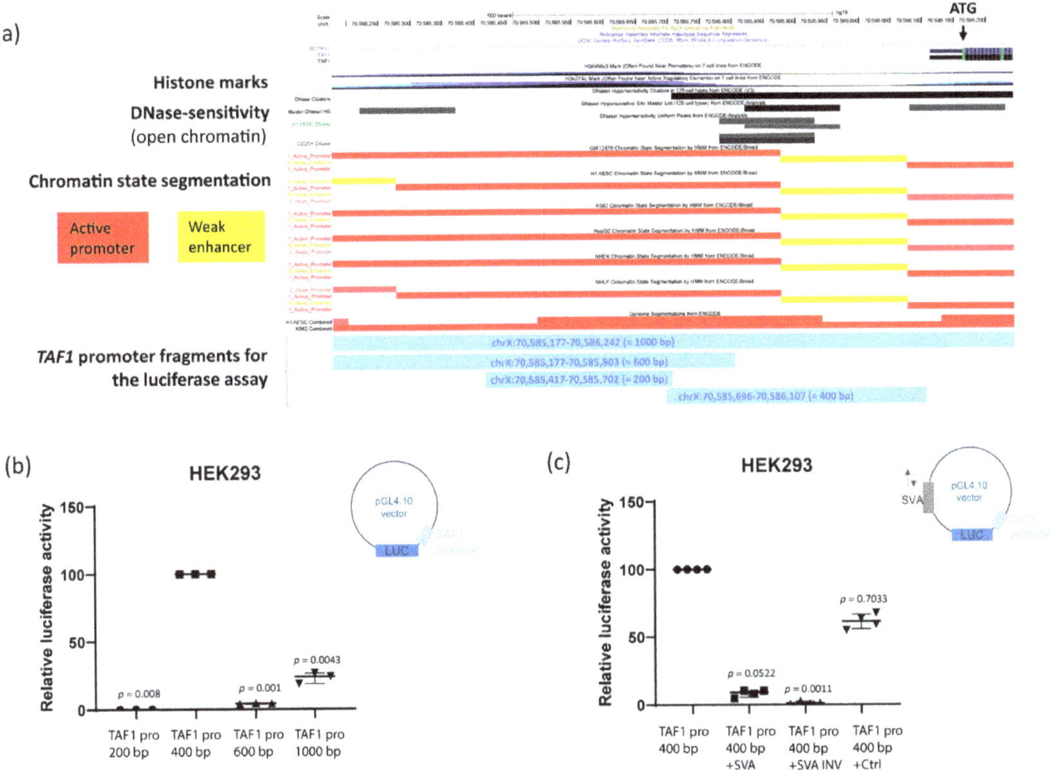

Figure 4. Regulatory potential of the SVA in vitro. (**a**) UCSC browser screenshot of the *TAF1* promoter region, showing DNase sensitivity, histone modifications and chromatin state segmentation in various cell types. In light blue (the lower part of the figure) are shown promoter fragments that were investigated in luciferase reporter assays, along with their genomic coordinates (hg19). Note that the most active *TAF1* promoter region (fragment ≈ 400 bp) overlaps with the DNase-sensitive region, characteristic of open chromatin. (**b**) Relative luciferase activity (Firefly counts/Renilla TK counts) of different *TAF1* promoter regions define the most active region (Fragment ≈ 400 bp). (**c**) Relative luciferase activity of the XDP-specific SVA-inserted sense or antisense (INV-inverted), as compared to a size-matched control. Maximum activity (100%) was exerted by the vector containing only the most active promoter region (*TAF1* pro 400 bp), and *p* values are calculated relative to this sample using Kruskal–Wallis and Dunn's multiple comparison tests.

3. Discussion

TAF1 is the largest subunit of TFIID, the initial basal transcription factor that recognizes and binds to the core promoter, and is thus essential for the subsequent formation of the functional preinitiation complex that positions RNA polymerase II at transcription start sites (reviewed in [32]). Furthermore, *TAF1* is extremely intolerant to loss-of-function (LoF) mutations (probability of LoF Intolerance, pLI = 1 in gnomAD database), and a *taf1* knockout zebrafish model shows embryonic lethality, pointing to its crucial role in development [33]. Together, this implies that the expression levels and function of *TAF1*/TAF1 must be tightly regulated, and any dysregulation could have a plethora of different consequences. In addition, this is a large gene with numerous transcript variants (currently, 27 annotated in Ensembl) that contributes to both proteomic diversity and to the tissue-specific gene regulatory network. While the canonical and neuron-specific *TAF1* isoforms differ in only 6 bp that determine the tissue distribution (e.g., neuronal commitment), it is still unclear whether the alternative transcript variant containing the deep intronic region performs a specific role in the cell. Given its increased amounts in XDP cells, it is tempting to speculate that it might exert a dominant-negative effect if it is being translated into a protein or RNA-induced toxicity and accumulation/deposition of RNA-binding proteins.

Our results show that although the level of the *TAF1-32i* transcript differs significantly between XDP patients and controls, the retained region within intron 32 is identical in all individuals. However, we do not know where the transcript terminates, or whether it even includes a part of the SVA in XDP patients. SVA retrotransposons have been reported to cause aberrant splicing, altering the canonical transcripts [34–36]. Instead, the XDP-specific SVA seems to enhance the synthesis of an already existing transcript, as we demonstrated that low levels of the *TAF1-32i* transcript can be detected in various cell types and that it undergoes degradation by NMD. Recent evidence indicates that coordinated action between alternative splicing and NMD functions to achieve the proper expression level of a given gene and/or protein, and that intron retention may be used to regulate a specific differentiation event, as shown in the hematopoietic system [37,38]. Thus, it seems plausible that levels of the *TAF1-32i* transcript might fine-tune expression, cellular differentiation, or cellular decisions, and that altered levels could contribute to disease manifestation. Of note, our experiments suggest that the levels of *TAF1-32i* in non-XDP cells might be higher than previously estimated when analyzed directly from cDNA (i.e., without prior preamplification).

In accordance with its role in marking transcribed regions, we detected a weak H3K36me3 signal in the deep intronic region included in the *TAF1-32i* transcript. This histone mark is also associated with the binding of PTBP (polypyrimidine tract-binding-protein), one of the major regulators of splicing that was shown to bind to silencing elements and regulate whether or not an alternative exon will be included in a transcript [21]. Although there were no differences in H3K36me3 levels between patients and controls along the transcribed regions, we cannot conclude that there is no signal within the SVA. Namely, with short-read sequencing technologies, it is challenging to map a putative H3K36me3 signal coming from any SVA, even if it is being transcribed (at least partially). That is, because there are >2700 SVA elements in the human genome, they are being filtered out during the short-read bioinformatical analysis that includes only the regions that can be mapped to the reference sequence. Conversely, for estimating DNA methylation, long-read sequencing technologies exist and enable the measuring of DNA methylation along the region spanning the SVA insertion of interest. Therefore, we chose to use the nanopore technology that detects native DNA modifications. Subsequently, we applied single-nucleotide-specific pyrosequencing, which is methodologically different and relies on bisulfite conversion prior to measurement. Although differences in *TAF1-32i* amounts are prominent between controls and XDP lines in iPSC (Figure 1), these differences are not caused by changes in DNA methylation, as demonstrated by our results (Figure 3a,b). Specifically, these differences in levels of the *TAF1-32i* transcript are visible in XDP ΔSVA cells (Figure 1a), which do not coincide with any alterations in DNA methylation at these two CpG sites (Figure 3b). DNA methylation levels at the position chrX:70,659,134 appear to vary drastically among tissues, with the lowest levels in the cerebellum, potentially indicating a tissue-specific regulatory effect. Although we have not observed any alterations in DNA methylation and H3K36me3 levels in the intronic region within *TAF1-32i* in XDP-derived samples, further investigations of other histone marks and other CpG sites are warranted. For instance, recent work on XDP-derived cells reported local changes in H3 acetylation (AcH3), affecting an exon proximal to the SVA insertion. This decrease in AcH3 level was normalized by CRISPR/Cas9-excision of the SVA, suggesting that the SVA alters epigenetic marks in the region [39]. In addition, a significant increase in histone H3 citrullination (H3R2R8R17cit3) was reported in the XDP post-mortem prefrontal cortex [40]. When considering XDP-relevant epigenetic changes beyond those potentially introduced by the SVA, the three disease-specific single-nucleotide changes introduce or abolish CpG sites of DNA methylation, introducing a possible additional mechanism that might modulate *TAF1* expression in XDP in addition to the SVA [41].

Transposable elements comprise a large portion of the human genome, and in healthy cells, they are silenced and usually inactive. However, they have been reported to become active and mobile in aging mammalian tissues [42] and are regarded as a source of genomic variation or even as "controlling elements" [43]. Currently, it is well known that transposable elements can influence gene expression by acting as promoters, enhancers, repressors, or insulators (reviewed in [44]). Our results showing that the XDP-specific SVA retrotransposon represses *TAF1* promoter activity (Figure 4c) suggest that it could function as a transcriptional repressor. This is in line with a previous report investigating an SVA element inserted upstream of the *FUS* gene and is thus associated with amyotrophic lateral sclerosis and frontotemporal dementia. Namely, this SVA exerts a repressive function on the SV40 minimal promoter [45], indicating that this property is universal rather than sequence-dependent. Conversely, the XDP-specific SVA was shown to act as a promoter in a different experimental setup [15], also leading to the conclusion that there are transcription factor binding sites (TFBS) within the SVA. However, due to its repetitive sequence and genome-wide distribution, it is still experimentally challenging to prove the exact transcription factor(s) and their binding sites within the SVA. Although our study was limited in some aspects, such as small sample size, that did not always allow for statistical testing, it adds to the growing body of evidence that transposable elements affect gene expression. Together, our results show transcriptional alterations in XDP caused by the SVA retrotransposon, suggesting that it contains binding sites for transcription factors and possibly splicing regulators.

4. Materials and Methods

4.1. Study Participants

We analyzed biomaterials from a total of 83 individuals (52 XDP patients carrying the SVA insertion and 31 (21 males) healthy ethnicity-matched controls with wild-type genotype). The median age at sample collection was 40 (interquartile range (IQR: 35.0–47.8, range: 30–60) years for the XDP patients and 35 (IQR: 30.0–42.0, range: 18–54) years for controls. In XDP patients, the median repeat number was 43 (IQR: 40.0–45.8, range: 35–53), median AAO was 37 (IQR: 31.0–41.8, range: 26–51) years, and median disease duration at the time of sample collection was 3 (IQR: 2–4, range: −2–19) years.

DNA was available for all samples. With respect to RNA, for 41 XDP patients, only RNA from blood was available and used in experiments, while for two patients, only RNA from fibroblasts was available. For the remaining patients, RNA from the blood and/or fibroblasts or iPSCs was available. For two XDP patients and 3 controls, postmortem brain tissue was available. The XDP patients died at the age of 36 and 38 years. Two of the controls were male and one female, and they were 68 years old at the time of death, with no neurodegenerative findings at the time of pathological examination. They were of German ethnicity, as no postmortem tissue from Filipino individuals was available. The study was conducted according to the guidelines of the Declaration of Helsinki and approved by the Ethics Committee of the University of Lübeck (AZ12-219). All autopsies had been performed either as clinical autopsies with first-line relatives, next of kin or their legally authorized representatives giving informed consent or as legal autopsies on behalf of investigating authorities. The use of specimens obtained at autopsies for research upon anonymization is in accordance with local ethical standards and regulations at the University Hospital Schleswig–Holsten (the "Gesetz uber das Leichen–, Bestattungs– und Friedhofswesen (Bestattungsgesetz) des Landes Schleswig–Holstein vom 04.02.2005, Abschnitt II, 9 (Leichen offnung, anatomisch)") or University Medical Center Hamburg–Eppendorf ("Hamburgisches Krankenhausgesetz vom 17.04.1991, §12, Abs. 1").

4.2. Nucleic Acid Extraction and Reverse Transcription

The genomic DNA was routinely extracted from peripheral blood leukocytes using the salting-out method. QIAamp DNA Mini Kit (Qiagen, Hilden, Germany) was used for DNA extraction from iPSCs, while the Blood and cell culture DNA midi kit (Qiagen) was used to extract high-molecular-weight DNA from brain tissue (i.e., cerebellum and frontal cortex). RNA was extracted from the whole blood using the PAXgene Blood RNA Kit (Qiagen), and from cells using the RNeasy Mini Kit (Qiagen), according to the manufacturer's instructions. Only the RNA samples with an RNA integrity number (RIN) of >6 were included in the analyses. Maxima First Strand cDNA Synthesis Kit for RT-qPCR with dsDNase (Thermo Scientific, Waltham, MA, USA) was used for reverse transcription, starting with 500 ng total RNA.

4.3. Quantitative PCR (qPCR)

Maxima SYBR Green/Fluorescein qPCR Master Mix (Thermo Scientific) was used for qPCR, in a 10 µL reaction volume, on the Light Cycler 96 Instrument (Roche, Basel, Switzerland). *TAF1-32i* primers target exon 32 (5′-GTATAATGATTCAGGAAGTTGCAAG-3′) and intron 32 (5′-GTAATGTACCAATATAAATTTCCTGGTTT-3′). *GAPDH* primers target exon 1 (5′-GTCAGCCGCATCTTCTTTTG-3′) and exon 3 (5′-GCGCCCAATACGACCAAATC-3′). Cycling conditions for the 3-step amplification are as follows: 95° for 15 s; 57° for 30 s; 72° for 30 s. The analysis of each sample was performed in triplicate. Statistical analyses were performed on dCt values (Ct target–Ct reference gene) using an unpaired *t* test since they are logarithmic and thus normally distributed, which allows for one to perform a *t* test.

4.4. Sanger Sequencing

PCR products using cDNA as a template and the above-given sequences of the *TAF1* in32 primers were purified by Exonuclease I and Fast AP Thermosensitive Alkaline Phosphatase (Thermo Scientific). Subsequently, the sequencing reaction was performed using only one of the primers and the BigDye Terminator v3.1 Cycle Sequencing Kit (Applied Biosystems, Waltham, MA, USA). Samples were purified by Sodium Acetate/Ethanol precipitation, dissolved in Hi-Di Formamide (Applied Biosystems), and loaded on the 3500xL Genetic Analyzer (Applied Biosystems). Electropherograms were visualized using Chromas Lite (Technelysium Pty Ltd., South Brisbane, Australia).

4.5. Cell Culture

Fibroblast lines were established from skin biopsies, and these cells, as well as HEK293, were grown in DMEM medium (Thermo Scientific, Waltham, MA, USA), supplemented with 10% fetal bovine serum (Thermo Scientific, Waltham, MA, USA) and 1% Penicillin–Streptomycin (Thermo Scientific, Waltham, MA, USA). Generation and characterization of the iPS cell lines from XDP patients and ethnically matched controls was performed previously (https://www.wicell.org/home/stem-cells/catalog-of-stem-cell-lines/collections/massachusetts-general-hospital.cmsx (accessed on 10 December 2021)), and gene-edited lines have been examined in an earlier study [10]. Here, they were grown on Matrigel-coated plates in mTeSR medium (StemCell Technologies, Vancouver, BC, Canada).

4.6. Cycloheximide Treatment

iPSCs from healthy controls and XDP patients were grown to 70% confluency in 6-well plates, prior to cycloheximide treatment (C4859, Sigma-Aldrich, St. Louis, MO, USA). On the day of the treatment, the medium was removed from the cells, and fresh medium containing cycloheximide to a final concentration of 50 µg/mL was added to each well. The cells were incubated overnight and pelleted the next day for further experiments (i.e., RNA extraction and qPCR).

4.7. Chromatin Immunoprecipitation (ChIP)

Chromatin immunoprecipitation was performed according to Lee et al. [46]. Briefly, iPSCs were fixed for 10 min on ice with 1% formaldehyde in mTeSR. The reaction was quenched with 2.5 M glycine, followed by extraction of the nuclear lysate and chromatin sonication on Diagenode Bioruptor Pico (15 cycles; 30 s pulse–30 s pause). For ChIP, 15 µg of chromatin was incubated with 5 µg of the H3K36me3 antibody (ab9050, Abcam, Cambridge, UK) overnight. The next day, blocked magnetic beads were added to the chromatin–antibody complexes and incubated overnight, followed by 7 washes with RIPA buffer and 1 with TE buffer. The immunoprecipitated DNA was extracted with phenol–chloroform, washed with ethanol, eluted in ultra-pure nuclease-free water, and used for subsequent experiments. For qPCR, the primers were designed to cover both the enriched H3K36me3 signal (Figure 2a) and the region included in the *TAF1-32i* transcript (forward: 5′-GCTCATGAATGTATTCTGATCC-3′; reverse: 5′-GTACAGCTATGTAAGATATTGCC-3′). For NGS, library preparation was performed with NEBNext Ultra II DNA Library Prep with Sample Purification Beads (E7103, NEB), and the sequencing was performed on NextSeq 2000 (Illumina, San Diego, CA, USA). Reads were mapped using Bowtie2, while parsing to bigwig format was performed using DROMPAplus and/or MACS2 (as described in [47]).

4.8. DNA Methylation Analyses by Nanopore Sequencing

Cas9-targeted sequencing from Oxford Nanopore Technologies was performed to enrich the target region and to obtain the epigenetic information. CRISPR RNAs (crRNAs) were designed with CHOPCHOP (https://chopchop.cbu.uib.no (accessed on 10 December 2021)). Four crRNAs were used upstream of the *TAF1* SVA insertion, and four crRNAs were used downstream. Two libraries were prepared per sample. The enriched DNA was prepared with the Nanopore Ligation Sequencing Kit (SQK-LSK109), loaded on a R9.4.1 flow cell and sequenced with MinION or GridION. For methylation analysis, all sequencing data obtained were combined to maximize coverage depth. Methylation was called with the software Nanopolish (v0.13.2) (Oxford Nanopore Technologies, Oxford, UK), which can detect 5′-methylcytosine (5mC) in a CpG context. To counteract potential off-target effects of the CRISPR/Cas9 enrichment, the BAM file was filtered for reads with an alignment length >3kb in the patient- or >1.5kb in control-derived samples. Only CpG sites covered by >10 reads were included in the analysis.

4.9. DNA Methylation Analyses by Pyrosequencing

Bisulfite conversion of genomic DNA was performed with the EpiTect Fast DNA Bisulfite Kit (Qiagen), according to the manufacturer's instructions. Converted DNA was PCR amplified using primers specific for the converted sequence (forward: 5′-ATAATTTTTAA TTTGGGTTTAATGGGG-3′; reverse: 5′-[BIO]CTACCTAACAAAAATATAAATAATAAA TTAA-3′). Samples were sequenced on PyroMark Q48 Autoprep (Qiagen) using two different sequencing primers (70659134: 5′-GTATTAATATTATTTAGTAGTT-3′; 70659225: 5′-GTTTATATTATATTTTGTTTAG-3′). Data were tested for normal distribution using the Kolmogorov–Smirnov test and analyzed with an unpaired *t* test.

4.10. Luciferase Assay

To define the most active *TAF1* promoter region, various fragments were inserted into the pGL4.10(*luc2*) vector (Promega, Madison, WI, USA) using the Gibson Assembly cloning strategy (E2621, NEB). In the next step, either the full-length SVA or a size-matched control were inserted in the vector containing the most active *TAF1* promoter region. Primers used for cloning are available upon request. To improve the cloning efficiency of the repetitive DNA regions found within the SVA (e.g., hexanucleotide repeats), OneShot Stbl3 chemically competent *E. coli* were used (Invitrogen, Waltham, MA, USA), and bacteria were grown at 30 °C. In addition to Sanger sequencing, inserts were verified with fragment analysis targeting the hexanucleotide repeats, as described previously [14,15]. HEK293 cells were transfected with different constructs, using FuGENE HD (Promega). We co-transfected cells with a thymidine kinase promoter-Renilla luciferase reporter plasmid (pRL-TK) as an internal control. After 24 h, the cells were lysed, and the activity of Firefly and Renilla luciferase was determined with the Dual Luciferase Reporter Assay (Promega) in a TriStar2 LB Multidetection Microplate Reader (Berthold, Bad Wildbad, Germany). All measurements were verified in at least three independent experiments and as triplicates in each experiment. Firefly luciferase signals were corrected for transfection efficiency using Renilla signals of the co-transfected control vector. Relative light units were then normalized relative to the *TAF1* pro 400 bp-containing plasmid. Statistical analysis was performed using the Kruskal–Wallis test with Dunn's multiple comparison test.

Supplementary Materials: The following supporting information can be downloaded at: https://www.mdpi.com/article/10.3390/ijms23042231/s1.

Author Contributions: Conceptualization, J.P., F.J.K. and A.W.; methodology, J.P., S.M.A., J.N.C., K.G., S.S., R.T., V.Y., N.K., K.H. and A.R.; software, J.L., K.S. and V.K.A.S.; validation, J.P., J.T., J.L., H.K. and A.W.; formal analysis, J.P., S.M.A., J.N.C., J.T., J.L., K.S., H.K., F.J.K. and A.W.; investigation, J.P., S.M.A., J.N.C., J.T., J.L., K.S., M.S., F.J.K. and A.W.; resources, N.B., R.L.R., R.D.G.J., C.C.E.D., J.M., M.G., P.S., A.R., M.S., F.J.K., C.K. and A.W.; data curation, J.P., S.M.A., J.N.C., S.S., R.T., V.Y. and H.K.; writing—original draft preparation, J.P. and A.W.; writing—review and editing, J.P., S.M.A., J.N.C., J.T., N.B., J.L., K.G., R.L.R., R.D.G.J., C.C.E.D., J.M., H.K., K.H., P.S., A.R., M.S., F.J.K., C.K. and A.W.; visualization, J.P., J.L. and H.K.; supervision, H.K., M.S., F.J.K., C.K. and A.W.; project administration, J.P., F.J.K., C.K. and A.W.; funding acquisition, J.P., A.R., M.S., F.J.K., C.K. and A.W. All authors have read and agreed to the published version of the manuscript.

Funding: This study was funded by the Deutsche Forschungsgemeinschaft (DFG; FOR 2488 to J.T., N.B., P.S., A.R., C.K. and A.W.; KI-1887/2-1 to H.K; and SP1532/3-1, SP1532/4-1, and SP1532/5-1, to M.S.). J.P. was supported by a research grant from the University of Lübeck, Germany (J14-2021). N.B., P.S., A.R., C.K. and A.W. were funded by the Collaborative Center for X-Linked Dystonia Parkinsonism. M.S. is supported by the Max Planck Society and the Deutsches Zentrum für Luft- und Raumfahrt (DLR 01GM1925).

Institutional Review Board Statement: The study was conducted according to the guidelines of the Declaration of Helsinki and approved by the Ethics Committee of the University of Lübeck (AZ12-219; approval date: 18 December 2012).

Informed Consent Statement: Informed consent was obtained from all subjects involved in the study.

Data Availability Statement: The data presented in this study are available from the corresponding authors upon reasonable request.

Acknowledgments: We thank our patients and their families for participating in this study. We also thank Björn-Hergen Laabs for his help with the statistical analysis of qPCR data from blood-derived samples.

Conflicts of Interest: The authors report no conflict of interest. N.B. received honoraria from Abbott, Abbvie, Biogen, Biomarin, Bridgebio, Centogene GmbH and Zambon. He is funded by the DFG (BR4328.2-1, GRK1957). C.K. serves as a medical advisor for genetic testing reports to Centogene GmbH in the fields of movement disorders and dementia, excluding Parkinson's disease, and is a member of the Scientific Advisory Boards of Retromer Therapeutics and Klink. A.W. serves as a consultant for medical writing to CENTOGENE GmbH.

References

1. Lee, L.V.; Rivera, C.; Teleg, R.A.; Dantes, M.B.; Pasco, P.M.D.; Jamora, R.D.G.; Arancillo, J.; Villareal-Jordan, R.F.; Rosales, R.L.; Demaisip, C.; et al. The Unique Phenomenology of Sex-Linked Dystonia Parkinsonism (XDP, DYT3, "Lubag"). *Int. J. Neurosci.* **2010**, *121*, 3–11. [CrossRef]
2. Pauly, M.G.; López, M.R.; Westenberger, A.; Saranza, G.; Brüggemann, N.; Weissbach, A.; Rosales, R.L.; Diesta, C.C.; Jamora, R.D.; Reyes, C.J.; et al. Expanding Data Collection for the MDSGene Database: X-linked Dystonia-Parkinsonism as Use Case Example. *Mov. Disord.* **2020**, *35*, 1933–1938. [CrossRef]
3. Kawarai, T.; Morigaki, R.; Kaji, R.; Goto, S. Clinicopathological phenotype and genetics of X-linked dystonia-parkinsonism (XDP; DYT3; Lubag). *Brain Sci.* **2017**, *7*, 72. [CrossRef]

4.   Pasco, P.M.D.; Ison, C.V.; Muñoz, E.L.; Magpusao, N.S.; Cheng, A.E.; Tan, K.T.; Lo, R.W.; Teleg, R.A.; Dantes, M.B.; Borres, R.; et al. Understanding XDP Through Imaging, Pathology, and Genetics. *Int. J. Neurosci.* **2010**, *121*, 12–17. [CrossRef]

5.   Hanssen, H.; Heldmann, M.; Prasuhn, J.; Tronnier, V.; Rasche, D.; Diesta, C.C.; Domingo, A.; Rosales, R.L.; Jamora, R.D.; Klein, C.; et al. Basal ganglia and cerebellar pathology in X-linked dystonia-parkinsonism. *Brain* **2018**, *141*, 2995–3008. [CrossRef]

6.   Nolte, D.; Niemann, S.; Müller, U. Specific sequence changes in multiple transcript system DYT3 are associated with X-linked dystonia parkinsonism. *Proc. Natl. Acad. Sci. USA* **2003**, *100*, 10347–10352. [CrossRef]

7.   Makino, S.; Kaji, R.; Ando, S.; Tomizawa, M.; Yasuno, K.; Goto, S.; Matsumoto, S.; Tabuena, M.D.; Maranon, E.; Dantes, M.; et al. Reduced Neuron-Specific Expression of the TAF1 Gene Is Associated with X-Linked Dystonia-Parkinsonism. *Am. J. Hum. Genet.* **2007**, *80*, 393–406. [CrossRef]

8.   Domingo, A.; Westenberger, A.; Lee, L.V.; Brænne, I.; Liu, T.; Vater, I.; Rosales, R.; Jamora, R.D.; Pasco, P.M.; Paz, E.M.C.-D.; et al. New insights into the genetics of X-linked dystonia-parkinsonism (XDP, DYT3). *Eur. J. Hum. Genet.* **2015**, *23*, 1334–1340. [CrossRef]

9.   Aneichyk, T.; Hendriks, W.T.; Yadav, R.; Shin, D.; Gao, D.; Vaine, C.A.; Collins, R.L.; Domingo, A.; Currall, B.; Stortchevoi, A.; et al. Dissecting the Causal Mechanism of X-Linked Dystonia-Parkinsonism by Integrating Genome and Transcriptome Assembly. *Cell* **2018**, *172*, 897–909.e21. [CrossRef]

10. Rakovic, A.; Domingo, A.; Grütz, K.; Msc, L.K.; Capetian, P.; Cowley, S.; Lenz, I.; Brüggemann, N.; Rosales, R.; Jamora, D.; et al. Genome editing in induced pluripotent stem cells rescues TAF1 levels in X-linked dystonia-parkinsonism. *Mov. Disord.* **2018**, *33*, 1108–1118. [CrossRef]

11. Domingo, A.; Amar, D.; Grütz, K.; Lee, L.V.; Rosales, R.; Brüggemann, N.; Jamora, R.D.; Paz, E.C.-D.; Rolfs, A.; Dressler, D.; et al. Evidence of TAF1 dysfunction in peripheral models of X-linked dystonia-parkinsonism. *Cell. Mol. Life Sci.* **2016**, *73*, 3205–3215. [CrossRef]

12. Ito, N.; Hendriks, W.T.; Dhakal, J.; Vaine, C.A.; Liu, C.; Shin, D.; Shin, K.; Wakabayashi-Ito, N.; Dy, M.; Multhaupt-Buell, T.; et al. Decreased N-TAF1 expression in X-Linked Dystonia-Parkinsonism patient-specific neural stem cells. *Dis. Model. Mech.* **2016**, *9*, 451–462. [CrossRef]

13. Al Ali, J.; Vaine, C.A.; Shah, S.; Ms, L.C.; Hakoum, A.; Supnet, M.L.; Acuña, P.; Ms, G.A.; Ms, T.M.; Bs, N.G.G.; et al. TAF1 Transcripts and Neurofilament Light Chain as Biomarkers for X-linked Dystonia-Parkinsonism. *Mov. Disord.* **2020**, *36*, 206–215. [CrossRef]

14. Westenberger, A.; Reyes, C.J.; Saranza, G.; Dobricic, V.; Hanssen, H.; Domingo, A.; Laabs, B.; Schaake, S.; Pozojevic, J.; Rakovic, A.; et al. A hexanucleotide repeat modifies expressivity of X-linked dystonia parkinsonism. *Ann. Neurol.* **2019**, *85*, 812–822. [CrossRef]

15. Bragg, D.C.; Mangkalaphiban, K.; Vaine, C.A.; Kulkarni, N.J.; Shin, D.; Yadav, R.; Dhakal, J.; Ton, M.-L.; Cheng, A.; Russo, C.T.; et al. Disease onset in X-linked dystonia-parkinsonism correlates with expansion of a hexameric repeat within an SVA retrotransposon in TAF1. *Proc. Natl. Acad. Sci. USA* **2017**, *114*, E11020–E11028, Erratum in *Proc. Natl. Acad. Sci. USA* **2020**, *117*, 6277. [CrossRef]

16. Wassarman, D.A.; Sauer, F. TAF II 250: A transcription toolbox. *J. Cell Sci.* **2001**, *114*, 2895–2902. [CrossRef]

17. Thomas, M.C.; Chiang, C.M. The general transcription machinery and general cofactors. *Crit. Rev. Biochem. Mol. Biol.* **2006**, *41*, 105–178. [CrossRef]

18. Nickless, A.; Bailis, J.M.; You, Z. Control of gene expression through the nonsense-mediated RNA decay pathway. *Cell Biosci.* **2017**, *7*, 26. [CrossRef]

19. García-Moreno, J.F.; Romão, L. Perspective in alternative splicing coupled to nonsense-mediated mrna decay. *Int. J. Mol. Sci.* **2020**, *21*, 9424. [CrossRef]

20. The ENCODE Project Consortium. An Integrated Encyclopedia of DNA Elements in the Human Genome. *Nature* **2012**, *489*, 57–74. [CrossRef]

21. Luco, R.F.; Pan, Q.; Tominaga, K.; Blencowe, B.J.; Pereira-Smith, O.M.; Misteli, T. Regulation of Alternative Splicing by Histone Modifications. *Science* **2010**, *327*, 996–1000. [CrossRef] [PubMed]

22. Pradeepa, M.M.; Sutherland, H.; Ule, J.; Grimes, G.R.; Bickmore, W.A. Psip1/Ledgf p52 Binds Methylated Histone H3K36 and Splicing Factors and Contributes to the Regulation of Alternative Splicing. *PLoS Genet.* **2012**, *8*, e1002717. [CrossRef] [PubMed]

23. Chantalat, S.; Depaux, A.; Héry, P.; Barral, S.; Thuret, J.-Y.; Dimitrov, S.; Gérard, M. Histone H3 trimethylation at lysine 36 is associated with constitutive and facultative heterochromatin. *Genome Res.* **2011**, *21*, 1426–1437. [CrossRef] [PubMed]

24. Sun, Z.; Zhang, Y.; Jia, J.; Fang, Y.; Tang, Y.; Wu, H.; Fang, D. H3K36me3, message from chromatin to DNA damage repair. *Cell Biosci.* **2020**, *10*, 9. [CrossRef]

25. Maor, G.L.; Yearim, A.; Ast, G. The alternative role of DNA methylation in splicing regulation. *Trends Genet.* **2015**, *31*, 274–280. [CrossRef]

26. Greenberg, M.V.C.; Bourc'His, D. The diverse roles of DNA methylation in mammalian development and disease. *Nat. Rev. Mol. Cell Biol.* **2019**, *20*, 590–607. [CrossRef]

27. Wang, H.; Xing, J.; Grover, D.; Hedges, D.J.; Han, K.; Walker, J.A.; Batzer, M.A. SVA Elements: A Hominid-specific Retroposon Family. *J. Mol. Biol.* **2005**, *354*, 994–1007. [CrossRef]

28. Lüth, T.; Laß, J.; Schaake, S.; Wohlers, I.; Pozojevic, J.; Jamora, R.D.G.; Rosales, R.L.; Brüggemann, N.; Saranza, G.; Diesta, C.C.E.; et al. Elucidating Hexanucleotide Repeat Number and Methylation within the X-Linked Dystonia-Parkinsonism (XDP)-SVA Retrotransposon in TAF1 with Nanopore Sequencing. *Genes* **2022**, *13*, 126. [CrossRef]
29. Elbarbary, R.A.; Lucas, B.A.; Maquat, L.E. Retrotransposons as regulators of gene expression. *Science* **2016**, *351*, aac7247. [CrossRef]
30. Estécio, M.R.; Gallegos, J.; Dekmezian, M.; Lu, Y.; Liang, S.; Issa, J.-P. SINE Retrotransposons Cause Epigenetic Reprogramming of Adjacent Gene Promoters. *Mol. Cancer Res.* **2012**, *10*, 1332–1342. [CrossRef]
31. Sasaki, T.; Nishihara, H.; Hirakawa, M.; Fujimura, K.; Tanaka, M.; Kokubo, N.; Kimura-Yoshida, C.; Matsuo, I.; Sumiyama, K.; Saitou, N.; et al. Possible involvement of SINEs in mammalian-specific brain formation. *Proc. Natl. Acad. Sci. USA* **2008**, *105*, 4220–4225. [CrossRef] [PubMed]
32. Bhuiyan, T.; Timmers, H.M. Promoter Recognition: Putting TFIID on the Spot. *Trends Cell Biol.* **2019**, *29*, 752–763. [CrossRef] [PubMed]
33. Gudmundsson, S.; Wilbe, M.; Filipek-Górniok, B.; Molin, A.-M.; Ekvall, S.; Johansson, J.; Allalou, A.; Gylje, H.; Kalscheuer, V.M.; Ledin, J.; et al. TAF1, associated with intellectual disability in humans, is essential for embryogenesis and regulates neurodevelopmental processes in zebrafish. *Sci. Rep.* **2019**, *9*, 1–11. [CrossRef] [PubMed]
34. Taniguchi-Ikeda, M.; Kobayashi, K.; Kanagawa, M.; Yu, C.-C.; Mori, K.; Oda, T.; Kuga, A.; Kurahashi, H.; Akman, H.O.; DiMauro, S.; et al. Pathogenic exon-trapping by SVA retrotransposon and rescue in Fukuyama muscular dystrophy. *Nature* **2011**, *478*, 127–131. [CrossRef] [PubMed]
35. Kim, J.; Hu, C.; Moufawad El Achkar, C.; Black, L.E.; Douville, J.; Larson, A.; Pendergast, M.K.; Goldkind, S.F.; Lee, E.A.; Kuniholm, A.; et al. Patient-Customized Oligonucleotide Therapy for a Rare Genetic Disease. *N. Engl. J. Med.* **2019**, *381*, 1644–1652. [CrossRef] [PubMed]
36. Hancks, D.C.; Ewing, A.D.; Chen, J.E.; Tokunaga, K.; Kazazian, H.H. Exon-trapping mediated by the human retrotransposon SVA. *Genome Res.* **2009**, *19*, 1983–1991. [CrossRef] [PubMed]
37. Lewis, B.P.; Green, R.; Brenner, S.E. Evidence for the widespread coupling of alternative splicing and nonsense-mediated mRNA decay in humans. *Proc. Natl. Acad. Sci. USA* **2002**, *100*, 189–192. [CrossRef] [PubMed]
38. Wong, J.J.-L.; Ritchie, W.; Ebner, O.A.; Selbach, M.; Wong, J.W.; Huang, Y.; Gao, D.; Pinello, N.; Gonzalez, M.; Baidya, K.; et al. Orchestrated Intron Retention Regulates Normal Granulocyte Differentiation. *Cell* **2013**, *154*, 583–595. [CrossRef]
39. Petrozziello, T.; Dios, A.M.; Mueller, K.A.; Vaine, C.A.; Hendriks, W.T.; Glajch, K.E.; Mills, A.N.; Mangkalaphiban, K.; Penney, E.B.; Ito, N.; et al. SVA insertion in X-linked Dystonia Parkinsonism alters histone H3 acetylation associated with TAF1 gene. *PLoS ONE* **2020**, *15*, e0243655. [CrossRef]
40. Petrozziello, T.; Mills, A.N.; Vaine, C.A.; Penney, E.B.; Fernandez-Cerado, C.; Legarda, G.P.A.; Velasco-Andrada, M.S.; Acuña, P.J.; Ang, M.A.; Muñoz, E.L.; et al. Neuroinflammation and histone H3 citrullination are increased in X-linked Dystonia Parkinsonism post-mortem prefrontal cortex. *Neurobiol. Dis.* **2020**, *144*, 105032. [CrossRef]
41. Krause, C.; Schaake, S.; Grütz, K.; Sievert, H.; Msc, C.J.R.; König, I.R.; Msc, B.L.; Jamora, R.D.; Rosales, R.L.; Diesta, C.C.E.; et al. DNA Methylation as a Potential Molecular Mechanism in X-linked Dystonia-Parkinsonism. *Mov. Disord.* **2020**, *35*, 2220–2229. [CrossRef]
42. De Cecco, M.; Criscione, S.W.; Peterson, A.L.; Neretti, N.; Sedivy, J.M.; Kreiling, J.A. Transposable elements become active and mobile in the genomes of aging mammalian somatic tissues. *Aging* **2013**, *5*, 867–883. [CrossRef]
43. McClintock, B. Intranuclear systems controlling gene action and mutation. *Brookhaven Symp. Biol.* **1956**, *8*, 58–74.
44. Chuong, E.; Elde, N.C.; Feschotte, E.B.C.N.C.E.C. Regulatory activities of transposable elements: From conflicts to benefits. *Nat. Rev. Genet.* **2016**, *18*, 71–86. [CrossRef]
45. Savage, A.L.; Wilm, T.P.; Khursheed, K.; Shatunov, A.; Morrison, K.E.; Shaw, P.J.; Shaw, C.E.; Smith, B.; Breen, G.; Al-Chalabi, A.; et al. An Evaluation of a SVA Retrotransposon in the FUS Promoter as a Transcriptional Regulator and Its Association to ALS. *PLoS ONE* **2014**, *9*, e90833. [CrossRef] [PubMed]
46. Lee, T.I.; E Johnstone, S.; A Young, R. Chromatin immunoprecipitation and microarray-based analysis of protein location. *Nat. Protoc.* **2006**, *1*, 729–748. [CrossRef]
47. Nakato, R.; Sakata, T. Methods for ChIP-seq analysis: A practical workflow and advanced applications. *Methods* **2021**, *187*, 44–53. [CrossRef]

International Journal of
Molecular Sciences

Article

Comparative Transcriptome Profiling of Young and Old Brown Adipose Tissue Thermogenesis

Yumin Kim [1], Baeki E. Kang [2], Dongryeol Ryu [2], So Won Oh [3,*] and Chang-Myung Oh [1,*]

[1] Department of Biomedical Science and Engineering, Gwangju Institute of Science and Technology, Gwangju 61005, Korea; dbals123@gm.gist.ac.kr
[2] Department of Molecular Cell Biology, Sungkyunkwan University (SKKU) School of Medicine, Suwon 16419, Korea; baekikang@gmail.com (B.E.K.); freefall@skku.edu (D.R.)
[3] Department of Nuclear Medicine, Seoul National University Boramae Medical Center, Seoul 03080, Korea
* Correspondence: mdosw@snu.ac.kr (S.W.O.); cmoh@gist.ac.kr (C.-M.O.); Tel.: +82-62-715-5377 (C.-M.O.); Fax: +82-62-715-5309 (C.-M.O.)

Abstract: Brown adipose tissue (BAT) is a major site for uncoupling protein 1 (UCP1)-mediated non-shivering thermogenesis. BAT dissipates energy via heat generation to maintain the optimal body temperature and increases energy expenditure. These energetic processes in BAT use large amounts of glucose and fatty acid. Therefore, the thermogenesis of BAT may be harnessed to treat obesity and related diseases. In mice and humans, BAT levels decrease with aging, and the underlying mechanism is elusive. Here, we compared the transcriptomic profiles of both young and aged BAT in response to thermogenic stimuli. The profiles were extracted from the GEO database. Intriguingly, aging does not cause transcriptional changes in thermogenic genes but upregulates several pathways related to the immune response and downregulates metabolic pathways. Acute severe CE upregulates several pathways related to protein folding. Chronic mild CE upregulates metabolic pathways, especially related to carbohydrate metabolism. Our findings provide a better understanding of the effects of aging and metabolic responses to thermogenic stimuli in BAT at the transcriptome level.

Keywords: brown adipose tissue; transcriptome; cold exposure; aging

Citation: Kim, Y.; Kang, B.E.; Ryu, D.; Oh, S.W.; Oh, C.-M. Comparative Transcriptome Profiling of Young and Old Brown Adipose Tissue Thermogenesis. *Int. J. Mol. Sci.* **2021**, *22*, 13143. https://doi.org/10.3390/ijms222313143

Academic Editors: Amelia Casamassimi, Alfredo Ciccodicola and Monica Rienzo

Received: 12 October 2021
Accepted: 3 December 2021
Published: 5 December 2021

Publisher's Note: MDPI stays neutral with regard to jurisdictional claims in published maps and institutional affiliations.

1. Introduction

Brown adipose tissue (BAT) is a specialized site for uncoupling protein 1 (UCP1)-mediated non-shivering thermogenesis [1]. BAT dissipates chemical energy via heat generation to maintain the optimal body temperature against cold exposure and increases energy expenditure in response to excessive feeding [2]. Recent technical advances in the field of energy metabolism have revealed that the thermogenesis in BAT uses large amounts of intracellular triglycerides and glucose as the energy source [3,4], and thus activating the thermogenesis of BAT is a promising target for the treatment of obesity and related diseases, such as diabetes, dyslipidemia, and cardiovascular diseases [1,5]. Several human studies reported that BAT activities were affected by weather and climate [6–8]. There were strong associations between weather and BAT activity [6,8], and the Inuit, who live in the Arctic region, have genetic variants related to heat generation in BAT [7].

The amount of BAT and its thermogenic activity decrease with aging in both mice and humans [9]. Brown-like adipocytes in white adipose tissue (WAT), as well as brown adipocytes in BAT lose their thermogenic characteristics with aging [10]. Imaging studies using 18-fluorodeoxyglucose (18F-FDG) positron emission tomography (PET-CT) have revealed that young people have a higher stimulated/non-stimulated BAT ratio in the cervical-supraclavicular region than aged people [11]. Interscapular BAT abundantly exists in children aged <10 years but is dispersed in adults [12]. In humans, >90% metabolically active BAT is lost in their 50s and 60s [13].

Several approaches, such as cold exposure (CE), exercise, and beta 3-adrenergic receptor (β3-AR) agonists have been tried to activate the thermogenic activities of BAT [14–18]. Although their sample sizes were small, β3-AR stimulation has shown clinical benefits in clinical trials. Through acute administration of the mirabegron, the β3-AR agonist activated BAT metabolic activity and white adipose tissue (WAT) lipolysis in humans [16]. Chronic mirabegron therapy increased BAT activity and improved glucose homeostasis in both healthy and obese humans [17,18].

Although some human and animal studies have reported the metabolic benefits of activating BAT through these approaches [8,14,17,18], most clinical trials have been performed on young adults (ClinicalTrials.gov: NCT03793127, NCT03049462, and NCT02236962). In aged people, thermogenic stimuli do not activate the thermogenic activities of BAT [13]. Takeshi et al. identified cold-activated BAT by using PET-CT and 162 adult healthy volunteers aged 20–73 years [13]. In the same study, the incidence of activated BAT after 2 h of exposure to 19 °C was found to be 53% (44/83), 12.5% (1/8), and 0% (0/7) in humans during their 20s, 50s, and 60s, respectively. The underlying mechanism of this decline has not yet been elucidated.

Most chronic metabolic diseases, such as type-2 diabetes, develop with age [19]. Thus, the decline in the thermogenic activity and response to thermogenic stimuli with age is the major hurdle in harnessing BAT thermogenesis as a novel therapeutic strategy against metabolic disease. In this study, we compared the transcriptomic profiles of the BAT in both young and old mice in response to CE to find the molecular mechanisms underlying age-related dysfunctions in BAT. In addition, we compared the transcriptomic changes in BAT response to thermogenic stimuli such as acute severe CE (ACE), chronic mild CE (CCE), high-fat diet (HFD), and β3-adrenergic receptor (β3-AR) agonist *CL316243* treatment in young and old mice to determine which thermogenic stimulus is better.

2. Results

2.1. Comparison of BAT Transcriptome Profiles in Aging and Adaptive Thermogenesis

The gene expression profile of BAT was analyzed using transcriptome datasets of ACE studies (GSE135391), CCE study (GSE172021), HFD induced thermogenesis study (GSE112740), and β3-AR stimulation (GSE98132). First, we analyzed the differentially expressed genes (DEGs) between young and aged BAT at room temperature (RT). We identified 438 upregulated and 366 downregulated genes (Adjusted *p*-value < 0.05, Figure 1A). Figure 1B shows the top 10 upregulated and top 10 downregulated genes ranked by fold change. *Cyp2b10* (cytochrome P450 2B10), *Peg3* (paternally expressed gene 3), and *Mfsd2a* (major facilitator superfamily domain-containing 2A) are highly upregulated in aged BAT compared with the levels in young BAT. *Ttn* (titin), *Neb* (nebulin), and *Ttc25* (tetratricopeptide repeat protein 25) genes are mostly downregulated in aged BAT compared with the levels in young BAT. The *Cyp2b10, Peg3, Mfsd2a,* and *Ttn* genes have previously been reported to play critical roles in adipocyte identity and metabolism [20–23]. However, the *Neb* and *Ttc25* genes are novel genes associated with BAT aging. Further studies are needed to investigate the role of these two genes in BAT.

In young BAT, ACE (4 °C for 24 h) upregulated 444 genes and downregulated 266 genes (Figure 1C). CCE (gradual decrease from 23 °C to 10 °C, then 2 weeks of exposure) upregulated 514 genes and downregulated 369 genes (Figure 1D). Among the 444 upregulated genes upon ACE, only 33 are also upregulated upon CCE. Among the 266 downregulated genes upon ACE, only 25 are also downregulated upon CCE (Supplementary Figure S1). These small numbers of common genes suggest that each stress might trigger quite different signal responses for activating the BAT activity.

A total of 788 genes were found to be differentially expressed, with 408 upregulated and 380 downregulated genes, between young and aged BAT upon ACE (Figure 1E). In old mouse BAT, ACE was found to upregulate 510 genes and downregulate 437 genes (Figure 1F). When we compared these DEGs with those between young and aged BAT at RT, 13 downregulated genes in aged BAT compared with young BAT at RT were found to be upregulated in aged BAT upon ACE. Additionally, 29 upregulated genes in old BAT compared with young BAT at RT were found to be downregulated in aged BAT upon ACE (Supplementary Figure S2).

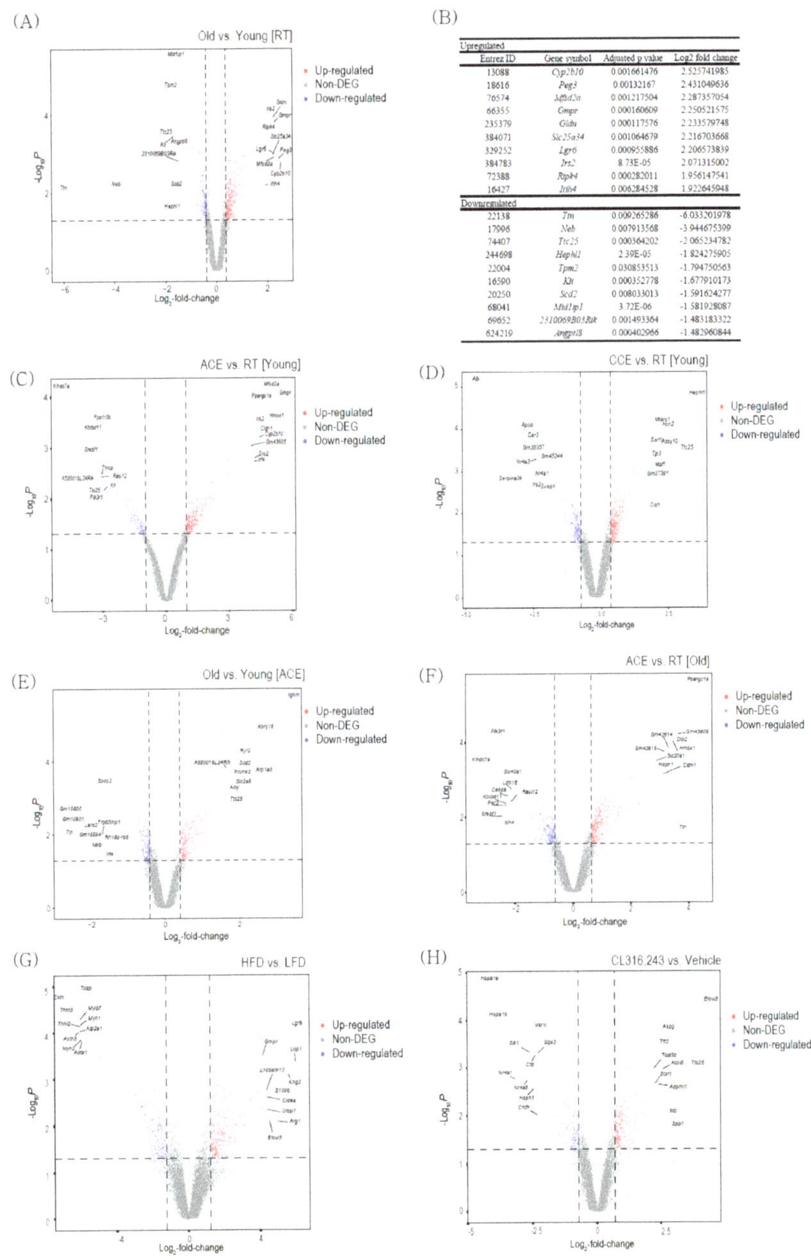

(B)

Upregulated			
Entrez ID	Gene symbol	Adjusted p value	Log2 fold change
13088	Cyp2b10	0.001661476	2.525741985
18616	Peg3	0.00132167	2.431049636
76574	Mfsd2a	0.001217504	2.287357054
66355	Gmpr	0.000160609	2.250521575
235379	Gldn	0.000117576	2.233579748
384071	Slc25a34	0.001064679	2.216703668
329252	Lgr6	0.000955896	2.206573839
384783	Irs2	8.73E-05	2.071315002
72388	Rtpk4	0.000282011	1.956147541
16427	Itih4	0.006284528	1.922645948
Downregulated			
22138	Tm	0.009265286	-6.033201978
17996	Neb	0.007913568	-3.944675399
74407	Ttc25	0.000364202	-2.065234782
244698	Hephl1	2.39E-05	-1.824275905
22004	Tpm2	0.030853513	-1.794750563
16590	Kit	0.000352778	-1.677910173
20250	Scd2	0.008033013	-1.591624277
68041	Mid1ip1	3.72E-06	-1.581929087
69652	2310069B03Rik	0.001493364	-1.483183322
624219	Angptl8	0.000402966	-1.482960844

Figure 1. Differentially expressed genes (DEGs) between young and old brown adipose tissues (BAT). (**A**) Volcano plot of the DEGs between old vs. young BAT (GSE135391). (**B**) The list of top 10 upregulated and downregulated DEGs between old vs. young BAT. (**C**) Volcano plot of the DEGs

between acute severe cold exposure (ACE) vs. room temperature (RT) in young BAT (GSE135391). (**D**) Volcano plot of the DEGs between chronic mild cold exposure (ACE) vs. room temperature (RT) in young BAT (GSE172021). (**E**) Volcano plot of the DEGs between old vs. young BAT upon ACE (GSE135391). (**F**) Volcano plot of the DEGs between ACE vs. RT in old BAT (GSE135391). (**G**) Volcano plot of the DEGs between high-fat diet (HFD) vs. low-fat diet (LFD) in young BAT (GSE112740). (**H**) Volcano plot of the DEGs between CL316243 treatment vs. vehicle treatment in young BAT (GSE98132).

A total of 948 genes were found to be differentially expressed, with 399 upregulated and 549 downregulated genes, between HFD and low-fat diet in young BAT (Figure 1G). The thermogenic genes such as *Ucp1, Cidea,* and *Elovl3* were upregulated after HFD feeding in young BAT (Figure 1G). After β3-AR treatment, 381 genes were upregulated, and 326 genes were downregulated (Figure 1H). β3-AR treatment also increased the thermogenic gene *Elovl3* (Figure 1H).

2.2. Pathway Alterations in BAT

We aimed to determine the biological characteristics of the DEGs in BAT that are associated with aging and/or other thermogenic stimuli. Thus, we performed gene ontology (GO) and Kyoto Encyclopedia of Genes and Genomes (KEGG) pathway analysis (Figure 2) using three ACE (GSE135391, GSE86590, and GSE119452), CCE (GSE172021), HFD (GSE112740), and β3-AR stimulation (GSE98132) transcriptomes.

ACE activates several pathways related to 'response to cold', 'response to stress', and 'brown fat cell differentiation' (Figure 2A). Both ACE and CCE did not induce significant changes in diet-induced thermogenesis (Figure 2A). ACE also activated endoplasmic reticulum (ER)-related protein folding and pathways related to the unfolded protein response in both young and old BAT (Figure 2B). This observation suggests that ACE increases the protein quality control related to cellular stress, and thus ACE cannot be a good candidate therapeutic strategy against the metabolic dysfunctions associated with aging. Figure 2C showed pathway changes related to metabolism. Metabolic pathways related to 'fatty acid metabolism', 'cholesterol metabolic process', and 'insulin signaling pathway' were upregulated in CCE, HFD, and CL316243 treatment. Figure 2D showed changes in signaling pathways. ACE upregulates 'apoptotic process' and 'MAPK signaling pathway'.

BAT also plays a role as an endocrine organ that controls whole-body glucose and lipid metabolism by secreting adipokines, which are called 'batokines' [24]. To assess batokine secretion by thermogenic stimulation, we analyzed the expressions of 'batokine' genes (Figure 2E). Bone morphogenetic protein 8B (*BMB8B*) gene was increased after ACE, CCE, HFD, and β3-AR agonist stimulation. Fibroblast growth factor 21 (*FGF21*) was increased only 48 h after acute CE in young BAT. In old BAT, most genes did not show significant changes after thermogenic stimulation (Figure 2E).

2.3. Mitochondrial Gene Expression in Brown Adipose

To evaluate the mitochondrial changes with aging and CE in BAT, we analyzed the expressions of genes related to mitochondrial proteome in the Mitocarta 3.0 gene list [25]. Figure 3 shows heatmaps composed of the DEGs related to mitochondria in BAT. The rows of each heatmap represent mitochondrion-related genes with significantly changed expression levels based on fold change, and the columns are the comparative result of each group.

Figure 3A showed DEGs related to protein homeostasis and Figure 3B showed the DEGs related to mitochondrial dynamics. Figure 3C showed the DEGs related to nucleotide metabolism. Both acute CE and chronic CE induces various changes in the expressions of genes related to mitochondrial proteostasis, dynamics, and nucleotide metabolism.

Figure 3D–F shows DEGs related nutrients metabolism. Chronic CE upregulated genes related to carbohydrate metabolism and amino acid metabolism (Figure 3D,E). Lipid metabolism-related genes were upregulated in both acute CE and chronic CE (Figure 3F). Thioesterase superfamily member 4 (*Them4*) and glycerol-3-phosphate acyltransferase

(*Gpam*) were commonly increased in the acute CE dataset (Figure 3F). Regarding the Fe-S cluster, only a few genes were differentially expressed in both acute CE and chronic CE (Figure 3G).

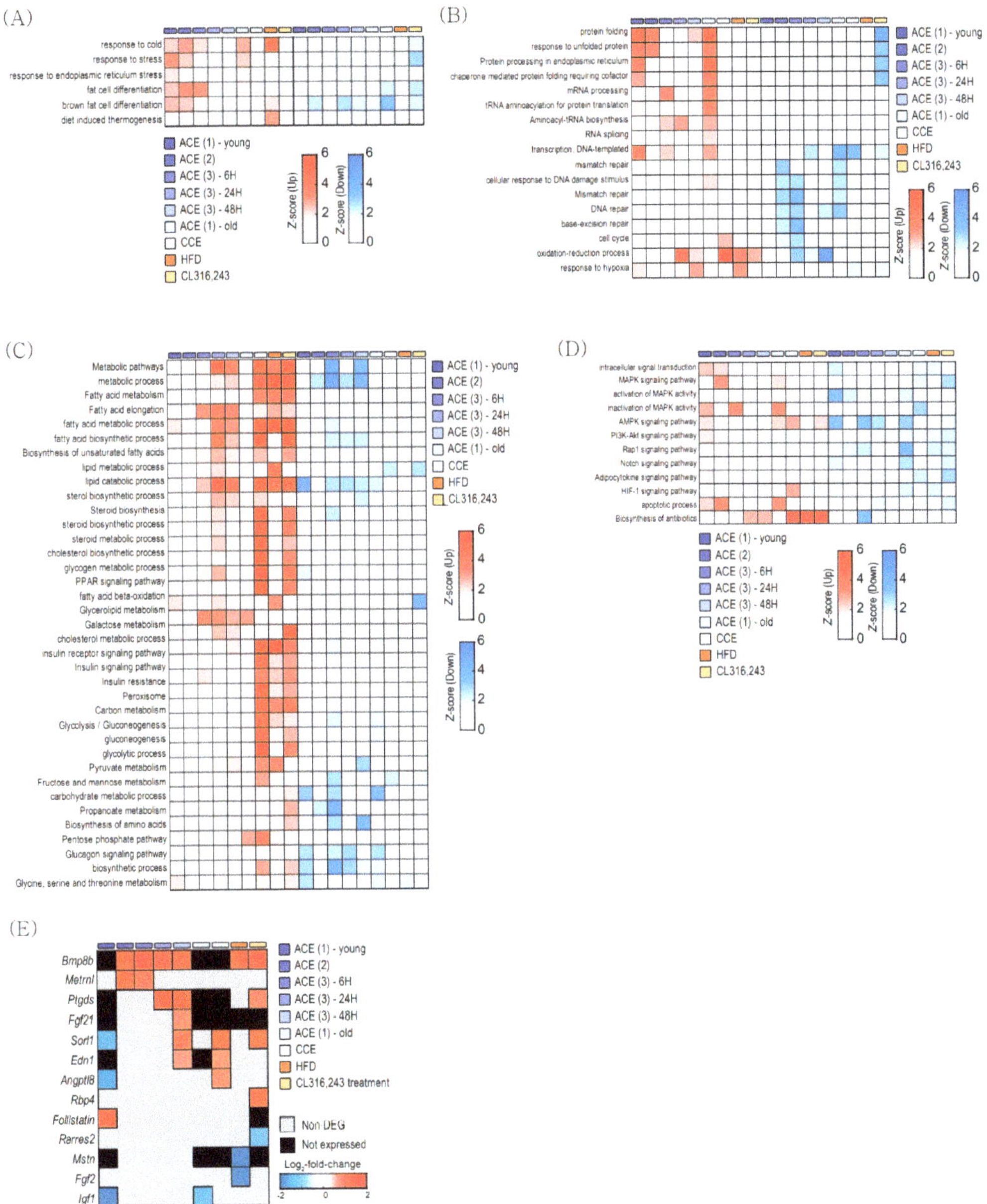

Figure 2. Clustered heatmap of functional enrichment analysis. (**A**) Pathways related to brown fat activity and stress. (**B**) Pathways related to protein processing. (**C**) Pathways related to metabolism. (**D**) Pathways related to signaling. (**E**) Genes related to adipokine secretion in brown adipose tissue. ACE, acute severe cold exposure; CCE, chronic mild cold exposure; HFD, high-fat diet.

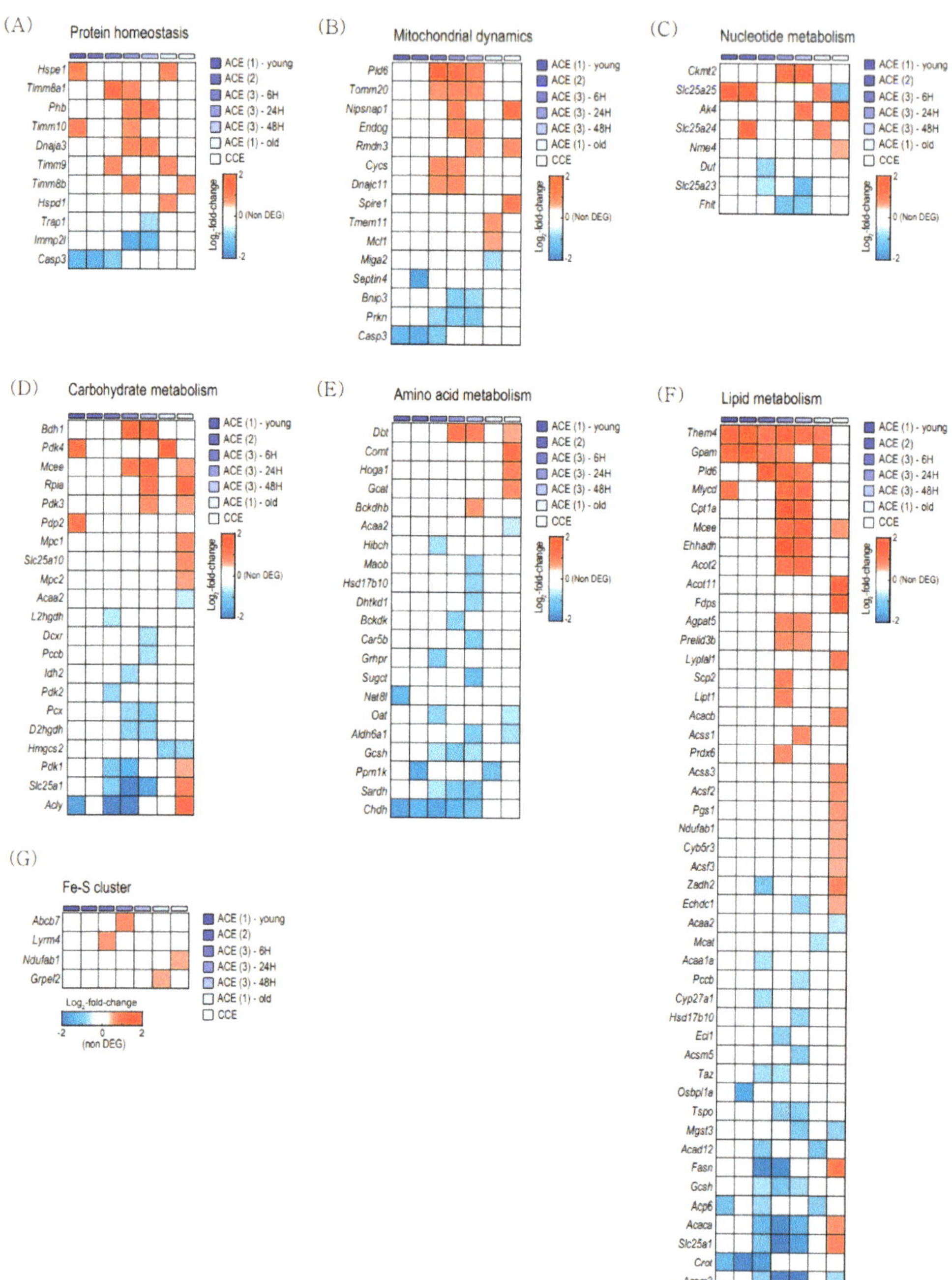

Figure 3. Heatmap visualization of differentially expressed mitochondrial genes (DEMGs) in brown adipose tissue (BAT). (**A**) DEMGs related to protein homeostasis. (**B**) DEMGs related to mitochondrial dynamics. (**C**) DEMGs related to nucleotide metabolism. (**D**) DEMGs related to carbohydrate metabolism. (**E**) DEMGs related to amino acid metabolism. (**F**) DEMGs related to lipid metabolism. (**G**) DEMGs related to the Fe-S cluster. ACE, acute severe cold exposure; CCE, chronic mild cold exposure.

2.4. Cold-Induced Changes in BAT

To determine specific changes between young and aged BAT after CE, we compared DEGs between ACE and RT in young BAT with DEGs in old BAT (GSE13591) (Figure 4). A total of 284 genes are commonly upregulated, and 132 genes are commonly downregulated in both DEGs (Figure 4A). Figure 4B shows the top genes related to protein folding in both DEGs. Heat shock protein family H (*Hsp110*) Member 1 (*Hsph1*), heat shock protein 90 alpha family class A Member 1 (*Hsp90aa1*), and heat shock protein family A Member 8 (*Hspa8*) are the top three upregulated genes in common DEGs.

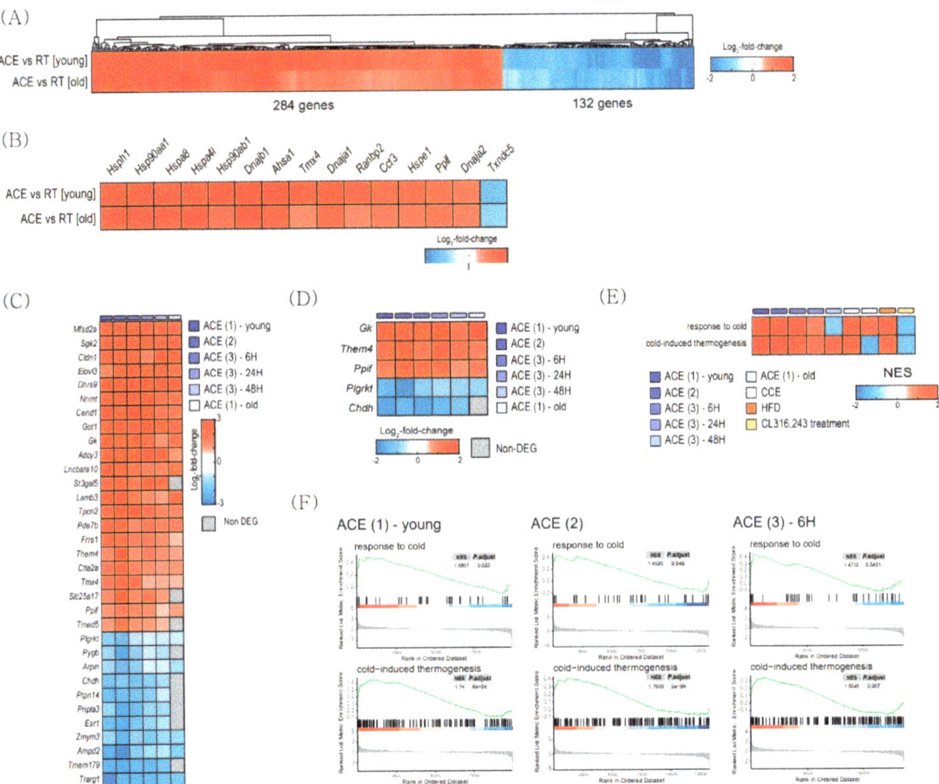

Figure 4. Cold-exposure-induced changes in gene expression in brown adipose tissue (BAT). (**A**) Heatmap of the common differentially expressed genes (DEGs) between acute severe cold exposure (ACE) vs. room temperature (RT) in young BAT, and DEGs between in old BAT (GSE135391). (**B**) Heatmap of the common DEGs related to protein folding pathway between ACE vs. RT in young BAT and DEGs between ACE vs. RT in aged BAT. (**C**) Common DEGs after CE in both young and old BAT from 3 datasets (GSE13591, GSE86590, and GSE119452). (**D**) Common mitochondrial DEGs from 4 datasets. (**E,F**) Gene set enrichment analysis (GSEA) result using all datasets. Heatmap (**E**) and enrichment plots (**F**) plots related to 'response to cold' and 'cold-induced thermogenesis.

To analyze the general features related to CE, we compared DEGs from the other two datasets, GSE86590 and GSE119452. Figure 4C,D shows common DEGs and mitochondrial DEGs, respectively. Three mitochondrial genes, glycerol kinase (*Gk*), *Them4*, and peptidyl-prolyl isomerase F (*Ppif*) gene were upregulated after CE in young and old BAT (Figure 4D). The functional enrichment analysis showed that pathways related to 'response to cold' and 'cold-induced thermogenesis' were upregulated in both ACE and CCE (Figure 4E,F).

2.5. Chronic Mild Cold Exposure Changes Metabolic Pathways in Brown Adipose Tissue

ACE and CCE have demonstrated beneficial effects in mice and humans by activating BAT thermogenesis [8,26,27]. Interestingly, DEG analysis (Figure 1) and pathway analysis (Figure 2) suggest that ACE and CCE use different signaling pathways for BAT activation. Thus, we next compared gene expressions and pathways between ACE and CCE using GSE135391 and GSE1127140 datasets (Figure 5). Figure 5A shows common genes in DEGs between ACE and RT in young BAT, DEGs between ACE and RT in old BAT, and DEGs between CCE and RT in young BAT. HSPs such as Hsph1, Hspa4l (heat shock protein family A (Hsp70) Member 4-like) and Hspb8 (heat shock protein family B (small) Member 8) are common upregulated DEGs (Figure 5A). Ebf2 (*early B-cell factor 2*) is a commonly downregulated DEG (Figure 5A). This gene is a specific marker gene for brown preadipocytes and drives brown adipocyte differentiation [28]. C/EBPα (CCAAT/enhancer-binding protein alpha) is also a commonly decreased DEG in CE, which triggers differentiation of white preadipocytes in mature white adipocytes [29].

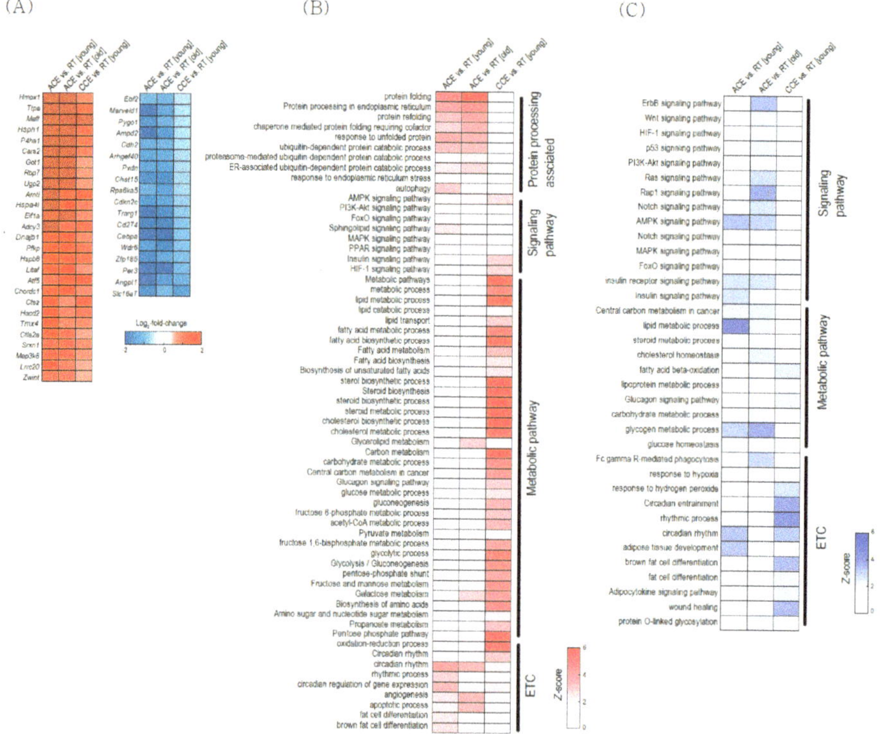

Figure 5. Transcriptional changes after chronic mild cold exposure (CCE) in brown adipose tissue (BAT). (**A**) Heatmap visualization of common differentially expressed genes (DEGs) between acute severe cold exposure (ACE) vs. room temperature (RT) in young BAT and DEGs between ACE vs. RT in aged BA and DEGs between CCE vs. RT in young BAT. (**B**,**C**) Gene Ontology Biologic Process (GOBP) and Kyoto Encyclopedia of Genes and Genomes (KEGG) pathway analysis result. Clustered heatmap of upregulated (**B**) and downregulated (**C**) pathways.

Figure 5B shows common upregulated pathways and Figure 5C shows common downregulated pathways. ACE activates pathways related to protein processing, such as protein folding, protein ubiquitination, and unfolded protein response. However, CCE did not upregulate protein processing associated pathways and CCE upregulates metabolic pathways such as glucose and lipid metabolism.

3. Discussion

In this study, we analyzed transcriptomic profiles of BAT with aging and/or thermogenic stimuli such as CE, HTN, and β3-AR agonist. Old BAT showed decreased gene expression related to the lipid metabolism, such as stearoyl-CoA desaturase 2 (*Scd2*) and angiopoietin-like 8 (*Angptl8*), and increased gene expression related to obesity in WAT, such as *Peg3* [30]. Interestingly, 118 genes upregulated with aging are also upregulated in young BAT after ACE (Supplementary Figure S3). Even among the top ten upregulated genes in aged BAT compared with young BAT at RT (Figure 1B), three genes (*Mfsd2a*, GMP reductase (*GMPR*) [21], solute carrier family 25 member 34 (*Slc25a34*) [31]) are known as elevated genes related to CE in BAT. The upregulation of genes related to CE in old BAT might be a compensatory response to metabolic dysfunctions with aging, or this observation means that age-related stress might trigger a similar signaling pathway related to that induced by CE.

The *Ttn* gene, which is downregulated in old BAT at RT, is one of the most downregulated genes in old BAT upon ACE (Figure 1B). This gene encodes a large protein called titin, which is an essential component of sarcomeres and plays an important role in muscle development [32]. BAT and the skeletal muscle arise from a common precursor (myf5-expressing precursor) cell. Accordingly, BAT and the skeletal muscle have been shown to share many genes as key regulators of their structures and functions [33,34].

Several studies have already reported the possible role of *Ttn* in adipose tissue. *Ttn* is significantly upregulated in the visceral adipose tissue of obese people, compared with the level in lean people [35]. Additionally, it is significantly downregulated in the BAT of obesity-prone rats, compared with the level in wild-type rats [36]. This finding suggests that *Ttn* might be a novel regulator of BAT. Further studies are needed to reveal the exact role of *Ttn* in the age-related changes of BAT.

Intriguingly, thermogenic genes, such as uncoupling protein 1 (*Ucp1*), cell death-inducing DNA fragmentation factor alpha-like effector A (*Cidea*), cytochrome C oxidase subunit 8B (*Cox8b*), and ELOVL fatty acid elongase 3 (*Elovl3*), did not show significant differences in expression level between young and old BAT at RT. This observation suggests that decreased thermogenic activity with aging might result from the dysfunction of other core genes or the post-transcriptional changes of thermogenic genes in BAT.

ACE significantly increased both *UCP1* and *PGC-1α* (Supplementary Figure S4). Intriguingly, CCE did not upregulate *PGC-1α* and CL316,243 treatment did not increase both *UCP1* and *PGC-1α*. *Cidea* increased only in HFD-induced thermogenesis. These results suggest that each stimulation may activate BAT through different thermogenic pathways.

The functional enrichment analysis revealed that CE increased several pathways related to response to cold and stress in both young and old BAT (Figure 2). Both ACE and CCE also induced many changes related to mitochondrial functions in young and old BAT (Figure 3). However, pathways related to metabolism were increased in young BAT but not old BAT (Figure 2C). In old BAT, ACE increased only one gene related to carbohydrate metabolism, no gene related to amino acid metabolism, and two genes related to lipid metabolism (Figure 3D–F). CCE showed more upregulated pathways related to the carbohydrate and amino acid metabolism than ACE in young BAT. These findings suggest CCE might be an effective therapeutic strategy for improving metabolic dysfunction in old BAT.

Aging did not cause transcriptional changes in thermogenic genes. CE activates thermogenesis-related genes in both young and aged BAT (Figure 4A,F). This observation means that a decrease in thermogenic gene expression is not the underlying cause of the reduced thermogenesis in aged BAT. Recently, Kazuki et al. also reported that the age-related impairment of BAT thermogenesis is not significantly associated with thermogenic genes and suggested post-translational regulated mitochondrial impairment, especially related to Fe-S cluster formation as a new underlying mechanism of BAT dysfunction with aging [37]. Among Fe-S cluster formation-related genes, ACE increased GrpE-like

2 (*Grpel2*) expression in old BAT (Figure 3G), which is a redox-sensitive protein against oxidative stress [38].

Pathway analysis revealed that ACE activated protein-folding-related pathways in both young and old BAT (Figures 2 and 5). *Hsph1*, *Hsp90aa1* and *Hspa8* are the top three commonly upregulated genes in young and old BAT after ACE (Figure 4B). *Hsph1* encodes a member of the heat shock protein 70 family of proteins and this protein is a known marker of both human and mouse brown adipocytes [39]. *Hsp90aa1* encodes heat shock protein 90α, which is the isoform of the molecular chaperone Hsp90 [40]. This protein plays a role in lipid metabolism [41]. *Hspa8* encodes a member of the heat shock protein 70 family, which interacts with negative charged phospholipids and acts as a membrane chaperone [42]. These increases might be the result of protective responses to cold-induced stress in BAT [43], or these HSPs may participate in the BAT metabolism directly, because HSPs are specific molecular chaperones that play various roles in metabolism [44] as well as protein quality control [45].

Interestingly, circadian-rhythm–related pathways are changed according to aging and CE (Figure 5). Recently, many studies have reported that the circadian rhythm regulates energy metabolism, and chronodisruption by chronic desynchronization of circadian rhythms has detrimental effects on adipose tissue function and differentiation [46,47]. Thus, our finding suggests that normalization of the circadian disruption can be an effective strategy to treat the BAT dysfunctions associated with aging.

Our study has several limitations. First, we did not obtain the transcriptome profile of aged BAT upon CCE. Comparison of the metabolic effects of ACE and CCE in aged BAT may provide us with a better understanding of the reduced thermogenic response to CE in aged BAT. Second, we used only transcriptome data for the evaluation of metabolic changes related to aging and CE. Integrated approaches using transcriptome with proteomics and metabolomics are needed to understand the age-related changes in BAT and confirm our analysis. Third, we used publicly available transcriptome for our study. Our results need to be verified by other methods such as real-time PCR. Further validation studies are needed.

In conclusion, our findings provided a better understanding of the various effects of thermogenic stimuli on BAT. Metabolic pathways, especially carbohydrate metabolism-related pathways, are more upregulated in BAT under CCE than other stimuli. Thus, CCE might be a better strategy for increasing metabolic activities and improving glucose homeostasis than other activating strategies in BAT. Further studies are needed to investigate the role of CCE in old BAT.

4. Material and Methods

4.1. RNA-Seq Analysis of NCBI Gene Omnibus (GEO) Datasets

Both RNA-seq data of ACE and CCE are deposited in GSE135391, GSE86590, GSE119452, and GSE172021. We used the GSE1127440 dataset for HFD-induced thermogenesis in BAT and GSE98133 dataset for β3-AR agonist-stimulated thermogenesis (Figure 6).

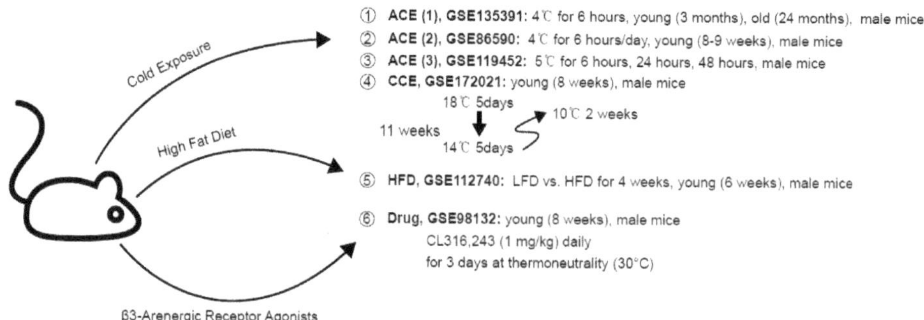

Figure 6. Information about transcriptome data that are used in this study.

4.2. Identification of DEGs

Raw data were processed to calculate counts per million (cpm) through 'edgeR' [48] package in R software package (version 4.0.0 for Windows; http://cran.r-project.org, accessed on 11 October 2021), and those were converted to $\log_2$ scale followed by normalization using quantile normalization. Then, to identify DEGs from several datasets, we applied the integrative statistical method to the normalized $\log_2$-cpm. For each gene, to calculate the observed T value and $\log_2$-median-ratio between two conditions, Student's *t*-test and $\log_2$-median-ratio were conducted, respectively. Then, random sampling 1000 times was performed to generate empirical null distributions for T value and $\log_2$-median-ratio.

To generate the overall *p*-value for each gene, we calculate adjusted *p*-values by applying a two-tailed test for the measured T value and $\log_2$-median-ratio through their corresponding empirical distributions. Then, adjusted *p* values were combined into an overall *p*-value using Stouffer's method [49]. For each comparison, we selected DEGs through two criteria: its overall *p*-value < 0.05 and the absolute $\log_2$-median-ratio > the mean of 2.5th and 97.5th percentiles of the empirical distribution for the $\log_2$-median-ratio (Supplementary Table S1).

Venn diagrams which showed the comparison of DEGs between datasets were made through Oliveros, J.C. (2007–2015) Venny, an interactive tool for comparing listed data with Venn diagrams at 20-08-2021 (https://bioinfogp.cnb.csic.es/tools/venny/index.html, accessed on 11 October 2021).

4.3. Pathway Analysis: Kyoto Encyclopeida of Genes and Genomes (KEGG) and Gene Ontology Biologic Process (GOBP) Analysis

We conducted the functional enrichment analysis of the DEGs by using DAVID software [50]. Our GOBPs, KEGGs were selected by applying two criteria: its *p*-value < 0.05 and count of genes > 3.

4.4. Gene Set Enrichment Analysis (GSEA)

To assess the enrichment in our DEGs upon cold exposure in young and aged, we used GSEA through "clusterProfiler" in R software (version 4.0) [51]. We conducted GSEA analysis with 10,000 permutations, minGSSize = 3, maxGSSize = 800 and pvaluecutoff = 0.05.

Supplementary Materials: The following are available online at https://www.mdpi.com/article/10.3390/ijms222413143/s1.

Author Contributions: Data search, Y.K. and B.E.K.; conceptualization and writing, Y.K., D.R. and C.-M.O.; supervision, S.W.O. and C.-M.O. All authors have read and agreed to the published version of the manuscript.

Funding: This research was supported by the Basic Science Research Program through the National Research Foundation of Korea (NRF), funded by the Ministry of Education (2020R1C1C1004999 to C.-M.O.), and supported by Seoul National University Hospital Research Fund (03-2016-0130 to S.W.O.)

Institutional Review Board Statement: Not applicable.

Informed Consent Statement: Not applicable.

Data Availability Statement: Data analyzed in this study were a re-analysis of existing data, which are openly available at locations cited in the reference section.

Conflicts of Interest: The authors declare no conflict of interest.

References

1. Kajimura, S.; Saito, M. A new era in brown adipose tissue biology: Molecular control of brown fat development and energy homeostasis. *Ann. Rev. Physiol.* **2014**, *76*, 225–249. [CrossRef] [PubMed]
2. Saito, M. Brown adipose tissue as a regulator of energy expenditure and body fat in humans. *Diabetes Metab. J.* **2013**, *37*, 22–29. [CrossRef] [PubMed]

3. Carpentier, A.C.; Blondin, D.P.; Virtanen, K.A.; Richard, D.; Haman, F.; Turcotte, É.E. Brown adipose tissue energy metabolism in humans. *Front. Endocrinol.* **2018**, *9*, 447. [CrossRef]
4. Zhang, F.; Hao, G.; Shao, M.; Nham, K.; An, Y.; Wang, Q.; Zhu, Y.; Kusminski, C.M.; Hassan, G.; Gupta, R.K. An adipose tissue atlas: An image-guided identification of human-like BAT and beige depots in rodents. *Cell Metab.* **2018**, *27*, 252–262. e253. [CrossRef] [PubMed]
5. Becher, T.; Palanisamy, S.; Kramer, D.J.; Eljalby, M.; Marx, S.J.; Wibmer, A.G.; Butler, S.D.; Jiang, C.S.; Vaughan, R.; Schöder, H. Brown adipose tissue is associated with cardiometabolic health. *Nat. Med.* **2021**, *27*, 58–65. [CrossRef] [PubMed]
6. Au-Yong, I.T.; Thorn, N.; Ganatra, R.; Perkins, A.C.; Symonds, M.E. Brown adipose tissue and seasonal variation in humans. *Diabetes* **2009**, *58*, 2583–2587. [CrossRef] [PubMed]
7. Fumagalli, M.; Moltke, I.; Grarup, N.; Racimo, F.; Bjerregaard, P.; Jørgensen, M.E.; Korneliussen, T.S.; Gerbault, P.; Skotte, L.; Linneberg, A. Greenlandic Inuit show genetic signatures of diet and climate adaptation. *Science* **2015**, *349*, 1343–1347. [CrossRef] [PubMed]
8. van der Lans, A.A.; Wierts, R.; Vosselman, M.J.; Schrauwen, P.; Brans, B.; van Marken Lichtenbelt, W.D. Cold-activated brown adipose tissue in human adults: Methodological issues. *Am. J. Physiol. Regul. Integr. Comp. Physiol.* **2014**, *307*, R103–R113. [CrossRef] [PubMed]
9. Cannon, B.; Nedergaard, J. Brown adipose tissue: Function and physiological significance. *Physiol. Rev.* **2004**, *84*, 277–359. [CrossRef] [PubMed]
10. Graja, A.; Schulz, T.J. Mechanisms of aging-related impairment of brown adipocyte development and function. *Gerontology* **2015**, *61*, 211–217. [CrossRef] [PubMed]
11. Saito, M.; Okamatsu-Ogura, Y.; Matsushita, M.; Watanabe, K.; Yoneshiro, T.; Nio-Kobayashi, J.; Iwanaga, T.; Miyagawa, M.; Kameya, T.; Nakada, K. High incidence of metabolically active brown adipose tissue in healthy adult humans: Effects of cold exposure and adiposity. *Diabetes* **2009**, *58*, 1526–1531. [CrossRef]
12. Heaton, J.M. The distribution of brown adipose tissue in the human. *J. Anat.* **1972**, *112*, 35.
13. Yoneshiro, T.; Aita, S.; Matsushita, M.; Okamatsu-Ogura, Y.; Kameya, T.; Kawai, Y.; Miyagawa, M.; Tsujisaki, M.; Saito, M. Age-related decrease in cold-activated brown adipose tissue and accumulation of body fat in healthy humans. *Obesity* **2011**, *19*, 1755–1760. [CrossRef]
14. da Silva, C.P.V.; Hernández-Saavedra, D.; White, J.D.; Stanford, K.I. Cold and exercise: Therapeutic tools to activate brown adipose tissue and combat obesity. *Biology* **2019**, *8*, 9. [CrossRef] [PubMed]
15. Cypess, A.M.; Weiner, L.S.; Roberts-Toler, C.; Elía, E.F.; Kessler, S.H.; Kahn, P.A.; English, J.; Chatman, K.; Trauger, S.A.; Doria, A.; et al. Activation of human brown adipose tissue by a β3-adrenergic receptor agonist. *Cell Metab.* **2015**, *21*, 33–38. [CrossRef]
16. Baskin, A.S.; Linderman, J.D.; Brychta, R.J.; McGehee, S.; Anflick-Chames, E.; Cero, C.; Johnson, J.W.; O'Mara, A.E.; Fletcher, L.A.; Leitner, B.P. Regulation of human adipose tissue activation, gallbladder size, and bile acid metabolism by a β3-adrenergic receptor agonist. *Diabetes* **2018**, *67*, 2113–2125. [CrossRef] [PubMed]
17. O'Mara, A.E.; Johnson, J.W.; Linderman, J.D.; Brychta, R.J.; McGehee, S.; Fletcher, L.A.; Fink, Y.A.; Kapuria, D.; Cassimatis, T.M.; Kelsey, N. Chronic mirabegron treatment increases human brown fat, HDL cholesterol, and insulin sensitivity. *J. Clin. Investig.* **2020**, *130*, 2209–2219. [CrossRef]
18. Finlin, B.S.; Memetimin, H.; Zhu, B.; Confides, A.L.; Vekaria, H.J.; El Khouli, R.H.; Johnson, Z.R.; Westgate, P.M.; Chen, J.; Morris, A.J. The β3-adrenergic receptor agonist mirabegron improves glucose homeostasis in obese humans. *J. Clin. Investig.* **2020**, *130*, 2319–2331. [CrossRef]
19. Spinelli, R.; Parrillo, L.; Longo, M.; Florese, P.; Desiderio, A.; Zatterale, F.; Miele, C.; Raciti, G.A.; Beguinot, F. Molecular basis of ageing in chronic metabolic diseases. *J. Endocrinol. Investig.* **2020**, *10*, 1373–1389. [CrossRef]
20. Harms, M.J.; Ishibashi, J.; Wang, W.; Lim, H.-W.; Goyama, S.; Sato, T.; Kurokawa, M.; Won, K.-J.; Seale, P. Prdm16 is required for the maintenance of brown adipocyte identity and function in adult mice. *Cell Metab.* **2014**, *19*, 593–604. [CrossRef]
21. Angers, M.; Uldry, M.; Kong, D.; Gimble, J.M.; Jetten, A.M. Mfsd2a encodes a novel major facilitator superfamily domain-containing protein highly induced in brown adipose tissue during fasting and adaptive thermogenesis. *Biochem. J.* **2008**, *416*, 347–355. [CrossRef]
22. Curley, J.; Pinnock, S.; Dickson, S.; Thresher, R.; Miyoshi, N.; Surani, M.; Keverne, E.B. Increased body fat in mice with a targeted mutation of the paternally expressed imprinted gene Peg3. *FASEB J.* **2005**, *19*, 1302–1304. [CrossRef] [PubMed]
23. Miyano, C. Severe thermoregulatory deficiencies in mice with a gene deletion in titin. Ph.D Thesis, Northern Arizona University, Flagstaff, AZ, USA, 2018.
24. Villarroya, F.; Cereijo, R.; Villarroya, J.; Giralt, M. Brown adipose tissue as a secretory organ. *Nat. Rev. Endocrinol.* **2017**, *13*, 26–35. [CrossRef] [PubMed]
25. Rath, S.; Sharma, R.; Gupta, R.; Ast, T.; Chan, C.; Durham, T.J.; Goodman, R.P.; Grabarek, Z.; Haas, M.E.; Hung, W.H. MitoCarta3. 0: An updated mitochondrial proteome now with sub-organelle localization and pathway annotations. *Nucleic Acids Res.* **2021**, *49*, D1541–D1547. [CrossRef]
26. Lu, X.; Solmonson, A.; Lodi, A.; Nowinski, S.M.; Sentandreu, E.; Riley, C.L.; Mills, E.M.; Tiziani, S. The early metabolomic response of adipose tissue during acute cold exposure in mice. *Sci. Rep.* **2017**, *7*, 3455. [CrossRef] [PubMed]
27. Wang, Z.; Ning, T.; Song, A.; Rutter, J.; Wang, Q.A.; Jiang, L. Chronic cold exposure enhances glucose oxidation in brown adipose tissue. *EMBO Rep.* **2020**, *21*, e50085. [CrossRef]

28. Rajakumari, S.; Wu, J.; Ishibashi, J.; Lim, H.-W.; Giang, A.-H.; Won, K.-J.; Reed, R.R.; Seale, P. EBF2 determines and maintains brown adipocyte identity. *Cell Metab.* **2013**, *17*, 562–574. [CrossRef]
29. Linhart, H.G.; Ishimura-Oka, K.; DeMayo, F.; Kibe, T.; Repka, D.; Poindexter, B.; Bick, R.J.; Darlington, G.J. C/EBPα is required for differentiation of white, but not brown, adipose tissue. *Proc. Natl. Acad. Sci. USA* **2001**, *98*, 12532–12537. [CrossRef] [PubMed]
30. Moraes, R.; Blondet, A.; Birkenkamp-Demtroeder, K.; Tirard, J.; Orntoft, T.; Gertler, A.; Durand, P.; Naville, D.; Bégeot, M. Study of the alteration of gene expression in adipose tissue of diet-induced obese mice by microarray and reverse transcription-polymerase chain reaction analyses. *Endocrinology* **2003**, *144*, 4773–4782. [CrossRef]
31. Hao, Q.; Yadav, R.; Basse, A.L.; Petersen, S.; Sonne, S.B.; Rasmussen, S.; Zhu, Q.; Lu, Z.; Wang, J.; Audouze, K.; et al. Transcriptome profiling of brown adipose tissue during cold exposure reveals extensive regulation of glucose metabolism. *Am. J. Physiol. Endocrinol. Metab.* **2015**, *308*, E380–E392. [CrossRef]
32. Labeit, S.; Kolmerer, B.; Linke, W.A. The giant protein titin: Emerging roles in physiology and pathophysiology. *Circ. Res.* **1997**, *80*, 290–294. [CrossRef] [PubMed]
33. Kong, X.; Yao, T.; Zhou, P.; Kazak, L.; Tenen, D.; Lyubetskaya, A.; Dawes, B.A.; Tsai, L.; Kahn, B.B.; Spiegelman, B.M. Brown adipose tissue controls skeletal muscle function via the secretion of myostatin. *Cell Metab.* **2018**, *28*, 631–643. [CrossRef]
34. Sanchez-Gurmaches, J.; Guertin, D.A. Adipocyte lineages: Tracing back the origins of fat. *Biochim Biophys. Acta Mol. Basis Dis.* **2014**, *1842*, 340–351. [CrossRef] [PubMed]
35. Gómez-Ambrosi, J.; Catalán, V.; Diez-Caballero, A.; Martínez-Cruz, L.A.; Gil, M.J.; García-Foncillas, J.; Cienfuegos, J.A.; Salvador, J.; Mato, J.M.; Frühbeck, G. Gene expression profile of omental adipose tissue in human obesity. *FASEB J.* **2004**, *18*, 215–217. [CrossRef] [PubMed]
36. Joo, J.I.; Yun, J.W. Gene expression profiling of adipose tissues in obesity susceptible and resistant rats under a high fat diet. *Cell. Physiol. Biochem.* **2011**, *27*, 327–340. [CrossRef] [PubMed]
37. Tajima, K.; Ikeda, K.; Chang, H.-Y.; Chang, C.-H.; Yoneshiro, T.; Oguri, Y.; Jun, H.; Wu, J.; Ishihama, Y.; Kajimura, S. Mitochondrial lipoylation integrates age-associated decline in brown fat thermogenesis. *Nat. Metab.* **2019**, *1*, 886–898. [CrossRef]
38. Konovalova, S.; Liu, X.; Manjunath, P.; Baral, S.; Neupane, N.; Hilander, T.; Yang, Y.; Balboa, D.; Terzioglu, M.; Euro, L. Redox regulation of GRPEL2 nucleotide exchange factor for mitochondrial HSP70 chaperone. *Redox Biol.* **2018**, *19*, 37–45. [CrossRef]
39. Shinoda, K.; Luijten, I.H.; Hasegawa, Y.; Hong, H.; Sonne, S.B.; Kim, M.; Xue, R.; Chondronikola, M.; Cypess, A.M.; Tseng, Y.-H. Genetic and functional characterization of clonally derived adult human brown adipocytes. *Nat. Med.* **2015**, *21*, 389–394. [CrossRef]
40. Zuehlke, A.D.; Beebe, K.; Neckers, L.; Prince, T. Regulation and function of the human HSP90AA1 gene. *Gene* **2015**, *570*, 8–16. [CrossRef]
41. Kuan, Y.-C.; Hashidume, T.; Shibata, T.; Uchida, K.; Shimizu, M.; Inoue, J.; Sato, R. Heat shock protein 90 modulates lipid homeostasis by regulating the stability and function of sterol regulatory element-binding protein (SREBP) and SREBP cleavage-activating protein. *J. Biol. Chem.* **2017**, *292*, 3016–3028. [CrossRef]
42. Dores-Silva, P.R.; Cauvi, D.M.; Coto, A.L.; Silva, N.S.; Borges, J.C.; De Maio, A. Human heat shock cognate protein (HSC70/HSPA8) interacts with negatively charged phospholipids by a different mechanism than other HSP70s and brings HSP90 into membranes. *Cell Stress Chaperones* **2021**, *4*, 671–684. [CrossRef] [PubMed]
43. Colinet, H.; Lee, S.F.; Hoffmann, A. Temporal expression of heat shock genes during cold stress and recovery from chill coma in adult Drosophila melanogaster. *FEBS J.* **2010**, *277*, 174–185. [CrossRef]
44. Archer, A.E.; Von Schulze, A.T.; Geiger, P.C. Exercise, heat shock proteins and insulin resistance. *Philos. Trans. R. Soc. B Biol. Sci.* **2018**, *373*, 20160529. [CrossRef]
45. Ciechanover, A.; Kwon, Y.T. Protein quality control by molecular chaperones in neurodegeneration. *Front. Neurosci.* **2017**, *11*, 185. [CrossRef] [PubMed]
46. Froy, O.; Garaulet, M. The circadian clock in white and brown adipose tissue: Mechanistic, endocrine, and clinical aspects. *Endocr. Rev.* **2018**, *39*, 261–273. [CrossRef]
47. Froy, O. Metabolism and circadian rhythms—Implications for obesity. *Endocr. Rev.* **2010**, *31*, 1–24. [CrossRef] [PubMed]
48. Robinson, M.D.; McCarthy, D.J.; Smyth, G.K. edgeR: A Bioconductor package for differential expression analysis of digital gene expression data. *Bioinformatics* **2010**, *26*, 139–140. [CrossRef]
49. Hwang, D.; Rust, A.G.; Ramsey, S.; Smith, J.J.; Leslie, D.M.; Weston, A.D.; De Atauri, P.; Aitchison, J.D.; Hood, L.; Siegel, A.F.; et al. A data integration methodology for systems biology. *Proc. Natl. Acad. Sci. USA* **2005**, *102*, 17296–17301. [CrossRef]
50. Huang, D.W.; Sherman, B.T.; Lempicki, R.A. Systematic and integrative analysis of large gene lists using DAVID bioinformatics resources. *Nat. Protoc.* **2009**, *4*, 44–57. [CrossRef]
51. Subramanian, A.; Tamayo, P.; Mootha, V.K.; Mukherjee, S.; Ebert, B.L.; Gillette, M.A.; Paulovich, A.; Pomeroy, S.L.; Golub, T.R.; Lander, E.S.; et al. Gene set enrichment analysis: A knowledge-based approach for interpreting genome-wide expression profiles. *Proc. Natl. Acad. Sci. USA* **2005**, *102*, 15545–15550. [CrossRef]

International Journal of
Molecular Sciences

Article

Transcriptomic Profiling of Adult-Onset Asthma Related to Damp and Moldy Buildings and Idiopathic Environmental Intolerance

Hille Suojalehto [1], Joseph Ndika [2], Irmeli Lindström [1], Liisa Airaksinen [1], Kirsi Karvala [1,3], Paula Kauppi [4], Antti Lauerma [4], Sanna Toppila-Salmi [4], Piia Karisola [2] and Harri Alenius [2,5,*]

[1] Occupational Medicine, Finnish Institute of Occupational Health, 00032 Helsinki, Finland; Hille.Suojalehto@ttl.fi (H.S.); irmeli.lindstrom@ttl.fi (I.L.); Liisa.Airaksinen@ttl.fi (L.A.); kirsi.karvala@varma.fi (K.K.)

[2] Human Microbiome (HUMI) Research Program, Faculty of Medicine, University of Helsinki, 00014 Helsinki, Finland; joseph.ndika@helsinki.fi (J.N.); Piia.Karisola@helsinki.fi (P.K.)

[3] Varma, 00098 Helsinki, Finland

[4] Skin and Allergy Hospital, Helsinki University Hospital, 00250 Helsinki, Finland; paula.kauppi@hus.fi (P.K.); antti.lauerma@hus.fi (A.L.); sanna.salmi@helsinki.fi (S.T.-S.)

[5] Institute of Environmental Medicine (IMM), Karolinska Institutet, 171 77 Stockholm, Sweden

* Correspondence: Harri.Alenius@helsinki.fi; Tel.: +358-50-4489526

Citation: Suojalehto, H.; Ndika, J.; Lindström, I.; Airaksinen, L.; Karvala, K.; Kauppi, P.; Lauerma, A.; Toppila-Salmi, S.; Karisola, P.; Alenius, H. Transcriptomic Profiling of Adult-Onset Asthma Related to Damp and Moldy Buildings and Idiopathic Environmental Intolerance. *Int. J. Mol. Sci.* **2021**, *22*, 10679. https://doi.org/10.3390/ijms221910679

Academic Editors: Amelia Casamassimi, Alfredo Ciccodicola and Monica Rienzo

Received: 4 September 2021
Accepted: 28 September 2021
Published: 1 October 2021

Publisher's Note: MDPI stays neutral with regard to jurisdictional claims in published maps and institutional affiliations.

Abstract: A subset of adult-onset asthma patients attribute their symptoms to damp and moldy buildings. Symptoms of idiopathic environmental intolerance (IEI) may resemble asthma and these two entities overlap. We aimed to evaluate if a distinct clinical subtype of asthma related to damp and moldy buildings can be identified, to unravel its corresponding pathomechanistic gene signatures, and to investigate potential molecular similarities with IEI. Fifty female adult-onset asthma patients were categorized based on exposure to building dampness and molds during disease initiation. IEI patients (n = 17) and healthy subjects (n = 21) were also included yielding 88 study subjects. IEI was scored with the Quick Environmental Exposure and Sensitivity Inventory (QEESI) questionnaire. Inflammation was evaluated by blood cell type profiling and cytokine measurements. Disease mechanisms were investigated via gene set variation analysis of RNA from nasal biopsies and peripheral blood mononuclear cells. Nasal biopsy gene expression and plasma cytokine profiles suggested airway and systemic inflammation in asthma without exposure to dampness (AND). Similar evidence of inflammation was absent in patients with dampness-and-mold-related asthma (AAD). Gene expression signatures revealed a greater degree of similarity between IEI and dampness-related asthma than between IEI patients and asthma not associated to dampness and mold. Blood cell transcriptome of IEI subjects showed strong suppression of immune cell activation, migration, and movement. QEESI scores correlated to blood cell gene expression of all study subjects. Transcriptomic analysis revealed clear pathomechanisms for AND but not AAD patients. Furthermore, we found a distinct molecular pathological profile in nasal and blood immune cells of IEI subjects, including several differentially expressed genes that were also identified in AAD samples, suggesting IEI-type mechanisms.

Keywords: adult-onset asthma; building dampness and molds; endotypes; environmental intolerance; transcriptomics; pathobiological mechanisms

1. Introduction

Asthma is a heterogeneous disease driven by interactions between airway epithelium, the immune system, and environmental exposure. It can be sub-classified into several pheno- and endotypes based on clinical, functional, and inflammatory features [1]. Adult-onset asthma is more common in females, is typically nonallergic, and has less favorable prognosis, when compared with early-onset asthma [2,3]. Contrary to early-onset allergic asthma, there is limited understanding of the biology that underlies the adult-onset asthma phenotype.

Some studies have reported an increase of asthma incidence in adults living or working in damp and moldy buildings [4]. Adult-onset asthma initiated during exposure to building dampness and molds does not demographically differ from the adult-onset asthma phenotype; it is more common in females and predominantly non-allergic [5,6]. These patients use asthma medication excessively, suggesting poor prognosis. The underlying mechanisms are poorly known; however, it seems immunoglobin E (IgE)-mediated sensitization to molds is not an essential factor for asthma onset in these environments [5].

Idiopathic environmental intolerance (IEI) (or multiple chemical sensitivity) patients have recurring, non-specific symptoms in multiple organ systems attributed to environmental factors with no medical and exposure-related explanation [7]. IEI is considered a functional somatic disorder of biopsychosocial nature rather than a toxicological response [8,9]. In Finland and other Northern European countries, these patients often attribute their symptoms to buildings [10]. IEI and adult-onset asthma patient groups are demographically similar, they report similar respiratory symptoms, and there is comorbidity between these conditions [11,12].

We aimed to study if a specific clinically identified subset of patients with adult-onset asthma associated with damp and moldy buildings can be distinguished by assessing transcriptomic profiles and corresponding pathobiological mechanisms in nasal mucosa and peripheral blood mononuclear cells (PBMC). In addition, we evaluated if these patients have similar mechanisms of disease to IEI.

2. Results

2.1. Demographic and Clinical Data

Figure 1 shows a workflow of the study.

All participants were nonsmoking women. BMI, previous smoking, or atopy did not differ significantly between the groups, and the controls were younger than other groups ($p = 0.021$) (Table 1). Six IEI patients had concomitant asthma. Symptom duration, or the proportion of severe asthma or the ICS dose, did not differ significantly between the asthma and IEI groups. Chronic rhinitis was less common in controls than in other groups ($p = 0.004$). QEESI chemical intolerance ($p < 0.001$) and symptom and life impact scores ($p < 0.001$) were highest in the IEI group and lowest in the control group, with the asthma groups' scores in between. The IEI group reported most disabling symptoms related to muscles or joints, heart or chest, and ability to think. Total IgE, blood eosinophils, fractional exhaled nitric oxide, FEV1, and FEV1/FVC did not differ significantly between the groups. The asthma and IEI patients had more bronchial hyperresponsiveness than the controls ($p = 0.029$).

2.2. Inflammatory Cytokine and Gene Expression Profile of Patients and Controls

At the protein level, proinflammatory cytokines IL-8, IL-12(p70), and IL-17A and regulatory cytokine IL-10 did not differ from the controls in any of the tested groups. However, IL-6 was slightly elevated in the plasma of the asthma not associated with dampness and molds (AND) patients (Figure S1).

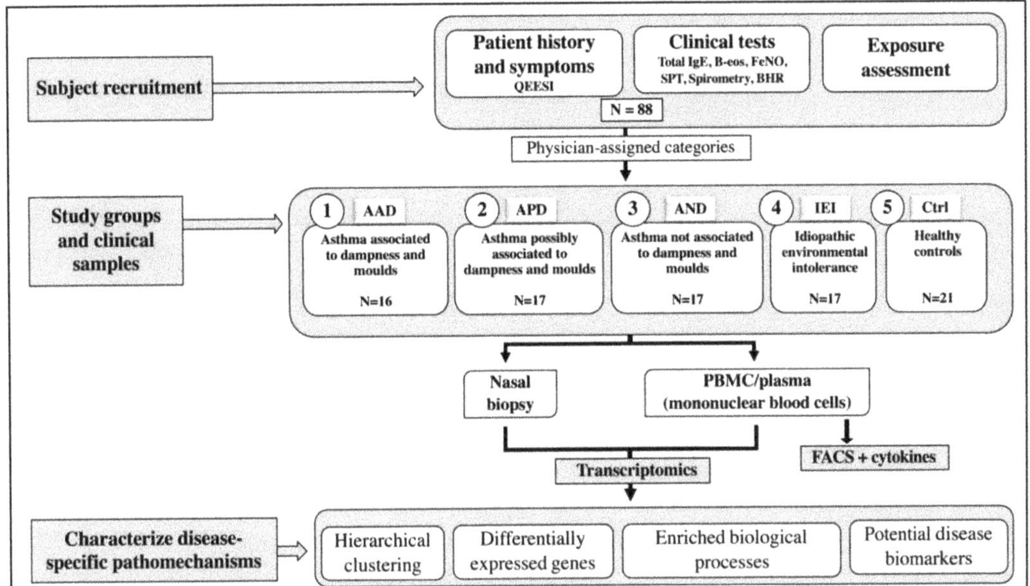

Figure 1. Flowchart outlining sample categorization and study workflow. Clinical tests included spirometry, assessment of non-specific bronchial hyperresponsiveness (BHR), fractional exhaled nitric oxide (FeNO), and blood eosinophil (B-eos) counts. Allergy to common environmental allergens was determined via skin prick testing (SPT). The Quick Environmental Exposure and Sensitivity Inventory questionnaire (QEESI) was used to assess idiopathic environmental intolerance. Nasal biopsies and peripheral blood mononuclear cell (PBMC) samples were obtained. Disease mechanisms were investigated with gene expression profiling, cytokine analysis (ELISA), and immune cell subtype profiling (FACS).

Based on the global gene expression profiles, there was considerable overlap between the different asthma subgroups (Figure S2). Nasal biopsy and PBMC transcriptomes separated the controls from the asthma patients. However, the distinction of controls from the asthma subgroups was different for the two tissue types. In the nasal biopsy, all the control samples were entirely separated from all the other patient samples, whereas in the PBMC samples, the controls only separated clearly from the asthma possibly associated with dampness and molds (APD) and IEI patient subgroups. When the individual patient groups were compared with the controls, 409 genes in the nasal biopsies and 266 genes in the PBMCs were identified as significantly different. In the nasal biopsy samples, AND appeared to be the most distinct of the asthma groups with 202 differentially expressed genes (DEGs). Within the IEI category, 101 DEGs were identified in the nasal biopsies and 222 DEGs in PBMCs (Figure 2A). DEGs from all patient/control contrasts in each sample type are provided in Table S1 (PBMC DEGs) and Table S2 (nasal biopsy DEGs).

Table 1. Demographic and clinical characteristics during the sample collection of the study subjects. All subjects were nonsmoking women.

	Asthma Associated with Dampness and Molds (AAD) n = 16	Asthma Possibly Associated with Dampness and Molds (APD) n = 17	Asthma Not Associated with Dampness and Molds (AND) n = 17	Idiopathic Environmental Intolerance (IEI) [a] n = 17	Healthy Controls (Ctrl) n = 21	p-Value
Age, years, md (IQR)	47.3 (40.9–53.4)	50.5 (43.7–55.7)	51.8 (38.5–57.4)	50.3 (44.4–55.1)	41.2 (32.8–47.0)	0.021
Body mass index, kg/m^2, md (IQR)	24.9 (24.1–30.4)	27.3 (24.0–32.9)	29.0 (26.0–34.5)	25.1 (23.7–27.3)	25.5 (21.1–28.7)	0.060
Ex-smoker, n (%)	3 (19)	6 (35)	4 (24)	3 (18)	1 (10)	0.399
Atopy [b], n (%)	3 (19)	4 (24)	1 (6)	6 (35)	7 (33)	0.238
Duration of respiratory symptoms, years, md (IQR)	3.0 (2.0–5.0)	5.0 (1.5–12.5)	8.0 (2.0–16.5)	5.0 (2.5–10.5)	NA	0.122
Use of inhaled steroid within one month, n (%)	16 (100)	16 (94)	17 (100)	10 (59)	0 (0)	<0.001
Inhaled steroid dose, n (%) [c]						0.786
Low	5 (31)	2 (13)	2 (12)	2 (20)	NA	
Medium	6 (37)	6 (38)	7 (41)	3 (30)	NA	
High	5 (31)	8 (50)	8 (47)	5 (50)	NA	
Daily use of short-acting β2-agonist, n (%)	0 (0)	1 (6)	2 (12)	1 (17)	NA	0.696
Severe asthma [d], n (%)	3 (19)	4 (24)	5 (29)	1 (17)	NA	0.875
Chronic rhinitis, n (%)	11 (69)	11 (65)	10 (59)	10 (59)	3 (14)	0.004
QEESI chemical intolerance score, md (IQR), n = 87	39 (24–53)	34 (19–55)	29 (9–71)	76 (39–85)	5 (3–15)	<0.001
QEESI symptom score, md (IQR)	42 (16–61)	35 (15–44)	29 (17–47)	51 (32–78)	6 (3–13)	<0.001
QEESI life impact score, md (IQR)	23 (4–52)	22 (9–40)	15 (15–38)	64 (43–88)	2 (0–7)	<0.001
Total IgE kU/L, md (IQR)	27 (15–78)	14 (7–41)	26 (13–69)	20 (12–52)	22 (16–35)	0.690
Blood eosinophils /μL, md (IQR)	175 (93–215)	130 (95–220)	180 (120–250)	120 (120–175)	160 (110–210)	0.918
Fractional exhaled nitric oxide, md (IQR)	13.5 (11.5–22.1)	17.5 (12.3–23.5)	14.5 (9.0–30.5)	11.5 (9.1–20.6)	12.0 (9.3–16.3)	0.445
FEV$_1$ % pred, md (IQR)	91 (86–97)	84 (78–100)	85 (70–95)	95 (88–103)	89 (83–98)	0.093
FVC % pred, md (IQR)	93 (86–101)	92 (79–102)	86 (78–81)	96 (94–103)	95 (86–105)	0.033
FEV$_1$/FVC, md (IQR)	78 (75–84)	78 (73–83)	80 (75–81)	81 (76–84)	79 (76–82)	0.671
Nonspecific bronchial hyperresponsiveness, n (%), n = 84						0.029
Moderate	4 (29)	3 (18)	1 (7)	1 (6)	0 (0)	
Mild	2 (14)	5 (29)	3 (20)	4 (24)	0 (0)	
No	8 (57)	9 (53)	11 (73)	12 (71)	21 (100)	

FEV$_1$ = forced expiratory volume in one second; FVC = forced vital capacity; IQR = interquartile range; IgE = immunoglobin E; GINA = Global Initiative for Asthma; md = median; NA = not available; SPT = skin prick test; QEESI = the Quick Environmental Exposure and Sensitivity Inventory; [a] six subjects had asthma diagnosis; [b] positive skin prick test to at least one environmental allergen; [c] proportion of cases using inhaled steroid within one month; [d] ERS/ATS criteria [13].

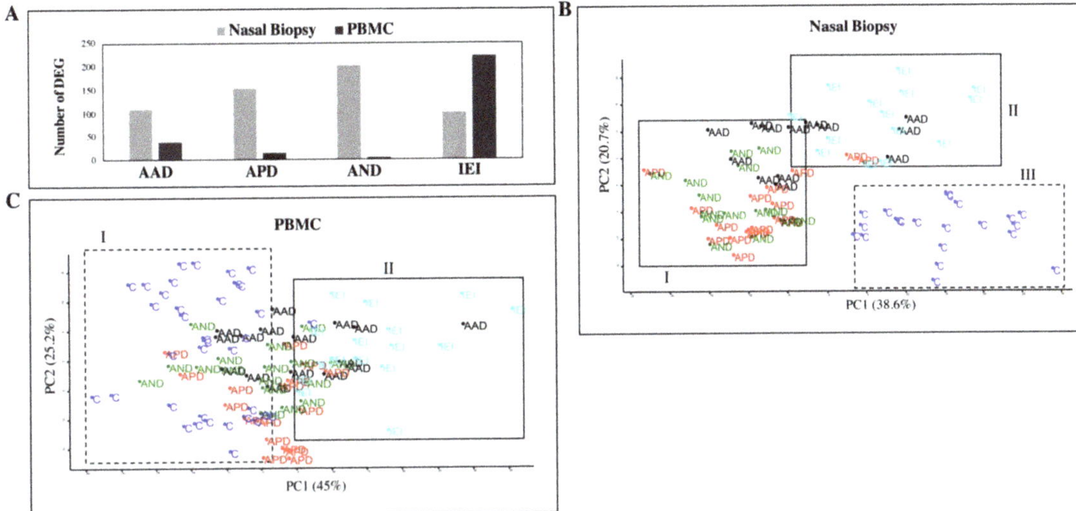

Figure 2. Comparison of differentially expressed genes (DEGs) identified in patient versus controls (**A**). Patient groups with asthma that is *associated* to dampness and molds, *possibly associated* to dampness and molds, and *not associated* with dampness and molds are denoted as AAD, APD, and AND, respectively. Patients diagnosed with idiopathic environmental intolerance are denoted as IEI, and controls are denoted as (**C**). Principal component analysis (PCA) shows top two components represented by genes identified as differentially expressed in the nasal biopsies (**B**) and blood cells (PBMCs) (**C**) of patients versus controls. In the nasal biopsies (**B**), the top 2 components of the PCA plot separated all samples into three main clusters. Cluster I consisted of all AND patients, 90% of all APD patients, and 50% of all AAD patients. Cluster II consisted of all IEI patients and a couple of APD and half of all AAD patients. All control samples grouped together in Cluster III. In blood cells (**C**), clearly distinct asthma patient or control subgroup clusters cannot be identified from the top 2 components of the PCA plot. However, the control samples (Cluster I) do cluster separately from IEI individuals (Cluster II).

Principal component analysis (PCA) based on log2-transformed gene expression intensities of identified DEGs from nasal biopsy or PBMC, separated the study samples into three and two main clusters, respectively (Figure 2B,C). When disease and ICS dose are incorporated into a PCA plot constructed from the patient/control DEGs in the nasal biopsies, the clustering of the samples is not explained by differences in ICS dose (Figure S3). The systemic effects of ICS were taken into account during analysis of differentially expressed genes in PBMCs. A subtle separation of the asthma groups from the controls was observed from PCA analysis of DEGs. The APD and AND groups clustered closest to the controls, and just like in the nasal biopsies, asthma associated with dampness and molds (AAD) patients were closest (50% overlap) to the IEI cluster (Figure 2C). This IEI cluster (100% IEI + 5% Ctrl) was clearly distinct from the Ctrl cluster (95% Ctrl) (Figure 2C).

2.3. Biological Significance of Differentially Expressed Genes

Venn comparisons of identified DEGs suggested very little overlap (96% of DEGs were unique to each asthma patient subgroup) in disease pathobiological mechanisms in blood cells (Figure 3A). Only AAD and IEI patients had common DEGs—12 in total. On the other hand, the degree of overlap was notably higher in the patient/control DEGs identified from nasal biopsies (Figure 3B). With 27% DEGs in common, the AND and APD asthma subtypes were the most closely related asthma groups in this study. AAD had more DEGs in common with IEI (20%) than either APD (8%) or AND (6%). Nineteen DEGs were commonly shared between all three asthma groups.

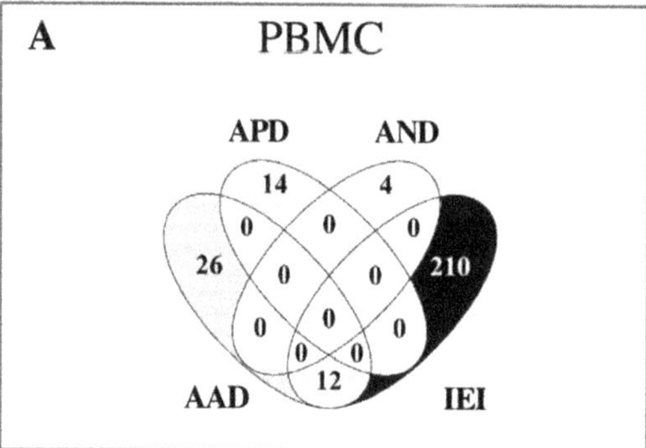

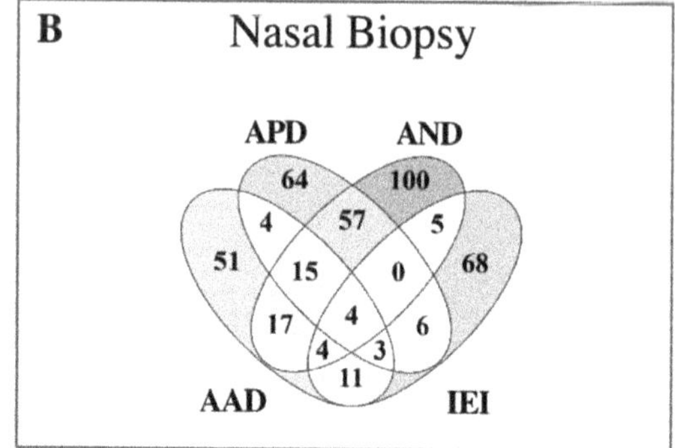

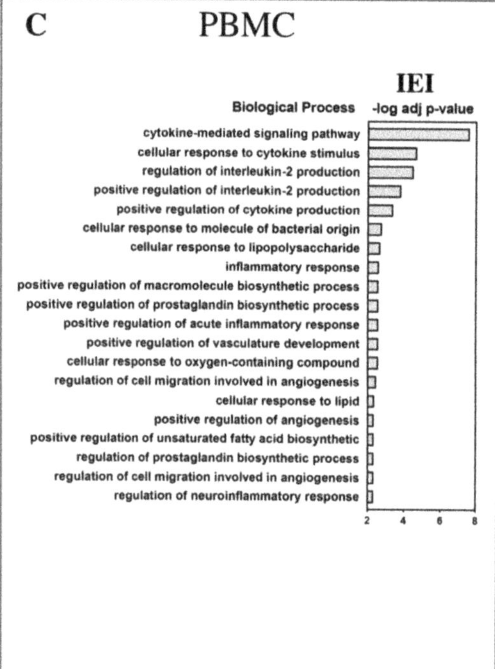

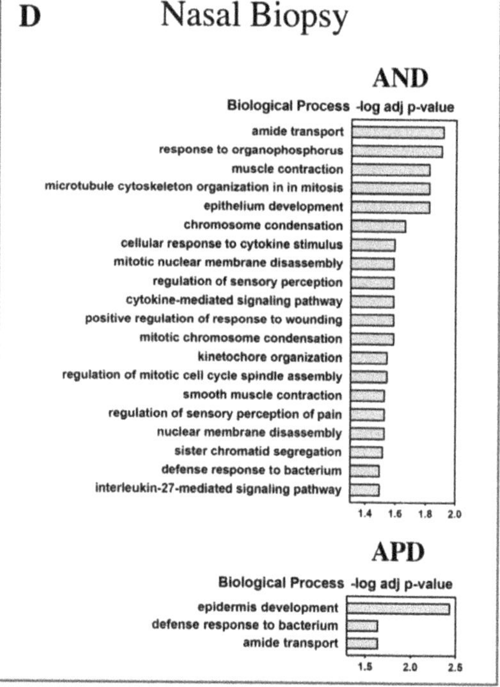

Figure 3. Venn diagram of differentially expressed genes (DEGs) identified in patient versus controls in PBMC (**A**) and nasal biopsy (**B**). Patient groups with asthma that is *associated* to dampness and molds, *possibly associated* to dampness and molds, and *not associated* with dampness and molds are denoted as AAD, APD, and AND, respectively. Patients diagnosed with idiopathic environmental intolerance are denoted as IEI and controls are denoted as *C. AND* and *APD* had the greatest number of shared DEGs in nasal biopsy, while AAD and IEI have the most DEGs in common in blood cells (PBMC). Gene ontology-based analysis of enriched biological processes in different patient groups reveals several significantly overrepresented pathways. The top pathways, ranked by −log adjusted *p* value (cutoff, 1.3) are shown for each differentially expressed gene set identified in PBMC (**C**) and nasal biopsy (**D**).

In nasal biopsy samples, due to the limited number of shared DEGs between the different asthma subtypes, the unique DEGs for each asthma group were analyzed for enrichment of potentially unique pathobiological mechanisms. Although 51 (AAD), 64 (APD), and 100 (AND) genes were uniquely differentially expressed in the associated patient subgroup, these genes were not significantly associated with any known biological processes or pathways. Therefore, we proceeded with functional enrichment analysis of all the identified DEGs in each asthma subtype. Analysis of the identified DEGs for functional enrichment based on gene ontology (GO) biological processes revealed clear differences between the patient groups and the target tissues. Due to the low number of DEGs, no enriched biological processes were observed in any of the asthma groups in blood cells. In sharp contrast, the IEI group demonstrated very significant (adjusted *p*-value, 1×10^{-1} to 1×10^{-7}) enrichment of biological processes such as *"cytokine-mediated signaling pathway"*, "cellular *response to cytokine stimulus"*, *"regulation of IL2 production"*, and "cellular *response to lipopolysaccharide"*, *"inflammatory response"*, etc. The top 20 overrepresented GO biological processes and their corresponding *p*-values are shown in Figure 3C. On the other hand, the IEI group DEGs from the nasal biopsy transcriptome did not show any enrichment of biological processes. In the nasal biopsy samples, significantly enriched (adjusted *p*-value < 0.05) GO biological processes were only identified from DEGs in the AND and APD groups. The top enriched biological processes in the AND group were *"amide transport"*, "response to organophosphorous" and "muscle contraction". The top 20 GO biological processes overrepresented in the AND subgroup DEGs are shown in Figure 3D, upper panel. Only three biological processes (*"epidermis development"* and *"defense response to bacterium"* and *"amide transport"*) were enriched in the APD group (Figure 3D, lower panel). No significant enrichments of biological processes were observed in the AAD asthma group.

For the patient subgroups demonstrating DEGs with significantly enriched GO biological processes (i.e., IEI for PBMC and APD, AND for nasal biopsies), we next sought to identify specific co-expressed gene networks (modules) for each clinical asthma subtype using the INfORM tool for module prioritization and Ingenuity's pathway analysis (IPA) for identification of up and downregulated downstream biological functions within each co-expressed gene network. In PBMC, three gene modules from IEI DEGs were identified. Significantly enriched pathways were identified from modules 1 (M1) and 3 (M3), all of which were predicted to be downregulated (Z-score < 0). The most significant biological processes represented by co-expressed genes in M1 were decreased recruitment, migration, and activation of blood cells/leukocytes (B-H adjusted *p*-value < 1×10^{-12}), and in M3, decreased proliferation and migration of blood cells/lymphocytes (B-H *p*-value < 1×10^{-7}). The interaction between all IEI-associated gene network modules and their corresponding downstream biological processes are depicted in Figure 4.

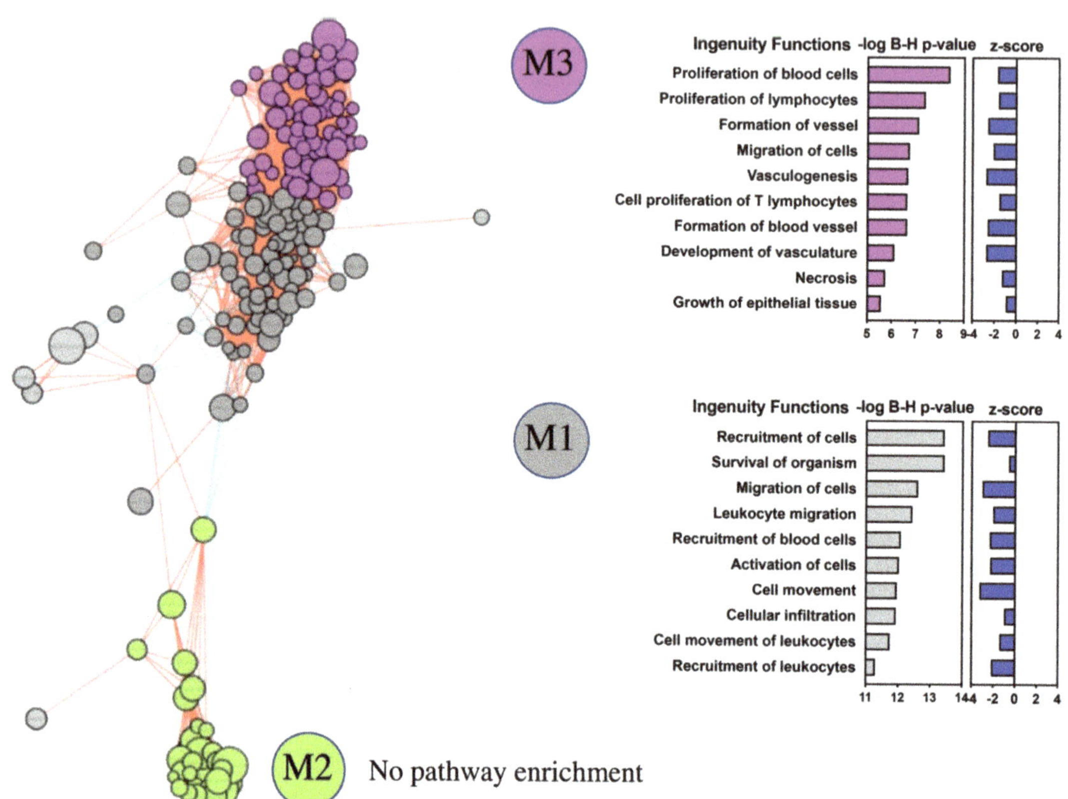

Figure 4. Prediction of the disease relevant pathways represented by differentially expressed genes (DEGs) in peripheral mononuclear cells of patients with idiopathic environmental intolerance (IEI). Three co-expressed gene networks (modules; M1, M2, and M3) were identified from genes that are differentially expressed between IEI patients and controls. The interactions between different inferred co-expressed gene network modules, as well as their corresponding top enriched biological functions and predicted downstream activation states are shown for modules M1 and M3. Biological functions are ranked by −log of Benjamini–Hochberg corrected p values (implemented filter: exclude cancer pathways; include only pathways with ≥ 5 differentially expressed genes). Positive (+) or negative (-) activation scores correspond to downstream activated or inhibited disease/functions, respectively. An activation Z score > |2| is highly predictive of an activated or inhibited disease/function. No enriched biological functions were identified from the genes in module M2.

From the nasal biopsy DEGs, five modules (M1–M5) associated with the AND asthma subtype were identified. All five gene modules consisted of significantly enriched biological processes, some of which were predicted as activated (Z-score > 0) or inhibited (Z-score < 0). The most significant biological processes were identified in modules M2 (decreased *segregation of chromosomes*; B-H p-value $< 1 \times 10^{-13}$, *mitosis*; B-H p-value $< 1 \times 10^{-12}$) and M4 (increased *smooth muscle contraction*; B-H p-value $< 1 \times 10^{-9}$, *formation of filaments*; B-H p-value $< 1 \times 10^{-4}$). The interaction between gene network modules associated with the AND asthma subgroup, their corresponding overrepresented biological functions, and predicted activation states are shown in Figure 5. The most significantly enriched function in module M1 was *secretion of triacyl glycerol* (B-H p-value $< 1 \times 10^{-3}$), *metabolism of prostaglandin* (B-H p-value $< 1 \times 10^{-3}$) in M3, and *quantity of MHC Class I on cell surface* (B-H p-value $< 1 \times 10^{-2}$) in module M5.

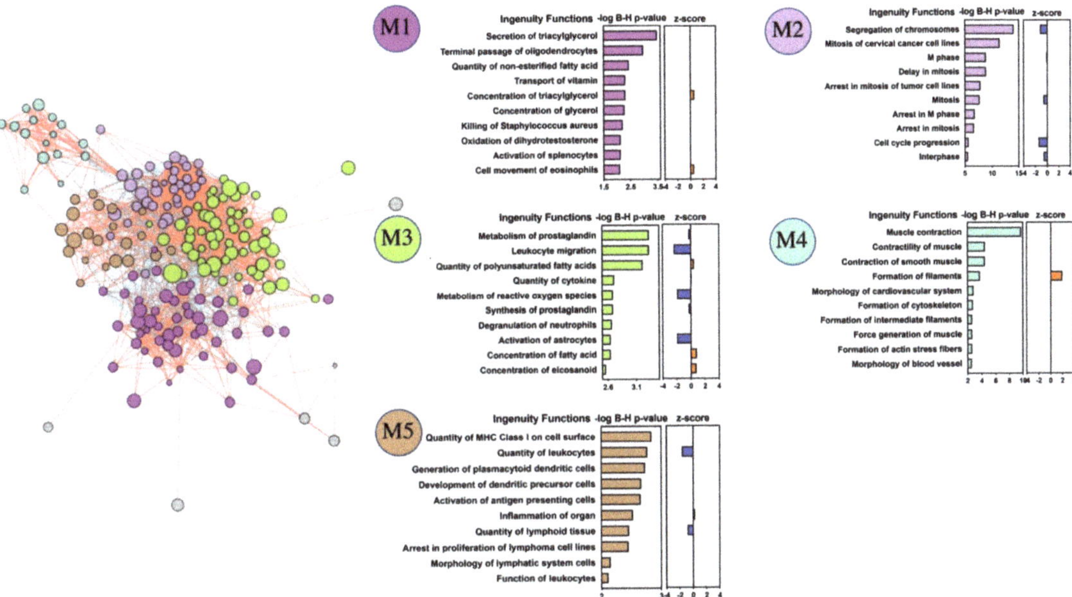

Figure 5. Prediction of the disease relevant pathways represented by differentially expressed genes (DEGs) in nasal biopsy of patients with asthma that is not associated to damp and moldy buildings (*AND*). Five co-expressed gene networks (modules; M1, M2, M3, M4, and M5) were identified from genes that are differentially expressed between *AND* patients and controls. The interactions between different inferred co-expressed gene network modules, as well as their corresponding biological processes and predicted downstream activation states are shown. Biological functions are ranked by −log of Benjamini–Hochberg corrected p values (implemented filter: exclude cancer pathways; include only pathways with ≥5 differentially expressed genes). Positive (+) or negative (-) activation scores correspond to downstream activated or inhibited disease/functions, respectively. An activation Z score > |2| is highly predictive of an activated or inhibited disease/function.

Four modules were also found to be associated with nasal biopsy DEGs in the APD asthma subtype. Module interaction and their overrepresented biological functions are depicted in Figure S4. Enriched biological functions were identified from three (M2, M3, and M4) of the four co-expressed gene networks. The most significant functions were transport of vitamin/folic acid (M2), interphase/mitosis (M3), and keratinization/differentiation of keratinocytes (M4).

2.4. Cell Type Analysis Based on Gene Expression Signature and Flow Cytometry Indicate Macrophages Are Key Mechanistic Players in IEI Pathobiology

Because the functional indications from DEGs in the blood cells of the IEI patients were decreased cell migration, proliferation, and movement, we sought to answer whether these can be substantiated by alternative enrichment analysis. We performed cell type prediction analysis based on DEGs identified in PBMCs of the IEI subjects, as well as flow cytometry cell type profiling. Genes identified as differentially expressed in the IEI subjects relative to the controls were submitted to a publicly available database of RNA-seq data (ARCHS[4]) [14]. In this study, 1716 macrophage samples were identified to be highly enriched (adj. *p* value 4.3×10^{-25}) for 40% (88/222 genes) of the IEI/Ctrl DEGs (Figure 6A). Interestingly, elevated monocyte levels in circulation of the IEI patients was the most significant finding from flow cytometry analysis (Figure 6B). A heatmap based on the IEI/Ctrl DEGs identified as signature macrophage genes (88 genes) shows that 94% of these genes were downregulated in the IEI patients (Figure 6C).

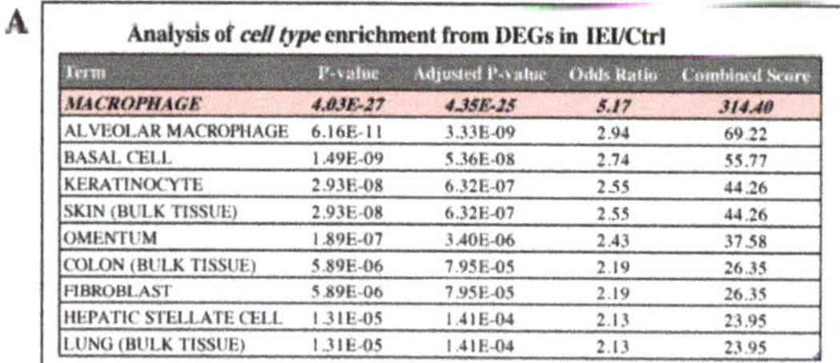

Analysis of *cell type* enrichment from DEGs in IEI/Ctrl

Term	P-value	Adjusted P-value	Odds Ratio	Combined Score
MACROPHAGE	*4.03E-27*	*4.35E-25*	*5.17*	*314.40*
ALVEOLAR MACROPHAGE	6.16E-11	3.33E-09	2.94	69.22
BASAL CELL	1.49E-09	5.36E-08	2.74	55.77
KERATINOCYTE	2.93E-08	6.32E-07	2.55	44.26
SKIN (BULK TISSUE)	2.93E-08	6.32E-07	2.55	44.26
OMENTUM	1.89E-07	3.40E-06	2.43	37.58
COLON (BULK TISSUE)	5.89E-06	7.95E-05	2.19	26.35
FIBROBLAST	5.89E-06	7.95E-05	2.19	26.35
HEPATIC STELLATE CELL	1.31E-05	1.41E-04	2.13	23.95
LUNG (BULK TISSUE)	1.31E-05	1.41E-04	2.13	23.95

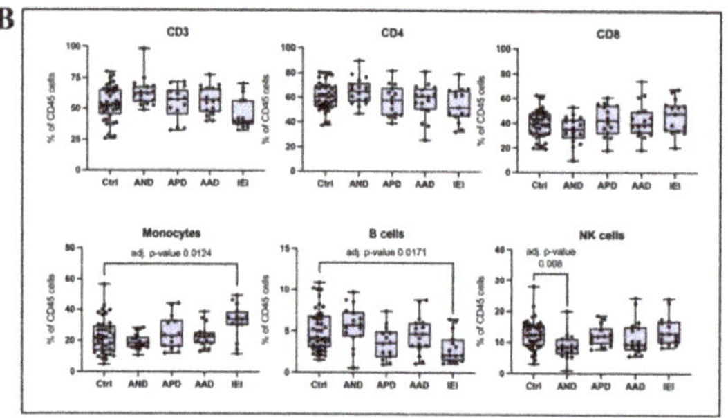

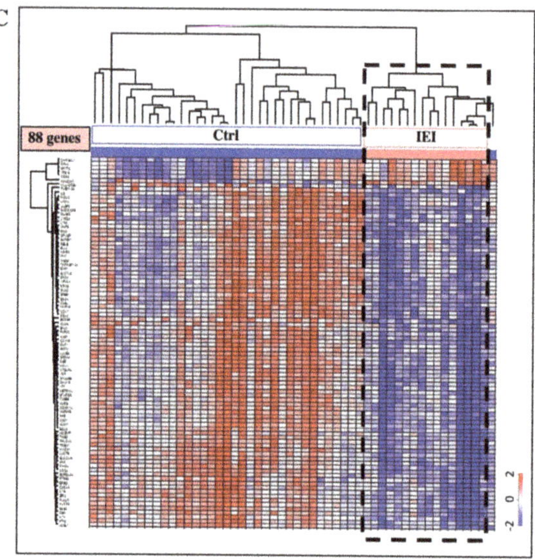

Figure 6. Identified differentially expressed genes (DEGs) in blood cells of idiopathic environmental intolerance (*IEI*) patients relative to controls (Ctrl) were submitted to cell type enrichment analysis. The most significant cell type represented by the expression signature of these genes were macrophages (**A**). The relative expression of 88 genes identified to be differentially expressed between IEI patients and controls was consistent with genes enriched in RNA-seq expression profiles (1716 samples) of macrophages in ARCHS database (overlapping not shown). Cell type enrichment analysis was also performed via FACS analysis. Significantly elevated levels of monocytes were observed in IEI subjects' blood. Patient groups with asthma that is *associated* to dampness and molds, *possibly associated* to dampness and molds, and *not associated* with dampness and molds are denoted as AAD (n = 16), APD (n = 17), and AND (n = 17), respectively. Patients diagnosed with idiopathic environmental intolerance are denoted as IEI (n = 16) and controls are denoted as *Ctrl* (n = 21). Analysis was performed using the Kruskal–Wallis test with Dunn's post-hoc correction (**B**). Despite having elevated monocyte levels, macrophage signature genes were found to be predominantly downregulated in the blood cells of individuals diagnosed with IEI. A heatmap showing the relative expression (z-score normalized) of these genes in control and IEI blood cell samples is shown in (**C**). Controls comprise individuals who did not take any inhaled corticosteroid (n = 21) plus a subset of controls that used inhaled corticosteroid for 4 weeks (n = 14). The dashed bold rectangle highlights a cluster of IEI patients with downregulated expression of these macrophage-associated genes.

2.5. QEESI Score and Disease Duration Have the Strongest Correlations to DEGs Identified in Blood Cells

Since the most significantly enriched biological processes were observed from blood cell DEGs, we next sought to investigate whether the DE genes in PBMC are significantly associated to any of the asthma-relevant clinical parameters. To identify genes with specific clinical relevance, Pearson's correlation analysis was carried out between DEGs identified in blood cells and several clinical parameters in this cohort. A heatmap of genes having a correlation score R > |0.5| to at least one clinical finding is shown in Figure 7A. In total, 49 genes were identified with an R median of −0.52 (QEESI life impact), −0.51 (QEESI chemical intolerance), and −0.33 (symptom duration). These 49 genes (Table S3) were strongly predicted (Z-score > |2|) to cause decreased immune cell proliferation and migration, decreased cellular homeostasis, decreased cytotoxicity of T-lymphocytes, and decreased activation of antigen-presenting cells (Figure 7B).

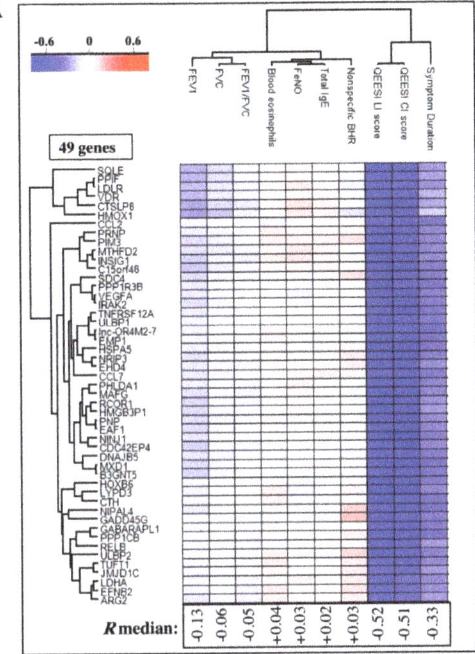

B

Diseases or Functions Annotation	B-H FDR	Z-score	Prediction	# Genes
Proliferation of blood cells	1.2E-09	-1.9		24
Proliferation of lymphatic system cells	8.4E-09	-2.0		22
Proliferation of lymphocytes	8.8E-09	-2.2	Decreased	21
Cell proliferation of T lymphocytes	8.8E-07	-2.3	Decreased	17
Migration of cells	5.8E-06	-2.3	Decreased	30
Cell movement	1.6E-05	-2.5	Decreased	31
Migration of monocytes	2.8E-05	-2.0		7
Cytotoxicity of lymphocytes	4.6E-05	-2.4	Decreased	8
Quantity of cells	4.6E-05	-1.9		26
Activation of microglia	4.6E-05	-1.7		7
Apoptosis	4.6E-05	0.8		32
Cellular homeostasis	5.5E-05	-2.6	Decreased	24
Quantity of blood cells	5.5E-05	-1.3		19
Cell proliferation of tumor cell lines	6.4E-05	-2.5	Decreased	25
Activation of leukocytes	6.9E-05	-2.8	Decreased	15
Formation of blood vessel	6.9E-05	-2.5	Decreased	8
Cell movement of monocytes	7.1E-05	-2.0		8
Quantity of connective tissue	7.1E-05	-0.6		14
Leukocyte migration	7.4E-05	-2.3	Decreased	18
Cytotoxicity of T lymphocytes	7.5E-05	-2.2	Decreased	6
Fibrosis	7.5E-05	0.5		14
Growth of epithelial tissue	7.6E-05	-1.8		15
Uptake of Ca2+	7.6E-05	-0.4		6
Activation of lymphocytes	8.4E-05	-2.4	Decreased	12
Chronic psoriasis	8.7E-05			7
Necrosis	8.7E-05	-0.5		31
Activation of antigen presenting cells	9.3E-05	-2.4	Decreased	10
Concentration of lipid	9.3E-05	1.5		16
Function of leukocytes	9.3E-05			12
Psoriasis	9.3E-05			13

Figure 7. Regression analyses of differentially expressed genes (DEGs) and selected clinical parameters relevant to diagnosis of asthma and/or idiopathic environmental intolerance identified 49 genes with a Pearson's correlation coefficient of R > |0.5| to at least 1 clinical parameter. A hierarchical cluster of correlated genes (cutoff R > |0.5|) and clinical parameters is shown in (**A**). The correlation coefficients of these 49 genes clusters the clinical parameters into 2 main clusters. The strongest correlations were identified in both the QEESI chemical intolerance (CI) and life impact (LI) scores, wherein expression of all 49 genes was negatively correlated with QEESI scores across the entire study cohort. Pathway analysis of these 49 genes predicted highly significant enrichment of genes involved in suppression of immune cell activation, proliferation, and/or migration (**B**). BHR is non-specific bronchial hyperresponsiveness, CI is chemical intolerance, FeNO is fractional exhaled nitric oxide, FEV1 is forced expiratory volume in one second, FVC is forced vital capacity, IgE is immunoglobin E, LI is life impact, and QEESI is Quick Environmental Exposure and Sensitivity Inventory.

The top anti-correlated genes were identified as *NINJ1* (R = −0.59, *p* < 0.0001) and *SQLE* (R = −0.62, *p* < 0.0001). NINJ1 (Ninjurin1) and SQLE1 (squalene epoxidase) were significantly anti-correlated to QEESI life impact and chemical intolerance findings (Figure S5A). QEESI life impact and chemical intolerance scores were lowest in the controls and expression of NINJ1 and SQLE were highest in the controls (Figure S5B).

3. Discussion

There is a lack of knowledge on the mechanisms of adult-onset asthma and IEI [1,15,16]. We used a combination of clinical assessment and global transcriptomic analysis of blood and airway epithelium to investigate whether specific pathobiological mechanisms of adult-onset asthma associated with dampness and molds could be identified. The asthma patients were divided into three groups based on exposure and symptoms to building dampness and molds during disease initiation. Furthermore, we included a group of IEI patients to investigate potential molecular similarities with asthma patients.

The study population represented nonsmoking women, and the clinical characteristics of asthma patients were similar to female and obesity-related asthma phenotypes identified earlier in the cluster analysis of adult-onset asthma [17,18]. Asthma was mainly mild or moderate [13]. Previous tests had confirmed asthma diagnosis of asthma patients and regular ICS use could explain near-normal lung function results during sampling. Atopy and markers of eosinophilic inflammation (blood eosinophils, fractional exhaled nitric oxide, total IgE) did not differ between asthma patients and healthy controls, suggesting non-T2 type disease [1]. Because IEI and asthma commonly overlap [12], it was not possible to completely exclude subjects with an asthma diagnosis from the IEI group. Nevertheless, asthma did not explain the symptoms in that group and QEESI scales showed a clear difference between asthma and IEI groups. To assess IEI, we used chemical intolerances and life impact scales of the QEESI questionnaire, which have shown the largest discriminatory validity of QEESI scales [19]. The IEI patients had experienced symptoms for several years and reported high QEESI scores suggesting chronic and severe IEI [20]. A previous study showed that asthma does not substantially change QEESI results [19]. Higher QEESI scales of asthma groups than the control group suggest more IEI-type symptomatology among asthma patients. This is in line with previous findings showing an overlap between these two entities [12].

We selected nasal epithelia as a local sampling site because in previous studies the quality of nasal epithelia samples has been sufficient for transcriptomic analysis [21], whereas quality of sputum samples varies [22]. Previous studies have identified IL-17, IL-8, and IL-6 as key cytokines in non-T2 asthma [23]. Increased levels of IL-6 have also been found in serum [24] of asthmatic patients and in BALF from patients with nonallergic asthma compared with that from patients with allergic asthma [25]. In the present study, none of these cytokines were significantly different on an mRNA level, and on the protein level, only IL-6 was found to be mildly elevated in the AND subgroup suggesting a weak systemic inflammation. Previous studies have failed to show consistent findings related to immunological parameters [26]. Luca et al. reported that IFN-gamma, IL-8, IL-10, MCP-1, PDGFbb, and VEGF were significantly increased in the plasma compared to healthy controls [27]. Dantoft et al. found that plasma levels of IL-1-beta, IL-2, IL-4, and IL-6 were significantly increased in IEI, whereas IL-13 was downregulated [28]. We did not find significant differences in any of the cytokines studied between IEI and the controls. It is therefore possible that larger sample sizes than that used in the present study are needed to reach statistical significance in the plasma cytokine analysis in IEI.

Gene expression profiling of whole blood of the U-BIOPRED cohort, identified 1693 DEGs (referred as severe asthma disease signature—SADS) between patients with severe asthma and healthy controls, whereas the number of DEGs between mild/moderate asthma and healthy controls were fewer [29]. These results demonstrate that the disease severity has major impact on the blood transcriptome and the number of DEGs. We found a relatively low number of DEGs in the blood cells of the adult-onset asthma patient groups, suggesting weak systemic evidence of disease. Compared to PBMC, six times more DEGs were identified from nasal biopsies of our asthma patients suggesting that disease pathology is primarily localized to airway epithelia. Indeed, several biological processes with previously published relevance to asthma, such as contraction of (airway) smooth muscle [30,31] and altered cell division [32–34], were identified as highly enriched functions in the AND group. In addition, enrichment of genes involved in leukocyte migration and cytokine production are the major drivers of the immune dysregulation and clinical manifestations of asthma. Given that clinical and transcriptomics findings in the AND group are well in line with published pathomechanisms associated with asthma, we propose that the AND represents the phenotype of non-IgE-mediated adult-onset asthma. The APD group was the closest to the AND group, in terms of shared DEG and affected biological functions, wherein we observed altered mitosis as one of the most significantly affected airway epithelial functions. In contrast, there was no molecular evidence of localized or systemic inflammation in the AAD group. We cannot exclude the possibility that the use of ICS, which may attenuate inflammation, could explain partially our findings of modest to no systemic or localized inflammation in the airway epithelium.

Although the AAD group did not show evidence of active airway inflammation, based on negative findings in GO term enrichment analysis, it revealed modest clustering together with the IEI patients in PCA. This PCA clustering was due to that 33% (12/38) of the DEGs in AAD were shared with IEI in the blood cell transcriptome and 20% (22/109) of the DEGs in AAD were shared with IEI in the nasal biopsies. Thus, although we did not detect a specific molecular endotype for AAD, partial transcriptomic clustering together with IEI may indicate IEI-type disease mechanisms in this group. In our previous study, we did not find significant differences in nasal mucosa or in the blood transcriptome in subjects with respiratory symptoms associated with moisture damage when compared to non-exposed or non-symptomatic persons [35].

Unlike asthma groups, the IEI patients showed strong enrichment of biological processes related to cytokine stimulus or cytokine signaling in GO term analysis of the blood transcriptome. In pathway analysis, all enriched biological functions related to the immune system were markedly suppressed in the IEI patients. Our findings emphasized the possible role for macrophages in IEI pathology since the macrophage signature genes were suppressed in the blood cells. Our results also show that no active inflammation can be detected in the airway epithelia of the IEI patients. Due to the absence of internationally accepted biological or physiological criteria for diagnosing IEI patients, it has been difficult to reconcile pathological mechanisms underlying IEI. Moreover, the physiological mechanisms of functional somatic disorders including IEI are under debate. In several models, stress-related physiology has been considered a key element in symptom perception [36]. Stress modulates immune function depending on the duration of stress [37,38]. Immunomodulatory mechanisms related to chronic stress could be involved in detected immunosuppression of our IEI patients.

The QEESI life impact and chemical intolerance scores across all patients in this cohort were most correlated to a subset of 49 DEGs identified in blood cells. Enrichment of cellular functions with this gene set is consistent with suppressed immune cell activation and underscores the biological relevance of QEESI-linked genes. The top two genes that were significantly correlated to the subject's QEESI scores are NINJ1 and SQLE. NINJ1 (Ninjurin1; synonym—nerve injury-induced protein) is involved in macrophage activation and migration [39–41]. Squalene epoxidase (SQLE) is an essential enzyme for cholesterol biosynthesis, and a clinical target for treatment of hypercholesteremia, cancer, and fungal infections [42]. In future studies, these genes could be used for more specific phenotyping in larger IEI cohorts, in order to extensively investigate disease mechanisms and identify potential therapeutic targets.

As a conclusion, gene signatures for adult-onset asthma without exposure to dampness revealed clear evidence of inflammation in the airways. In contrast, a clear endotype of asthma related to damp and moldy buildings could not be distinguished from existing biological pathways and co-expressed gene networks. Interestingly, blood cell transcriptomics of IEI subjects showed heavy suppression of immune cell functions. Twenty to thirty percent of the DEGs identified in IEI nasal and blood samples were also detected in samples from asthma patients exposed to dampness and molds, suggesting IEI-type mechanisms.

4. Materials and Methods

The study methods are briefly described below. Where applicable, additional details are provided in the manuscripts' online Supplementary Materials.

4.1. Exposure Assessment

Participants' exposure to indoor dampness and mold during the onset of asthma symptoms was evaluated independently by two experts (KK, HS). The assessment was based on indoor air and building evaluation reports, medical records, and workplace reports. The person's own perception of exposure was also considered. Exposure was classified in three categories: (1) *Substantial exposure to dampness and molds* was found if at home or at the workplace a significant water damage and related microbial growth in the building materials was verified. In this category, a reliable description of a wide-ranging water-damaged structure (with a size of more than 1 m^2) was available. Fungal spores were found in material samples or in indoor air samples exceeding national limit values [43]. (2) *Possible exposure to dampness and molds* was found when evidence of damage was sparse or unsure. Subjects were categorized in this group also if they reported dampness or mold damage at home or at the workplace during asthma onset, but there was no evidence of the damage. (3) *No exposure to dampness and molds* group had no evidence of exposure to dampness or mold growth at home or at the workplace and the subjects did not report suspicions of these damages.

4.2. Study Population and Groups

All subjects were nonsmoking women aged 18–65 years. Asthma and IEI patients were recruited from tertiary hospital asthma and occupational disease outpatient clinics at Helsinki University Hospital and the Finnish Institute of Occupational Health during 2015–2018, and the control group was recruited using advertisements and social media. A medical doctor (HS) went through medical files to verify that the asthma diagnosis was confirmed according to standard procedure, and at least one of the diagnostic criteria were fulfilled: (1) marked $\geq$12% and 200 mL postbronchodilator increase in FEV$_1$ in spirometry, (2) recurrent significant reversibility in diagnostic peak flow monitoring, or (3) non-specific bronchial hyperresponsiveness in histamine or methacholine provocation test. Age of asthma onset, and allergic and asthma symptoms were evaluated from medical files and patients' interviews. In addition to gender and age, the inclusion criteria of asthma groups were (1) adult-onset asthma (onset of asthma symptoms $\geq$18 years of age), and (2) no allergic or asthma symptoms related to common environmental allergens. The subjects were classified in the following groups:

1. *Asthma associated with dampness and molds (AAD)* had '*substantial exposure to dampness and molds*'—during the asthma onset their symptoms deteriorated at the damaged building.
2. *Asthma possibly associated with dampness and molds (APD)* had '*possible exposure to dampness and molds*'—during the asthma onset their symptoms deteriorated at the workplace or at home.
3. *Asthma not associated with dampness and molds (AND)* had '*no exposure to dampness and molds*' or deterioration of asthma symptoms at home or at the workplace during asthma onset.
4. *Idiopathic environmental intolerance (IEI)* group inclusion criteria were modified from WHO's IEI criteria [16] and that of Lacour et al. [44]: (1) respiratory symptoms, (2) symptoms associated environmental factors in low exposure levels which do not cause symptoms to the majority of people, (3) symptoms from several organ systems, (4) recurrent symptoms related to certain environments that diminish or end when the factor related to symptoms is removed, (5) the condition had lasted at least 6 months and causes major living restrictions in ability of function, (6) symptoms cannot be explained by any known somatic or psychiatric disease, (7) the person experienced that the symptoms are caused by environmental factors. Asthma patients were not excluded.
5. *Healthy controls (Ctrl)* had no allergic symptoms related to common environmental allergens, asthma, or chronic cough or dyspnea. They had no known long-term working or living in a building with dampness or mold damage. They did not fulfill the criteria of IEI.

Methods of clinical assessments at the time of sampling are presented in the online Supplementary Materials. Chemical intolerances and life impact scales of the Quick Environmental Exposure and Sensitivity Inventory (QEESI) were used to gauge IEI [45].

The study was approved by Helsinki University Hospital Coordinating Ethical Committee (80/13/03/00/15). Each participant gave written informed consent.

4.3. Inhaled Steroid Course for Healthy Controls

For safety reasons, inhaled corticosteroids (ICS) were continued during sample collection, and nasal steroids were ceased one month earlier when possible. To control the systemic ICS effect in the analysis, we assessed the ICS effect in the healthy controls. They used high dose ICS (budesonide DPI 800 ug/day) for 5–6 weeks and a PBMC sample was taken before and after the treatment.

4.4. Sampling

Nasal biopsies taken from mucosa of inferior concha were stored in RNAlater solution at −80 °C. PBMCs were isolated from venous blood (8 mL) and collected into CPT tubes (BD).

4.5. PBMC Separation and Flow Cytometry

Separated plasma was aliquoted and stored at −20 °C. The extracted PBMCs were frozen in cell freezing medium (Gibco) and stored in a deep freezer (−80 °C). The number of total cells was counted, and the relative proportions of cell populations was determined by flow cytometry using surface markers for T cells, B cells, NK cells, and monocytes.

4.6. Cytokine Profiling by Luminex Assay

The soluble inflammatory markers including IL-1ra, IL-6, IL-8, IL-10, IL-12(p70), and IL-17A were measured from plasma by Bio-Plex Pro Assays (Bio-Rad Corporation, Hercules, CA, USA) in a Luminex (Bio-Plex 200, Bio-Rad) system according to the provided instructions.

4.7. Transcriptomics

Total RNA was isolated from biopsy samples on the RNeasy Plus Mini Kit (QIAGEN, Hilden, Germany), and RNA from the white blood cells was isolated by RNA AllPrep DNA/RNA/miRNA Universal Kit (QIAGEN). The amount and quality of RNA was checked and confirmed. Briefly, 100 ng of total RNA was amplified, labelled with Cy3 and Cy5 dyes, and hybridized to human microarrays (SurePrint G3, Agilent Technologies, Canta Clara, CA, USA). Raw data were monitored for quality, and quantile normalized (Bioconductor package Limma).

4.8. Data Analysis

We used SPSS version 25.0 (IBM Corporation, Armonk, NY, USA) for analysis of demographic and clinical parameters. The differences between the groups were analyzed using the Kruskal–Wallis test for continuous variables and chi-squared tests for categorical variables. The Kruskal–Wallis test was used also for Luminex cytokine analysis followed by Dunn's test for multiple comparisons when applicable.

For transcriptomics, preprocessing and differential gene analysis were analyzed with *eUTOPIA* bioinformatics packages [46]. Differential gene expression analysis between different test groups was performed by *Limma Model* analysis [47]. Age (biopsy and PBMC), body mass index (BMI) (biopsy and PBMC), and ICS use (PBMC only) were used as co-factors. The multivariate correction of false discovery rate was performed by the *Benjamini–Hochberg* method. A minimum log2 difference of 0.58, together with a maximum adjusted p value of 0.05 was implemented as a cut-off to consider a gene as significantly differentially expressed. Based on our experience in toxicogenomics and clinical transcriptomics/proteomics, a combination of 1.5-fold change threshold, FDR correction, and identification of overrepresented biological pathways is a reliable and validated approach to identify biologically relevant differentially expressed genes. *Perseus* and *Chipster* were used to generate gene clusters and heatmaps [48,49]. The top 2 principal components were used to depict clustering of the samples in PCA plots. Heatmap clustering parameters used were as follows; distance: Euclidean, linkage: average and cluster preprocessing: k-means, with the number of clusters set to exceed the total number of samples—default was 300. Gene-phenotype correlations were studied with a Pearson's correlation. Analysis details are given in Supplementary Materials. All raw data associated with this study, as well as the normalized data matrix, have been submitted into the GEO repository with the accession number: GSE182797, GSE182798.

4.9. Pathway Enrichment Analysis

A three-step process was used to infer the biological relevance of differentially expressed genes. First, the list of differentially expressed genes (DEG) from each patient/control contrast was submitted into Enrichr for identification of overrepresented gene ontology biological processes (GO Biological Process 2021). Fisher's exact test with a maximum FDR threshold of 5% (adjusted $p < 0.05$) was implemented to consider a biological process as significantly enriched. Next, each set of patient/control DEGs consisting of significantly enriched GO biological processes was submitted to INfORM (Inference of NetwOrk Response Modules) [50]—an R-shiny application wherein gene level expression, fold changes, and differential p-values are used to detect, evaluate, and select gene modules with high statistical and biological significance. Finally, the activated or inhibited biological functions represented by each module gene set was determined using a z-scoring algorithm and Fischer's exact test (FET) implemented in the IPA pathway analysis tool (Ingenuity IPA version 65367011, QIAGEN). In IPA, significant overlaps between INfORM-identified response modules and biological processes (functions) relative to the whole Ingenuity knowledgebase were determined (adjusted p-value threshold of 0.05). Z-scoring assessment using the corresponding patient/control expression fold-change of each module gene was then used to predict whether the identified significantly enriched functions were activated or inhibited. Where applicable, we referred to a significantly enriched downstream biological process as activated if the z-score > 0 or inhibited if the z-score < 0. Z-score $= 0$ implies the direction of the effect is unknown or ambiguous. Z-scores that exceed ± 2 are 'highly predictive' of an activated or inhibited biological process.

Supplementary Materials: The following are available online at https://www.mdpi.com/article/10.3390/ijms221910679/s1.

Author Contributions: Study Design, H.A., H.S.; Clinical Samples, H.S., I.L., L.A., P.K. (Paula Kauppi), S.T.-S.; Microarrays, P.K. (Piia Karisola); Bioinformatic Analysis, J.N., H.A., P.K. (Piia Karisola); Statistical Analysis for Clinical Parameters, H.S.; Visualization, J.N., H.A.; Supervision, H.A., H.S., P.K. (Piia Karisola); Writing—Original Draft, H.A., H.S., J.N., P.K. (Piia Karisola); Writing—Review and Editing, H.S., J.N., I.L., L.A., K.K., P.K. (Paula Kauppi), A.L., S.T.-S., P.K. (Piia Karisola), H.A.; Project Administration, H.A., H.S. All authors have read and agreed to the published version of the manuscript.

Funding: This research was funded by The Finnish Work Environment Fund (grant no.114350). Open access funding provided by University of Helsinki.

Institutional Review Board Statement: The study was approved by Helsinki University Hospital Coordinating Ethical Committee (80/13/03/00/15). Each participant gave written informed consent.

Informed Consent Statement: Written informed consent has been obtained from the patient(s) to publish this paper.

Data Availability Statement: The transcriptomics datasets associated with the current study have been deposited in the GEO Omnibus database and are publicly available with accession numbers GSE182797 and GSE182798.

Acknowledgments: We thank Sirpa Hyttinen, Tanja Katovich, and the study nurses for their assistance in this study. We thank Terhi Vesa, Varpu Alenius, and Mikko Asiala for their excellent help in processing cell samples.

Conflicts of Interest: The authors declare no conflict of interest.

References

1. Kaur, R.; Chupp, G. Phenotypes and endotypes of adult asthma: Moving toward precision medicine. *J. Allergy Clin. Immunol.* **2019**, *144*, 1–12. [CrossRef]
2. De Nijs, S.B.; Venekamp, L.N.; Bel, E.H. Adult-onset asthma: Is it really different? *Eur. Respir. Rev.* **2013**, *22*, 44–52. [CrossRef]
3. Tuomisto, L.E.; Ilmarinen, P.; Kankaanranta, H. Prognosis of new-onset asthma diagnosed at adult age. *Respir. Med.* **2015**, *109*, 944–954. [CrossRef]
4. Wang, J.; Pindus, M.; Janson, C.; Sigsgaard, T.; Kim, J.-L.; Holm, M.; Sommar, J.; Orru, H.; Gislason, T.; Johannessen, A.; et al. Dampness, mould, onset and remission of adult respiratory symptoms, asthma and rhinitis. *Eur. Respir. J.* **2019**, *53*, 1801921. [CrossRef] [PubMed]
5. Karvala, K.; Toskala, E.; Luukkonen, R.; Lappalainen, S.; Uitti, J.; Nordman, H. New-onset adult asthma in relation to damp and moldy workplaces. *Int. Arch. Occup. Environ. Health* **2010**, *83*, 855–865. [CrossRef] [PubMed]
6. Karvala, K.; Uitti, J.; Luukkonen, R.; Nordman, H. Quality of life of patients with asthma related to damp and moldy work environments. *Scand. J. Work Environ. Health* **2012**, *39*, 96–105. [CrossRef] [PubMed]
7. Lessof, M. Report of Multiple Chemical Sensitivities (MCS) Workshop, Berlin, Germany, 21–23 February 1996. PCS/96.29 IPCS, Geneva, Switzerland. *Hum. Exp. Toxicol.* **1997**, *16*, 233–234. [CrossRef] [PubMed]
8. Hetherington, L.H.; Battershill, J.M. Review of evidence for a toxicological mechanism of idiopathic environmental intolerance. *Hum. Exp. Toxicol.* **2012**, *32*, 3–17. [CrossRef]
9. Nordin, S. Mechanisms underlying nontoxic indoor air health problems: A review. *Int. J. Hyg. Environ. Health* **2020**, *226*, 113489. [CrossRef]
10. Karvala, K.; Sainio, M.; Palmquist, E.; Nyback, M.H.; Nordin, S. Prevalence of various environmental intolerances in a Swedish and Finnish general population. *Environ. Res.* **2018**, *161*, 220–228. [CrossRef]
11. Karvala, K.; Sainio, M.; Palmquist, E.; Claeson, A.S.; Nyback, M.H.; Nordin, S. Building-Related Environmental Intolerance and Associated Health in the General Population. *Int. J. Environ. Res. Public Health* **2018**, *15*, 2047. [CrossRef]
12. Lind, N.; Söderholm, A.; Palmquist, E.; Andersson, L.; Millqvist, E.; Nordin, S. Comorbidity and Multimorbidity of Asthma and Allergy and Intolerance to Chemicals and Certain Buildings. *J. Occup. Environ. Med.* **2017**, *59*, 80–84. [CrossRef]
13. Chung, K.F.; Wenzel, S.E.; Brozek, J.L.; Bush, A.; Castro, M.; Sterk, P.J.; Adcock, I.M.; Bateman, E.D.; Bel, E.H.; Bleecker, E.R.; et al. International ERS/ATS guidelines on definition, evaluation and treatment of severe asthma. *Eur. Respir. J.* **2014**, *43*, 343–373. [CrossRef]
14. Lachmann, A.; Torre, D.; Keenan, A.B.; Jagodnik, K.M.; Lee, H.J.; Wang, L.; Silverstein, M.C.; Ma'ayan, A. Massive mining of publicly available RNA-seq data from human and mouse. *Nat. Commun.* **2018**, *9*, 1366. [CrossRef] [PubMed]
15. Sze, E.; Bhalla, A.; Nair, P. Mechanisms and therapeutic strategies for non-T2 asthma. *Allergy* **2020**, *75*, 311–325. [CrossRef] [PubMed]
16. IPCS/WHO. Conclusions and recommendations of a workshop on multiple chemical sensitivities (MCS). *Regul. Toxicol. Pharm.* **1996**, *24*, 188–189. [CrossRef]
17. Ilmarinen, P.; Tuomisto, L.E.; Niemela, O.; Tommola, M.; Haanpaa, J.; Kankaanranta, H. Cluster Analysis on Longitudinal Data of Patients with Adult-Onset Asthma. *J. Allergy Clin. Immunol. Pract.* **2017**, *5*, 967–978.e963. [CrossRef] [PubMed]
18. Kuruvilla, M.E.; Lee, F.E.; Lee, G.B. Understanding Asthma Phenotypes, Endotypes, and Mechanisms of Disease. *Clin. Rev. Allergy Immunol.* **2019**, *56*, 219–233. [CrossRef] [PubMed]
19. Skovbjerg, S.; Berg, N.D.; Elberling, J.; Christensen, K.B. Evaluation of the quick environmental exposure and sensitivity inventory in a Danish population. *J. Environ. Public Health* **2012**, *2012*, 304314. [CrossRef] [PubMed]
20. Nordin, S.; Andersson, L. Evaluation of a Swedish version of the Quick Environmental Exposure and Sensitivity Inventory. *Int. Arch. Occup. Environ. Health* **2010**, *83*, 95–104. [CrossRef]
21. Drizik, E.; Corbett, S.; Zheng, Y.; Vermeulen, R.; Dai, Y.; Hu, W.; Ren, D.; Duan, H.; Niu, Y.; Xu, J.; et al. Transcriptomic changes in the nasal epithelium associated with diesel engine exhaust exposure. *Environ. Int.* **2020**, *137*, 105506. [CrossRef]
22. Matsuoka, H.; Niimi, A.; Matsumoto, H.; Ueda, T.; Takemura, M.; Yamaguchi, M.; Jinnai, M.; Chang, L.; Otsuka, K.; Oguma, T.; et al. Patients' characteristics associated with unsuccessful sputum induction in asthma. *J. Allergy Clin. Immunol.* **2008**, *121*, 774–776. [CrossRef]
23. Laan, M.; Lotvall, J.; Chung, K.F.; Linden, A. IL-17-induced cytokine release in human bronchial epithelial cells in vitro: Role of mitogen-activated protein (MAP) kinases. *Br. J. Pharm.* **2001**, *133*, 200–206. [CrossRef]
24. Yokoyama, A.; Kohno, N.; Fujino, S.; Hamada, H.; Inoue, Y.; Fujioka, S.; Ishida, S.; Hiwada, K. Circulating interleukin-6 levels in patients with bronchial asthma. *Am. J. Respir. Crit. Care Med.* **1995**, *151*, 1354–1358. [CrossRef]
25. Virchow, J.C., Jr.; Kroegel, C.; Walker, C.; Matthys, H. Inflammatory determinants of asthma severity: Mediator and cellular changes in bronchoalveolar lavage fluid of patients with severe asthma. *J. Allergy Clin. Immunol.* **1996**, *98*, S27–S33, discussion S33–S40. [CrossRef]
26. Mitchell, C.S.; Donnay, A.; Hoover, D.R.; Margolick, J.B. Immunologic parameters of multiple chemical sensitivity. *Occup. Med.* **2000**, *15*, 647–665. [PubMed]
27. De Luca, C.; Scordo, M.G.; Cesareo, E.; Pastore, S.; Mariani, S.; Maiani, G.; Stancato, A.; Loreti, B.; Valacchi, G.; Lubrano, C.; et al. Biological definition of multiple chemical sensitivity from redox state and cytokine profiling and not from polymorphisms of xenobiotic-metabolizing enzymes. *Toxicol. Appl. Pharm.* **2010**, *248*, 285–292. [CrossRef] [PubMed]

28. Dantoft, T.M.; Elberling, J.; Brix, S.; Szecsi, P.B.; Vesterhauge, S.; Skovbjerg, S. An elevated pro-inflammatory cytokine profile in multiple chemical sensitivity. *Psychoneuroendocrinology* **2014**, *40*, 140–150. [CrossRef] [PubMed]
29. Bigler, J.; Boedigheimer, M.; Schofield, J.P.R.; Skipp, P.J.; Corfield, J.; Rowe, A.; Sousa, A.R.; Timour, M.; Twehues, L.; Hu, X.; et al. A Severe Asthma Disease Signature from Gene Expression Profiling of Peripheral Blood from U-BIOPRED Cohorts. *Am. J. Respir. Crit. Care Med.* **2017**, *195*, 1311–1320. [CrossRef] [PubMed]
30. Doeing, D.C.; Solway, J. Airway smooth muscle in the pathophysiology and treatment of asthma. *J. Appl. Physiol.* **2013**, *114*, 834–843. [CrossRef]
31. Zuyderduyn, S.; Sukkar, M.B.; Fust, A.; Dhaliwal, S.; Burgess, J.K. Treating asthma means treating airway smooth muscle cells. *Eur. Respir. J.* **2008**, *32*, 265–274. [CrossRef]
32. Alcala, S.E.; Benton, A.S.; Watson, A.M.; Kureshi, S.; Reeves, E.M.; Damsker, J.; Wang, Z.; Nagaraju, K.; Anderson, J.; Williams, A.M.; et al. Mitotic asynchrony induces transforming growth factor-beta1 secretion from airway epithelium. *Am. J. Respir. Cell Mol. Biol.* **2014**, *51*, 363–369. [CrossRef] [PubMed]
33. Freishtat, R.J.; Watson, A.M.; Benton, A.S.; Iqbal, S.F.; Pillai, D.K.; Rose, M.C.; Hoffman, E.P. Asthmatic airway epithelium is intrinsically inflammatory and mitotically dyssynchronous. *Am. J. Respir. Cell Mol. Biol.* **2011**, *44*, 863–869. [CrossRef] [PubMed]
34. Wang, Z.N.; Su, R.N.; Yang, B.Y.; Yang, K.X.; Yang, L.F.; Yan, Y.; Chen, Z.G. Potential Role of Cellular Senescence in Asthma. *Front. Cell Dev. Biol.* **2020**, *8*, 59. [CrossRef]
35. Ndika, J.; Suojalehto, H.; Taubel, M.; Lehto, M.; Karvala, K.; Pallasaho, P.; Sund, J.; Auvinen, P.; Jarvi, K.; Pekkanen, J.; et al. Nasal mucosa and blood cell transcriptome profiles do not reflect respiratory symptoms associated with moisture damage. *Indoor Air* **2018**, *28*, 721–731. [CrossRef]
36. Van den Bergh, O.; Witthoft, M.; Petersen, S.; Brown, R.J. Symptoms and the body: Taking the inferential leap. *Neurosci. Biobehav. Rev.* **2017**, *74*, 185–203. [CrossRef]
37. Segerstrom, S.C.; Miller, G.E. Psychological stress and the human immune system: A meta-analytic study of 30 years of inquiry. *Psychol. Bull.* **2004**, *130*, 601–630. [CrossRef]
38. Tian, R.; Hou, G.; Li, D.; Yuan, T.F. A possible change process of inflammatory cytokines in the prolonged chronic stress and its ultimate implications for health. *Sci. World J.* **2014**, *2014*, 780616. [CrossRef] [PubMed]
39. Choi, S.; Woo, J.K.; Jang, Y.-S.; Kang, J.-H.; Hwang, J.-I.; Seong, J.K.; Yoon, Y.S.; Oh, S.H. Ninjurin1 Plays a Crucial Role in Pulmonary Fibrosis by Promoting Interaction between Macrophages and Alveolar Epithelial Cells. *Sci. Rep.* **2018**, *8*, 17542. [CrossRef]
40. Ifergan, I.; Kebir, H.; Terouz, S.; Alvarez, J.I.; Lécuyer, M.-A.; Gendron, S.; Bourbonnière, L.; Dunay, I.R.; Bouthillier, A.; Moumdjian, R.; et al. Role of ninjurin-1 in the migration of myeloid cells to central nervous system inflammatory lesions. *Ann. Neurol.* **2011**, *70*, 751–763. [CrossRef] [PubMed]
41. Jennewein, C.; Sowa, R.; Faber, A.C.; Dildey, M.; von Knethen, A.; Meybohm, P.; Scheller, B.; Dröse, S.; Zacharowski, K. Contribution of Ninjurin1 to Toll-Like Receptor 4 Signaling and Systemic Inflammation. *Am. J. Respir. Cell Mol. Biol.* **2015**, *53*, 656–663. [CrossRef]
42. Padyana, A.K.; Gross, S.; Jin, L.; Cianchetta, G.; Narayanaswamy, R.; Wang, F.; Wang, R.; Fang, C.; Lv, X.; Biller, S.A.; et al. Structure and inhibition mechanism of the catalytic domain of human squalene epoxidase. *Nat. Commun.* **2019**, *10*, 97. [CrossRef] [PubMed]
43. Tahtinen, K.; Lappalainen, S.; Karvala, K.; Remes, J.; Salonen, H. Association between Four-Level Categorisation of Indoor Exposure and Perceived Indoor Air Quality. *Int. J. Environ. Res. Public Health* **2018**, *15*, 679. [CrossRef]
44. Lacour, M.; Zunder, T.; Schmidtke, K.; Vaith, P.; Scheidt, C. Multiple chemical sensitivity syndrome (MCS)–suggestions for an extension of the U.S. MCS-case definition. *Int. J. Hyg. Environ. Health* **2005**, *208*, 141–151. [CrossRef]
45. Miller, C.S.; Prihoda, T.J. The Environmental Exposure and Sensitivity Inventory (EESI): A standardized approach for measuring chemical intolerances for research and clinical applications. *Toxicol. Ind. Health* **1999**, *15*, 370–385. [CrossRef] [PubMed]
46. Marwah, V.S.; Scala, G.; Kinaret, P.A.S.; Serra, A.; Alenius, H.; Fortino, V.; Greco, D. eUTOPIA: SolUTion for Omics data PreprocessIng and Analysis. *Source Code Biol. Med.* **2019**, *14*, 1. [CrossRef] [PubMed]
47. Ritchie, M.E.; Phipson, B.; Wu, D.; Hu, Y.; Law, C.W.; Shi, W.; Smyth, G.K. limma powers differential expression analyses for RNA-sequencing and microarray studies. *Nucleic Acids Res.* **2015**, *43*, e47. [CrossRef]
48. Kallio, M.A.; Tuimala, J.T.; Hupponen, T.; Klemelä, P.; Gentile, M.; Scheinin, I.; Koski, M.; Käki, J.; Korpelainen, E.I. Chipster: User-friendly analysis software for microarray and other high-throughput data. *BMC Genom.* **2011**, *12*, 507. [CrossRef]
49. Tyanova, S.; Temu, T.; Sinitcyn, P.; Carlson, A.; Hein, M.Y.; Geiger, T.; Mann, M.; Cox, J. The Perseus computational platform for comprehensive analysis of (prote)omics data. *Nat. Methods* **2016**, *13*, 731–740. [CrossRef]
50. Marwah, V.; Kinaret, P.; Serra, A.; Scala, G.; Lauerma, A.; Fortino, V.; Greco, D. INfORM: Inference of NetwOrk Response Modules. *Bioinformatic* **2018**, *34*, 2136–2138. [CrossRef]

International Journal of
Molecular Sciences

Review

Nuclear mRNA Export and Aging

Hyun-Sun Park [1,†], Jongbok Lee [2,†], Hyun-Shik Lee [3], Seong Hoon Ahn [4,*,‡] and Hong-Yeoul Ryu [3,*,‡]

[1] Department of Biochemistry, Inje University College of Medicine, Busan 50834, Korea; hspark@inje.ac.kr
[2] Department of Biological and Chemical Engineering, Hongik University, 2639, Sejong-ro, Jochiwon-eup, Sejong-si 30016, Korea; jlee0917@hongik.ac.kr
[3] BK21 FOUR KNU Creative BioResearch Group, School of Life Sciences, College of National Sciences, Kyungpook National University, Daegu 41566, Korea; leeh@knu.ac.kr
[4] Department of Molecular and Life Science, College of Science and Convergence Technology, ERICA Campus, Hanyang University, Ansan 15588, Korea
* Correspondence: hoon320@hanyang.ac.kr (S.H.A.); rhr4757@knu.ac.kr (H.-Y.R.);
  Tel.: +82-31-400-5518 (S.H.A.); +82-53-950-6352 (H.-Y.R.)
† These authors contributed equally to this work.
‡ These authors jointly supervised.

Abstract: The relationship between transcription and aging is one that has been studied intensively and experimentally with diverse attempts. However, the impact of the nuclear mRNA export on the aging process following its transcription is still poorly understood, although the nuclear events after transcription are coupled closely with the transcription pathway because the essential factors required for mRNA transport, namely TREX, TREX-2, and nuclear pore complex (NPC), physically and functionally interact with various transcription factors, including the activator/repressor and pre-mRNA processing factors. Dysregulation of the mediating factors for mRNA export from the nucleus generally leads to the aberrant accumulation of nuclear mRNA and further impairment in the vegetative growth and normal lifespan and the pathogenesis of neurodegenerative diseases. The optimal stoichiometry and density of NPC are destroyed during the process of cellular aging, and their damage triggers a defect of function in the nuclear permeability barrier. This review describes recent findings regarding the role of the nuclear mRNA export in cellular aging and age-related neurodegenerative disorders.

Keywords: mRNA export; TREX; TREX-2; NPC; lifespan; neurodegenerative diseases

Citation: Park, H.-S.; Lee, J.; Lee, H.-S.; Ahn, S.H.; Ryu, H.-Y. Nuclear mRNA Export and Aging. *Int. J. Mol. Sci.* **2022**, *23*, 5451. https://doi.org/10.3390/ijms23105451

Academic Editors: Amelia Casamassimi, Alfredo Ciccodicola and Monica Rienzo

Received: 12 April 2022
Accepted: 12 May 2022
Published: 13 May 2022

Publisher's Note: MDPI stays neutral with regard to jurisdictional claims in published maps and institutional affiliations.

1. Introduction

Eukaryotic transcription is a complex stepwise process comprised of the transcription initiation, elongation, and termination and requires multiple factors, including transcription machinery (RNA polymerase and general transcription factors), transcription cofactors (coactivator or corepressor), and chromatin regulators [1]. During the process of transcription, the nascent pre-mRNA is processed by 5′-end capping, removal of intron via splicing, and 3′-end cleavage and polyadenylation. The mature mRNA is then exported from the nucleus to the cytoplasm to undergo protein translation, and aberrantly processed pre-mRNAs and mRNAs are eliminated via the RNA surveillance system. Despite the distinct factors that carry out each of the steps in the pathway of gene expression, each factor interacts both physically and functionally with other proteins in the different pathways, coupling among the gene expression machineries [2].

Aging is a process that is accompanied by the progressive impairment at the molecular, cellular, and organ levels, eventually leading to the decay of biological and physiological functions and the increased risk of diverse aging-related diseases such as cancer, cardiovascular, and neurodegenerative diseases [3,4]. The rate and progression of aging are influenced by highly complex and diverse genetic and environmental factors, and the transcription process is linked closely to aging and its related disorders [5].

The transcription of genes triggers an increased opportunity for damage or mutation to affect DNA because the transcription machinery-mediated unwinding of the DNA double helix leads to exposure of single strand DNA to mutagenic agents [6]. The transcription's fidelity is remarkably impaired with aging, contributing to genotoxicity and proteotoxicity and the eventual reduction of cellular longevity [7]. The transcription errors are not always random or temporary and often mimic DNA mutations, frequently inducing genetic diseases. For example, the transcription errors in the genes encoding UBB and APP lead to translation of the toxic forms of ubiquitin-B and amyloid precursor proteins in patients with Alzheimer's disease [8,9], and the 8-Oxoguanine-mediated transcription errors in the *RAS* gene can induce the oncogenic pathway in mammalian cells [10]. Furthermore, such errors occasionally cause proteins to be misfolded, which can escape recognition by protein quality control machinery and survive inside cells for extended periods of time [11]. In various species, tissues, and cell types, aging is associated with alterations in the expression of diverse genes involved in signaling pathways, genetics, translational mechanisms, and metabolism, and its maintenance is critical for normal functioning to continue [5]. Although transcription is an essential process for life and survival, transcription itself and its misregulation cause genome instability and premature aging.

Similarly, the correlation between transcription and aging has been studied intensively through various experiments. However, extraordinarily little is known about the effects exhibited on the aging process as a result of nuclear events after transcription takes place, and such research has not been focused up to date. Thus, this review provides an initial overview of the current studies and recent progress in elucidating the role of the nuclear RNA export in cellular aging and the pathogenesis of neurodegenerative disorders.

2. Nuclear mRNA Export Pathway

In eukaryotic cells, the nuclear export of mRNA transcripts requires multiple cellular events including transcription, maturation of pre-mRNA, and the assembly of mature mRNA with specific RNA binding proteins, an establishing messenger ribonucleoprotein (mRNP) complex, and the mRNA transport through the nuclear pore complexes (NPCs) into the cytoplasm [12].

The process of mRNA processing is cotranscriptionally coupled to the mRNA transport pathway. In particular, $3'$-end mRNA processing is clearly linked with the mRNA export, and the involved factors are evolutionary conserved from yeast to humans [13,14] (Figure 1). In yeast, polyadenylation leads to the recruitment of the poly(A) binding protein Nab2 (ZC3H14 in human) and its binding partner Yra1 (ALY in human), interacting directly with the essential mRNA export receptor Mex67-Mtr2 (NXF1-NXT1 in human) [15–18]. In addition, the recruitment of Yra1 is dependent on the interaction with Pcf11, an essential component of cleavage and the polyadenylation factor IA, which then transfers Yra1 to the transcription/export (TREX) complex with the aid of the Sub2 helicase (UAP56 in human) [19–22]. The THO proteins (Tho2, Hpr1, Mft1, Thp2, and Tex1), a core member of the TREX complex, and the Sub2 helicase are required for efficient polyadenylation by the Pap1 Poly(A) polymerase, indicating coupling among polyadenylation, dissociation of the polyadenylation proteins, and the release of the mRNP from the transcription unit [23–25]. Recently, it was also reported that two distinct ALY-interacting factors, NXF1 and TREX, prefer selectively to export different transcript groups depending on exon architecture and G/C content in human cells [26]. Additionally, mammalian SR proteins known as an alternative pre-mRNA splicing factor promote NXF1 recruitment to mRNA, and this interaction suggests a link between alternative splicing and the mRNA export, thereby controlling the cytoplasmic abundance of transcripts with alternative $3'$ ends [27].

The transcription and export complex-2 (TREX-2), composed of Sac3, Thp1, Cdc31, Sem1, and Sus1, physically and functionally interacts with both the Spt-Ada-Gcn5 acetyltransferase (SAGA) transcription coactivator complex and NPC to link the transcription, mRNA export, and targeting of active genes to NPC [28] (Figure 2A). The N-terminus in Sac3 acts as a scaffold for association with Thp1 and Sem1, creating an mRNA-binding module, and with the mRNA exporter Mex67-Mtr2, whereas its C-terminus binds to Sus1, Cdc31, and Nup1 nucleoporin, providing a docking platform at NPC [29–33]. TREX-2 shares one subunit Sus1 with the DUB module for deubiquitination of H2B in the SAGA complex, and Sus1 simultaneously associates with the promoter and coding regions of some SAGA-dependent genes, offering a functional link of the transcription activation to the mRNA export [34–37]. Specifically, the Sus1, Sac3, and Thp1 subunits facilitate the post-transcriptional anchoring of transcribed genes to NPC upon the activation of transcription [38–40]. Both Cdc31 and Sem1 also contribute synergistically to mediate the association of TREX-2 with NPC for promoting the mRNA export process [31,41]. In human TREX-2, a germinal center associated nuclear protein (GANP), known as Sac3 orthologue, is associated with the RNA polymerase II and Nxf1 (Mex67 in yeast), facilitating the movement of mRNP to NPC [42]. Inhibition of the processing of mRNA leads to the redistribution of GANP from NPC into nuclear foci, suggesting that TREX-2 mediates the transportation of the mRNP from active genes to NPC [43].

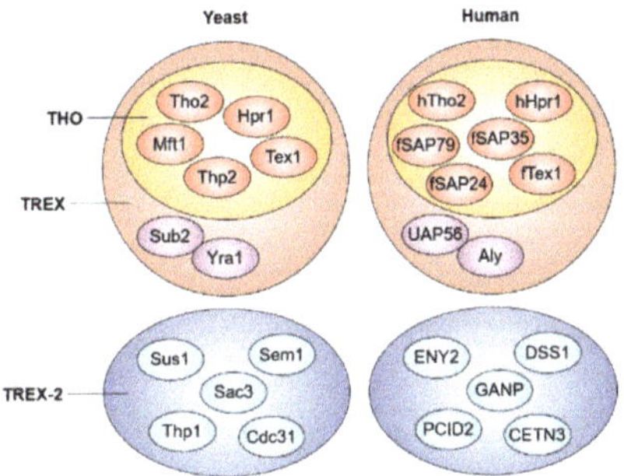

Figure 1. The conserved TREX and TREX-2 complexes. Both TREX and TREX-2 complexes are conserved between yeast and human. The TREX complex includes the multi-subunit THO complex and mRNA export proteins.

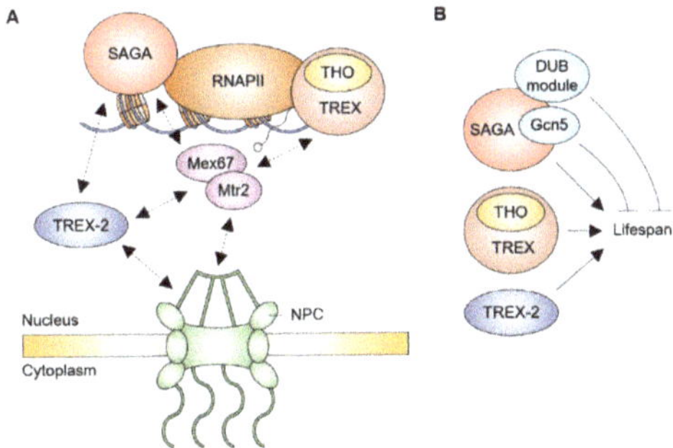

Figure 2. The effects of SAGA, TREX, and TREX-2 complexes on the yeast lifespan. (**A**) Model of SAGA, TREX, and TREX-2 complexes-mediated mRNA export pathway. At the stage of transcription initiation, SAGA is recruited to RNA polymerase II (RNAPII) machinery and mediates activation of transcription. TREX is co-transcriptionally recruited and associates with nascent transcripts. The mRNA export receptor Mex67-Mtr2 interacts with SAGA, TREX, and TREX-2 complexes and NPC, which facilitates passage of mature mRNP to cytoplasm. TREX-2 shares a component with SAGA and promotes anchoring of mRNP to NPC. Illustration reflects the relevant location of proteins but not precise physical association. (**B**) Many subunits in SAGA, TREX, and TREX-2 are required for blocking a shortened lifespan, whereas Gcn5 and DUB module (except for Sus1) in SAGA limit an abnormal extension of the lifespan.

3. NPC and mRNA Export

The eukaryotic NPC is composed of about 30 nucleoporin proteins and has a radial symmetry of eightfold. Its structure is composed of three main parts, a central core spanning the nuclear envelope (NE) membrane, a nuclear basket, and long cytoplasmic filaments, and selectively allows most of the mRNPs to disperse in and out of the nucleus in order to maintain a barrier of nuclear permeability [44]. The transmembrane nucleoporins physically tether NPC to the NE membrane, while structural nucleoporins, embedded in the NPC, serve as a platform for the other nucleoporins and FG-nucleoporins containing phenylalanine–glycine (FG)-repeats, such as FG, FXFG, and GLFG [45,46]. The symmetrical FG-nucleoporins are located on the both sides of the NPC, while the asymmetrical FG-nucleoporins are observed exclusively on one side of the NPC [46].

The mRNP anchors to the NPC by interacting directly between the mRNA export factors and basket nucleoporins located in the nucleoplasmic region [12]. In yeast, the FG-repeats of Nup49, Nup57, Nup1, and Nup2 nucleoporins provide the first docking sites for mRNP to NPC via their interaction with Mex67 [47]. However, when the transport route of single native mRNA particles was monitored in insect and human cells, 60–75% of them are able to return to the interchromatin region after association with the basket [48,49]. Furthermore, when the export procedure was inhibited by the treatment of wheat germ agglutinin (WGA) in human cells, an accumulation of mRNPs at the nuclear periphery was found, suggesting that the interaction between the mRNP and NPC is independent of the export process [50]. Another plausible explanation is that such nucleoplasmic mRNP flux may function as a rate-limiting step at the NPC basket by spending for a long duration before reaching the mRNP to the NPC. When the imaging study revealed single native mRNA particles moving across the NE in insect cells, only 25% of the encounter particles with NE were successfully sent to the cytoplasm [48]. Additionally, monitoring the actual flow of the β-actin mRNA revealed that the rate-limiting steps for the nucleocytoplasmic transport of the mRNP are both the access and the release from the NPC [51]. Therefore, the quality control and surveillance mechanisms for mRNA are estimated to be important pathways for the rate-limiting step [12].

4. mRNA Export Factors and Aging

The connection between gene expression and aging is reflected in the diverse transcription factors that can operate as the key factors in regulating the various cellular processes [52–60]. Among such factors involved in the regulation of lifespan, the SAGA complex, a physical and functional partner of TREX-2, has multiple roles depending on its independent modules, HAT module (histone acetylation), DUB module (deubiquitination of H2B), TAF module (coactivator architecture), and SPT module (assembly of the preinitiation complex), in yeast aging pathway [61] (Figure 2B). The presence of a HAT inhibitor, inducing a low level of histone acetylation, leads to an extended replicative lifespan (RLS) which is completely abolished upon the loss of Gcn5, a catalytic subunit of the SAGA HAT module [62]. A RLS is significantly also extended in the presence of the heterozygous mutant *gcn5* or *ngg1*, a gene encoding a linking protein between Gcn5 and SAGA [62,63], whereas each loss of other components in the SAGA HAT module does not lead to an increase in the yeast lifespan [64,65]. A loss of Ubp8, Sgf73, or Sgf11 in the SAGA DUB module greatly extends a RLS in a Sir2-dependent mechanism for maintaining telomeric silencing and rDNA stability, the most representative pathway for controlling the lifespan of yeast [65], while both a RLS and the chronological lifespan (CLS) are mostly decreased in the cells lacking each component in the SAGA SPT module [64]. In addition, SAGA promotes anchoring the non-chromosomal DNA circles to the NPC and concomitantly leads to confinement of such circles in the mother nucleus, which is a characteristic feature of aged nucleus [66]. Although it is still unclear how a single complex has multiple functions that ensure a normal lifespan, SAGA is a good example of how aging is finely tuned by regulators in a complex network.

The THO complex is required for the environmental stress response and maintaining a normal fly lifespan. Mutations in the THO complex resulted in a shortened lifespan and strong sensitivity to certain environmental stressors. This is suppressed by the upregulation of c-Jun N-terminal kinase signaling which regulates stress tolerance and longevity [67]. Genome-wide transcriptomic analyses revealed that the gene expression of TREX and other factors that are required for trafficking nucleocytoplasma were globally downregulated in five distinct types of senescent cells, representing replicative senescence, tumor cell senescence, oncogene-induced senescence, stem cell senescence, and progeria and endothelial cell senescence. Such a similar enrichment pattern was observed in two large human tissue genomic databases: Genotype-Tissue Expression and The Cancer Genome Atlas [68]. Furthermore, the enrichment patterns of TREX and NPC-related factors were conversely upregulated during the process of tumorogenesis, suggesting that the failure of age-related changes in gene expression profile of TREX and related factors may lead to an increased risk for aging-related cancer [68].

A very recent study revealed that TREX-2 is also involved in the maintenance of a normal lifespan in yeast [69]. The loss of two major structural components of TREX-2, Thp1 and Sac3, and a linker protein Sus1 between the SAGA DUB module and TREX-2 impaired the normal lifespan and vegetative growth. In particular, TREX-2 regulates the RLS in a Sir2-independent manner, and the growth and lifespan defects by the loss of Sus1 were the fault of TREX-2 rather than the SAGA DUB. Moreover, the growth defect, shortened lifespan, and nuclear accumulation of poly(A)$^+$ RNA in cells lacking Sus1 were rescued by an increased dosage of the mRNA export factors Mex67 and Dbp5, whose association with the nuclear rim was affected by Sus1, suggesting that boosting the mRNA export process restores the defect of mRNA transport and further damage in the growth and lifespan by lack of Sus1 (Figure 3). In short, an abnormal accumulation of nuclear RNA is a negative factor for ensuring a normal lifespan.

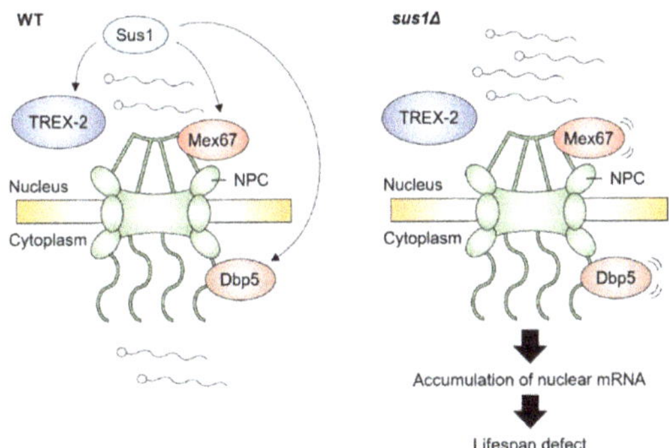

Figure 3. Sus1-mediated mRNA export pathway is required for maintaining a normal lifespan in yeast. In WT, Sus1, a component of TREX-2 complex, facilitates the proper association of Mex67 and Dbp5 with NPC, which requires efficient mRNA transport from nucleus to cytoplasm. In contrast, deletion of *SUS1* leads to mislocalization of Mex67 and Dbp5 and accumulation of nuclear mRNA, resulting in a further defect in the lifespan. Illustration reflects the relevant location of proteins but not their precise physical association.

5. NPC and Aging

The age-dependent deterioration of nucleoporins accelerates the damages in the structure and function of the NPC, leading to the loss of the barrier of nuclear permeability and a leakage of cytoplasmic proteins into the nucleus [70]. In differentiated rat brain cells, nucleoporins are oxidized and long-lived without a turnover of the NPC via the degradation of old proteins and a new synthesis which results in definite harmful effects [70,71]. In yeast, the correlation between NPCs and the lifespan of cells was directly analyzed by a

RLS measurement method [72]. The RLS was impaired by lacking the GLFG domain of Nup116, while such a shortened lifespan is rescued by the overexpression of Gsp1, the small GTPase that facilitates the karyopherin Kap121-mediated transport. However, the Nup100-mediated control of the tRNA life cycle potentially limits the yeast lifespan [72,73].

The optimal stoichiometry and density of NPC are disrupted during the process of aging [74]. The senescent human fibroblasts exhibit several characteristic features, such as hypo-responsiveness either to growth factors or to apoptotic signals that are induced by diverse stimuli [75–78] and a decreased cellular level of nucleocytoplasmic transport factors, Nup88, Nup107, Nup155, Nup50, karyopherin, Ran (Ras-related GTPase), and Ran-regulating factors, suggesting that senescence-associated hypo-responsiveness would be the result from a reduction in the nuclear translocation by the loss of the stoichiometry of nucleocytoplasmic transporters [79]. Such an alteration in the optimal level of the NPC is similarly observed in older yeast cells [80,81]. Although another senescent phenotype is the changed distribution and density of the NPC at NE, reflecting irregular nuclear organization and function [82], it is unclear how the density of the NPC increases its effects on the longevity and the process of aging.

6. MRNA Turnover and Aging

The mRNA surveillance process ensures that the properly processed transcripts are present within the cell and are coupled to the mRNA export pathway [83], and its defects often drive cellular senescence [84]. For instance, a decrease in the human RNA turnover rate via the declined activity of the RNA exosome or oxidative stress triggers cellular senescence [85]. The expression of the *PHO84* gene is repressed by the corresponding antisense RNA in chronologically aged yeast cells, and stabilization of the antisense RNA is facilitated by the Rrp6/exosome complex and histone acetylation [86]. In addition, unspliced or malformed transcripts are identified and degraded during the quality control step involved with certain nucleoporins, endonuclease Swt1, and protease Ulp1 upon the docking of the mRNP to the basket of NPC in yeast [87,88]. Therefore, the mRNA turnover mechanism inhibits nuclear accumulation and the abnormal export of misprocessed RNA species and is important in preventing pathophysiological cell senescence and cell death.

7. MRNA Export and Age-Related Neurodegenerative Disorders

A defective mRNA export is implicated in diverse neurodegenerative disorders [89,90]. The mislocalization of the THO complex subunit two (THOC2) to the cytoplasm was detected in HEK293T cells that were transfected with Htt96Q or TDP-43 associated with Huntington's disease (HD) and amyotrophic lateral sclerosis (ALS), respectively [91]. The mutations of THOC4 act as a potential neurodegeneration suppressor in the fly ALS model [92], while a knockout of THOC5 in mouse dopaminergic neurons leads to a defect in the nuclear export of synaptic transcripts and degeneration of the neurons, leading to the death of the animal [93]. Matrin3, a protein associated with ALS, interacts physically with multiple TREX proteins, and its mutations cause the nuclear mRNA export defects of both the global mRNA and ALS-related transcripts in particular [94]. In addition, TDP-43 itself binds thousands of introns and $3'$ UTRs of pre-mRNAs, and its mutations lead to abnormal localization of nucleoporins and the nuclear retention of poly(A)$^+$ RNA [95,96]. Similar to TDP-43, FUS, whose mutations cause ALS, also associates with thousands of mRNAs, and it appears to promote mRNA export in neural dendrites [97–99].

The abnormal aggregation of two scaffold nucleoporins Nup205 and Nup107, mislocalization of Nup62 at NE, and mutations in the GLE1 gene encoding the nuclear mRNA export factor that physically interacts with NPC were found in patients with ALS [100–102]. A loss of Nup358 (also called E3 SUMO-protein ligase RanBP2) in murine motoneurons drives the ALS-like syndrome, suggesting that the irregular composition and distribution of nucleoporin might play an important role in ALS pathophysiology [103]. Similar to ALS, abnormal localization of Gle1, Nup62, and RanGAP1 (the binding partner of Nup358) in multiple models of HD and Nup62 in the hippocampus and neocortex of Alzheimer's disease patients was previously reported [104–106]. Parkin, whose mutations are considered to be one of the most common causes of the familial Parkinson's disease, selectively binds to Nup358 and promotes its degradation [107,108]. In addition, defects in the export of the mitochondrial mRNA through NE budding, a distinct pathway with the NPC-mediated nucleocytoplasmic transport, displayed progressive mitochondrial disruption, resulting in accelerated aging [109]. Taken together, these studies indicate that the disruption of the nucleocytoplasmic transport is a central feature of neurodegenerative diseases. Studying the function of the mRNA export in aging may provide clues for developing new therapies that can block neurodegeneration.

8. Conclusions

The link between gene transcription and aging has been well characterized in diverse studies. The gene expression profile is extremely changed in senescent cells, indicating the various biological events that occur during the process of aging [110], and the transcription itself is able to accelerate the rate of damage to DNA, leading to genomic instability and further premature aging [6]. The transcription error rates are increased with aging, inducing the aggregation of peptides that characterize age-associated disorders [11]. The nuclear events that occur after the process of transcription are also an important element related to cellular aging, and this process is closely coupled with the transcription progress. The essential factors required for the mRNA export, TREX, TREX-2, and NPC, are dynamically interacted with a number of transcription factors, including SAGA and pre-mRNA processing factors [14,19,20,22,24,28,42], and the mutation of such factors involved in the transportation of mRNA generally triggers the accumulation of the nuclear RNA by blocking a release of the mRNA into the cytoplasm and further shortened lifespan [15,17–20,29–33,36,47,69,72,73,93]. However, because the study of a defect in the mRNA export was focused on monitoring a single native mRNA molecule or poly(A)$^+$ RNA, genome-wide analysis approaches may assist in uncovering whether nucleoplasmic trafficking of specific RNA transcript(s) affects cellular lifespan. In addition, the induction of the smooth mRNA transport by an increased dosage of Mex67 or Dbp5 rescues the decreased lifespan in *sus1Δ* cells, implying that the prevention of nuclear RNA accumulation plays an important role in cellular aging [69]. Therefore, a change in the localization, stoichiometry, and density of the mRNA export factors may potentially hold value as a new marker for the detection of cellular aging and the study of longevity.

The histones are subject to multiple PTMs, including acetylation, methylation, phosphorylation, ubiquitination, and sumoylation, and such patterns of PTM constitute codes that regulate elaborate chromatin-based processes [111–115]. Not only SAGA-mediated regulation of deubiquitination and acetylation on histones but also the diverse modifications on histones is able to influence the pathway of the mRNA export. For example, yeast Mog1, a Ran GTPase-binding protein required for the nuclear protein import, maintains normal levels of the H2B ubiquitination and H3K4 methylation, and the mRNA export defect in *mog1Δ* is aggravated by the additional loss of factors for H2B ubiquitylation [116]. Direct interaction between the Setd2 H3K36 methyltransferase and Spt6/Iws1 transcription elongation complex may facilitate kinetics of the mRNA transport in human cell lines [117]. Additionally, because evidence has provided insight into the connection between histone modifications and aging [118], histone modifications may have a potential role in linking the nuclear RNA export to the lifespan. However, except for the SAGA complex having the activity of histone deubiquitination and acetylation, there are no available reports concerning the effects of the histone modification-mediated control of the mRNA export on the aging pathway. Hence, the better characterization of how histone modifications modulate the mRNA export from the nucleus to the cytoplasm may be a promising avenue for future research exploring the prevention of premature aging and the development of a new therapy for neurodegenerative disorders.

Author Contributions: Conceptualization, H.-S.P. and J.L.; writing, H. S.P. and H.-Y.R.; review and editing, J.L., H.-S.L., S.H.A. and H.-Y.R.; funding acquisition, S.H.A. and H.-Y.R. All authors have read and agreed to the published version of the manuscript.

Funding: This study was supported by a National Research Foundation of Korea (NRF) grant funded by the South Korean government (MSIT) (no. 2021R1F1A1051082) to S.H.A. and (nos. 2020R1C1C1009367, and 2020R1A4A1018280) to H.-Y.R.

Institutional Review Board Statement: Not applicable.

Informed Consent Statement: Not applicable.

Data Availability Statement: Not applicable.

Conflicts of Interest: The authors declare that they have no conflict of interest.

References

1. Kornberg, R.D. The molecular basis of eukaryotic transcription. *Proc. Natl. Acad. Sci. USA* **2007**, *104*, 12955–12961. [CrossRef] [PubMed]
2. Maniatis, T.; Reed, R. An extensive network of coupling among gene expression machines. *Nature* **2002**, *416*, 499–506. [CrossRef] [PubMed]
3. Fontana, L.; Partridge, L.; Longo, V.D. Extending Healthy Life Span—From Yeast to Humans. *Science* **2010**, *328*, 321–326. [CrossRef] [PubMed]
4. Lai, R.W.; Lu, R.; Danthi, P.S.; Bravo, J.I.; Goumba, A.; Sampathkumar, N.K.; Benayoun, B.A. Multi-level remodeling of transcriptional landscapes in aging and longevity. *BMB Rep.* **2019**, *52*, 86–108. [CrossRef] [PubMed]
5. Stegeman, R.; Weake, V.M. Transcriptional Signatures of Aging. *J. Mol. Biol.* **2017**, *429*, 2427–2437. [CrossRef] [PubMed]
6. Callegari, A.J. Does transcription-associated DNA damage limit lifespan? *DNA Repair* **2016**, *41*, 1–7. [CrossRef]
7. Verheijen, B.M.; Van Leeuwen, F.W. Commentary: The landscape of transcription errors in eukaryotic cells. *Front. Genet.* **2017**, *8*, 219. [CrossRef]
8. Van Leeuwen, F.W.; Hol, E.M.; Burbach, J.P. Mutations in RNA: A first example of molecular misreading in Alzheimer's disease. *Trends Neurosci.* **1998**, *21*, 331–335. [CrossRef]
9. van Leeuwen, F.W.; de Kleijn, D.P.V.; Van Den Hurk, H.H.; Neubauer, A.; Sonnemans, M.A.F.; Sluijs, J.A.; Köycü, S.; Ramdjielal, R.D.J.; Salehi, A.; Martens, G.J.M.; et al. Frameshift Mutants of β Amyloid Precursor Protein and Ubiquitin-B in Alzheimer's and Down Patients. *Science* **1998**, *279*, 242–247. [CrossRef]
10. Saxowsky, T.T.; Meadows, K.L.; Klungland, A.; Doetsch, P.W. 8-Oxoguanine-mediated transcriptional mutagenesis causes Ras activation in mammalian cells. *Proc. Natl. Acad. Sci. USA* **2008**, *105*, 18877–18882. [CrossRef]
11. Vermulst, M.; Denney, A.S.; Lang, M.; Hung, C.-W.; Moore, S.; Moseley, M.A.; Thompson, W.J.; Madden, V.; Gauer, J.; Wolfe, K.J.; et al. Transcription errors induce proteotoxic stress and shorten cellular lifespan. *Nat. Commun.* **2015**, *6*, 8065. [CrossRef] [PubMed]
12. Björk, P.; Wieslander, L. Mechanisms of mRNA export. *Semin. Cell Dev. Biol.* **2014**, *32*, 47–54. [CrossRef] [PubMed]
13. Schmid, M.; Jensen, T.H. Quality control of mRNP in the nucleus. *Chromosoma* **2008**, *117*, 419–429. [CrossRef] [PubMed]
14. Qu, X.; Lykke-Andersen, S.; Nasser, T.; Saguez, C.; Bertrand, E.; Jensen, T.H.; Moore, C. Assembly of an Export-Competent mRNP Is Needed for Efficient Release of the 3′-End Processing Complex after Polyadenylation. *Mol. Cell. Biol.* **2009**, *29*, 5327–5338. [CrossRef]
15. Strässer, K.; Hurt, E. Yra1p, a conserved nuclear RNA-binding protein, interacts directly with Mex67p and is required for mRNA export. *EMBO J.* **2000**, *19*, 410–420. [CrossRef]
16. Batisse, J.; Batisse, C.; Budd, A.; Böttcher, B.; Hurt, E. Purification of Nuclear Poly(A)-binding Protein Nab2 Reveals Association with the Yeast Transcriptome and a Messenger Ribonucleoprotein Core Structure. *J. Biol. Chem.* **2009**, *284*, 34911–34917. [CrossRef]
17. Green, D.M.; Marfatia, K.A.; Crafton, E.B.; Zhang, X.; Cheng, X.; Corbett, A.H. Nab2p Is Required for Poly(A) RNA Export in Saccharomyces cerevisiae and Is Regulated by Arginine Methylation via Hmt1p. *J. Biol. Chem.* **2002**, *277*, 7752–7760. [CrossRef]
18. Hector, R.E.; Nykamp, K.R.; Dheur, S.; Anderson, J.T.; Non, P.J.; Urbinati, C.R.; Wilson, S.; Minvielle-Sebastia, L.; Swanson, M.S. Dual requirement for yeast hnRNP Nab2p in mRNA poly(A) tail length control and nuclear export. *EMBO J.* **2002**, *21*, 1800–1810. [CrossRef]
19. Strässer, K.; Hurt, E. Splicing factor Sub2p is required for nuclear mRNA export through its interaction with Yra1p. *Nature* **2001**, *413*, 648–652. [CrossRef]
20. Strässer, K.; Masuda, S.; Mason, P.; Pfannstiel, J.; Oppizzi, M.; Rodriguez-Navarro, S.; Rondón, A.G.; Aguilera, A.; Struhl, K.; Reed, R.; et al. TREX is a conserved complex coupling transcription with messenger RNA export. *Nature* **2002**, *417*, 304–308. [CrossRef]
21. Johnson, S.A.; Kim, H.; Erickson, B.; Bentley, D.L. The export factor Yra1 modulates mRNA 3′ end processing. *Nat. Struct. Mol. Biol.* **2011**, *18*, 1164–1171. [CrossRef] [PubMed]
22. Johnson, S.A.; Cubberley, G.; Bentley, D.L. Cotranscriptional Recruitment of the mRNA Export Factor Yra1 by Direct Interaction with the 3′ End Processing Factor Pcf11. *Mol. Cell* **2009**, *33*, 215–226. [CrossRef] [PubMed]

23. Saguez, C.; Schmid, M.; Olesen, J.R.; Ghazy, M.A.E.-H.; Qu, X.; Poulsen, M.B.; Nasser, T.; Moore, C.; Jensen, T.H. Nuclear mRNA Surveillance in THO/sub2 Mutants Is Triggered by Inefficient Polyadenylation. *Mol. Cell* **2008**, *31*, 91–103. [CrossRef] [PubMed]
24. Chávez, S.; Beilharz, T.; Rondón, A.G.; Erdjument-Bromage, H.; Tempst, P.; Svejstrup, J.; Lithgow, T.; Aguilera, A. A protein complex containing Tho2, Hpr1, Mft1 and a novel protein, Thp2, connects transcription elongation with mitotic recombination in Saccharomyces cerevisiae. *EMBO J.* **2000**, *19*, 5824–5834. [CrossRef]
25. Hur, J.K.; Chung, Y.D. A novel model of THO/TREX loading onto target RNAs in metazoan gene expression. *BMB Rep.* **2016**, *49*, 355–356. [CrossRef]
26. Zuckerman, B.; Ron, M.; Mikl, M.; Segal, E.; Ulitsky, I. Gene Architecture and Sequence Composition Underpin Selective Dependency of Nuclear Export of Long RNAs on NXF1 and the TREX Complex. *Mol. Cell* **2020**, *79*, 251–267.e6. [CrossRef]
27. Müller-McNicoll, M.; Botti, V.; Domingues, A.M.D.J.; Brandl, H.; Schwich, O.D.; Steiner, M.C.; Curk, T.; Poser, I.; Zarnack, K.; Neugebauer, K.M. SR proteins are NXF1 adaptors that link alternative RNA processing to mRNA export. *Genes Dev.* **2016**, *30*, 553–566. [CrossRef]
28. Rodríguez-Navarro, S. Insights into SAGA function during gene expression. *EMBO Rep.* **2009**, *10*, 843–850. [CrossRef]
29. Ellisdon, A.; Dimitrova, L.; Hurt, E.; Stewart, M. Structural basis for the assembly and nucleic acid binding of the TREX-2 transcription-export complex. *Nat. Struct. Mol. Biol.* **2012**, *19*, 328–336. [CrossRef]
30. Fischer, T.; Strässer, K.; Racz, A.; Rodriguez-Navarro, S.; Oppizzi, M.; Ihrig, P.; Lechner, J.; Hurt, E. The mRNA export machinery requires the novel Sac3p-Thp1p complex to dock at the nucleoplasmic entrance of the nuclear pores. *EMBO J.* **2002**, *21*, 5843–5852. [CrossRef]
31. Jani, D.; Lutz, S.; Marshall, N.J.; Fischer, T.; Köhler, A.; Ellisdon, A.M.; Hurt, E.; Stewart, M. Sus1, Cdc31, and the Sac3 CID Region Form a Conserved Interaction Platform that Promotes Nuclear Pore Association and mRNA Export. *Mol. Cell* **2009**, *33*, 727–737. [CrossRef] [PubMed]
32. Jani, D.; Valkov, E.; Stewart, M. Structural basis for binding the TREX2 complex to nuclear pores, GAL1 localisation and mRNA export. *Nucleic Acids Res.* **2014**, *42*, 6686–6697. [CrossRef] [PubMed]
33. Lei, E.P.; Stern, C.A.; Fahrenkrog, B.; Krebber, H.; Moy, T.I.; Aebi, U.; Silver, P.A. Sac3 Is an mRNA Export Factor That Localizes to Cytoplasmic Fibrils of Nuclear Pore Complex. *Mol. Biol. Cell* **2003**, *14*, 836–847. [CrossRef] [PubMed]
34. García-Molinero, V.; García-Martínez, J.; Reja, R.; Furió-Tarí, P.; Antúnez, O.; Vinayachandran, V.; Conesa, A.; Pugh, B.F.; Pérez-Ortín, J.E.; Rodríguez-Navarro, S. The SAGA/TREX-2 subunit Sus1 binds widely to transcribed genes and affects mRNA turnover globally. *Epigenetics Chromatin* **2018**, *11*, 13. [CrossRef]
35. Köhler, A.; Garcia, P.P.; Llopis, A.; Zapater, M.; Posas, F.; Hurt, E.; Rodríguez-Navarro, S. The mRNA Export Factor Sus1 Is Involved in Spt/Ada/Gcn5 Acetyltransferase-mediated H2B Deubiquitinylation through Its Interaction with Ubp8 and Sgf11. *Mol. Biol. Cell* **2006**, *17*, 4228–4236. [CrossRef]
36. Rodriguez-Navarro, S.; Fischer, T.; Luo, M.-J.; Antúnez, O.; Brettschneider, S.; Lechner, J.; Pérez-Ortín, J.E.; Reed, R.; Hurt, E. Sus1, a Functional Component of the SAGA Histone Acetylase Complex and the Nuclear Pore-Associated mRNA Export Machinery. *Cell* **2004**, *116*, 75–86. [CrossRef]
37. Schneider, M.; Hellerschmied, D.; Schubert, T.; Amlacher, S.; Vinayachandran, V.; Reja, R.; Pugh, B.F.; Clausen, T.; Köhler, A. The Nuclear Pore-Associated TREX-2 Complex Employs Mediator to Regulate Gene Expression. *Cell* **2015**, *162*, 1016–1028. [CrossRef]
38. Cabal, G.; Genovesio, A.; Rodriguez-Navarro, S.; Zimmer, C.; Gadal, O.; Lesne, A.; Buc, H.; Feuerbach-Fournier, F.; Olivo-Marin, J.-C.; Hurt, E.C.; et al. SAGA interacting factors confine sub-diffusion of transcribed genes to the nuclear envelope. *Nature* **2006**, *441*, 770–773. [CrossRef]
39. Chekanova, J.A.; Abruzzi, K.C.; Rosbash, M.; Belostotsky, D.A. Sus1, Sac3, and Thp1 mediate post-transcriptional tethering of active genes to the nuclear rim as well as to non-nascent mRNP. *RNA* **2008**, *14*, 66–77. [CrossRef]
40. Kurshakova, M.M.; Krasnov, A.; Kopytova, D.V.; Shidlovskii, Y.V.; Nikolenko, J.V.; Nabirochkina, E.N.; Spehner, D.; Schultz, P.; Tora, L.; Georgieva, S.G. SAGA and a novel Drosophila export complex anchor efficient transcription and mRNA export to NPC. *EMBO J.* **2007**, *26*, 4956–4965. [CrossRef]
41. Faza, M.B.; Kemmler, S.; Jimeno, S.; González-Aguilera, C.; Aguilera, A.; Hurt, E.; Panse, V.G. Sem1 is a functional component of the nuclear pore complex–associated messenger RNA export machinery. *J. Cell Biol.* **2009**, *184*, 833–846. [CrossRef] [PubMed]
42. Wickramasinghe, V.O.; McMurtrie, P.I.; Mills, A.D.; Takei, Y.; Penrhyn-Lowe, S.; Amagase, Y.; Main, S.; Marr, J.; Stewart, M.; Laskey, R.A. mRNA Export from Mammalian Cell Nuclei Is Dependent on GANP. *Curr. Biol.* **2010**, *20*, 25–31. [CrossRef] [PubMed]
43. Jani, D.; Lutz, S.; Hurt, E.; Laskey, R.A.; Stewart, M.; Wickramasinghe, V.O. Functional and structural characterization of the mammalian TREX-2 complex that links transcription with nuclear messenger RNA export. *Nucleic Acids Res.* **2012**, *40*, 4562–4573. [CrossRef] [PubMed]
44. Wente, S.R.; Rout, M.P. The Nuclear Pore Complex and Nuclear Transport. *Cold Spring Harb. Perspect. Biol.* **2010**, *2*, a000562. [CrossRef] [PubMed]
45. Aitchison, J.D.; Rout, M. The Yeast Nuclear Pore Complex and Transport Through It. *Genetics* **2012**, *190*, 855–883. [CrossRef]
46. Terry, L.J.; Wente, S.R. Flexible Gates: Dynamic Topologies and Functions for FG Nucleoporins in Nucleocytoplasmic Transport. *Eukaryot. Cell* **2009**, *8*, 1814–1827. [CrossRef]
47. Terry, L.J.; Wente, S.R. Nuclear mRNA export requires specific FG nucleoporins for translocation through the nuclear pore complex. *J. Cell Biol.* **2007**, *178*, 1121–1132. [CrossRef]

48. Siebrasse, J.P.; Kaminski, T.; Kubitscheck, U. Nuclear export of single native mRNA molecules observed by light sheet fluorescence microscopy. *Proc. Natl. Acad. Sci. USA* **2012**, *109*, 9426–9431. [CrossRef]

49. Ma, J.; Liu, Z.; Michelotti, N.; Pitchiaya, S.; Veerapaneni, R.; Androsavich, J.R.; Walter, N.G.; Yang, W. High-resolution three-dimensional mapping of mRNA export through the nuclear pore. *Nat. Commun.* **2013**, *4*, 2414. [CrossRef]

50. Mor, A.; Suliman, S.; Ben-Yishay, R.; Yunger, S.; Brody, Y.; Shav-Tal, Y. Dynamics of single mRNP nucleocytoplasmic transport and export through the nuclear pore in living cells. *Nat. Cell Biol.* **2010**, *12*, 543–552. [CrossRef]

51. Grünwald, D.; Singer, R.H. In vivo imaging of labelled endogenous β-actin mRNA during nucleocytoplasmic transport. *Nature* **2010**, *467*, 604–607. [CrossRef] [PubMed]

52. Alic, N.; Giannakou, M.E.; Papatheodorou, I.; Hoddinott, M.P.; Andrews, T.D.; Bolukbasi, E.; Partridge, L. Interplay of dFOXO and Two ETS-Family Transcription Factors Determines Lifespan in Drosophila melanogaster. *PLoS Genet.* **2014**, *10*, e1004619. [CrossRef] [PubMed]

53. Cai, L.; McCormick, M.A.; Kennedy, B.K.; Tu, B.P. Integration of Multiple Nutrient Cues and Regulation of Lifespan by Ribosomal Transcription Factor Ifh1. *Cell Rep.* **2013**, *4*, 1063–1071. [CrossRef] [PubMed]

54. Ghazi, A.; Henis-Korenblit, S.; Kenyon, C. A Transcription Elongation Factor That Links Signals from the Reproductive System to Lifespan Extension in Caenorhabditis elegans. *PLoS Genet.* **2009**, *5*, e1000639. [CrossRef]

55. Mittal, N.; Guimaraes, J.C.; Gross, T.; Schmidt, A.; Vina-Vilaseca, A.; Nedialkova, D.D.; Aeschimann, F.; Leidel, S.; Spang, A.; Zavolan, M. The Gcn4 transcription factor reduces protein synthesis capacity and extends yeast lifespan. *Nat. Commun.* **2017**, *8*, 457. [CrossRef]

56. Oh, S.W.; Mukhopadhyay, A.; Svrzikapa, N.; Jiang, F.; Davis, R.J.; Tissenbaum, H.A. JNK regulates lifespan in *Caenorhabditis elegans* by modulating nuclear translocation of forkhead transcription factor/DAF-16. *Proc. Natl. Acad. Sci. USA* **2005**, *102*, 4494–4499. [CrossRef]

57. Postnikoff, S.D.L.; Malo, M.E.; Wong, B.; Harkness, T.A.A. The Yeast Forkhead Transcription Factors Fkh1 and Fkh2 Regulate Lifespan and Stress Response Together with the Anaphase-Promoting Complex. *PLoS Genet.* **2012**, *8*, e1002583. [CrossRef]

58. Rodriguez-Lopez, M.; Gonzalez, S.; Hillson, O.; Tunnacliffe, E.; Codlin, S.; Tallada, V.A.; Bähler, J.; Rallis, C.; Rodriguez-Lopez, M.; Gonzalez, S.; et al. The GATA Transcription Factor Gaf1 Represses tRNAs, Inhibits Growth, and Extends Chronological Lifespan Downstream of Fission Yeast TORC1. *Cell Rep.* **2020**, *30*, 3240–3249.e4. [CrossRef]

59. Ryu, H.-Y.; Rhie, B.-H.; Ahn, S.H. Loss of the Set2 histone methyltransferase increases cellular lifespan in yeast cells. *Biochem. Biophys. Res. Commun.* **2014**, *446*, 113–118. [CrossRef]

60. Yamamoto, R.; Tatar, M. Insulin receptor substrate chico acts with the transcription factor FOXO to extend Drosophila lifespan. *Aging Cell* **2011**, *10*, 729–732. [CrossRef]

61. Samara, N.L.; Wolberger, C. A new chapter in the transcription SAGA. *Curr. Opin. Struct. Biol.* **2011**, *21*, 767–774. [CrossRef] [PubMed]

62. Huang, B.; Zhong, D.; Zhu, J.; An, Y.; Gao, M.; Zhu, S.; Dang, W.; Wang, X.; Yang, B.; Xie, Z. Inhibition of histone acetyltransferase GCN5 extends lifespan in both yeast and human cell lines. *Aging Cell* **2020**, *19*, e13129. [CrossRef] [PubMed]

63. Horiuchi, J.; Silverman, N.; Marcus, G.A.; Guarente, L. ADA3, a putative transcriptional adaptor, consists of two separable domains and interacts with ADA2 and GCN5 in a trimeric complex. *Mol. Cell. Biol.* **1995**, *15*, 1203–1209. [CrossRef] [PubMed]

64. Lim, S.; Ahn, H.; Duan, R.; Liu, Y.; Ryu, H.-Y.; Ahn, S.H. The Spt7 subunit of the SAGA complex is required for the regulation of lifespan in both dividing and nondividing yeast cells. *Mech. Ageing Dev.* **2021**, *196*, 111480. [CrossRef] [PubMed]

65. McCormick, M.A.; Mason, A.G.; Guyenet, S.J.; Dang, W.; Garza, R.M.; Ting, M.K.; Moller, R.M.; Berger, S.L.; Kaeberlein, M.; Pillus, L.; et al. The SAGA Histone Deubiquitinase Module Controls Yeast Replicative Lifespan via Sir2 Interaction. *Cell Rep.* **2014**, *8*, 477–486. [CrossRef]

66. Denoth-Lippuner, A.; Krzyzanowski, M.K.; Stober, C.; Barral, Y. Role of SAGA in the asymmetric segregation of DNA circles during yeast ageing. *eLife* **2014**, *3*, e03790. [CrossRef] [PubMed]

67. Kim, H.; Cho, B.; Moon, S.; Chung, Y.D. The THO complex is required for stress tolerance and longevity in Drosophila. *Genes Genom.* **2011**, *33*, 291–297. [CrossRef]

68. Kim, S.Y.; Yang, E.J.; Lee, S.B.; Lee, Y.-S.; Cho, K.A.; Park, S.C. Global transcriptional downregulation of TREX and nuclear trafficking machinery as pan-senescence phenomena: Evidence from human cells and tissues. *Exp. Mol. Med.* **2020**, *52*, 1351–1359. [CrossRef]

69. Lim, S.; Liu, Y.; Rhie, B.-H.; Kim, C.; Ryu, H.-Y.; Ahn, S.H. Sus1 maintains a normal lifespan through regulation of TREX-2 complex-mediated mRNA export. *bioRxiv* **2022**. [CrossRef]

70. D'Angelo, M.A.; Raices, M.; Panowski, S.H.; Hetzer, M.W. Age-Dependent Deterioration of Nuclear Pore Complexes Causes a Loss of Nuclear Integrity in Postmitotic Cells. *Cell* **2009**, *136*, 284–295. [CrossRef]

71. Savas, J.N.; Toyama, B.H.; Xu, T.; Yates, J.R., 3rd; Hetzer, M.W. Extremely Long-Lived Nuclear Pore Proteins in the Rat Brain. *Science* **2012**, *335*, 942. [CrossRef] [PubMed]

72. Lord, C.; Timney, B.L.; Rout, M.; Wente, S.R. Altering nuclear pore complex function impacts longevity and mitochondrial function in S. cerevisiae. *J. Cell Biol.* **2015**, *208*, 729–744. [CrossRef] [PubMed]

73. Lord, C.L.; Ospovat, O.; Wente, S.R. Nup100 regulates Saccharomyces cerevisiae replicative life span by mediating the nuclear export of specific tRNAs. *RNA* **2017**, *23*, 365–377. [CrossRef] [PubMed]

74. Cho, U.H.; Hetzer, M.W. Nuclear Periphery Takes Center Stage: The Role of Nuclear Pore Complexes in Cell Identity and Aging. *Neuron* **2020**, *106*, 899–911. [CrossRef] [PubMed]

75. Ryu, S.J.; Cho, K.A.; Oh, Y.S.; Park, S.C. Role of Src-specific phosphorylation site on focal adhesion kinase for senescence-associated apoptosis resistance. *Apoptosis* **2006**, *11*, 303–313. [CrossRef] [PubMed]

76. Cho, K.A.; Ryu, S.J.; Park, J.S.; Jang, I.S.; Ahn, J.S.; Kim, K.T.; Park, S.C. Senescent Phenotype Can Be Reversed by Reduction of Caveolin Status. *J. Biol. Chem.* **2003**, *278*, 27789–27795. [CrossRef]

77. Park, W.-Y.; Park, J.-S.; Cho, K.-A.; Kim, D.-I.; Ko, Y.-G.; Seo, J.-S.; Park, S.C. Up-regulation of Caveolin Attenuates Epidermal Growth Factor Signaling in Senescent Cells. *J. Biol. Chem.* **2000**, *275*, 20847–20852. [CrossRef]

78. Seluanov, A.; Gorbunova, V.; Falcovitz, A.; Sigal, A.; Milyavsky, M.; Zurer, I.; Shohat, G.; Goldfinger, N.; Rotter, V. Change of the Death Pathway in Senescent Human Fibroblasts in Response to DNA Damage Is Caused by an Inability To Stabilize p53. *Mol. Cell. Biol.* **2001**, *21*, 1552–1564. [CrossRef]

79. Kim, S.Y.; Ryu, S.J.; Ahn, H.J.; Choi, H.R.; Kang, H.T.; Park, S.C. Senescence-related functional nuclear barrier by down-regulation of nucleo-cytoplasmic trafficking gene expression. *Biochem. Biophys. Res. Commun.* **2010**, *391*, 28–32. [CrossRef]

80. Janssens, G.; Meinema, A.C.; González, J.; Wolters, J.C.; Schmidt, A.; Guryev, V.; Bischoff, R.; Wit, E.C.; Veenhoff, L.M.; Heinemann, M. Protein biogenesis machinery is a driver of replicative aging in yeast. *eLife* **2015**, *4*, e08527. [CrossRef]

81. Rempel, I.L.; Crane, M.M.; Thaller, D.J.; Mishra, A.; Jansen, D.P.; Janssens, G.; Popken, P.; Akşit, A.; Kaeberlein, M.; Van Der Giessen, E.; et al. Age-dependent deterioration of nuclear pore assembly in mitotic cells decreases transport dynamics. *eLife* **2019**, *8*, e48186. [CrossRef] [PubMed]

82. Maeshima, K.; Yahata, K.; Sasaki, Y.; Nakatomi, R.; Tachibana, T.; Hashikawa, T.; Imamoto, F.; Imamoto, N. Cell-cycle-dependent dynamics of nuclear pores: Pore-free islands and lamins. *J. Cell Sci.* **2006**, *119*, 4442–4451. [CrossRef] [PubMed]

83. Hieronymus, H.; Yu, M.C.; Silver, P.A. Genome-wide mRNA surveillance is coupled to mRNA export. *Genes Dev.* **2004**, *18*, 2652–2662. [CrossRef] [PubMed]

84. Son, H.G.; Lee, S.-J.V. Longevity regulation by NMD-mediated mRNA quality control. *BMB Rep.* **2017**, *50*, 160–161. [CrossRef]

85. Mullani, N.; Porozhan, Y.; Mangelinck, A.; Rachez, C.; Costallat, M.; Batsché, E.; Goodhardt, M.; Cenci, G.; Mann, C.; Muchardt, C. Reduced RNA turnover as a driver of cellular senescence. *Life Sci. Alliance* **2021**, *4*, e202000809. [CrossRef]

86. Camblong, J.; Iglesias, N.; Fickentscher, C.; Dieppois, G.; Stutz, F. Antisense RNA Stabilization Induces Transcriptional Gene Silencing via Histone Deacetylation in S. cerevisiae. *Cell* **2007**, *131*, 706–717. [CrossRef]

87. Vinciguerra, P.; Iglesias, N.; Camblong, J.; Zenklusen, D.; Stutz, F. Perinuclear Mlp proteins downregulate gene expression in response to a defect in mRNA export. *EMBO J.* **2005**, *24*, 813–823. [CrossRef]

88. Niño, C.A.; Hérissant, L.; Babour, A.; Dargemont, C. mRNA Nuclear Export in Yeast. *Chem. Rev.* **2013**, *113*, 8523–8545. [CrossRef]

89. Cookson, M.R. Aging-RNA in development and disease. *Wiley Interdiscip. Rev. RNA* **2012**, *3*, 133–143. [CrossRef]

90. Sakuma, S.; D'Angelo, M.A. The roles of the nuclear pore complex in cellular dysfunction, aging and disease. *Semin. Cell Dev. Biol.* **2017**, *68*, 72–84. [CrossRef]

91. Woerner, A.C.; Frottin, F.; Hornburg, D.; Feng, L.R.; Meissner, F.; Patra, M.; Tatzelt, J.; Mann, M.; Winklhofer, K.F.; Hartl, F.U.; et al. Cytoplasmic protein aggregates interfere with nucleocytoplasmic transport of protein and RNA. *Science* **2016**, *351*, 173–176. [CrossRef] [PubMed]

92. Freibaum, B.D.; Lu, Y.; Lopez-Gonzalez, R.; Kim, N.C.; Almeida, S.; Lee, K.-H.; Badders, N.; Valentine, M.; Miller, B.L.; Wong, P.C.; et al. GGGGCC repeat expansion in C9orf72 compromises nucleocytoplasmic transport. *Nature* **2015**, *525*, 129–133. [CrossRef] [PubMed]

93. Maeder, C.I.; Kim, J.-I.; Liang, X.; Kaganovsky, K.; Shen, A.; Li, Q.; Li, Z.; Wang, S.; Xu, X.S.; Li, J.B.; et al. The THO Complex Coordinates Transcripts for Synapse Development and Dopamine Neuron Survival. *Cell* **2018**, *174*, 1436–1449.e20. [CrossRef] [PubMed]

94. Boehringer, A.; Garcia-Mansfield, K.; Singh, G.; Bakkar, N.; Pirrotte, P.; Bowser, R. ALS Associated Mutations in Matrin 3 Alter Protein-Protein Interactions and Impede mRNA Nuclear Export. *Sci. Rep.* **2017**, *7*, 14529. [CrossRef]

95. Tollervey, J.R.; Curk, T.; Rogelj, B.; Briese, M.; Cereda, M.; Kayikci, M.; König, J.; Hortobágyi, T.; Nishimura, A.L.; Župunski, V.; et al. Characterizing the RNA targets and position-dependent splicing regulation by TDP-43. *Nat. Neurosci.* **2011**, *14*, 452–458. [CrossRef]

96. Chou, C.-C.; Zhang, Y.; Umoh, M.E.; Vaughan, S.W.; Lorenzini, I.; Liu, F.; Sayegh, M.; Donlin-Asp, P.; Chen, Y.H.; Duong, D.; et al. TDP-43 pathology disrupts nuclear pore complexes and nucleocytoplasmic transport in ALS/FTD. *Nat. Neurosci.* **2018**, *21*, 228–239. [CrossRef]

97. Kapeli, K.; Pratt, G.A.; Vu, A.Q.; Hutt, K.R.; Martinez, F.J.; Sundararaman, B.; Batra, R.; Freese, P.D.; Lambert, N.J.; Huelga, S.C.; et al. Distinct and shared functions of ALS-associated proteins TDP-43, FUS and TAF15 revealed by multisystem analyses. *Nat. Commun.* **2016**, *7*, 12143. [CrossRef]

98. Fujii, R.; Takumi, T. TLS facilitates transport of mRNA encoding an actin-stabilizing protein to dendritic spines. *J. Cell Sci.* **2005**, *118*, 5755–5765. [CrossRef]

99. Fujii, R.; Okabe, S.; Urushido, T.; Inoue, K.; Yoshimura, A.; Tachibana, T.; Nishikawa, T.; Hicks, G.; Takumi, T. The RNA Binding Protein TLS Is Translocated to Dendritic Spines by mGluR5 Activation and Regulates Spine Morphology. *Curr. Biol.* **2005**, *15*, 587–593. [CrossRef]

100. Zhang, K.; Donnelly, C.J.; Haeusler, A.R.; Grima, J.C.; Machamer, J.B.; Steinwald, P.; Daley, E.; Miller, S.J.; Cunningham, K.; Vidensky, S.; et al. The C9orf72 repeat expansion disrupts nucleocytoplasmic transport. *Nature* **2015**, *525*, 56–61. [CrossRef]
101. Kinoshita, Y.; Ito, H.; Hirano, A.; Fujita, K.; Wate, R.; Nakamura, M.; Kaneko, S.; Nakano, S.; Kusaka, H. Nuclear Contour Irregularity and Abnormal Transporter Protein Distribution in Anterior Horn Cells in Amyotrophic Lateral Sclerosis. *J. Neuropathol. Exp. Neurol.* **2009**, *68*, 1184–1192. [CrossRef] [PubMed]
102. Kaneb, H.M.; Folkmann, A.W.; Belzil, V.V.; Jao, L.-E.; Leblond, C.S.; Girard, S.L.; Daoud, H.; Noreau, A.; Rochefort, D.; Hince, P.; et al. Deleterious mutations in the essential mRNA metabolism factor, hGle1, in amyotrophic lateral sclerosis. *Hum. Mol. Genet.* **2015**, *24*, 1363–1373. [CrossRef] [PubMed]
103. Cho, K.-I.; Yoon, D.; Qiu, S.; Danziger, Z.; Grill, W.M.; Wetsel, W.C.; Ferreira, P.A. Loss of Ranbp2 in motor neurons causes the disruption of nucleocytoplasmic and chemokine signaling and proteostasis of hnRNPH3 and Mmp28, and the development of amyotrophic lateral sclerosis (ALS)-like syndromes. *Dis. Model. Mech.* **2017**, *10*, 559–579. [CrossRef]
104. Gasset-Rosa, F.; Chillon-Marinas, C.; Goginashvili, A.; Atwal, R.S.; Artates, J.W.; Tabet, R.; Wheeler, V.C.; Bang, A.G.; Cleveland, D.W.; Lagier-Tourenne, C. Polyglutamine-Expanded Huntingtin Exacerbates Age-Related Disruption of Nuclear Integrity and Nucleocytoplasmic Transport. *Neuron* **2017**, *94*, 48–57.e4. [CrossRef] [PubMed]
105. Grima, J.C.; Daigle, J.G.; Arbez, N.; Cunningham, K.C.; Zhang, K.; Ochaba, J.; Geater, C.; Morozko, E.; Stocksdale, J.; Glatzer, J.C.; et al. Mutant Huntingtin Disrupts the Nuclear Pore Complex. *Neuron* **2017**, *94*, 93.e6–107.e6. [CrossRef] [PubMed]
106. Sheffield, L.G.; Miskiewicz, H.B.; Tannenbaum, L.B.; Mirra, S.S. Nuclear Pore Complex Proteins in Alzheimer Disease. *J. Neuropathol. Exp. Neurol.* **2006**, *65*, 45–54. [CrossRef] [PubMed]
107. Um, J.W.; Min, D.S.; Rhim, H.; Kim, J.; Paik, S.R.; Chung, K.C. Parkin Ubiquitinates and Promotes the Degradation of RanBP2. *J. Biol. Chem.* **2006**, *281*, 3595–3603. [CrossRef] [PubMed]
108. Mata, I.F.; Lockhart, P.J.; Farrer, M.J. Parkin genetics: One model for Parkinson's disease. *Hum. Mol. Genet.* **2004**, *13*, R127–R133. [CrossRef]
109. Li, Y.; Hassinger, L.; Thomson, T.; Ding, B.; Ashley, J.; Hassinger, W.; Budnik, V. Lamin Mutations Accelerate Aging via Defective Export of Mitochondrial mRNAs through Nuclear Envelope Budding. *Curr. Biol.* **2016**, *26*, 2052–2059. [CrossRef]
110. Thakur, M.; Oka, T.; Natori, Y. Gene expression and aging. *Mech. Ageing Dev.* **1993**, *66*, 283–298. [CrossRef]
111. Lennartsson, A.; Ekwall, K. Histone modification patterns and epigenetic codes. *Biochim. Biophys. Acta* **2009**, *1790*, 863–868. [CrossRef] [PubMed]
112. Ryu, H.-Y.; Hochstrasser, M. Histone sumoylation and chromatin dynamics. *Nucleic Acids Res.* **2021**, *49*, 6043–6052. [CrossRef] [PubMed]
113. Ryu, H.; Su, D.; Wilson-Eisele, N.R.; Zhao, D.; López-Giráldez, F.; Hochstrasser, M. The Ulp2 SUMO protease promotes transcription elongation through regulation of histone sumoylation. *EMBO J.* **2019**, *38*, e102003. [CrossRef] [PubMed]
114. Ryu, H.-Y.; Zhao, D.; Li, J.; Su, D.; Hochstrasser, M. Histone sumoylation promotes Set3 histone-deacetylase complex-mediated transcriptional regulation. *Nucleic Acids Res.* **2020**, *48*, 12151–12168. [CrossRef] [PubMed]
115. Choi, J.; Ryoo, Z.Y.; Cho, D.-H.; Lee, H.-S.; Ryu, H.-Y. Trans-tail regulation-mediated suppression of cryptic transcription. *Exp. Mol. Med.* **2021**, *53*, 1683–1688. [CrossRef] [PubMed]
116. Oliete-Calvo, P.; Serrano-Quílez, J.; Nuño-Cabanes, C.; Pérez-Martínez, M.E.; Soares, L.M.; Dichtl, B.; Buratowski, S.; Pérez-Ortín, J.E.; Rodríguez-Navarro, S. A role for Mog1 in H2Bub1 and H3K4me3 regulation affecting RNAPII transcription and mRNA export. *EMBO Rep.* **2018**, *19*, e45992. [CrossRef]
117. Yoh, S.M.; Lucas, J.S.; Jones, K.A. The Iws1:Spt6:CTD complex controls cotranscriptional mRNA biosynthesis and HYPB/Setd2-mediated histone H3K36 methylation. *Genes Dev.* **2008**, *22*, 3422–3434. [CrossRef]
118. Wang, Y.; Yuan, Q.; Xie, L. Histone Modifications in Aging: The Underlying Mechanisms and Implications. *Curr. Stem Cell Res. Ther.* **2018**, *13*, 125–135. [CrossRef]

International Journal of
Molecular Sciences

Epigenetic Biomarkers of Transition from Metabolically Healthy Obesity to Metabolically Unhealthy Obesity Phenotype: A Prospective Study

Carolina Gutiérrez-Repiso [1,2,†], Teresa María Linares-Pineda [1,†], Andres Gonzalez-Jimenez [3], Francisca Aguilar-Lineros [1], Sergio Valdés [4,5], Federico Soriguer [4,5], Gemma Rojo-Martínez [4,5], Francisco J. Tinahones [1,2,6,*] and Sonsoles Morcillo [1,2,*]

[1] Unidad de Gestión Clínica de Endocrinología y Nutrición del Hospital Virgen de la Victoria, Instituto de Investigación Biomédica de Málaga (IBIMA), 29010 Málaga, Spain; carogure@hotmail.com (C.G.-R.); teresamaria712@gmail.com (T.M.L.-P.); franciscaaguilarjara@gmail.com (F.A.-L.)
[2] Centro de Investigación Biomédica en Red de Fisiopatología de la Obesidad y la Nutrición (CIBERobn), Instituto de Salud Carlos III, 28029 Madrid, Spain
[3] ECAI Bioinformática Instituto de Investigación Biomédica de Málaga (IBIMA), 29010 Málaga, Spain; bioinformatica@ibima.eu
[4] Departamento de Endocrinología and Nutrición, Hospital Regional Universitario de Málaga, Instituto de Investigación Biomédica de Málaga (IBIMA), 29009 Málaga, Spain; sergio.valdes@hotmail.es (S.V.); federicosoriguer@gmail.com (F.S.); gemma.rojo.m@gmail.com (G.R.-M.)
[5] Centro de Investigación Biomédica en Red de Diabetes y Enfermedades Metabólicas Asociadas (CIBERDEM), Instituto de Salud Carlos III, 28029 Madrid, Spain
[6] Departamento de Medicina y Dermatología, Universidad de Málaga, 29010 Málaga, Spain
[*] Correspondence: fjtinahones@uma.es (F.J.T.); sonsoles.morcillo@ibima.eu (S.M.)
[†] These authors contributed equally to this work.

Citation: Gutiérrez-Repiso, C.; Linares-Pineda, T.M.; Gonzalez-Jimenez, A.; Aguilar-Lineros, F.; Valdés, S.; Soriguer, F.; Rojo-Martínez, G.; Tinahones, F.J.; Morcillo, S. Epigenetic Biomarkers of Transition from Metabolically Healthy Obesity to Metabolically Unhealthy Obesity Phenotype: A Prospective Study. *Int. J. Mol. Sci.* **2021**, *22*, 10417. https://doi.org/10.3390/ijms221910417

Academic Editors: Amelia Casamassimi, Alfredo Ciccodicola and Monica Rienzo

Received: 30 August 2021
Accepted: 23 September 2021
Published: 27 September 2021

Publisher's Note: MDPI stays neutral with regard to jurisdictional claims in published maps and institutional affiliations.

Abstract: Background: Identifying those parameters that could potentially predict the deterioration of metabolically healthy phenotype is a matter of debate. In this field, epigenetics, in particular DNA methylation deserves special attention. Results: The aim of the present study was to analyze the long-term evolution of methylation patterns in a subset of metabolically healthy subjects in order to search for epigenetic markers that could predict the progression to an unhealthy state. Twenty-six CpG sites were significantly differentially methylated, both at baseline and 11-year follow-up. These sites were related to 19 genes or pseudogenes; a more in-depth analysis of the methylation sites of these genes showed that *CYP2E1* had 50% of the collected CpG sites differently methylated between stable metabolically healthy obesity (MHO) and unstable MHO, followed by *HLA-DRB1* (33%), *ZBTB45* (16%), *HOOK3* (14%), *PLCZ1* (14%), *SLC1A1* (12%), *MUC2* (12%), *ZFPM2* (12.5%) and *HLA-DQB2* (8%). Pathway analysis of the selected 26 CpG sites showed enrichment in pathways linked to th1 and th2 activation, antigen presentation, allograft rejection signals and metabolic processes. Higher methylation levels in the cg20707527 (*ZFPM2*) could have a protective effect against the progression to unstable MHO (OR: 0.21, 95%CI (0.067–0.667), $p < 0.0001$), whilst higher methylation levels in cg11445109 (*CYP2E1*) would increase the progression to MUO; OR: 2.72, 95%CI (1.094–6.796), $p < 0.0014$; respectively). Conclusions: DNA methylation status is associated with the stability/worsening of MHO phenotype. Two potential biomarkers of the transition to an unhealthy state were identified and deserve further investigation (cg20707527 and cg11445109). Moreover, the described differences in methylation could alter immune system-related pathways, highlighting these pathways as therapeutic targets to prevent metabolic deterioration in MHO patients.

Keywords: metabolically healthy obesity; epigenetic biomarkers; metabolic syndrome; DNA methylation

1. Introduction

Worldwide, obesity has reached epidemic proportions and at least 2.8 million people die each year as result of overweight or obesity. According to the World Health Organization, the prevalence of obesity has nearly tripled since 1975 [1].

Obesity is associated with higher risk of developing metabolic syndrome, type 2 diabetes (T2D) and cardiovascular diseases, resulting in an increase in mortality. However, not all people with obesity present the typical pattern of metabolic complications. This phenotype has been defined as metabolically healthy obesity (MHO) and its prevalence rate varies widely, ranging from 10% to 35% depending on the criteria used and population studied [2,3].

The MHO phenotype can progress to an unhealthy state known as metabolically unhealthy obesity (MUO). It has been suggested that this progression could be a matter of time [4], although there is evidence that suggests that a relevant percentage of MHO individuals maintain their status over time [5].

Despite the growing interest in these groups of subjects, there is a great lack of knowledge concerning the factors that determine why some obese subjects are protected from developing metabolic complications. Different studies propose that higher insulin sensitivity, specific distribution of fat, reduced infiltration of immune cells into adipose tissue, and consequently, a metabolically beneficial cytokine and adipokine secretion pattern, could be some of the mechanisms involved in the genesis of MHO [6,7].

It is estimated the 40–70% of obesity and metabolic disease has a inherited component, but large genome-wide association studies (GWAS) have shown that only 20% of variants in genes related to obesity can explain the predisposition to this condition [8]. Therefore, it has been suggested that epigenetic processes may have a role in the regulation of metabolic diseases. DNA methylation is one of the main epigenetic mechanisms, and can alter gene expression without changing the DNA sequence by adding methyl groups at cytosine residues. This field is still young, but it is attracting interest in various areas such as oncology and metabolic disorders such as obesity.

Previous studies have evaluated the relationship between epigenetic variants and metabolic diseases such as obesity and T2D [9]. It has been suggested that obesity is related to different methylation levels in blood cells compared with those in healthy cohorts [10–12]. Also, DNA methylation data from adipose tissue show that epigenetic variation is involved in obesity-associated comorbidities and T2D [13,14].

The aim of the present study was to analyze the long-term evolution of methylation patterns in a subset of MHO subjects in order to search for epigenetic markers that could predict the progression of MHO to MUO.

2. Results

Table 1 presents the metabolic variables used to classify the patients included in the study. Briefly, patients were considered as MHO if they had abdominal obesity and <2 of the NCEP ATPIII metabolic syndrome criteria were present. At baseline, triglyceride levels were significantly higher in the unstable MHO group (p = 0.001). No statistical differences were found for the rest of the studied variables. At 11-year follow-up, fasting glucose (p = 0.01), diastolic (p = 0.0159) and systolic blood pressure values (p = 0.024) were significantly higher in the unstable MHO group.

2.1. Principal Component Analysis

Principal component analysis (PCA) analysis was carried out using the double selection of methylated CpG loci. Firstly, CpG sites that clearly discriminated the two populations at 11-year follow-up were selected; in this selection, both components explained around 58% of the variance (Figure 1A). In this step, a total of 8200 (1%) differentially methylated CpG loci were selected from 815,389 probes, based on component contribution criteria. These CpG sites were tested at baseline to determine markers that discriminated between the two populations both at baseline and at the end of follow-up (Figure 1B). Then, those CpG sites in each component whose contribution values were relevant, were selected.

At baseline, sites with a contribution higher than 0.04% (half the maximum contribution value of the best variable) in component 1 and 0.25% (half the maximum contribution) in component 2 were selected and used to establish the methylation changes during the follow-up in the study population.

Table 1. Anthropometric and biochemical characteristics of the subjects included in the study.

Title 1	Baseline			11-Year Follow-Up		
	Stable MHO (n = 9)	Unstable MHO (n = 9)	*p*-Value	Stable MHO (n = 9)	Unstable MHO (n = 9)	*p*-Value
Age	45 ± 11	53 ± 9	NS			
Gender (Male/Female)	2/7	3/6	NS			
Fasting glucose (mg/dl)	103.5 ± 13.3	106.6 ± 10.3	NS	90 ± 4.9	108.4 ± 16.6	0.01
BMI	28.2 ± 1.6	29 ± 4.3	NS	29.9 ± 3.4	31.1 ± 4.0	NS
Triglycerides (mg/dl)	52.7 ± 12.3	92.7 ± 37.8	0.01	82 ± 25.2	100.33 ± 53.0	NS
HDL-cholesterol	57.8 ± 12.3	51.4 ± 9.6	NS	60.7 ± 6.3	53.6 ± 8.4	0.06
DBP (mm Hg)	81.4 ± 7.5	88.6 ± 16.8	NS	75.7 ± 9.3	90.5 ± 11.5	0.015
SBP (mm Hg)	121 ± 16.9	138.6 ± 26.3	NS	126.3 ± 19.3	153.8 ± 23.2	0.024
HTA treatment (%)	0	22.2	NS	0	55.6	0.015

Data are expressed as the mean ± standard deviation, or as (percentage). *p*-values for continuous data were calculated using the Kruskal–Wallis test, and for categorical data they were calculated using the chi-square test or Fisher's exact test if the frequency was <5. BMI: body mass index. HDL cholesterol: high density lipoprotein cholesterol. DBP: diastolic blood pressure. SBP: systolic blood pressure. HTA treatment: arterial hypertension treatment.

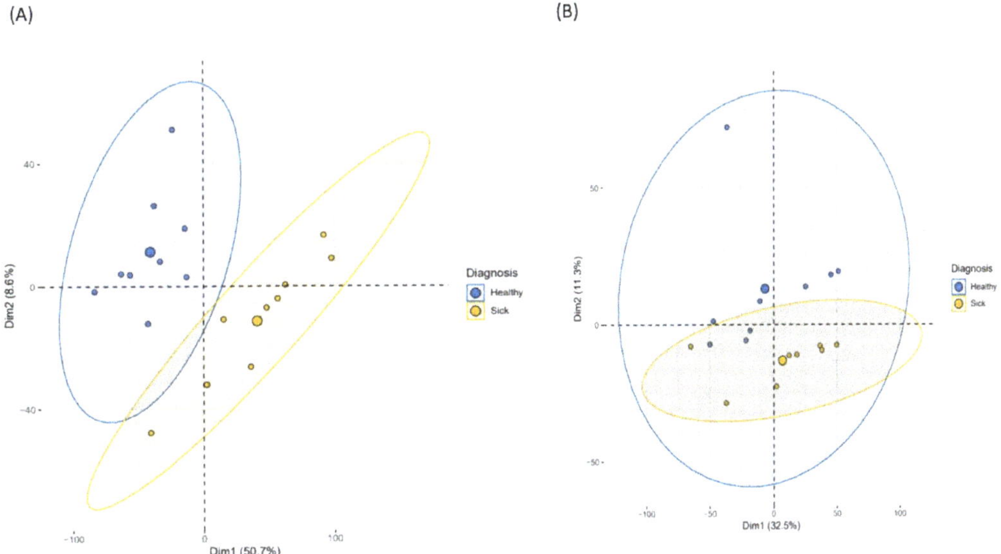

Figure 1. PCA at baseline and 11-year follow-up. (**A**) PCA performed on 11-year follow-up dataset. (**B**) PCA of the most important 11-year follow-up methylation sites at baseline.

Finally, 26 significantly differentially methylated CpG sites were selected for further analysis (Supplementary Table S1). Most of them (fifteen) were hypermethylated in stable MHO compared to the unstable MHO population both at baseline and 11-year follow-up, while 11 were hypomethylated. The differences between the mean methylation values in both populations at the two study points are shown in Figure 2.

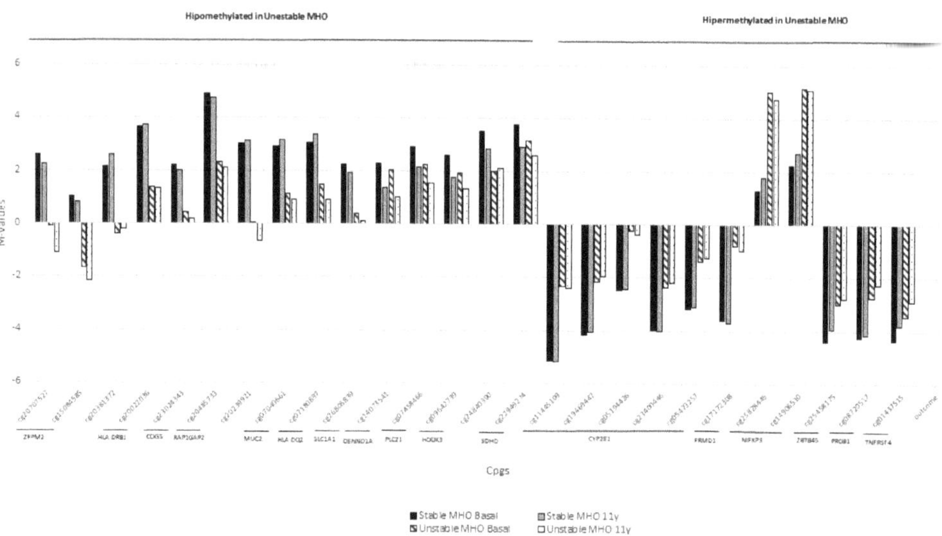

Figure 2. Methylation levels of 26 significantly differentially methylated CpG sites identified in the double PCA selection at baseline and 11-year follow-up.

2.2. Differentially Methylated Genes

A total of 17 genes and 2 pseudogenes were related to the 26 CpG sites identified in the double PCA selection. The top ten significantly differentially methylated CpG sites were associated with eight unique genes or pseudogenes; two were pseudogenes, namely, nucleolar protein interacting with the FHA domain (*NIPFK3*) and *DTX2P1-UPK3BP1-PMS2P11*. The rest of the sites were unique genes including zinc finger protein, FOG family member 2 (*ZFMP2*), cytochrome P450 family 2 subfamily E member 1 (*CYP2E1*), major histocompatibility complex, class II, DQ beta 1 and beta 2 (*HLA-DQB1* and *HLA-DQB2*), solute carrier family 1 (*SLC1A1*) and phospholipase C zeta 1 (*PLCZ1*). The characteristics of these CpG loci including probe ID, location, gene region or direction of methylation are shown in Table 2.

A more in-depth analysis was performed on these nineteen unique genes (seventeen genes and 2 pseudo genes). All the CpG sites in each of these genes, as well as flanked sequences were collected from the UCSC genome and checked as to whether they are detected in the Methylation EPIC Bead. The CpG sites described in each gene were analyzed to investigate to what extent these genes present multiple different CpG sites in our population.

Fourteen of the nineteen genes identified (73.6%) showed multiple, significant CpG sites. The gene with the largest difference in methylation was *CYP2E1* with 50% of the collected CpG sites differently methylated in the stable MHO and unstable MHO, followed by *HLA-DRB1* (33%), *ZBTB45* (16%), *HOOK3* (14%), *PLCZ1* (14%), *SLC1A1* (12%), *MUC2* (12%), *ZFPM2* (12.5%) and *HLA-DQB2* (8%), and several flanked sequences were identified as differentially methylated in *MUC2*. None of the flanked sequences were found to be significantly differentially methylated in the rest of the genes. The differentially methylated CpG sites in these genes are shown in Supplementary Table S2.

Table 2. The top ten significantly differentially methylated CpG sites in stable MHO and unstable MHO throughout the study.

Probe ID	Location	Gene Symbol	Gene Name	*p*-Value	Hypermethylated
cg20707527	Chr: 8 q23.1	*ZFPM2*	Zinc Finger Protein. FOG Family Member 2	0.0001	Stable MHO
cg15084585	Chr: 8 q23.1	*ZFPM2*	Zinc Finger Protein. FOG Family Member 2	0.0001	Stable MHO
cg20022036	Chr: 6 p21.32	*HLA-DRB1*	Major Histocompatibility Complex. Class II. DR Beta 1	0.0015	Stable MHO
cg20239921	Chr: 7 q 11.23	*DTX2P1-UPK3BP1-PMS2P11*	DTX2P1-UPK3BP1-PMS2P11	0.0015	Stable MHO
cg26805839	Chr: 9p24.2	*SLC1A1*	Solute Carrier Family 1 Member 1	0.0035	Stable MHO
cg11445109	Chr: 10 q26.3	*CYP2E1*	Cytochrome P450 Family 2 Subfamily E Member 1	0.0046	Unstable MHO
Cg07180987	Chr: 6 p21.32	*HLA-DQB2*	Major Histocompatibility Complex. Class II. DQ Beta 2	0.0057	Stable MHO
cg05194426	Chr: 10 q26.3	*CYP2E1*	Cytochrome P450 Family 2 Subfamily E Member 1	0.0057	Unstable MHO
cg25828445	Chr: 12 p13.31	*NIFKP3*	Nucleolar Protein Interacting with The FHA Domain	0.0067	Unstable MHO
cg07458466	Chr: 12 p 12.2	*PLCZ1*	Phospholipase C Zeta 1	0.0083	Stable MHO

2.3. Potential Biomarker of Transition to Unhealthy State

A backward stepwise logistic regression was performed using all the methylated sites to evaluate the prediction power of the different methylation in these sites. The final model selected two sites as the best markers to predict the deterioration of stable MHO to an unhealthy phenotype. So, a higher methylation level in the site cg20707527 in the gene *ZFPM2* could have a protective effect against progression to MUO (OR: 0.21, 95%CI (0.067–0.667), $p < 0.0001$); on the contrary, a higher methylation level of the site cg11445109 into the gene *CYP2E1* would increase the progression of the patient to MUO (OR: 2.72, 95%CI (1.094–6.796), $p < 0.0014$). As the baseline triglycerides levels were significantly different, this variable was also included in the model; however, they were not statistically significant.

2.4. Enrichment Analysis

The 26 differentially methylated CpG sites selected through double PCA selection were annotated by GO analysis and their functions were classified by biological processes, molecular function, and cellular components using an enrichment analysis. The top 10 GO terms categorized into biological processes, molecular functions and cellular components are illustrated in Supplementary Figure S1. Biological processes were shown to be linked to the metabolic process of a wide variety of substrates such as halogen compound, benzene, monoterpenoid, etc. Processes not related to metabolism were protein transport, antigen presentation and regulation of cytosolic calcium. Meanwhile, cellular components were mainly associated with transport between membranes, especially Golgi transport or coated-clathrin vesicles (Supplementary Table S3).

2.5. Pathway Analysis

Finally, pathway analyses were used to assess the biological pathways implicated in the differences between the methylation status in stable MHO and unstable MHO patients related to the 26 CpG sites identified in the double PCA selection. Immune-mediated processes could play a role in the progression to the unhealthy state considering that specific pathways such as Th1 and Th2 activation, antigen presentation, allograft rejection signalling were shown to be hypermethylated in stable MHO (Figure 3).

Figure 3. Main pathways related to the significantly differentially methylated CpG sites. Blue bars: biochemical pathways with CpGs sites hypermethylated in stable MHO. Green bars: biochemical pathways with CpGs sites hypermethylated in unstable MHO.A network analysis was performed to examine the inter-relationships between these genes. Almost half of them were linked in a unique network with transcription factors and transcription regulators (AHR, SIP1 or HNF4A) as the main nodes (Figure 4).

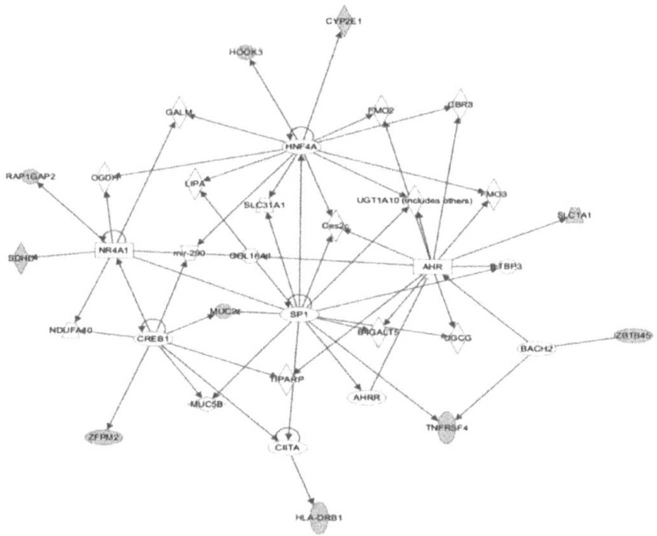

Figure 4. Gene network of interactions related to more relevant CPGs sites.

3. Discussion

Identifying those parameters that may predict the metabolic deterioration of MHO phenotype to unhealthy phenotype or the maintenance of metabolic healthy status over the course of time is currently a matter of debate. Among these factors, the role of epigenetics in the stability of MHO phenotype has attracted attention.

DNA methylation represents major epigenetic modification at the transcriptional regulation level. The function of DNA methylation seems to vary with the genomic context (transcriptional start sites, gene bodies, regulatory elements); in this way, DNA methylation of gene promoters is usually associated with transcriptional silencing, while gene body methylation has been associated with transcription enhancement [15].

Previous studies have investigated alterations in DNA methylation in adipose tissue in relation to obesity, insulin resistance and systemic inflammation [16,17], highlighting the relevance of this epigenetic mechanism in obesity and associated comorbidities. Additionally, modifications in the methylation profile of blood cells associated with obesity and metabolic syndrome have been described [18,19]. However, to the best of our knowledge, there is no previous study that evaluates the long-term methylation changes in patients with obesity according to their metabolic status.

Our results showed 26 CpG sites differentially methylated, both at baseline and 11-year follow-up, associated to 19 genes or pseudogenes, which deserve further investigation to decipher their potential role in the stability of MHO phenotype.

Among the pathways altered by these differences in methylation, immune-related pathways stand out as they could be involved in MHO progression to an unhealthy state. It is well-known that obesity is characterized by a chronic low-grade inflammatory state accompanied by macrophage infiltration in adipose tissue. It has been shown that both obesity and T2DM cause dysregulation of the immune system [20,21]. In our population, CpG sites located in *HLA-DRB1* and *HLA-DQB2* genes were shown to be hypermethylated, being higher than the methylation in stable MHO group. These genes belong to the human leukocyte antigen (HLA) class II complex, which is part of the antigen processing and presentation machinery, and a cornerstone of the adaptative immune system. In a previous study, components of HLA class II have shown increased expression in the adipose tissue of patients with obesity and metabolic syndrome [22]. In adipocytes of subjects with obesity, HLA class II has been shown to play a role in triggering inflammation. Indeed, adaptive immunity has been suggested to have a role in the onset and progression of inflammation and insulin resistance in obesity-associated adipose tissue [23]. SNP genotyping has indicated the role of *HLA-DRB1* in T2D [24]. Some *HLA-DRB1* polymorphisms have been suggested to be protective for T2DM; the hypothesized mechanism seems to be a protective role against autoimmune-mediated loss of insulin secretion [25]. Moreover, in obese adolescents, the development of insulin resistance was associated with a down-regulation of *HLA-DRB1* [26].

The rest of the genes associated with the methylation sites described are involved in a wide range of biological processes, highlighting the roles of potential biomarkers that could predict the progression to an unhealthy state at long-term follow-up. Our results showed that higher methylation in cg20707527 (*ZFPM2* gene) and lower methylation in cg11445109 (*CYP2E1* gene) could have a role in the stability of the healthy phenotype in obesity.

In our study, methylation in the *ZFPM2* gene showed a different tendency between groups; our results described two CpG sites that were hypermethylated in stable MHO, whilst in unstable MHO, these CpG sites were hypomethylated at both baseline and 11-year follow-up. *ZFPM2*, also known as *FOG2*, encodes a zinc finger transcription factor that regulates GATA protein activity, including GATA4, which is involved in cardiac function and modulation of angiogenesis [27]; however, it has also been suggested that *FOG2* develops other roles. Previous studies have associated genetic variants of *ZFPM2* with hypercholesterolemia and metabolic syndrome [28,29]. In animal models, triggering inflammation has been shown to lead to a decrease in *FOG2* expression in hepatocytes [30]. In another study, hepatic *FOG2* was shown to attenuate insulin sensitivity by promoting glycogenolysis [31].

The *CYP2E1* gene showed a high proportion of differentially methylated sites, and tended to be hypomethylated in both stable and unstable MHO. Moreover, the hypomethylation levels were higher in stable MHO. *CYP2E1* belongs to the superfamily of enzymes, cytochrome P450 (CYP), whose members are involved in the biotransformation of drugs, xenobiotics and endogenous substances [32]. The increased activity of *CYP2E1* may promote oxidative stress due to its ability to produce excessive reactive oxygen species [33]. This induction has been described at hepatic level in patients with non-alcoholic fatty liver disease [34]. Additionally, *CYP2E1* activity has been shown to be higher in patients with obesity [35] and an animal model of metabolic syndrome [36]. Although the results are contradictory, some studies have suggested an increase in *CYP2E1* activity in patients with T2D [37], and both glucose and insulin may modulate its activity [38]. All these data suggest that *CYP2E1* may have a role in metabolic alterations with an inflammatory component.

To the best of our knowledge, this is the first study to perform a longitudinal analysis of methylation status in an obese population with a 11-year follow-up. However, our study also presents some limitations. We used blood samples to assess differential DNA methylation, therefore further research on tissue-specific methylation patterns would be necessary. We could not perform RNA analysis to relate DNA methylation to gene expression. Due to the sample size, some relevant differences may not have been detected. Additionally, although the Infinium EPIC array is a very useful tool to interrogate CpGs sites, it only covers 30% of the human methylome. Finally, a validation cohort would be necessary to confirm our results although obtaining a cohort for long-term follow-up (11 years) makes the validation overly complicated.

For a better understanding of the MHO phenotype as well as the predictors factors in the transition of MHO to MUO, more longitudinal studies with a larger number of subjects will be needed. Epigenome-wide studies with samples from adipose tissue will be required in order to increase our knowledge of the mechanisms involved in the development of the MUO phenotype.

4. Materials and Methods

4.1. Design and Subjects

This study is part of the Pizarra study, the details of which have been previously published [4,39]. Briefly, the Pizarra study is a prospective, population-based cohort study of 1051 subjects aged 18–65 years from Pizarra, a town in the province of Malaga (Andalusia, southern Spain). The cohort was re-evaluated after 11 years, and a total of 547 individuals completed the follow-up. Blood samples at both baseline and 11-year follow-up were available from 276 of 547 individuals who completed the follow-up. Of them, 137 patients were obese, both at baseline and 11-year follow-up. Among 137 patients, 58 were classified as MHO at baseline. After matching by age, 18 patients were selected to be included in the study.

Informed consent was obtained from each participant, and the study was approved by the medical ethics committee of the Carlos Haya Regional University Hospital of Malaga.

4.2. Classification Criteria

The NCEP ATPIII criteria were used to classify the subjects according to their metabolic status [40]. They were considered as MHO if they had abdominal obesity (waist circumference >102 cm in men and >88 cm in women) and <2 of the NCEP ATPIII metabolic syndrome criteria were present: systolic blood pressure $\geq$135 mmHg or diastolic blood pressure $\geq$85 mmHg; fasting plasma glucose concentration $\geq$100 mg/dL; HDL-C concentration <40 mg/dL in men and <50 mg/dL in women; fasting plasma TG concentration $\geq$150 mg/dL; or treatment with antihypertensive, lipid lowering, or glucose-lowering medications.

For this study, a subset of 18 MHO subjects at baseline were selected for genome-wide DNA methylation analysis. Of these, 9 MHO subjects developed metabolic complications at 11-year follow-up (unstable MHO; n = 9), whilst the other sub-set of samples remained metabolically healthy at 11-year follow-up (stable MHO; n = 9).

4.3. Procedures

Weight and height measurements were made at baseline and 11-year follow-up. Body mass index (BMI) was calculated as: weight (kg)/height^2 (m^2). Blood pressure was measured twice with a sphygmomanometer with an interval of 5 min between measurements and the average of the two measurements was used in the analyses.

At baseline and 11-year follow-up, blood samples were collected after a 10–12 h fast. The serum was separated, and blood and serum samples were immediately frozen at $-80\,^\circ$C until analysis. Biochemical variables were measured in duplicate. Blood glucose was measured using the glucose oxidase method (Bayer, Leverkusen, Germany). Enzymatic methods were used to measure total cholesterol, triglycerides, and high-density lipoprotein cholesterol.

4.4. DNA Methylation Assay

DNA was extracted from peripheral blood using the QIAmp DNA Blood Mini Kit (Qiagen, Hilden, Germany) following the manufacturer's instructions. DNA concentration was quantified with a Qubit 3.0 Fluorometer (Thermo Fisher Scientific, Waltham, MA, USA) using Qubit dsDNA HS Assay Kit Fluorometer (Thermo Fisher Scientific, Waltham, MA, USA) After quantification, a total of 500 ng of genomic DNA was bisulfite-treated using a Zymo EZ-96 DNA Methylation™ Kit (Zymo Research Corp, Irvine, CA, USA) and was purified using a DNA-Clean-Up Kit (Zymo Research Corp, Irvine, CA, USA).

Over 850,000 methylation sites were interrogated with the Infinium Methylation EPIC Bead Chip Kit (Illumina, San Diego, CA, USA) following the Infinium HD Assay Methylation protocol, and raw data were obtained from iS (Illumina) software.

4.5. Methylation Data Analysis

We used statistical programming language R 3.5.1 (https://www.r-project.org/, accessed on 1 April 2021) to perform the methylation data analysis. Raw data files (idat files) were read with the minfi package [41] to calculate raw β-values. Normal-exponential out-of-band (NOOB) normalization [42] was used to correct the background. Probes located at sexual chromosomes or near SNPs were removed from the analysis. Low quality probes (those with a detection p-value > 0.01 in at least 10% of samples) were also removed. Finally, beta-mixture quantile (BMIQ) normalization [43] was applied to correct for the two different bead designs in the microarrays. For the differential methylation analysis, we transformed β-values to M-values.

4.6. Statistical Analysis

Statistical analysis and comparison were performed using R software (3.5.1) to study differences in anthropometric and biochemical variables with the Kruskall–Wallis test for continuous data and the chi-square test for categorial data. Data are expressed as the mean $\pm$ standard deviation, or as a percentage. Values were statistically significant when $p < 0.05$.

Principal Component Analysis (PCA)

Two complete datasets of normalized CpGs sites were obtained at baseline and the 11-year follow-up. Principal component analysis (PCA) was implemented using native R implementation through R Studio Software 1.2.5033 (version 3.5.1). Classical PCA can be considered as a projection-based approach to find the low-dimensional space that best represents a cloud of high-dimensional points [44]. Firstly, we performed PCA on the dataset of the 11-year follow-up and used the most important CpG sites in both components as subsets for the dataset. Around 1% of them were selected (8200 CpG sites). These sites were tested at baseline and those with a contribution higher than 0.04% (half the maximum contribution value of the best variable) in component 1 and 0.25% in component 2 were selected and used to establish the methylation changes through the follow-up of the study population.

To validate the importance of the selected CpG sites in the PCA, a comparative analysis was performed for each site. Differences between groups were established by using the Kruskal-Wallis test.

Differentially methylated CpG sites identified at both baseline and 11-year follow-up were used to perform a backward stepwise logistic regression to evaluate the prediction power of these sites for the progression to a metabolically unhealthy obesity state.

4.7. Gene Ontology and Pathway Testing

These CpG sites were studied using two different approaches; on one hand, Gene ontology (GO) was used to determine the main processes associated with the selected CpG sites by using AmiGO, a web application that allows users to query, browse and visualize ontologies and related gene product annotation (association) [45]. On the other hand, the selected CpG sites were analyzed through ingenuity pathway analysis software from QIA-GEN. This software allowed us to determine which canonical pathways were related with the selected CpG sites and to establish the more relevant processes altered in both groups at the follow-up. Finally, statistical analysis was performed using R software (3.5.1 version).

5. Conclusions

In conclusion, we described differentially methylated sites that could have a role in the stability/worsening of MHO phenotype; among them, two potential biomarkers have been suggested (cg20707527 and cg11445109). Moreover, the described differences in methylation could alter immune system-related pathways, suggesting these pathways as therapeutic targets to ameliorate metabolic deterioration in MHO patients.

Supplementary Materials: The following are available online at https://www.mdpi.com/article/10.3390/ijms221910417/s1.

Author Contributions: C.G.-R. and T.M.L.-P. contributed to the analysis and interpretation of the data and wrote the manuscript. A.G.-J. performed the statistical analysis of data and provided critical revision of the paper. F.A.-L. processed the samples and performed the experiments prior to the DNA methylation assay. S.V., F.S. and G.R.-M. contributed to the conception of the study. F.J.T. contributed to the conception and design of the study data and provided critical revision of the paper. S.M. contributed to conception/design, funding acquisition, interpretation of the data and has substantively revised the work. All authors have read and agreed to the published version of the manuscript.

Funding: T.M.L.-P. and C.G.-R. were supported by a grant from the Instituto de Salud Carlos III (FI19/00178 and CP20/00066, respectively). F.A.-L. is supported by a grant from "Programa Estatal de Promoción del Talento y su empleabilidad 2018" (PEJ2018-005156-A). S.M. and G.R.-M. are supported by Nicolas Monardes program of the Consejería de Salud de la Junta de Andalucía (C-0050-2017, C-0060-2012, respectively). This work was supported in part by a grant from the Instituto de Salud Carlos III (PI15-01350). This study has been co-funded by FEDER funds ("A way to make Europe"). CIBER Fisiopatología de la Obesidad y Nutrición (CIBEROBN) and CIBER de Diabetes y Enfermedades Metabólicas Asociadas (CIBERDEM) are ISCIII projects.

Institutional Review Board Statement: The study was conducted according to the guidelines of the Declaration of Helsinki and approved by the Ethics Committee of the Carlos Haya Regional University Hospital of Malaga.

Informed Consent Statement: Informed consent was obtained from each participant, and the study was approved by the Medical Ethics Committee of the Carlos Haya Regional University Hospital of Malaga.

Data Availability Statement: The datasets analyzed during the current study will be available on the GEO platform when the manuscript is accepted. Epigenome data from the Infinium Methylation EPIC Bead Chip and metadata including identifier and group of patients will be deposited.

Acknowledgments: We thank the participants in the Pizarra study for their important contribution.

Conflicts of Interest: The authors declare no conflict of interest.

References

1. World Health Organization Obesity and Overweight. Available online: https://www.who.int/news-room/fact-sheets/detail/obesity-and-overweight (accessed on 1 May 2021).
2. Kuk, J.L.; Ardern, C.I. Are metabolically normal but obese individuals at lower risk for all-cause mortality? *Diabetes Care* **2009**, *32*, 2297–2299. [CrossRef] [PubMed]
3. Rey-López, J.P.; de Rezende, L.F.; Pastor-Valero, M.; Tess, B.H. The prevalence of metabolically healthy obesity: A systematic review and critical evaluation of the definitions used. *Obes. Rev.* **2014**, *15*, 781–790. [CrossRef]
4. Soriguer, F.; Gutiérrez-Repiso, C.; Rubio-Martín, E.; García-Fuentes, E.; Almaraz, M.C.; Colomo, N.; De Antonio, I.E.; De Adana, M.S.R.; Chaves, F.J.; Morcillo, S.; et al. Metabolically healthy but obese, a matter of time? Findings from the prospective pizarra study. *J. Clin. Endocrinol. Metab.* **2013**, *98*, 2318–2325. [CrossRef]
5. Lin, L.; Zhang, J.; Jiang, L.; Du, R.; Hu, C.; Lu, J.; Wang, T.; Li, M.; Zhao, Z.; Xu, Y.; et al. Transition of metabolic phenotypes and risk of subclinical atherosclerosis according to BMI: A prospective study. *Diabetologia* **2020**, *63*, 1312–1323. [CrossRef] [PubMed]
6. Iacobini, C.; Pugliese, G.; Blasetti Fantauzzi, C.; Federici, M.; Menini, S. Metabolically healthy versus metabolically unhealthy obesity. *Metabolism* **2019**, *92*, 51–60. [CrossRef] [PubMed]
7. Stefan, N.; Häring, H.U.; Hu, F.B.; Schulze, M.B. Metabolically healthy obesity: Epidemiology, mechanisms, and clinical implications. *Lancet Diabetes Endocrinol.* **2013**, *1*, 152–162. [CrossRef]

8. Locke, A.E.; Kahali, B.; Berndt, S.I.; Justice, A.E.; Pers, T.H.; Day, F.R.; Powell, C.; Vedantam, S.; Buchkovich, M.L.; Yang, J.; et al. Genetic studies of body mass index yield new insights for obesity biology. *Nature* **2015**, *518*, 197–206. [CrossRef] [PubMed]
9. Ling, C.; Rönn, T. Cell Metabolism Review Epigenetics in Human Obesity and Type 2 Diabetes. *Cell Metab.* **2019**, *29*, 1028–1044. [CrossRef] [PubMed]
10. Wang, X.; Pan, Y.; Zhu, H.; Hao, G.; Huang, Y.; Barnes, V.; Shi, H.; Snieder, H.; Pankow, J.; North, K.; et al. An epigenome-wide study of obesity in African American youth and young adults: Novel findings, replication in neutrophils, and relationship with gene expression. *Clin. Epigenetics* **2018**, *10*, 3. [CrossRef]
11. Fradin, D.; Boëlle, P.Y.; Belot, M.P.; Lachaux, F.; Tost, J.; Besse, C.; Deleuze, J.F.; De Filippo, G.; Bougnères, P. Genome-Wide Methylation Analysis Identifies Specific Epigenetic Marks in Severely Obese Children. *Sci. Rep.* **2017**, *7*, 46311. [CrossRef]
12. Xu, X.; Su, S.; Barnes, V.A.; De Miguel, C.; Pollock, J.; Ownby, D.; Shi, H.; Zhu, H.; Snieder, H.; Wang, X. A genome-wide methylation study on obesity: Differential variability and differential methylation. *Epigenetics* **2013**, *8*, 522–533. [CrossRef]
13. Dahlman, I.; Sinha, I.; Gao, H.; Brodin, D.; Thorell, A.; Rydén, M.; Andersson, D.P.; Henriksson, J.; Perfilyev, A.; Ling, C.; et al. The fat cell epigenetic signature in post-obese women is characterized by global hypomethylation and differential DNA methylation of adipogenesis genes. *Int. J. Obes.* **2015**, *39*, 910–919. [CrossRef]
14. Nilsson, E.; Jansson, P.A.; Perfilyev, A.; Volkov, P.; Pedersen, M.; Svensson, M.K.; Poulsen, P.; Ribel-Madsen, R.; Pedersen, N.L.; Almgren, P.; et al. Altered DNA methylation and differential expression of genes influencing metabolism and inflammation in adipose tissue from subjects with type 2 diabetes. *Diabetes* **2014**, *63*, 2962–2976. [CrossRef] [PubMed]
15. Jones, P.A. Functions of DNA methylation: Islands, start sites, gene bodies and beyond. *Nat. Rev. Genet.* **2012**, *13*, 484–492. [CrossRef] [PubMed]
16. Pheiffer, C.; Willmer, T.; Dias, S.; Abrahams, Y.; Louw, J.; Goedecke, J.H. Ethnic and Adipose Depot Specific Associations Between DNA Methylation and Metabolic Risk. *Front. Genet.* **2020**, *11*, 967. [CrossRef]
17. Crujeiras, A.B.; Diaz-Lagares, A.; Moreno-Navarrete, J.M.; Sandoval, J.; Hervas, D.; Gomez, A.; Ricart, W.; Casanueva, F.F.; Esteller, M.; Fernandez-Real, J.M. Genome-wide DNA methylation pattern in visceral adipose tissue differentiates insulin-resistant from insulin-sensitive obese subjects. *Transl. Res.* **2016**, *178*, 13–24.e5. [CrossRef] [PubMed]
18. Chambers, J.C.; Loh, M.; Lehne, B.; Drong, A.; Kriebel, J.; Motta, V.; Wahl, S.; Elliott, H.R.; Rota, F.; Scott, W.R.; et al. Epigenome-wide association of DNA methylation markers in peripheral blood from Indian Asians and Europeans with incident type 2 diabetes: A nested case-control study. *Lancet Diabetes Endocrinol.* **2015**, *3*, 526–534. [CrossRef]
19. Van Otterdijk, S.D.; Binder, A.M.; Szarc Vel Szic, K.; Schwald, J.; Michels, K.B. DNA methylation of candidate genes in peripheral blood from patients with type 2 diabetes or the metabolic syndrome. *PLoS ONE* **2017**, *12*, e0180955. [CrossRef]
20. Richard, C.; Wadowski, M.; Goruk, S.; Cameron, L.; Sharma, A.M.; Field, C.J. Individuals with obesity and type 2 diabetes have additional immune dysfunction compared with obese individuals who are metabolically healthy. *BMJ Open Diabetes Res. Care* **2017**, *5*, e000379. [CrossRef]
21. Ip, B.C.; Hogan, A.E.; Nikolajczyk, B.S. Lymphocyte roles in metabolic dysfunction: Of men and mice. *Trends Endocrinol. Metab.* **2015**, *26*, 91–100. [CrossRef]
22. Klimčáková, E.; Roussel, B.; Márquez-Quiñones, A.; Kováčová, Z.; Kováčiková, M.; Combes, M.; Šiklová-Vítková, M.; Hejnová, J.; Šrámková, P.; Bouloumié, A.; et al. Worsening of obesity and metabolic status yields similar molecular adaptations in human subcutaneous and visceral adipose tissue: Decreased metabolism and increased immune response. *J. Clin. Endocrinol. Metab.* **2011**, *96*, E73–E82. [CrossRef]
23. Deng, T.; Lyon, C.J.; Minze, L.J.; Lin, J.; Zou, J.; Liu, J.Z.; Ren, Y.; Yin, Z.; Hamilton, D.J.; Reardon, P.R.; et al. Class II major histocompatibility complex plays an essential role in obesity-induced adipose inflammation. *Cell Metab.* **2013**, *17*, 411–422. [CrossRef]
24. Zhong, H.; Yang, X.; Kaplan, L.M.; Molony, C.; Schadt, E.E. Integrating Pathway Analysis and Genetics of Gene Expression for Genome-wide Association Studies. *Am. J. Hum. Genet.* **2010**, *86*, 581–591. [CrossRef] [PubMed]
25. Williams, R.C.; Muller, Y.L.; Hanson, R.L.; Knowler, W.C.; Mason, C.C.; Bian, L.; Ossowski, V.; Wiedrich, K.; Chen, Y.F.; Marcovina, S.; et al. HLA-DRB1 reduces the risk of type 2 diabetes mellitus by increased insulin secretion. *Diabetologia* **2011**, *54*, 1684–1692. [CrossRef]
26. Minchenko, D.O. Insulin resistance in obese adolescents affects the expression of genes associated with immune response. *Endocr. Regul.* **2019**, *53*, 71–82. [CrossRef]
27. Zhou, B.; Ma, Q.; Sek, W.K.; Hu, Y.; Campbell, P.H.; McGowan, F.X.; Ackerman, K.G.; Wu, B.; Zhou, B.; Tevosian, S.G.; et al. Fog2 is critical for cardiac function and maintenance of coronary vasculature in the adult mouse heart. *J. Clin. Investig.* **2009**, *119*, 1462–1476. [CrossRef] [PubMed]
28. Azimi-Nezhad, M.; Mirhafez, S.R.; Stathopoulou, M.G.; Murray, H.; Ndiaye, N.C.; Bahrami, A.; Varasteh, A.; Avan, A.; Bonnefond, A.; Rancier, M.; et al. The Relationship Between Vascular Endothelial Growth Factor Cis- and Trans-Acting Genetic Variants and Metabolic Syndrome. *Am. J. Med. Sci.* **2018**, *355*, 559–565. [CrossRef] [PubMed]
29. Salami, A.; El Shamieh, S. Association between SNPs of Circulating Vascular Endothelial Growth Factor Levels, Hypercholesterolemia and Metabolic Syndrome. *Medicina* **2019**, *55*, 464. [CrossRef]
30. Bagu, E.T.; Layoun, A.; Calvé, A.; Santos, M.M. Friend of GATA and GATA-6 modulate the transcriptional up-regulation of hepcidin in hepatocytes during inflammation. *BioMetals* **2013**, *26*, 1051–1065. [CrossRef]

31. Guo, Y.; Yu, J.; Deng, J.; Liu, B.; Xiao, Y.; Li, K.; Xiao, F.; Yuan, F.; Liu, Y.; Chen, S.; et al. A novel function of hepatic FOG2 in insulin sensitivity and lipid metabolism through PPARα. *Diabetes* **2016**, *65*, 2151–2163. [CrossRef]

32. Stipp, M.C.; Acco, A. Involvement of cytochrome P450 enzymes in inflammation and cancer: A review. *Cancer Chemother. Pharmacol.* **2020**, *87*, 295–309. [CrossRef]

33. Lu, Y.; Cederbaum, A.I. CYP2E1 and oxidative liver injury by alcohol. *Free Radic. Biol. Med.* **2008**, *44*, 723–738. [CrossRef]

34. Teufel, U.; Peccerella, T.; Engelmann, G.; Bruckner, T.; Flechtenmacher, C.; Millonig, G.; Stickel, F.; Hoffmann, G.F.; Schirmacher, P.; Mueller, S.; et al. Detection of carcinogenic etheno-DNA adducts in children and adolescents with non-alcoholic steatohepatitis (NASH). *Hepatobiliary Surg. Nutr.* **2015**, *4*, 426–435. [CrossRef] [PubMed]

35. Lucas, D.; Farez, C.; Bardou, L.G.; Vaisse, J.; Attali, J.R.; Valensi, P. Cytochrome P450 2E1 activity in diabetic and obese patients as assessed by chlorzoxazone hydroxylation. *Fundam. Clin. Pharmacol.* **1998**, *12*, 553–558. [CrossRef]

36. Bondarenko, L.B.; Shayakhmetova, G.M.; Voronina, A.K.; Kovalenko, V.M. Age-dependent features of CYP3A, CYP2C, and CYP2E1 functioning at metabolic syndrome. *J. Basic Clin. Physiol. Pharmacol.* **2016**, *27*, 603–610. [CrossRef]

37. Gravel, S.; Chiasson, J.L.; Turgeon, J.; Grangeon, A.; Michaud, V. Modulation of CYP450 Activities in Patients With Type 2 Diabetes. *Clin. Pharmacol. Ther.* **2019**, *106*, 1280–1289. [CrossRef]

38. Massart, J.; Begriche, K.; Fromenty, B. Cytochrome P450 2E1 should not be neglected for acetaminophen-induced liver injury in metabolic diseases with altered insulin levels or glucose homeostasis. *Clin. Res. Hepatol. Gastroenterol.* **2020**, *45*, 101470. [CrossRef]

39. Soriguer, F.; Rojo-Martínez, G.; Valdés, S.; Tapia, M.J.; Botas, P.; Morcillo, S.; Delgado, E.; Esteva, I.; Ruiz De Adana, M.S.; Almaraz, M.C.; et al. Factors determining weight gain in adults and relation with glucose tolerance. *Clin. Endocrinol.* **2013**, *78*, 858–864. [CrossRef] [PubMed]

40. Cleeman, J.I. Executive summary of the third report of the National Cholesterol Education Program (NCEP) expert panel on detection, evaluation, and treatment of high blood cholesterol in adults (adult treatment panel III). *J. Am. Med. Assoc.* **2001**, *285*, 2486–2497. [CrossRef]

41. Aryee, M.J.; Jaffe, A.E.; Corrada-Bravo, H.; Ladd-Acosta, C.; Feinberg, A.P.; Hansen, K.D.; Irizarry, R.A. Minfi: A flexible and comprehensive Bioconductor package for the analysis of Infinium DNA methylation microarrays. *Bioinformatics* **2014**, *30*, 1363–1369. [CrossRef]

42. Triche, T.J.; Weisenberger, D.J.; Van Den Berg, D.; Laird, P.W.; Siegmund, K.D. Low-level processing of Illumina Infinium DNA Methylation BeadArrays. *Nucleic Acids Res.* **2013**, *41*, e90. [CrossRef] [PubMed]

43. Teschendorff, A.E.; Marabita, F.; Lechner, M.; Bartlett, T.; Tegner, J.; Gomez-Cabrero, D.; Beck, S. A beta-mixture quantile normalization method for correcting probe design bias in Illumina Infinium 450 k DNA methylation data. *Bioinformatics* **2013**, *29*, 189–196. [CrossRef] [PubMed]

44. Song, Y.; Westerhuis, J.A.; Aben, N.; Michaut, M.; Wessels, L.F.A.; Smilde, A.K. Principal component analysis of binary genomics data. *Brief. Bioinform.* **2019**, *20*, 317–329. [CrossRef] [PubMed]

45. Udhaya Kumar, S.; Thirumal Kumar, D.; Siva, R.; George Priya Doss, C.; Zayed, H. Integrative bioinformatics approaches to map potential novel genes and pathways involved in ovarian cancer. *Front. Bioeng. Biotechnol.* **2019**, *7*, 391. [CrossRef]

International Journal of
Molecular Sciences

Article

The QseEF Two-Component System-GlmY Small RNA Regulatory Pathway Controls Swarming in Uropathogenic *Proteus mirabilis*

Wen-Yuan Lin [1], Yuan-Ju Lee [2,†], Ping-Hung Yu [3,†], Yi-Lin Tsai [1,†], Pin-Yi She [1], Tzung-Shian Li [1] and Shwu-Jen Liaw [1,4,*]

[1] Department and Graduate Institute of Clinical Laboratory Sciences and Medical Biotechnology, College of Medicine, National Taiwan University, Taipei 10048, Taiwan; f04424003@ntu.edu.tw (W.-Y.L.); joolliint@gmail.com (Y.-L.T.); R02424027@ntu.edu.tw (P.-Y.S.); qwerty7749@hotmail.com (T.-S.L.)

[2] Department of Urology, National Taiwan University Hospital, Taipei 10002, Taiwan; leeyuanju@hotmail.com

[3] Department of Nursing, National Taichung University of Science and Technology, Taichung City 404348, Taiwan; oxfordocean@gmail.com

[4] Department of Laboratory Medicine, National Taiwan University Hospital, Taipei 10002, Taiwan

* Correspondence: sjliaw@ntu.edu.tw; Tel.: +886-02-23123456 (ext. 6911)

† These authors contributed equally to this work.

Citation: Lin, W.-Y.; Lee, Y.-J.; Yu, P.-H.; Tsai, Y.-L.; She, P.-Y.; Li, T.-S.; Liaw, S.-J. The QseEF Two-Component System-GlmY Small RNA Regulatory Pathway Controls Swarming in Uropathogenic *Proteus mirabilis*. *Int. J. Mol. Sci.* **2022**, *23*, 487. https://doi.org/10.3390/ijms23010487

Academic Editors: Amelia Casamassimi, Alfredo Ciccodicola and Monica Rienzo

Received: 30 November 2021
Accepted: 28 December 2021
Published: 1 January 2022

Publisher's Note: MDPI stays neutral with regard to jurisdictional claims in published maps and institutional affiliations.

Abstract: Bacterial sensing of environmental signals through the two-component system (TCS) plays a key role in modulating virulence. In the search for the host hormone-sensing TCS, we identified a conserved *qseEGF* locus following *glmY*, a small RNA (sRNA) gene in uropathogenic *Proteus mirabilis*. Genes of *glmY-qseE-qseG-qseF* constitute an operon, and QseF binding sites were found in the *glmY* promoter region. Deletion of *glmY* or *qseF* resulted in reduced swarming motility and swarming-related phenotypes relative to the wild-type and the respective complemented strains. The *qseF* mutant had decreased *glmYqseEGF* promoter activity. Both *glmY* and *qseF* mutants exhibited decreased *flhDC* promoter activity and mRNA level, while increased *rcsB* mRNA level was observed in both mutants. Prediction by TargetRNA2 revealed *cheA* as the target of GlmY. Then, construction of the translational fusions containing various lengths of *cheA* 5′UTR for reporter assay and site-directed mutagenesis were performed to investigate the *cheA*-GlmY interaction in *cheA* activation. Notably, loss of *glmY* reduced the *cheA* mRNA level, and urea could inhibit swarming in a QseF-dependent manner. Altogether, this is the first report elucidating the underlying mechanisms for modulation of swarming motility by a QseEF-regulated sRNA GlmY, involving expression of *cheA*, *rcsB* and *flhDC* in uropathogenic *P. mirabilis*.

Keywords: *cheA*; *flhDC*; GlmY; *Proteus mirabilis*; QseEF; *rcsB*; swarming

1. Introduction

Proteus mirabilis is an important pathogen of the urinary tract, especially in patients with indwelling urinary catheters [1]. Common strategies of pathogenesis employed by *P. mirabilis* include adherence via fimbriae [2,3], biofilm formation, flagella-mediated motility, immune modulation and urease production [4]. *P. mirabilis* exhibits a form of multicellular behavior termed swarming [5]. Swarming motility are the results of complex signal transduction and gene regulation [5]. It is generally believed that signals could be sensed and transmitted by two-component systems (TCSs). It is worth noting that the ability of *P. mirabilis* to express virulence factors is coupled to swarming differentiation [5]. In general, flagella are thought to assist in colonization and dissemination during *P. mirabilis* catheter-associated UTIs, and FlhDC is a master regulator that controls the expression of flagellum-related genes [6,7]. During swarmer cell differentiation, the expression level of *flhDC* rises dramatically [8].

Among multiple strategies for many bacteria to respond rapidly to changing environments is the regulation by the very versatile and adaptable regulatory small non-coding RNAs (sRNAs) [9–11]. Numerous cellular processes, such as motility [12], various stress responses and virulence factor expression are subject to the post-transcriptional control of sRNAs [9–11,13]. In many cases, the trans-acting sRNA mediated-regulation requires the chaperone protein Hfq to facilitate sRNA-mRNA interaction [14,15]. Two well-known ways for an sRNA to regulate translation are sequestering of ribosome binding site (rbs) and melting a secondary structure in the $5'$ UTR to expose rbs, resulting in repression and activation of translation, respectively [16]. TCS and alternative σ factors have been shown to control sRNA expression [17,18]. In this regard, a PhoPQ TCS-regulated sRNA MgrR modulates expression of *eptB*, an LPS modification gene regulated by σ^E, to affect the sensitivity to antimicrobial peptides in response to low Mg^{2+} or the presence of antimicrobial peptides [19]. In addition, a σ^E-dependent sRNA, MicA, directly inhibited PhoPQ synthesis and consequently downregulated the PhoPQ regulons involved in pathogenicity, cell envelope composition and stress resistance [20].

Bacterial sensing of environmental signals plays a key role in regulating virulence and mediating bacterium host interactions. TCS is a common strategy used by bacteria to regulate gene expression in response to environmental cues. Enterohemorrhagic *E. coli* (EHEC) senses host hormones (epinephrine and norepinephrine) via the QseEF two-component system to activate actin polymerization and initiate formation of attaching and effacing (AE) lesions [21,22]. Deletion mutants of the *qseEF* are attenuated in virulence, as demonstrated for *E. coli*, *Citrobacter rodentium*, *Salmonella* and *Yersinia pseudotuberculosis* [23–25]. In addition, epinephrine can induce QseEF expression [21,24]. QseE is a sensor histidine kinase; QseF is a response regulator protein. The *qseE* and *qseF* genes are co-transcribed with *qseG*, a small outer membrane lipoprotein-encoded gene located between *qseE* and *qseF* [26]. The gene cluster *glmY-qseE-qseG-qseF* is conserved in *Enterobacteriaceae* [27]; however, previous work investigating QseEF or GlmY has been carried out mainly in EHEC and *Salmonella*. Although conserved across *E. coli* and *Salmonella*, the QseEF system has undergone specialization to regulate gene expression unique to each species.

Knowing the sensing of the stress hormones epinephrine (adrenaline) and norepinephrine (noradrenaline) through QseBC and QseEF TCSs plays an important role in modulating bacterial stress responses and virulence [24,28], we searched for the counterparts in uropathogenic *P. mirabilis* and identified QseEF homologue with high sequence identity but not QseBC. We then undertook an investigation to disclose the role of QseEF in *P. mirabilis*. Among the phenotypes screened, we found that *P. mirabilis qseF* mutant exhibited significantly reduced swarming ability and swarming-related phenotypic traits. Subsequently, the underlying mechanism of QseF-regulated swarming phenomenon was revealed to involve GlmY sRNA. For the first time, a pathway mediated by a two-component system through an sRNA was disclosed to be involved in swarming migration of uropathogenic *P. mirabilis*. The study provides a new insight into the underlying mechanisms of swarming motility in *P. mirabilis*.

2. Results

2.1. Identification of P. mirabilis qseEGF Gene Locus

Seeing that epinephrine and norepinephrine exist in urine, we sought QseBC and QseEF homologues in uropathogenic *P. mirabilis* N2. We identified a gene locus, *purL-qseE-qseG-qseF-nadE-glnB*, whose gene product with amino acid sequence identity of around 76, 60, 20, 84 and 90% to PurL, QseE, QseG, QseF and GlnB, respectively, with *E. coli* and *Salmonella* (both lacking *nadE*) (Figure 1A). We then tested whether epinephrine or norepinephrine has any effect on phenotypic traits of *P. mirabilis*. Neither hormone in the range 0.1–400 μM altered the phenotypes assayed (motility, biofilm formation, etc.) or showed any effect on expression of *qseEF* by real time RT-PCR.

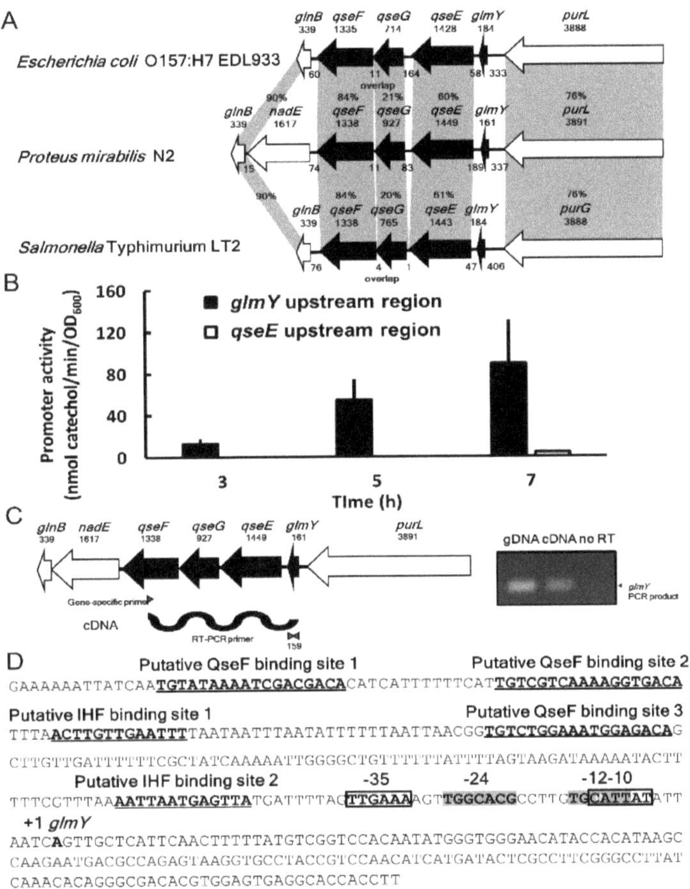

Figure 1. The *glmYqseEGF* gene locus in *P. mirabilis* N2. (**A**) The *P. mirabilis glmYqseEGF* gene locus corresponds to the similar locus in *E. coli* O157:H7 EDL933 and *Salmonella* Typhimurium LT2 with corresponding genes in shadows. An amino acid sequence analysis of the locus in these strains was performed using position-specific iterative BLAST. The percent amino acid identities of PurL, QseE, QseG, QseF and GlnB between *P. mirabilis* N2 and *E. coli* O157:H7 EDL933 or *Salmonella* Typhimurium LT2 are shown. The number above each arrow represents the gene length (bp). The intergenic space or overlap in terms of base pairs is also shown. (**B**) The promoter activities of *glmY* (−381 to −3 from putative *glmY* transcriptional start site) and *qseE* (−553 to −6 from *qseE* start codon) upstream region. (**C**) The *glmY-qseE-qseG-qseF* constitute an operon by gene-specific reverse transcription PCR. cDNA was synthesized by the *qseF*-specific primer and then the *glmY* DNA fragment was amplified by PCR. no RT, negative control; gDNA (genomic DNA), positive control. (**D**) The sequence of *glmYqseEGF* promoter region in *P. mirabilis* N2. The putative QseF and IHF binding sites are bold and underlined. The conserved σ70 and σ54 binding sites are indicated in boxes and shadows, respectively.

2.2. Identification of P. mirabilis glmY and Co-Transcription of glmY, qseE, qseG and qseF

In view of conservation of the gene cluster *glmY-glrK* (*qseE*)-*yfhG* (*qseG*)-*glrR* (*qseF*)-*glnB* in *Enterobacteriaceae* [27] and modulation of motility [12] and pathogenesis [29,30] by GlmY, we searched the bacterial small RNA database (BSRD) for *P. mirabilis* GlmY counterpart and found it located upstream of *qseE* and downstream of *purL* (Figure 1A). We first examined the promoter activity in the upstream 548-bp region of *qseE* (nucleotide −553 to −6 of QseE start codon) by the *xylE* reporter assay. No promoter activity was found (Figure 1B). Then, we tested the *glmY* promoter activity using the upstream 379-bp DNA fragment (nucleotide −381 to −3 from the putative transcription start site) of *glmY* and found the promoter activity at 3, 5 and 7 h after incubation (Figure 1B). Furthermore, we demonstrated that *glmY*, *qseE*, *qseG* and *qseF* belong to a transcript by RT (reverse transcription)-PCR using *qseF* specific primer to obtain the cDNA for amplifying the *glmY* fragment (Figure 1C). The *glmY* PCR product of 159 bp was observed, but no product was produced in the no RT control (Figure 1C). These data indicate that *glmY*, *qseE*, *qseG* and *qseF* can share the same promoter upstream of *glmY*. The lack of promoter activity in the 548-bp DNA fragment upstream *qseE* results from the fragment containing only partial *glmY* promoter DNA sequences (201/379 bp). In summary, *glmY*, *qseE*, *qseG* and *qseF* are co-transcribed from the *glmY* promoter in *P. mirabilis*. In addition, we identified one overlapped σ^{70}/σ^{54}, two IHF and three QseF putative binding sites in the upstream promoter region of *glmY* (Figure 1D).

2.3. Phenotypic Traits of glmY and qse Mutants

To investigate the roles of GlmY and QseEF TCS in *P. mirabilis*, we first generated isogenic mutants of *glmY* and *qseF*. Among the phenotypes assayed, both *glmY* and *qseF* mutant strains exhibited reduced swarming and swimming abilities relative to the wild-type and respective complemented strains (Figure 2A,D). Comparable growth of the wild-type and respective complemented strains was observed, and no growth detect was found in the mutant strains compared to the wild-type (data not shown). *glmY* and *qseF* mutants migrated much slower than the wild-type and respective complemented strains during the 8-h period after inoculation on the swarming plate. Mutants lacking *qseE*, *qseG* or *qseEGF* were also constructed with no growth defect compared to the wild-type bacteria. We found similar swarming and swimming patterns among *qseF*, *qseE*, *qseG* and *qseEGF* mutants (data not shown). Therefore, we investigated the QseEGF-mediated motility using *qseF* mutant to represent QseEF malfunction in the following experiments. Significantly reduced swarming-related phenotypes, including cell differentiation, hemolysin activity and flagellin level, were observed in the *glmY* and *qseF* mutants compared to the wild-type and respective complemented strains (Figure 2B,C,E). Transmission electron microscopy (TEM) also showed *glmY* and *qseF* mutant cells were shorter and had fewer flagella than the wild-type strain (Figure 2F).

2.4. Downregulation of glmYqseEGF Operon by qseF Deletion

Based on the presence of the putative QseF binding site in the *glmYqseEGF* promoter region (Figure 1D), we examined whether *glmY* expression is under the control of QseF by the reporter assay. Just as expected, *qseF* mutant had significantly lower *glmY* promoter activity compared to the wild-type and the *qseF*-complemented strain at 3, 5, 7 h after inoculation and incubation (Figure 3A).

2.5. Altered Expression of flhDC, rcsB and cheA in qseF and glmY Mutants

With the notion that disruption of *qseF*, encoding a TCS transcriptional regulator QseF, impaired the swarming motility and related phenotypes in *P. mirabilis*, we thus examined the effect of *qseF* loss on *flhDC* expression by the reporter assay. The loss of *qseF* incurred significant reduction in *flhDC* promoter activity compared to the wild-type and the complemented strains at 3, 5, 7 h after inoculation (Figure 3B). Since QseF is a transcriptional regulator, we then tried to find the consensus QseF binding site ($TGTCN_{10}GACA$) [31] in the promoter region of *flhDC* but failed. Since QseF could regulate *glmY* expression (Figure 3A) and both *qseF* and *glmY* mutants displayed decreased swarming and swimming motility (Figure 2A,D), we examined the promoter activity of *flhDC* in the *glmY* mutant. As is the case for *qseF* mutant, *glmY* mutant had a lower *flhDC* promoter activity than the wild-type and the complemented strains (Figure 3B). The mRNA level of *flhDC* in *qseF* and *glmY* mutants was consistent with the results of the reporter assay (Figure 3C).

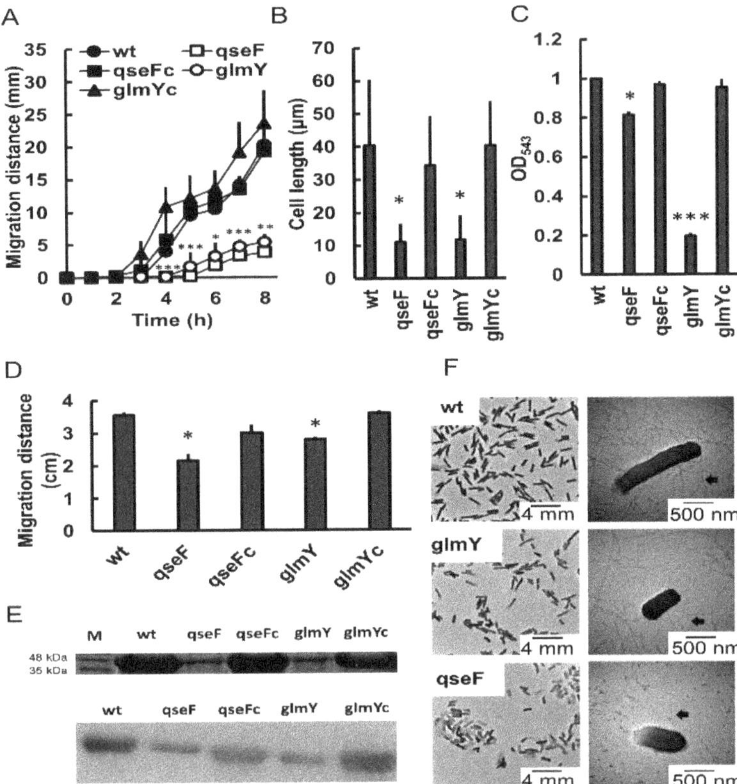

Figure 2. GlmY and QseF regulated swarming and related phenotypes. Swarming motility (**A**), cell length (**B**), hemolysin activity (**C**) and swimming motility (**D**) were determined in the wild-type, mutants of *qseF* and *glmY* and complemented strains. Swarming migration was monitored at 1-h intervals, and swimming motility was determined after incubation for 16 h. Cell length and hemolysin activity were determined at 5 h after incubation on the LB agar plate. The data are the averages and standard deviations of three independent experiments. Significant difference of *qseF* or *glmY* mutant from the wild-type at 4, 5, 6, 7 and 8 h is indicated (* $p < 0.05$; ** $p < 0.01$; *** $p < 0.001$ by the Student's *t* test) in (**A**). Significant difference from the wild-type is indicated (* $p < 0.05$; *** $p < 0.001$ by the Student's *t* test) in (**B**–**D**). (**E**) Analysis of flagellin expression by SDS-PAGE (upper panel) and Western blotting (lower panel). The flagellin level of the wild-type, mutants of *qseF* and *glmY* and complemented strains were examined at 5 h after seeding on the swarming plates by the SDS-PAGE and Western blotting as described in Materials and Methods. The representative picture of three independent experiments is shown. M, molecular weight marker. (**F**) TEM pictures of wild-type and mutants of *qseF* and *glmY*. Bacterial cultures were applied onto a carbon-coated grid, cells were stained with 1% PTA and TEM pictures were taken. Flagella are indicated by arrows. wt, wild-type; qseF, *qseF* mutant; glmY, *glmY* mutant; qseFc, *qseF*-complemented strain; glmYc, *glmY*-complemented strain.

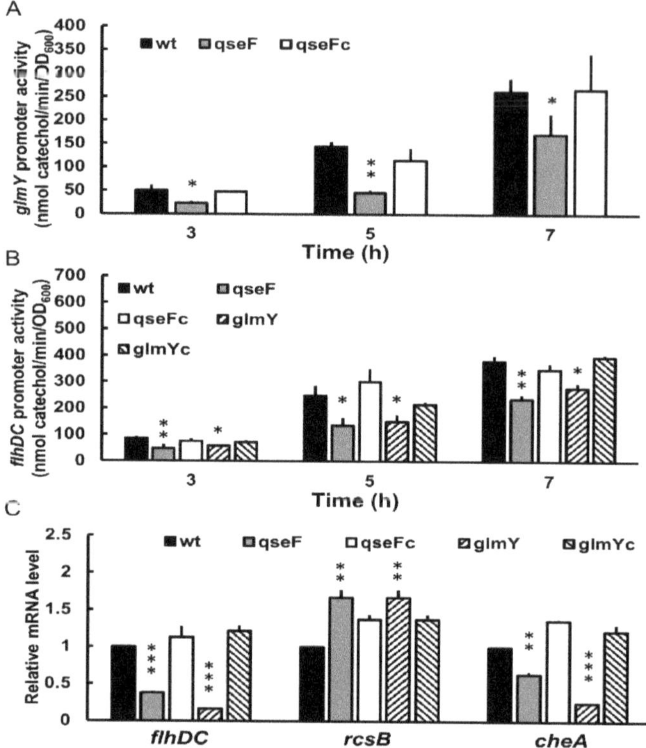

Figure 3. QseF-controlled GlmY regulated expression of *flhDC*, *rcsB* and *cheA*. (**A**) The *glmY* promoter activity in the wild-type, *qseF* mutant and *qseF*-complemented strain. (**B**) The *flhDC* promoter activity in the wild-type, mutants of *qseF* and *glmY* and the complemented strains. (**C**) The mRNA level of *flhDC*, *rcsB* and *cheA* in the wild-type, mutants of *qseF* and *glmY* and the complemented strains. In (**A,B**), bacteria cultures were spread on LB agar plate and XylE promoter activities were determined as described in Materials and Methods after incubation for 3, 5 and 7 h at 37 °C. In (**C**), the mRNA amount was measured at 5 h after inoculation on the LB agar plate. The value for the wild-type was set at 1, and other data are presented relative to this value. The data are the averages and standard deviations of three independent experiments. Significant difference from the wild-type is indicated (* $p < 0.05$; ** $p < 0.01$; *** $p < 0.001$ by the Student's *t* test). wt, wild-type; qseF, *qseF* mutant; glmY, *glmY* mutant; qseFc, *qseF*- complemented strain; glmYc, *glmY*-complemented strain.

Given the positive regulation of swarming and *flhDC* expression by both QseF and GlmY, the regulatory hierarchy from QseF to GlmY and the absence of QseF binding site in *flhDC* promoter region, we surmised that QseF may modulate swarming through GlmY to affect *flhDC* expression. How could GlmY, an sRNA, regulate both promoter activity and mRNA level of *flhDC* (Figure 3B,C)? In this respect, GlmY may modulate expression of a regulator which affects *flhDC* transcription directly and thus *flhDC* mRNA level. The direct regulation of *flhDC* mRNA level by GlmY may also contribute to the altered mRNA amount. As for *flhDC* regulator, *P. mirabilis* RcsB could modulate swarming by direct regulation of *flhDC* expression in a negative way [8,32] and *rcsB* overexpression led to a reduction of *flhDC* expression [33]. Therefore, we tested the effect of *glmY* mutation on *rcsB* mRNA level and found *glmY* mutant had increased *rcsB* mRNA level compared to the wild-type and *glmY*-complemented strain (Figure 3C). In view of regulation of *glmY* by QseF, *qseF* mutant exhibited a similar increase in *rcsB* mRNA level as the *glmY* mutant relative to the wild-type and the complemented strain (Figure 3C). On the other hand, we tried to identify whether GlmY could target *flhDC* or other mRNAs to modulate motility of *P. mirabilis*. We used the full sequence of GlmY as input to search for GlmY targets on the TargetRNA2 website. The tool did report a region in the 5′UTR of the *cheA* (but not *flhDC* or *rcsB*) as a candidate target with lower energy score and the *p*-value less than 0.05. The prediction shows interaction of GlmY (positions 17 to 28) with a 5′UTR region of *cheA* mRNA spanning −61 to −50 positions from the translation start site of *cheA* mRNA. CheA kinase, encoded by *cheA*, belongs to a family of two-component sensors responsible for bacterial chemotaxis [34]. As *P. mirabilis* GlmY positively regulated swarming (Figure 2A) and small RNAs exert positive regulation of virulence primarily at the level of mRNA stabilization [35], we assessed whether GlmY affects *cheA* mRNA level. As expected, the amount of *cheA* mRNA was reduced significantly in *glmY* and *qseF* mutants compared to the wild-type and respective complemented strains (Figure 3C).

2.6. GlmY Activates cheA Expression at the Post-Transcriptional Level

To investigate the role of interaction between GlmY and *cheA* 5′UTR in *cheA* expression and thus swarming, we generated plasmids carrying the individual *lac* promoter-driven *xylE* translational fusion containing different lengths (520, 253, 138 and 61 bp) of *cheA* 5′UTR and the first 27 bp of *cheA* ORF (Figure 4A). The individual fusion plasmid was transformed into the wild-type strain and *glmY* mutant to probe the essential region of *cheA* 5′UTR requiring GlmY for activating *cheA* translation by monitoring the *xylE* activity. While the *glmY* mutant carrying the plasmid of 520, 253 or 138 bp-5′UTR translational fusion exhibited significantly lower *xylE* activity than the wild-type strain carrying the respective fusion plasmid, albeit to a lesser extent for the 138 bp-fusion plasmid (Figure 4B), the *xylE* activity of the wild-type strain and the *glmY* mutant harboring the 61 bp-fusion plasmid was comparable (Figure 4B). The results indicate that there is no need for GlmY to activate *cheA* translation when only the 61 bp-5′UTR of *cheA* is present. The data indicate all the 520, 253 and 138-bp fragments of *cheA* 5′UTR had constrained structures requiring GlmY for releasing to activate *cheA* translation.

To further demonstrate whether GlmY affected *cheA* expression via direct base-pairing, we performed site-directed mutagenesis to inactivate the interaction of GlmY and *cheA* 5′UTR. We introduced an 8-bp mutation into the predicted pairing regions of the *cheA* 5′UTR (cheAm) and *glmY* (GlmYm) (Figure 4C) using the *cheA* 5′UTR (253 bp)-*xylE* translational fusion plasmid (pcheA) and *glmY*-containing pBAD33 (pBglmY) as templates to generate pcheAm and pBglmYm, respectively. The wild-type *P. mirabilis* harboring the pcheAm exhibited significantly lower XylE activity compared to that harboring the wild-type *cheA* 5′UTR-*xylE* plasmid (pcheA) (Figure 4D). The *glmY* mutant carrying pBglmY and pcheAm showed significantly lower XylE activity than the mutant carrying pBglmY and pcheA (0.4 vs. 1 in Figure 4D). GlmY with the 8-bp substitution (GlmYm) could not effectively enhance the expression of *cheA-xylE* fusion to the extent of wild-type GlmY (0.6 vs. 1) in the *glmY* mutant (Figure 4D). In addition, introduction of compensating mutations into the *cheA* 5′UTR could restore the ability of GlmYm to activate the *cheA-xylE* fusion to 80% of the wild-type GlmY and *cheA* interaction (pBglmYm-pcheAm vs. pBglmY-pcheA in Figure 4D).

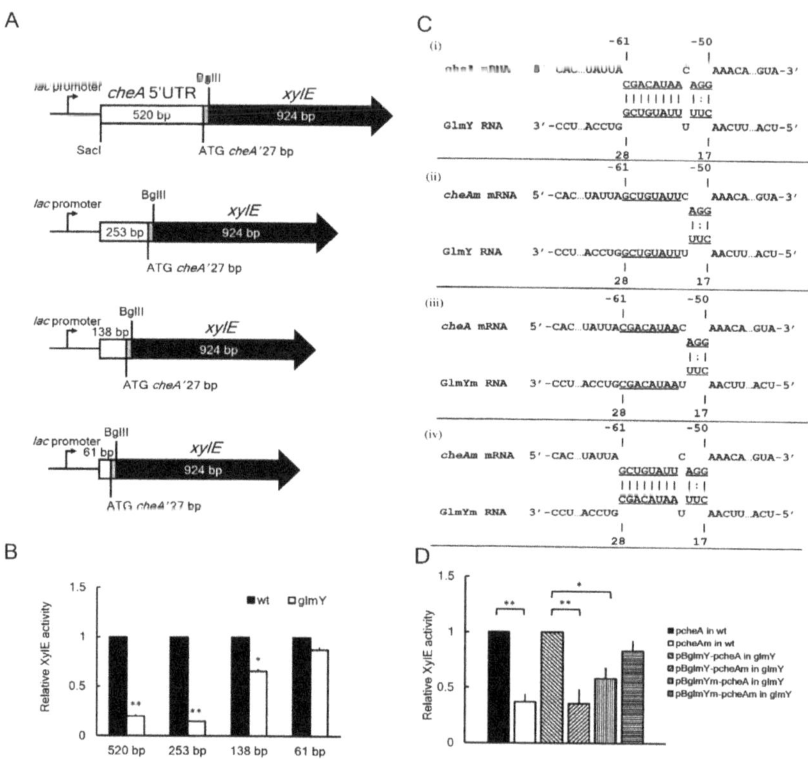

Figure 4. GlmY activated *cheA* expression at the post-transcriptional level. (**A**) Schematic representation of the different constructs used in the *cheA-xylE* translational reporter assay. The various lengths of *cheA* 5′UTR (520, 253, 138 and 61 bp before AUG plus 27 bp after AUG were in frame ligated to *xylE* gene to obtain the translational fusion under the control of *lac* promoter in the pGEM-T easy vector. (**B**) The effect of various lengths of *cheA* 5′UTR (as shown in (**A**)) on XylE activity of the *cheA* 5′UTR-*xylE* translational fusion in the wild-type and *glmY* mutant. The activity of XylE in the translational *cheA* 5′UTR-*xylE* reporter plasmid-transformed *P. mirabilis* strains was determined using the reporter assay at 5 h after incubation on an LB agar plate. The value obtained for the wild-type strain at a length of 5′UTR was set at 1. The data are the averages and standard deviations of three independent experiments. Significant difference from the wild-type is indicated (* $p < 0.05$; ** $p < 0.01$ by the Student's *t* test). (**C**) The predicted region of base-pairing between GlmY and the 5′ UTR of the *cheA* mRNA. The interaction site of *cheA* and GlmY are underlined and the substitutions present in GlmY (GlmYm) and *cheA* 5′ UTR (*cheA*m) are shown. (**i**). Wild-type GlmY and *cheA* 5′ UTR; (**ii**) wild-type GlmY and mutated *cheA* 5′ UTR; (**iii**) wild-type *cheA* 5′ UTR and mutated GlmY; (**iv**) mutated GlmY and the compensatory mutations in *cheA* 5′ UTR. (**D**) Analysis of GlmY interaction with *cheA* 5′ UTR for *cheA* expression. The *cheA* 5′UTR (253 bp)-*xylE* translational fusion plasmid (pcheA) and *glmY*-containing pBAD33 (pBglmY) were used as templates for introducing an 8-bp substitution (shown in (**C**)) into GlmY and *cheA* 5′ UTR by site-directed mutagenesis to produce pcheAm and pBglmYm, respectively. The pcheA and pcheAm were introduced into wild-type *P. mirabilis* separately, while *glmY* mutant was transformed with combinations of pBglmY-pcheA, pBglmY-pcheAm, pBglmYm-pcheA or pBglmYm-pcheAm. Then, the activities of XylE in the various *glmY* mutants (in the presence of 0.2% arabinose) and wild-types were determined at 5 h after incubation on an LB agar plate. The relative XylE activity was shown with the value of the pcheA-harbored wild-type or *glmY* mutant carrying pBglmY and pcheA set at 1, respectively. Significant difference from the wild-type or *glmY* mutant set at 1 is indicated (* $p < 0.05$; ** $p < 0.01$ by the Student's *t* test). wt, wild-type; glmY, *glmY* mutant.

These results prove the role of *cheA* 5′ UTR in interaction with GlmY and the 8-bp direct pairing between GlmY and *cheA* 5′ UTR critical for facilitating expression of *cheA*. For the first time, this work identifies *cheA* as a novel target of GlmY for modulating swarming motility of *P. mirabilis*.

2.7. Urea Inhibited Promoter Activity of the glmYqseEGF Operon in P. mirabilis

Based on the fact that *P. mirabilis* swarming was subject to positive regulation of the QseF-GlmY pathway, it was tempting to determine whether urea affect the expression of *glmYqseEGF* operon in the urinary tract, an environment containing a lot of urea. Therefore, the XylE activities of the *glmY-xylE* reporter plasmid-transformed wild-type and the *qseF* mutant were monitored in the presence of urea. The promoter activity of the *glmY* operon was reduced by urea at 50 mM in the wild-type but not the *qseF* mutant after incubation for 5 h on the LB agar plate (Figure 5A). We then examined the swarming motility of the wild-type and *qseF* mutant in the presence of urea or not. The swarming motility of the wild-type but not the *qseF* mutant was decreased by urea at 50 mM (Figure 5B). The results suggest that urea could be a negative signal for expression of *P. mirabilis glmYqseEGF* operon in the urinary tract.

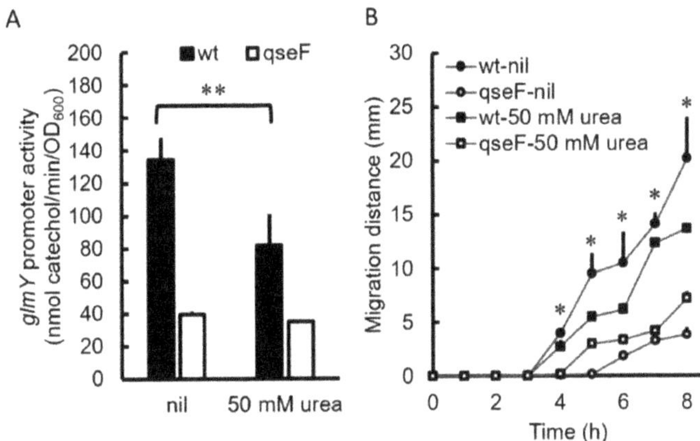

Figure 5. Urea inhibited *glmY* expression through QseF. (**A**) *glmY* promoter activity in the wild-type and *qseF* mutant in the presence and absence of urea. Bacterial cultures were spread on an LB agar plate with or without 50 mM of urea, and XylE promoter activities were determined at 5 h after incubation at 37 °C. (**B**) Swarming motility of the wild-type and *qseF* mutant in the presence and absence of 50 mM of urea was monitored at 1-h intervals. The data are the averages and standard deviations of three independent experiments. Significant difference between presence and absence of urea is indicated (* $p < 0.05$; ** $p < 0.01$ by the Student's *t* test). wt, wild-type; qseF, *qseF* mutant; nil, no urea.

3. Discussion

For the first time, in this study, a TCS regulator QseF participating in modulation of swarming motility through GlmY (an sRNA) and the underlying mechanisms were revealed in uropathogenic *P. mirabilis*. GlmY regulated swarming through direct GlmY-*cheA* interaction together with GlmY-*rcsB*- and/or GlmY-mediated *flhDC* expression (Figure 6), whereby affecting swarming-related chemotaxis system and flagellum production (Figure 6). The chemotaxis system plays an essential role in flagellar function and swarm cell differentiation, thus important for *P. mirabilis* to display a vigorous swarming pattern [34,36]. Given decreased not abolished GlmY expression upon loss of *qseF* (Figure 3A), this indicates GlmY is subject to QseF-independent control. Moreover, the finding that loss of *glmY* or *qseF* resulting in a similar motility ability, expression of *flhDC* and *rcsB* and *flhDC*-associated phenotypes (cell length and flagellin level) (Figures 2 and 3) indicates QseF could have alternative regulation of *flhDC* expression bypassing GlmY (Figure 6). In this regard, our preliminary data showed introduction of GlmY-expressing plasmid into *qseF* mutant could not restore *flhDC* mRNA to the wild-type level.

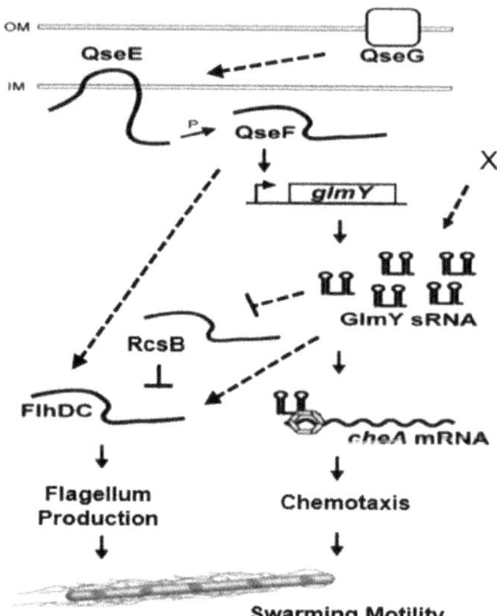

Figure 6. Summary of the swarming regulation by the QseEF-GlmY pathway involving expression of *rcsB*, *flhDC* and *cheA*. The sensor kinase QseE senses specific signals and transfers phosphoryl group to response regulator QseF. Then, QseF activates expression of GlmY, followed by *cheA* activation and *rcsB* inhibition. RcsB downregulation and CheA upregulation facilitate flagellum production and swarmer cell differentiation, respectively, thereby enhancing swarming motility. QseG could help activation of QseEF and unknown factor X facilitates GlmY expression. GlmY may interact with *flhDC* and modulate its expression. In addition, QseF could affect *flhDC* expression in a GlmY-independent way. Arrow, positive effect; line with a vertical bar, negative effect; dotted line, direct effect needed to be investigated.

An ever-increasing number and variety of sRNAs are being identified to serve regulatory functions for bacteria to respond to environmental cues and thrive in diverse habitats [9–11,37]. *E. coli* GlmY has been coopted to modulate the expression of virulence and be involved in cellular metabolism and architecture, including for biosynthesis of LPS [38–40], a permeability barrier and a major virulence determinant in pathogenic bacteria [41]. GlmY fine-tunes expression of type III secretion system and its effectors to promote bacterial attachment and subsequent actin rearrangement on host cells through post-transcriptional control of EspFu and the locus-of-enterocyte-effacement (LEE) [29]. In addition, GlmY and GlmZ participate in gene expression of curli adhesion, acid resistance and also tryptophan metabolism [42]. Despite similar genomic gene arrangement of *glmYqseEGF*, there are discrepancies of the QseF and GlmY-related regulation between *P. mirabilis* and *E. coli*. First, a Rho-independent terminator exists in the end of *E. coli* GlmY [43] but not in that of *P. mirabilis*. Second, neither promoter prediction nor reporter assay showed a promoter present in the intergenic region between *glmY* and *qseE* of *P. mirabilis*, which is not the case for *E. coli*, initiating *qseE* transcription in the *glmY-qseE* intergenic region [31]. Third, *E. coli* QseEF is involved in regulating genes required for pedestal formation but not motility [21]. There was no difference in motility and flagellar expression as seen by Western blotting between the wild-type *E. coli* strain and the *qseE* mutant [21], contrary to the similar defect of both *P. mirabilis qseF* and *qseE* mutants in swarming and swimming abilities. *E. coli glmY* and *qseE* are independently transcribed from different promoters [31], while we demonstrated that *glmY* and *qseEGF* constitute an operon by RT-PCR assay (Figure 1C). In this regard, the presence of the conserved RNase E-cleavage motif GCCUUAU in GlmY of *P. mirabilis* [38] indicates the *glmYqseEGF* transcript could be processed to produce effective GlmY for its function.

We found swarming, swimming and swarming-related phenotypes (cell length, haemolysin activity and flagellin level) were all reduced in both *qseF* and *glmY* mutants (Figure 2). Both *flhDC* promoter activity and mRNA level were downregulated in *qseF* and *glmY* mutants (Figure 3). Because no QseF binding site in the promoter region of *flhDC*, we speculated QseF may exert its effect on *flhDC* expression through GlmY. TargetRNA2 tool, focusing its search for an sRNA-mRNA interaction in a neighborhood around the rbs of the mRNA, revealed no GlmY binding site in *flhDC* and *rcsB* mRNAs. We then extended mRNA searching by IntaRNA for GlmY-interacting sites and binding sites of -101 to -92 and -284 to -271 from AUG (both outside rbs) for *flhDC* and *rcsB*, respectively, were identified. Based on the GlmY target prediction, there are two possibilities for upregulation of *flhDC* mRNA level by GlmY. One is by direct interaction with *flhDC* mRNA; the other is indirectly through *rcsB*. It also can not be ruled out that both modes of action coexist. Further studies are needed to investigate whether GlmY directly interacts with mRNAs of *flhDC* and *rcsB*. For example, the *flhDC* (or *rcsB*) mRNA amount of the *glmY* mutant harboring wild-type or interacting site-mutated *glmY* on a plasmid will be determined to assess the GlmY-*flhDC* (or *rcsB*) mRNA interaction. It is noteworthy that overexpression of *glmY* in the same *E. coli* strain either had no effect or resulted in motility repression [12,44]. It could be due to different plasmid vectors and assay conditions used.

Translational fusion assay indicated that regions of -520, -253 and -138 to $+27$ from AUG of *cheA* mRNA contain constrained secondary structures needed to be resolved by GlmY for *cheA* translation, albeit to a lesser extent for the region of -138 to $+27$ (Figure 4B). To elucidate how *cheA* 5′UTR interacts with GlmY, we uploaded sequences from -253 to -138 and -138 to $+27$ of *cheA* mRNA to the IntaRNA webserver and interaction between regions of -233 to -211 and -111 to -94 from AUG was revealed (Figure 7). Likewise, using sequences from -138 to -61 and -61 to $+27$ of *cheA* mRNA as input fragments showed interaction between regions of -88 to -82 and -15 to -9 by IntaRNA (Figure 7). Inspection of the sequence of -15 to -9 disclosed the existence of two overlapped putative ribosome binding sites (rbs), aagguga (gagguga in *E. coli*) and gaaugag (gaagga in *E. coli*) for translation of long *cheA* and short *cheA* of *E. coli* [45], respectively (Figure 7). Furthermore, IntaRNA revealed interaction of *cheA* -61 to -50 from AUG with GlmY $+17$ to $+28$ (Figure 7) by using full-length GlmY and -253 to $+27$ of *cheA* as the input fragments. These data suggest interaction between regions of -88 to -82 and -15 to -9 from AUG could inhibit *cheA* translation by hiding rbs. Hence GlmY is required for releasing the secondary structure to assist *cheA* translation. The reason for the less extent of translation affected in the absence of *glmY* using the translational fusion comprising -138 to $+27$ of *cheA* compared to the -253 to $+27$ of *cheA* fusion could be ascribed to the interaction between regions of -233 to -211 and -111 to -94 from AUG (Figure 7), thereby affecting the ease for GlmY to uncover the rbs for translation. It is interesting to know that GlmY not only activated *cheA* translation but also maintained *cheA* mRNA level. This is in line with that sRNA-mediated stability control is the crucial element of activation of *trans*-encoded mRNAs [35].

EHEC regulated pathogenesis and motility by sensing epinephrine or norepinephrine through QseBC and QseEF two-component signaling systems [21,22]. The membrane kinase QseC autophosphorylates and phosphorylates the QseB response regulator initiating a signaling cascade that activates QseEF to trigger expression of LEE genes, leading to AE lesions on intestinal epithelial cells. In *P. mirabilis* genome, only QseEF homologue of high similarity was found, with no homologue of QseBC existing. We found QseF participated in swarming regulation and deletion of *qseE* or *qseG* also had decreased swarming motility. Previous study has revealed the phosphorylation state of QseE and QseF is governed by interaction with QseG in response to epinephrine for post-transcriptional regulation of virulence genes through GlmY [46]. In addition, QseEGF has been shown to modulate transcription of *phoPQ*, linking to the virulence regulation [47]. The similarity of the transcriptome profiles of *qseE*, *qseF* and *qseG* mutants also indicates that these proteins work together [47]. Therefore, *P. mirabilis* QseEF likely should exert functions other than swarming to affect virulence. Accordingly, our preliminary data showing *qseF* mutant had a significantly impaired ability to colonize mouse bladders and kidneys compared to the wild-type. The investigation of virulence traits such as cytotoxicity, urothelial cell invasion and survival in macrophages is underway.

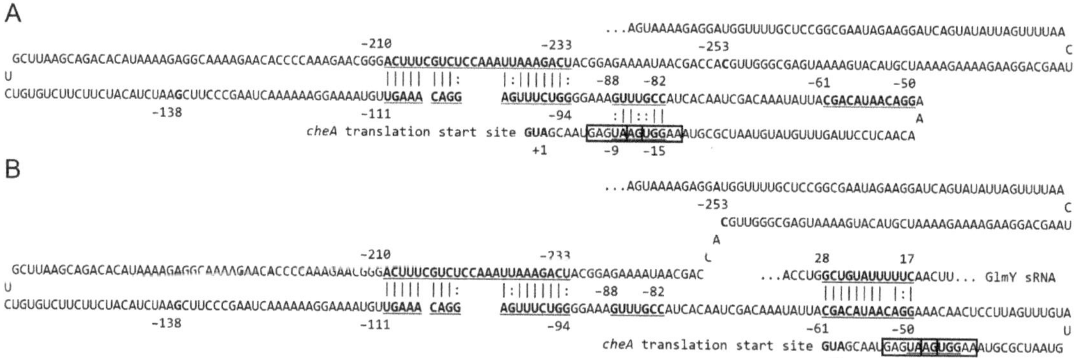

Figure 7. A model for regulation of *cheA* by GlmY sRNA. (**A**) The model structure for the 5′ UTR of *cheA* mRNA in the absence of GlmY. The expression of *cheA* is silenced at the post-transcriptional level by pairing −111 to −94 with −233 to −210 from AUG of *cheA* 5′ UTR and −88 to −82 from AUG with the putative ribosome binding site of *cheA*. (**B**) The putative structure for the 5′ UTR of *cheA* mRNA in the presence of GlmY. GlmY pairs with −61 to −50 from AUG of *cheA* 5′ UTR freeing the ribosome binding site and causing the *cheA* mRNA to be translated. Two overlapped putative ribosome binding sequences (AAGGUGA and GAAUGAG) are indicated in boxes, and the RNA-RNA base-pairing sequences predicted by intaRNA are underlined.

We found that urea could serve as a negative signal of QseEF for swarming (Figure 5). In this way, the expression of QseF will be inhibited in the urine (rich in urea) and the role of QseF in facilitating swarming will be neglected. In view of our preliminary data showing a significant difference between the wild-type and *qseF* mutant in colonization of mouse bladders and kidneys, it is tempting to surmise there is other cues in the urine to increase expression of *qseF*. In this aspect, our unpublished transcriptome data reveal mRNA levels of fatty acid synthetic genes increase but those of degradative genes decrease in the *qseF* mutant relative to the wild-type. This indicates QseF may sense fatty acid and be involved in fatty acid metabolism. Interestingly, we found oleic acid (0.01%) is a positive signal for *qseF* expression (data not shown) and swarming motility of *P. mirabilis* [48]. The oleic acid concentration used is an attainable concentration in the urine according to the Human Metabolome Database (https://hmdb.ca/metabolites?utf8=%E2%9C%93&quantified=1&urine=1&filter=true, accessed on 29 November 2021). Given QseE is also phosphatase [23], it is reasonable to infer that the phosphatase function of QseE prevents QseF activation under the presence of urea, whereas its kinase function triggers QseF regulon expression under appropriate concentrations of oleic acid. A finely tuned balance in these opposing activities should determine the regulon response of QseF as the case of the bifunctional DevS kinase responsive to environmental oxygen in *Mycobacterium tuberculosis* [49].

Since catheter-associated UTI (CAUTI) is a major health concern, research directed at understanding the pathogenesis is warranted and should lead to improved diagnosis, prevention and treatment. *P. mirabilis* is notorious for causing CAUTIs. In this work, we demonstrated a new regulatory pathway involving an sRNA, GlmY, participating in swarming motility of *P. mirabilis* during which expression of several virulence genes is increased. It is believed that *P. mirabilis* swarming up catheters is primed to infect the urinary tract [50], so elucidating the swarming mechanisms could provide new approaches in the development of intervention strategies and facilitate the discovery of novel therapeutics.

4. Materials and Methods

4.1. Bacterial Strains, Plasmids, Reagents and Growth Conditions

The bacterial strains and plasmids used in this study are listed in Table S1 in the Supplementary Materials. The bacterial strains used are a clinical isolate from a patient of UTI (the wild-type N2), its derived mutants and respective complemented strains. All chemicals were obtained from the Sigma-Aldrich unless otherwise indicated, and primer sequences are given in Table S2. Bacteria were stored at $-80\ ^{\circ}$C and routinely cultured in Luria-Bertani (LB) broth at 37 °C. The LSW⁻ agar plate [15] was used to prevent the phenotypic expression of swarming motility for selecting mutant clones and colony counting.

4.2. Construction of P. mirabilis Mutants and Complemented Strains

Sequences flanking the *qseF* gene was amplified by PCR using the primer pairs qseF-upF/XbaI-qseF-upR and XbaI-qseF-dnF/qseF-dnR for *qseF* mutant, and cloned into pGEM-T Easy (Promega) to generate pGqseF-up and pGqseF-dn. pGqseF-up was digested with SalI/XbaI, and the *qseF* upstream sequence-containing fragment was ligated to SalI/XbaI-digested pGqseF-dn to produce the pGqseF-updn plasmid, which contains both upstream and downstream sequences of *qseF*. A Km^r cassette was inserted in the XbaI-digested pGqseF-updn plasmid to generate pGqseF-updn-Km. The DNA fragment containing the Km^r cassette-disrupted combined upstream and downstream sequences of *qseF* was cleaved by SalI/SphI from pGqseF-updn-Km, and ligated into SalI/SphI-cleaved pUT-Km1 to generate pUTqseF-Km. For *glmY* mutant, pUTglmY-Km was constructed in a similar way except using primer pairs glmY-upF/XbaI-glmY-upR and glmY-dnF/glmY-dnR. For gene inactivation by homologous recombination, pUTqseF-Km or pUTglmY-Km was transferred to wild-type *P. mirabilis* N2 by conjugation. Transconjugants were spread on LSW⁻ plates containing tetracycline (20 µg/mL) and kanamycin (100 µg/mL), and confirmation of mutants with double-crossover events by colony PCR and Southern blot hybridization were performed. For complementation of mutants, the fragments containing full-length *qseF* gene or *glmY* was amplified by PCR using primer pairs qseF-comF/qseF-comR or glmY-comF/glmY-comR, and cloned into pGEM-T easy to generate the plasmid pGEM-qseF or pGEM-glmY. *qseF* is driven by *lac* promoter. *glmY* is driven by its own promoter (382 bp), divergent from *lac* promoter, to ensure expression of the correct transcript. pGEM-qseF or pGEM-glmY was then transformed into the respective mutant to generate the complemented strain.

4.3. Swarming and Swimming Assays

The swarming migration assays were performed as described previously [48]. The overnight bacterial cultures (5 µL) were inoculated onto the center of LB swarming plates containing 1.5% (wt/vol) agar, which were then incubated, and the swarming migration distance was measured by monitoring the swarm fronts of the bacterial cells at 1-h intervals. For swimming assays, the overnight culture was stabbed into the center of the swimming plates containing 0.3% agar and migration distance was recorded after incubation for 16 h at 37 °C.

4.4. Measurement of the Haemolysin Activity and Cell Length

The overnight bacterial cultures (120 µL) were inoculated onto the surface of LB swarming plates, which were then incubated at 37 °C for 5 h. The hemolysin activity and cell length were determined as described previously [48]. Cell-associated haemolytic activity was determined by incubation of a 20-µL cell suspension ($OD_{600nm} = 0.5$) in a 980-µL solution of 0.85% NaCl, 20 mM $CaCl_2$ and 2% washed sheep erythrocytes at 42 °C for 15 min. After centrifugation, the amount of haemoglobin released by lysis was measured by the increase of the optical density of the assay supernatant at 543 nm.

4.5. Measurement of the Flagellin Level

Flagellin levels were determined as described previously by SDS-PAGE and Coomassie brilliant blue staining [15]. Western blotting was performed to confirm the flagellin band. The flagellin samples on the SDS-PAGE gel were transferred to an Hybond-P membrane (GE Healthcare, Chicago, IL, USA). The blot was incubated with mouse polyclonal antiserum against FlaA, followed by sheep anti-mouse IgG conjugated with horseradish peroxidase (GE Healthcare, Chicago, IL, USA), and then developed using enhanced chemiluminescence detection reagents (PerkinElmer, Waltham, MA, USA).

4.6. Transmission Electron Microscopy

Transmission electron microscopy (TEM) was performed as previously described [3] by using 1% phosphotungstic acid (PTA)-stained bacteria on a carbon-coated grid and TEM pictures were obtained with a Hitachi H-7100 electron microscope (Hitachi High-Tech America, Pleasanton, CA, USA).

4.7. Real Time Reverse Transcription PCR (RT-PCR)

To study the effect of *qseF* or *glmY* deletion on the mRNA amount of motility-related genes, overnight LB cultures (100 µL) of the wild-type, mutants and the complemented strains were spread on the LB agar and incubated for 5 h at 37 °C. Total RNA was extracted, and real time RT-PCR was carried out as described previously [3] to measure the mRNA level using primer pairs listed in Table S2. The levels of RNAs were normalized against housekeeping gene *gyrB* mRNA.

4.8. Transcriptional and Translational Reporter Assays

For transcriptional reporter assay, the promoter region of the gene was amplified by SphI and PstI-included primers and cloned into pGEM-T Easy. These promoter-containing plasmids were cut by SphI and PstI, and the promoter-containing fragment was ligated, respectively to the *xylE*-containing pACYC184-xylE digested by SphI and PstI to construct the transcriptional reporter plasmid. The XylE activity of the transcriptional reporter plasmid-transformed wild-type and mutants was measured as described previously [3]. For translational reporter assay, the 5′UTR (520, 253, 138 or 61 bp) with 27 bp coding region (5′UTR-27) of *cheA* gene was amplified by SacI and BglII-included primers and cloned into pGEM-T Easy to generate the plasmid pGcheA-5′UTR. The fragment containing full-length *xylE* gene was amplified by PCR using primer pairs BglII-xylE-F and xylE-R, and cloned into pGEM-T Easy to generate the plasmid pGxylE. pGcheA-5′UTR was digested with SacI and BglII, and the *cheA* 5′UTR-27 sequence-containing fragment was ligated to SacI/BglII-digested pGxylE to produce the translational reporter plasmid pcheA of in-frame *cheA-xylE* fusion driven by *lac* promoter. The wild-type and mutants transformed with the translational reporter plasmid were grown overnight in LB broth containing ampicillin (100 µg/mL). Then, the cultures (100 µL) were spread onto the LB agar plate and incubated for 5 h at 37 °C before the XylE activity was measured.

4.9. Site-Directed Mutagenesis

The translational reporter plasmid pGEM-*cheA-xylE* (pcheA) and the *glmY*-harboring pBAD plasmid (pBglmY)) containing mutations in the putative GlmY- and *cheA* 5′UTR-interacting site, respectively, were generated (pcheAm and pBglmYm) by a KOD-Plus-mutagenesis kit (Toyobo, Osaka, Japan) using primers listed in Table S2 according to the manufacturer's protocol. The inverse PCR products of pcheA and pBglmY were digested by DpnI to remove the template plasmid DNA, self-ligation of PCR products was performed using T4 polynucleotide kinase and ligase and then DNA sequencing was performed to confirm the DNA sequence of the mutated sites. We introduced pcheA or pcheAm into the wild-type and compared the XylE activity to confirm the essential GlmY-interacting site in *cheA* 5′UTR. Additionally, we transformed the combinations of plasmids pBglmY-pcheA, pBglmY-pcheAm, pBglmYm-pcheA or pBglmYm-pcheAm into the *glmY* mutant and determined the XylE activity to evaluate the direct interaction of *cheA* 5′UTR and GlmY.

Supplementary Materials: The following supporting information can be downloaded at: https://www.mdpi.com/article/10.3390/ijms23010487/s1.

Author Contributions: Conceptualization, S.-J.L.; investigation, W.-Y.L., Y.-L.T., P.-Y.S. and T.-S.L.; formal analysis, W.-Y.L., Y.-L.T. and P.-H.Y.; resources, Y.-J.L.; data curation, W.-Y.L., Y.-L.T., Y.-J.L., P.-H.Y. and S.-J.L.; supervision, S.-J.L.; writing—original draft preparation, S.-J.L. and W.-Y.L.; writing—review and editing, S.-J.L.; project administration, S.-J.L.; funding acquisition, S.-J.L. and Y.-J.L. All authors have read and agreed to the published version of the manuscript.

Funding: This research was funded by the Ministry of Science and Technology (MOST), Taiwan (grant number: MOST 105-2320-B-002-050-MY3 to S.-J.L.).

Institutional Review Board Statement: Not applicable.

Informed Consent Statement: Not applicable.

Data Availability Statement: Not applicable.

Acknowledgments: We thank the Graduate Institute of Anatomy and Cell Biology, College of Medicine, National Taiwan University for providing assistance in TEM examination.

Conflicts of Interest: The authors declare no conflict of interest. The funders had no role in the design of the study, in the collection, analyses or interpretation of data, in the writing of the manuscript or in the decision to publish the results.

References

1. Norsworthy, A.N.; Pearson, M.M. From Catheter to Kidney Stone: The Uropathogenic Lifestyle of *Proteus mirabilis*. *Trends Microbiol.* **2017**, *25*, 304–315. [CrossRef]
2. Jansen, A.M.; Lockatell, V.; Johnson, D.E.; Mobley, H.L. Mannose-resistant *Proteus*-like fimbriae are produced by most *Proteus mirabilis* strains infecting the urinary tract, dictate the in vivo localization of bacteria, and contribute to biofilm formation. *Infect. Immun.* **2004**, *72*, 7294–7305. [CrossRef] [PubMed]
3. Tsai, Y.L.; Chien, H.F.; Huang, K.T.; Lin, W.Y.; Liaw, S.J. cAMP receptor protein regulates mouse colonization, motility, fimbria-mediated adhesion, and stress tolerance in uropathogenic *Proteus mirabilis*. *Sci. Rep.* **2017**, *7*, 7282. [CrossRef]
4. Nielubowicz, G.R.; Mobley, H.L. Host-pathogen interactions in urinary tract infection. *Nat. Rev. Urol.* **2010**, *7*, 430–441. [CrossRef] [PubMed]
5. Armbruster, C.E.; Mobley, H.L. Merging mythology and morphology: The multifaceted lifestyle of *Proteus mirabilis*. *Nat. Rev. Microbiol.* **2012**, *10*, 743–754. [CrossRef]
6. Jacobsen, S.M.; Stickler, D.J.; Mobley, H.L.; Shirtliff, M.E. Complicated catheter-associated urinary tract infections due to *Escherichia coli* and *Proteus mirabilis*. *Clin. Microbiol. Rev.* **2008**, *21*, 26–59. [CrossRef] [PubMed]
7. Fraser, G.M.; Hughes, C. Swarming motility. *Curr. Opin. Microbiol.* **1999**, *2*, 630–635. [CrossRef]
8. Clemmer, K.M.; Rather, P.N. Regulation of *flhDC* expression in *Proteus mirabilis*. *Res. Microbiol.* **2007**, *158*, 295–302. [CrossRef]
9. Gottesman, S.; Storz, G. Bacterial small RNA regulators: Versatile roles and rapidly evolving variations. *Cold Spring Harb. Perspect. Biol.* **2011**, *3*, a003798. [CrossRef]
10. Waters, L.S.; Storz, G. Regulatory RNAs in bacteria. *Cell* **2009**, *136*, 615–628. [CrossRef] [PubMed]
11. Guillier, M.; Gottesman, S.; Storz, G. Modulating the outer membrane with small RNAs. *Genes Dev.* **2006**, *20*, 2338–2348. [CrossRef]
12. Bak, G.; Lee, J.; Suk, S.; Kim, D.; Young Lee, J.; Kim, K.S.; Choi, B.S.; Lee, Y. Identification of novel sRNAs involved in biofilm formation, motility, and fimbriae formation in *Escherichia coli*. *Sci. Rep.* **2015**, *5*, 15287. [CrossRef] [PubMed]
13. Kulesus, R.R.; Diaz-Perez, K.; Slechta, E.S.; Eto, D.S.; Mulvey, M.A. Impact of the RNA chaperone Hfq on the fitness and virulence potential of uropathogenic *Escherichia coli*. *Infect. Immun.* **2008**, *76*, 3019–3026. [CrossRef]
14. Chao, Y.; Vogel, J. The role of Hfq in bacterial pathogens. *Curr. Opin. Microbiol.* **2010**, *13*, 24–33. [CrossRef] [PubMed]
15. Wang, M.C.; Chien, H.F.; Tsai, Y.L.; Liu, M.C.; Liaw, S.J. The RNA chaperone Hfq is involved in stress tolerance and virulence in uropathogenic *Proteus mirabilis*. *PLoS ONE* **2014**, *9*, e85626. [CrossRef]
16. Vogel, J.; Luisi, B.F. Hfq and its constellation of RNA. *Nat. Rev. Microbiol.* **2011**, *9*, 578–589. [CrossRef] [PubMed]
17. Valverde, C.; Haas, D. Small RNAs controlled by two-component systems. *Adv. Exp. Med. Biol.* **2008**, *631*, 54–79. [PubMed]
18. Vogel, J. A rough guide to the non-coding RNA world of *Salmonella*. *Mol. Microbiol.* **2009**, *71*, 1–11. [CrossRef]

19. Overgaard, M.; Kallipolitis, B.; Valentin-Hansen, P. Modulating the bacterial surface with small RNAs: A new twist on PhoP/Q-mediated lipopolysaccharide modification. *Mol. Microbiol.* **2009**, *74*, 1289–1294. [CrossRef]

20. Coornaert, A.; Lu, A.; Mandin, P.; Springer, M.; Gottesman, S.; Guillier, M. MicA sRNA links the PhoP regulon to cell envelope stress. *Mol. Microbiol.* **2010**, *76*, 467–479. [CrossRef]

21. Reading, N.C.; Torres, A.G.; Kendall, M.M.; Hughes, D.T.; Yamamoto, K.; Sperandio, V. A novel two-component signaling system that activates transcription of an enterohemorrhagic *Escherichia coli* effector involved in remodeling of host actin. *J. Bacteriol.* **2007**, *189*, 2468–2476. [CrossRef]

22. Njoroge, J.; Sperandio, V. Enterohemorrhagic *Escherichia coli* virulence regulation by two bacterial adrenergic kinases, QseC and QseE. *Infect. Immun.* **2012**, *80*, 688–703. [CrossRef]

23. Moreira, C.G.; Sperandio, V. Interplay between the QseC and QseE bacterial adrenergic sensor kinases in *Salmonella enterica* serovar Typhimurium pathogenesis. *Infect. Immun.* **2012**, *80*, 4344–4353. [CrossRef]

24. Moreira, C.G.; Russell, R.; Mishra, A.A.; Narayanan, S.; Ritchie, J.M.; Waldor, M.K.; Curtis, M.M.; Winter, S.E.; Weinshenker, D.; Sperandio, V. Bacterial adrenergic sensors regulate virulence of enteric pathogens in the gut. *mBio* **2016**, *7*, e00826-16. [CrossRef]

25. Cameron, E.A.; Gruber, C.C.; Ritchie, J.M.; Waldor, M.K.; Sperandio, V. The QseG lipoprotein impacts the virulence of enterohemorrhagic *Escherichia coli* and *Citrobacter rodentium* and regulates flagellar phase variation in *Salmonella enterica* serovar Typhimurium. *Infect. Immun.* **2018**, *86*, e00936-17. [CrossRef]

26. Reading, N.C.; Rasko, D.A.; Torres, A.G.; Sperandio, V. The two-component system QseEF and the membrane protein QseG link adrenergic and stress sensing to bacterial pathogenesis. *Proc. Natl. Acad. Sci. USA* **2009**, *106*, 5889–5894. [CrossRef] [PubMed]

27. Gopel, Y.; Luttmann, D.; Heroven, A.K.; Reichenbach, B.; Dersch, P.; Gorke, B. Common and divergent features in transcriptional control of the homologous small RNAs GlmY and GlmZ in Enterobacteriaceae. *Nucleic Acids Res.* **2011**, *39*, 1294–1309. [CrossRef]

28. Lustri, B.C.; Sperandio, V.; Moreira, C.G. Bacterial chat: Intestinal metabolites and signals in host-microbiota-pathogen interactions. *Infect. Immun.* **2017**, *85*, e00476-17. [CrossRef] [PubMed]

29. Gruber, C.C.; Sperandio, V. Posttranscriptional control of microbe-induced rearrangement of host cell actin. *mBio* **2014**, *5*, e01025-13. [CrossRef]

30. Bennett, A.M.; Shippy, D.C.; Eakley, N.; Okwumabua, O.; Fadl, A.A. Functional characterization of glucosamine-6-phosphate synthase (GlmS) in *Salmonella enterica* serovar Enteritidis. *Arch. Microbiol.* **2016**, *198*, 541–549. [CrossRef] [PubMed]

31. Reichenbach, B.; Gopel, Y.; Gorke, B. Dual control by perfectly overlapping sigma 54- and sigma 70- promoters adjusts small RNA GlmY expression to different environmental signals. *Mol. Microbiol.* **2009**, *74*, 1054–1070. [CrossRef]

32. Howery, K.E.; Clemmer, K.M.; Rather, P.N. The Rcs regulon in *Proteus mirabilis*: Implications for motility, biofilm formation, and virulence. *Curr. Genet.* **2016**, *62*, 775–789. [CrossRef]

33. Liu, M.C.; Lin, S.B.; Chien, H.F.; Wang, W.B.; Yuan, Y.H.; Hsueh, P.R.; Liaw, S.J. 10′(Z),13′(E)-heptadecadienylhydroquinone inhibits swarming and virulence factors and increases polymyxin B susceptibility in *Proteus mirabilis*. *PLoS ONE* **2012**, *7*, e45563. [CrossRef]

34. Burkart, M.; Toguchi, A.; Harshey, R.M. The chemotaxis system, but not chemotaxis, is essential for swarming motility in *Escherichia coli*. *Proc. Natl. Acad. Sci. USA* **1998**, *95*, 2568–2573. [CrossRef]

35. Podkaminski, D.; Vogel, J. Small RNAs promote mRNA stability to activate the synthesis of virulence factors. *Mol. Microbiol.* **2010**, *78*, 1327–1331. [CrossRef] [PubMed]

36. Verstraeten, N.; Braeken, K.; Debkumari, B.; Fauvart, M.; Fransaer, J.; Vermant, J.; Michiels, J. Living on a surface: Swarming and biofilm formation. *Trends Microbiol.* **2008**, *16*, 496–506. [CrossRef] [PubMed]

37. Valentin-Hansen, P.; Johansen, J.; Rasmussen, A.A. Small RNAs controlling outer membrane porins. *Curr. Opin. Microbiol.* **2007**, *10*, 152–155. [CrossRef]

38. Gopel, Y.; Papenfort, K.; Reichenbach, B.; Vogel, J.; Gorke, B. Targeted decay of a regulatory small RNA by an adaptor protein for RNase E and counteraction by an anti-adaptor RNA. *Genes Dev.* **2013**, *27*, 552–564. [CrossRef]

39. Gopel, Y.; Khan, M.A.; Gorke, B. Menage a trois: Post-transcriptional control of the key enzyme for cell envelope synthesis by a base-pairing small RNA, an RNase adaptor protein, and a small RNA mimic. *RNA Biol.* **2014**, *11*, 433–442. [CrossRef] [PubMed]

40. Urban, J.H.; Vogel, J. Two seemingly homologous noncoding RNAs act hierarchically to activate *glmS* mRNA translation. *PLoS Biol.* **2008**, *6*, e64. [CrossRef]

41. Klein, G.; Raina, S. Regulated assembly of LPS, its structural alterations and cellular response to LPS defects. *Int. J. Mol. Sci.* **2019**, *20*, 356. [CrossRef]

42. Gruber, C.C.; Sperandio, V. Global analysis of posttranscriptional regulation by GlmY and GlmZ in enterohemorrhagic *Escherichia coli* O157:H7. *Infect. Immun.* **2015**, *83*, 1286–1295. [CrossRef]

43. Urban, J.H.; Papenfort, K.; Thomsen, J.; Schmitz, R.A.; Vogel, J. A conserved small RNA promotes discoordinate expression of the *glmUS* operon mRNA to activate GlmS synthesis. *J. Mol. Biol.* **2007**, *373*, 521–528. [CrossRef]

44. De Lay, N.; Gottesman, S. A complex network of small non-coding RNAs regulate motility in *Escherichia coli*. *Mol. Microbiol.* **2012**, *86*, 524–538. [CrossRef] [PubMed]

45. McNamara, B.P.; Wolfe, A.J. Coexpression of the long and short forms of CheA, the chemotaxis histidine kinase, by members of the family Enterobacteriaceae. *J. Bacteriol.* **1997**, *179*, 1813–1818. [CrossRef] [PubMed]

46. Göpel, Y.; Görke, B. Interaction of lipoprotein QseG with sensor kinase QseE in the periplasm controls the phosphorylation state of the two-component system QseE/QseF in *Escherichia coli*. *PLoS Genet.* **2018**, *14*, e1007547. [CrossRef] [PubMed]

47. Reading, N.C.; Rasko, D.; Torres, A.G.; Sperandio, V. A transcriptome study of the QseEF two-component system and the QseG membrane protein in enterohaemorrhagic *Escherichia coli* O157:H7. *Microbiology* **2010**, *156*, 1167–1175. [CrossRef] [PubMed]
48. Liaw, S.J.; Lai, H.C.; Wang, W.B. Modulation of swarming and virulence by fatty acids through the RsbA protein in *Proteus mirabilis*. *Infect. Immun.* **2004**, *72*, 6836–6845. [CrossRef]
49. Kaur, K.; Kumari, P.; Sharma, S.; Sehgal, S.; Tyagi, J.S. DevS/DosS sensor is bifunctional and its phosphatase activity precludes aerobic DevR/DosR regulon expression in *Mycobacterium tuberculosis*. *FEBS J.* **2016**, *283*, 2949–2962. [CrossRef]
50. Schaffer, J.N.; Pearson, M.M. *Proteus mirabilis* and urinary tract infections. *Microbiol. Spectr.* **2015**, *3*, 1–39. [CrossRef]

International Journal of
Molecular Sciences

Article

Ultrasensitive Detection of *Bacillus anthracis* by Real-Time PCR Targeting a Polymorphism in Multi-Copy 16S rRNA Genes and Their Transcripts

Peter Braun, Martin Duy-Thanh Nguyen, Mathias C. Walter and Gregor Grass *

Bundeswehr Institute of Microbiology (IMB), 80937 Munich, Germany; peter3braun@bundeswehr.org (P.B.);
martin2nguyen@bundeswehr.org (M.D.-T.N.); mathias1walter@bundeswehr.org (M.C.W.)
* Correspondence: gregorgrass@bundeswehr.org; Tel.: +49-992692-3981

Abstract: The anthrax pathogen *Bacillus anthracis* poses a significant threat to human health. Identification of *B. anthracis* is challenging because of the bacterium's close genetic relationship to other *Bacillus cereus* group species. Thus, molecular detection is founded on species-specific PCR targeting single-copy genes. Here, we validated a previously recognized multi-copy target, a species-specific single nucleotide polymorphism (SNP) present in 2–5 copies in every *B. anthracis* genome analyzed. For this, a hydrolysis probe-based real-time PCR assay was developed and rigorously tested. The assay was specific as only *B. anthracis* DNA yielded positive results, was linear over 9 $\log_{10}$ units, and was sensitive with a limit of detection (LoD) of 2.9 copies/reaction. Though not exhibiting a lower LoD than established single-copy PCR targets (*dhp61* or *PL3*), the higher copy number of the *B. anthracis*–specific 16S rRNA gene alleles afforded ≤ 2 unit lower threshold (Ct) values. To push the detection limit even further, the assay was adapted for reverse transcription PCR on 16S rRNA transcripts. This RT-PCR assay was also linear over 9 $\log_{10}$ units and was sensitive with an LoD of 6.3 copies/reaction. In a dilution series of experiments, the 16S RT-PCR assay achieved a thousand-fold higher sensitivity than the DNA-targeting assays. For molecular diagnostics, we recommend a real-time RT-PCR assay variant in which both DNA and RNA serve as templates (thus, no requirement for DNase treatment). This can at least provide results equaling the DNA-based implementation if no RNA is present but is superior even at the lowest residual rRNA concentrations.

Keywords: anthrax; *Bacillus anthracis*; 16S rRNA; detection; identification; real-time PCR; RT-PCR

Citation: Braun, P.; Nguyen, M.D.-T.; Walter, M.C.; Grass, G. Ultrasensitive Detection of *Bacillus anthracis* by Real-Time PCR Targeting a Polymorphism in Multi-Copy 16S rRNA Genes and Their Transcripts. *Int. J. Mol. Sci.* **2021**, *22*, 12224. https://doi.org/10.3390/ijms222212224

Academic Editors: Amelia Casamassimi, Alfredo Ciccodicola and Monica Rienzo

Received: 14 October 2021
Accepted: 10 November 2021
Published: 12 November 2021

Publisher's Note: MDPI stays neutral with regard to jurisdictional claims in published maps and institutional affiliations.

1. Introduction

Within the genus *Bacillus*, the notorious anthrax pathogen *Bacillus anthracis* poses the greatest risk for humans, mammal livestock, and wildlife [1]. Other *Bacillus* spp. such *as B. cereus* or *B. thuringiensis*, which are typical soil bacteria, may also have pathogenic traits related to food poisoning, infections in immunocompromised persons, or production of insecticides [2]. Yet, only obligatory pathogenic *B. anthracis* (and a few *B. anthracis*-like bacilli) features a unique suite of pathogenicity factors rendering the endospore-forming bacterium a first-rate biothreat agent. These factors are encoded on two plasmids called pXO1 and pXO2. Plasmid pXO1 encodes the anthrax toxin genes producing the lethal toxin (gene products of *pagA* and *lef*) and edema toxin (gene products of *pagA* and *cya*) [1]. These toxins damage host cells on various levels [3]. Plasmid pXO2 harbors the capsule genes endowing the pathogen with a poly-glutamyl capsule which helps evade host immune response [1,4]. Phylogenetically, *B. anthracis* belongs to the very closely related *Bacillus cereus sensu lato* group. Besides the better-known species *B. cereus sensu stricto*, *B. anthracis*, or *B. thuringiensis*, the group also comprises several other familiar species such as *B. weihenstephanensis*, *B. mycoides*, *B. cytotoxicus*, and a variety of lesser-characterized members [5].

In the past, the high degree of genetic relatedness to several *B. cereus s.l.* strains has rendered molecular diagnostics of *B. anthracis* challenging (e.g., by polymerase chain reaction assays, PCR). One would think it should be straightforward to identify *B. anthracis* by detecting genetic marker genes (typically *pagA, lef, cya, capB,* or *capC*) [6–8] on one or both of its virulence plasmids. Identifying these genes comprising constituents of toxin or capsule biosynthesis (*cap*-genes), however, only verifies the presence of these plasmids. This is relevant because several *B. cereus s.l.* isolates are documented to possess very similar virulence plasmids, but not necessarily all of these belong to the species *B. anthracis.* Further, there are *B. anthracis* strains that lack one or both virulence plasmids. Species-specific molecular identification of *B. anthracis* is achieved by targeting a small number of validated chromosomal targets. These targets comprise sections of genes such as *dhp61* (*BA_5345*; [9]), *PL3* (*BA_5358*; [6]), or mutations characterized as single nucleotide polymorphisms (SNPs), e.g., in the *rpoB* [7] or the *plcR* [10] gene. A comprehensive overview of suitable and less ideal specific markers for *B. anthracis* has been provided previously [11]. Notwithstanding, the advantage of assaying for pXO1 or pXO2 markers over chromosomal ones is that the plasmid markers occur as multi-copy genes (since the virulence plasmids are present in more than one copy per cell) [12]. Large-scale genomic sequencing revealed that in *B. anthracis* plasmids, pXO1 and pXO2 (with their respective PCR-marker genes) are present on average in 3.86 and 2.29 copies, respectively [13]. Conversely, no multi-copy chromosomal marker has been employed for *B. anthracis* detection thus far.

Likewise, ribosomal RNA (particularly 16S rRNA) has not yet been routinely used for identification and detection of *B. anthracis* even though rRNA molecules are generally the most abundant ribonucleic acid entities in cells constituting up to approximately 80% of total RNA [14]. In fact, copies of 16S rRNA transcripts per cell as constituents of ribosomes number in many thousands (e.g., in *E. coli*, the number of ribosomes per cell ranges from 8×10^3 at a doubling time of 100 min to 7.3×10^4 at a doubling time of 20 min) [15]. Even in stationary culture, a single *E. coli* bacterium contains about 6.5×10^3 copies of ribosomes [16]. Phylogenetically closer to *B. anthracis* than *E. coli* is *Bacillus licheniformis.* For this bacillus, the average number of ribosomes per cell was calculated at 1.25×10^4, 3.44×10^4, or 9.2×10^4 in cultures growing at 37 °C with doubling times of 120, 60, and 35 min, respectively [17]. While these numbers are well in agreement, somewhat lower numbers of 9×10^3 ribosomes have been determined for exponentially growing cells of *Bacillus subtilis* [18]. While unexplored for *B. anthracis,* bacterial detection using rRNA genes and transcripts has been successfully harnessed to challenge previous limits of detection (LoD) for other pathogens [19–21].

In this study, we introduce a species-specific multi-copy chromosomal PCR marker of *B. anthracis.* This marker is represented by a unique SNP within a variable number of loci of the multi-copy 16S rRNA gene in this organism. Though the 16S rRNA gene sequences feature a very high degree of identity among the *B. cereus s.l.* group species [22], this SNP has previously been identified as unique and present in all publicly available *B. anthracis* genomic data [23–25]. Since all 16S rRNA gene copies harboring the SNP have 100% sequence identity, this specific sequence variation represents a distinct 16S rRNA gene allele named 16S-BA-allele. For simplification, all other 16S rRNA gene alleles lacking the sequence variation were named 16S-BC-allele. The relative abundance of these 16S-BA- and -BC-alleles were recently quantified in 959 *B. anthracis* isolates [25]. Here, we also harnessed this SNP to develop a *B. anthracis* specific reverse transcription (RT) real-time PCR assay. This approach brings the multi-copy marker concept for *B. anthracis* up to a new level owing to the excess numbers of ribosomes (and thus 16S rRNA moieties) in relation to chromosomes within a *B. anthracis* cell.

2. Results

2.1. Set-Up and Optimization of a New 16S rRNA Gene Allele-Specific PCR Assay

The "16S SNP BA probe" for hybridization to the *B. anthracis* specific sequence variation in 16S-BA-alleles in the *B. anthracis* genome was designed so that the SNP position was located centrally. In order to increase the fidelity of this probe, six locked nucleic acid (LNA) bases were introduced (Table 1). Similarly, five LNA positions were added to the alternative "16S SNP BC probe", recognizing the non-*B. anthracis* specific 16S-BC-alleles of *B. anthracis* (Table 1). The 16S SNP BA probe was verified in silico against the NCBI database to be highly specific for *B. anthracis*, i.e., all *B. anthracis* genomes showed a 100% match, and only genomes of a few other bacterial isolates exhibited identical sequences. Among these was, e.g., a small number of *Sphingomonas* spp. Others, such as a few genomes annotated as *Staphylococus aureus*, had the same one-base-pair mismatch at the SNP-position (relative to *B. anthracis*) and were thus identical to other *B. cereus s.l.* genomes, hybridizing perfectly against the alternative "16S SNP BC probe" (Figure S1).

Table 1. Primers and probes.

Oligonucleotide	Sequence (5′-3′)
16S SNP F *	CGAGCGCAACCCTTGA
16S SNP R *	CAGTCACCTTAGAGTGCCC
16S SNP BA probe	6FAM-CTT+AGTT+A+C+C+AT+CATT–BHQ1
16S SNP BC probe	HEX-CTT+AGTT+G+C+C+ATCATT–BHQ1
Dark 16S SNP BC probe	CTT+AGTT+G+C+C+ATCATT-C3-spacer **

Locked nucleic acids are designated by prepositioned (+); 6FAM—6-Fluorescein phosphoramidite; HEX—Hexachloro-fluorescein; BHQ1—Black Hole Quencher-1. * The expected amplicon length of the PCR reaction is 57 bp. ** blocked with a C3-spacer in 3′-position. Hairpin Tm: Primer 16S SNP R: 37.7, else: none; self-dimer Tm: Primer 16S SNP F: 11.5, else: none.

Initially, the 16S SNP BC probe, which deviates only by the one central SNP base from the 16S SNP BA probe, also carried a fluorescent dye/quencher pair. However, since this probe was found to be not entirely specific for recognizing 16S rRNA fragments of *B. cereus s.l.* members, we decided to additionally design this SNP-competing probe as a fluorescently "dark" probe in order to reduce costs of synthesis (Table 1). Thus, the 16S rRNA SNP-PCR may be considered a pseudo-duplex assay (see below for details). All PCR runs were performed with both probes, typically with the 6FAM-labeled 16S SNP BA probe and the dark 16S SNP BC probe.

In silico analysis against the NCBI nt database confirmed that the PCR amplification primers 16S SNP F and 16S SNP R (Table 1) were not species-specific for *B. anthracis*. Indeed, besides DNA from other members of the *B. cereus s.l.* group, these primers would also amplify genome sequences of various other bacteria, such as *Paenibacillus* spp., or the reverse primer would bind to sequences of *Alkalihalobacillus clausii* or *Bacillus licheniformis*, among others. This ambiguity is not surprising for primers hybridizing against 16S rRNA gene sequences. Conversely, the pivotal factor for the detection assay introduced here is that only the 16S SNP BA probe hybridizes without any mismatch against 16S-BA-allele in *B. anthracis* (Figure S1). Thus, the specificity of the PCR assay is uniquely and entirely governed by the LNA-enhanced 16S SNP BA probe.

The 16S rRNA SNP-PCR was robust for deviations from the optimum annealing temperature (62 °C; Table S1). Additionally, primer (Table S2), probe (Table S3), and MgCl$_2$ (Table S4) concentrations and pipetting errors (Table S5) were tolerated quite well. Intra- and inter-assay (Tables S6 and S7) variability was determined with positive, weakly positive, and negative template DNA. The average PCR variations were at 0.0–1.1% (intra-assay) and 1.1–1.2% (inter-assay), respectively (Tables S6 and S7), indicating high precision of the PCR. Melt point analysis of the 16S-BA-allele PCR product vs. the 16S-BC-allele PCR product (Figure S2) indicated specific amplification of each allele fragment.

2.2. Competitive Amplification-Inhibition of the 16S-BA-Allele Fragment-PCR by Excess of the Alternative 16S-BC-Allele

Though the new 16S rRNA SNP-PCR assay was tested very robust and precise, we wondered to which degree the assay would be inhibited by large excesses of the alternative 16S-BC-allele fragment featuring a single mismatch at the SNP located centrally in the hybridizing 16S-BA-allele-specific PCR probe (Figure S1). For testing this, we first evaluated which probe ratio (16S rRNA SNP BA vs. BC probe) would yield the lowest residual fluorescence values (in the 6FAM-channel of the 16S-BA-allele-specific probe) when providing only 16S-BC-allele containing DNA as PCR template. In these tests, the concentration of the 16S-BA-allele-specific probe was kept constant at 0.25 μM. The resulting 6FAM-fluorescence values were very low compared to regular amplification (Table S8), signals were weakly linearly increasing, and no Ct values were detected. The lowest fluorescence, barely above the negative control level, was recorded at a ratio of 0.25/0.75 μM (16S rRNA SNP-BA probe/-BC probe). Thus, this ratio was used for all following tests.

Next, a constant 100 template copies of the 16S-BA-allele fragment per reaction were titrated against increasing copy numbers of the alternative 16S-BC–allele fragment. Figure S3 and Table S9 show that an excess of 16S-BC-allele to BA-allele fragments of 10^6, 10^5, 10^4, or 10^3 to 1 (Table S9; assay #1–4) inhibits detection of the 16S-BA-allele fragment. This is because there was neither any *bona fide* sigmoidal PCR amplification nor were there any fluorescence signals with values meaningfully above the 16S-BC-allele-only controls (assays #11 and #12). Starting with 7.5×10^4 copies of competing 16S-BC-alleles (vs. 100 16S-BA-allele copies, i.e., 750 to 1; assay #5), both a regular Ct value was provided, and fluorescence started to markedly increase above the base level. At a ratio of 500 to 1 (16S-BC- to BA-alleles), *B. anthracis* detection became possible (assays #6 vs. #12; #7). Latest at a surplus of equal or less than 100 to 1 (assay #8), detection of 16S-BA-allele among BC-alleles was robustly possible. Thus, at the very least, a single copy of 16S-BA-allele can be detected in the presence of 100 BC-alleles.

2.3. Sensitivity and Specificity of the 16S rRNA SNP-PCR Assay

Similar to earlier work [26], we sought to harness the specificity of SNP-interrogation without assaying the alternative SNP state (i.e., the 16S-BC-allele here). Because detecting the 16S-BC-allele was not of interest for the assay at hand, the respective labeled 16S SNP BC probe was replaced by an unlabeled, fluorescently "dark" probe (i.e., a BA allele SNP-competitor probe; Table 1). In effect, primers would still amplify both alleles; however, the fluorescent probe for the 16S-BA-allele would be outcompeted by the dark probe on 16S-BC-allele targets, and the fluorescent 16S rRNA BA SNP probe would only generate signals in the presence of cognate 16S-BA-allele sequences. Thus, this approach using a dark competing probe would diminish the inadvertent generation of unspecific fluorescence generated by mishybridization of 16S rRNA BA SNP probes to 16S-BC-allele sequences.

To formally validate the sensitivity of the 16S rRNA SNP-PCR assay, a panel of 14 different *B. anthracis* DNAs was employed. These *B. anthracis* strains represent all major branches A, B, and C [27], including prominent sub-branches [28] of the global *B. anthracis* phylogeny (Table S10). All DNAs produced positive PCR results. Similarly, we tested a "specificity panel" of potentially cross-reacting organisms (Table S11). This panel included 13 DNAs of non-*anthracis B. cereus s.l.* strains. Additionally included were DNAs of common animal host organisms such as cattle, goat, sheep, and human. Neither of these DNAs yielded any positive PCR results. Finally, DNAs of organisms relevant for differential diagnostics and other prominent microbial pathogens were also assayed by the new *B. anthracis* specific 16S rRNA SNP-PCR (Table S12). Again, none of these DNAs resulted in false-positive PCR results. Of note, *Sphingomonas zeae* JM-791 [29] harboring 16S rRNA genes 100% identical in the region of the 16S SNP BA probe but different in the primer binding sites yielded negative PCR results. These results clearly indicated that the new PCR is both sensitive and specific for *B. anthracis*.

2.4. Linear Dynamic Range, Efficiency, and Limit of Detection of the B. anthracis Specific 16S rRNA SNP-PCR Assay

The linear dynamic range of the new PCR was determined based on measurements of serial DNA dilutions using recombinant 16S-BA-allele fragments or genomic DNA of *B. anthracis* Ames, respectively, as templates (Figure 1). Linearity was observed over a range from 10^1 to 10^9 copies per reaction for cloned template DNA (Figure 1A; Table S13). In nine out of nine PCR replicates, positive signals were obtained down to 10^1 copies per reaction. At 10^0, two out of nine reactions were negative, thus defining the lower limit of the linear dynamic range. The coefficient of determination (R^2) was calculated as >0.999. From the slope of the linear regression, the efficacy of the PCR was derived as 2.0 (which is 100.1% of the theoretical optimum). Thus, the 16S rRNA SNP-PCR assay performed very well over a wide 9 $\log_{10}$ concentration range of template DNA.

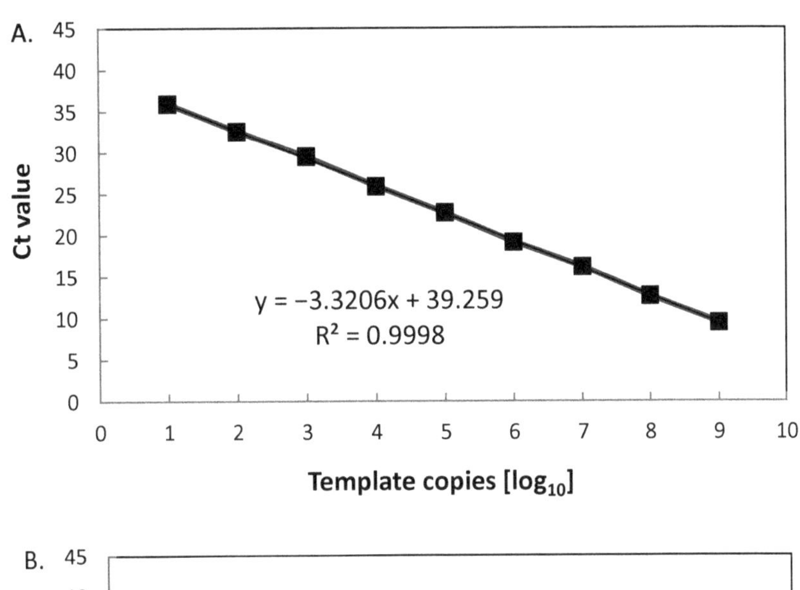

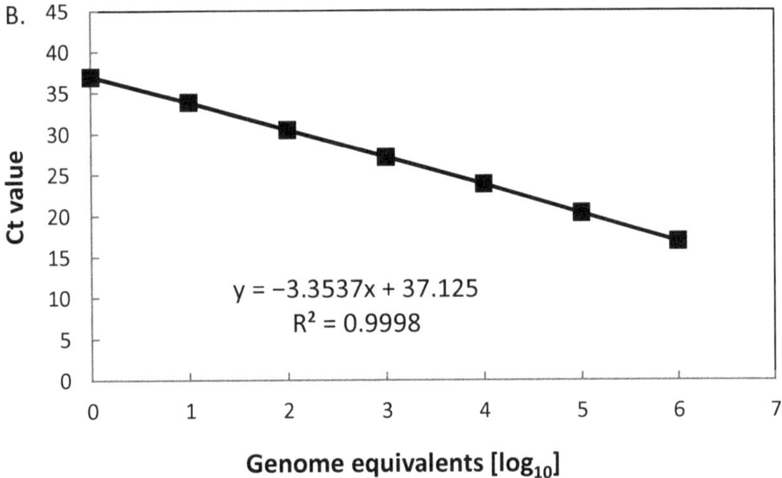

Figure 1. Linearity of the 16S rRNA SNP-PCR. Serial dilutions of DNA of (**A**) a fragment comprising the 16S-BA-allele or (**B**) *B. anthracis* strain Ames were serially diluted 1:10, PCR-tested, and template copies (**A**) or genome equivalents (**B**) plotted against Ct values. Indicated in the graphs are the slopes of the linear regressions and the coefficients of determination (R^2). Individual data points represent average values from $n = 3 \times 3$ PCR-tests.

The *B. anthracis* Ames genome harbors four copies of the 16S-BA-allele and seven copies of the BC-allele. Linear range parameters were very similar to that of cloned 16S-BA-allele DNA fragment (Figure 1B; Table S13). Because of the upper concentration limit of our *B. anthracis* Ames DNA preparations, the highest value in the linear range was 10^6 genome copies. Thus, here the linear range covered target concentrations from 10^0 to 10^6 copies per reaction. R^2 was determined as >0.999 and the efficacy of the PCR as 1.99 (which is 98.7% of the theoretical optimum). This indicated that the 16S rRNA SNP PCR assay yielded very similar results in these experiments, whether recombinant target DNA or authentic *B. anthracis* DNA was used as templates. Note, though, a single *B. anthracis* Ames genome carries four copies of the 16S-BA-allele. This explains why all PCRs yielded positive signals with DNA template at 10^0 copies (genome equivalents), whereas PCRs using single copy recombinant template did not.

Next, we determined the LoD for the 16S rRNA SNP-PCR assay by probit analysis (Figure 2; numerical data in Table S14). The assay had a limit of detection of 2.9 copies per reaction. This calculates to about 0.6 copies/µL with a probability of success of 95% with a confidence interval of 2.4–4.5 copies/assay.

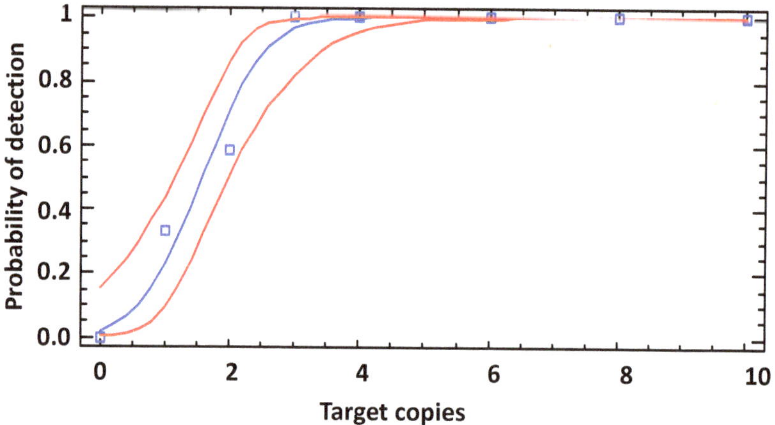

Figure 2. Limit of detection (LoD) of the 16S rRNA SNP-PCR (analytical sensitivity). DNA fragments comprising the 16S-BA-allele were diluted to the indicated copies per reaction (numerical data in Table S14) and subjected to real-time PCR (12 replicates for each data point). Probit analysis (plot of fitted model) was performed to determine the LoD by fitting template copies against the cumulative fractions of positive PCR observations (blue squares and line) and used for calculating the lower and upper 95% confidence limits (red lines).

2.5. Comparison of the New 16S rRNA SNP-PCR Assay with Existing PCR Assays

In order to further assess the performance of the 16S rRNA SNP-PCR assay, we compared it with other established PCR assays for *B. anthracis* identification currently used in our laboratory. These assays target the single-copy genes *dhp61* [9] or *PL3* [6] that have been individually validated before and compared to other commonly used *B. anthracis* PCRs [11]. Using $\log_{10}$ dilutions of *B. anthracis* Ames DNA, the 16S rRNA SNP-PCR exhibited markedly, at least three units, lower Ct values (27.9 ± 0.4; 31.7 ± 0.1; 35.4 ± 0.7) than *dhp61* (32.1 ± 0.0; 35.4 ± 0.6; 38.9 ± 1.5) or *PL3* (31.8 ± 0.2; 36.1 ± 0.7; >40) at 1000, 100, or 10 genome equivalents, respectively (Figure 3). *B. cereus* DNA did not result in amplification by any PCR assay. This result strongly suggested that the multi-copy 16S rRNA SNP-PCR assay performs competitively when compared back-to-back with established PCR assays for the detection of *B. anthracis*.

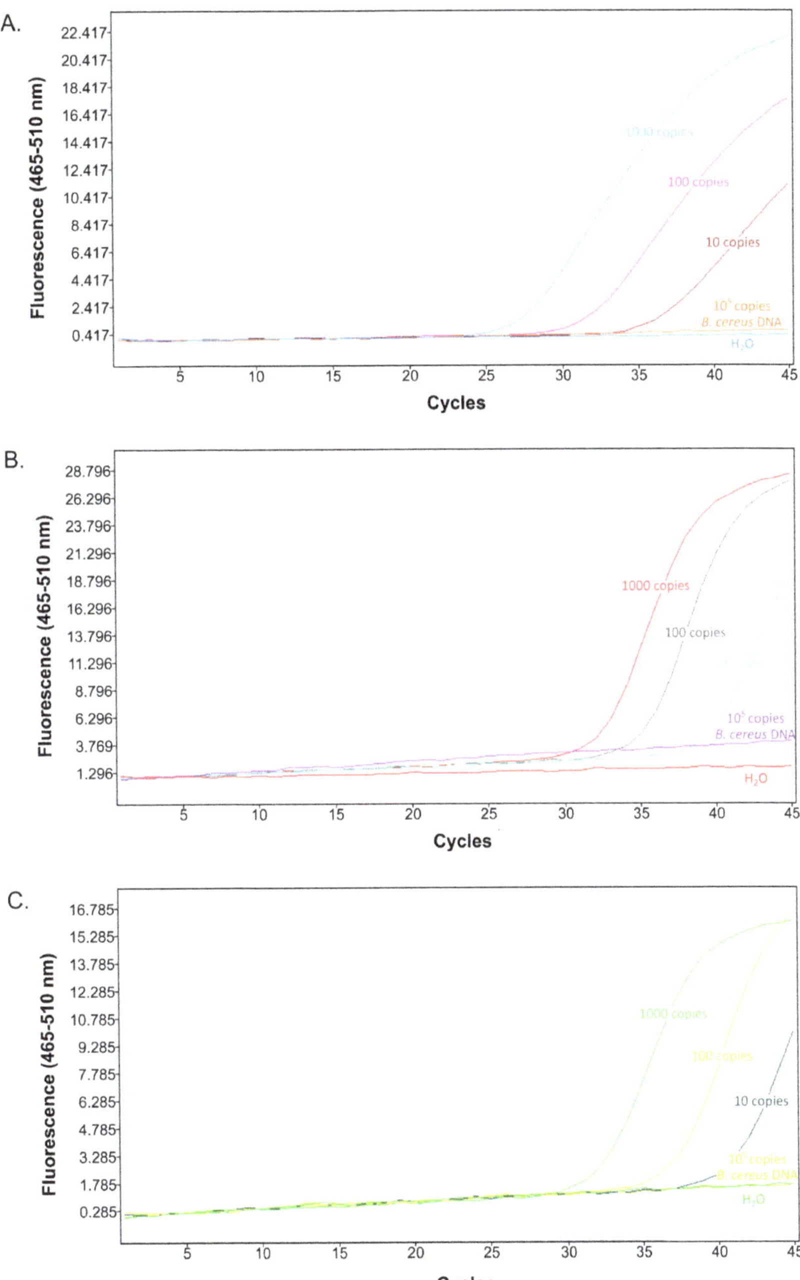

Figure 3. Comparison of the new 16S rRNA SNP-PCR assay with existing PCR assays. Different quantities of *B. anthracis* Ames template DNA (1000, 100, or 10 genome equivalents per reaction), non-target DNA (10^5 templates of *B. cereus* DNA), or water (negative) control were subjected to real-time PCR using the new 16S rRNA SNP assay (**A**), published *dhp61* gene assay [9] (**B**) or published *PL3* gene assay [6] (**C**). Representative amplification curves (from *n* = 3 with similar results) are shown.

2.6. Challenge of the New 16S rRNA SNP-PCR Assay with Samples from a Ring Trial

Along this line of reasoning, we next challenged the 16S rRNA SNP-PCR assay with samples from a previous ring trial for *B. anthracis* nucleic acid detection [30]. Again, the test was performed in comparison with the established PCR assays for *B. anthracis* identification, *dhp61* [9], and *PL3* [6]. Each of the assays was able to correctly identify the two positives out of four samples (Figure S4). Similar to evaluating known concentrations (Figure 3), the 16S rRNA SNP-PCR assay performed the best. It yielded the lowest Ct values (Figure S4), about two units lower than that of *dhp61* or *PL3* PCR. The 16S rRNA SNP-PCR assay may thus be ideally suited for this kind of analysis in which low target DNA quantities can be expected.

2.7. Challenge of the New 16S rRNA SNP-PCR Assay with Total DNA from Spiked Soil Samples

Since the 16S rRNA SNP-PCR assay performed well thus far, even in the presence of *E. coli* and human (Figure S4) or competing *B. cereus* (Figure S3) DNA, we evaluated to what extent the assay would be able to detect target DNA in spiked soil samples. These samples were spiked with cells of *E. coli* and *F. tularensis* and cells or endospores of *B. anthracis* and/or *B. thuringiensis* and were subjected to DNA purification. As above, the 16S rRNA SNP-PCR assay was conducted in comparison with the established PCR assays for *B. anthracis* identification *dhp61* [9] and *PL3* [6]. Figure S5 shows the PCR amplification curves. Samples #1, #2, and #4 were samples spiked with *B. anthracis*; sample #3 only contained *E. coli* and *B. thuringiensis*. Sample #4 had a large excess of *B. thuringiensis* over *B. anthracis* (a factor of 10^4). The 16S rRNA SNP-PCR assay detected *B. anthracis* in samples #1 and #2 but not in #4. Conversely, *dhp61* or *PL3* assays detected all three positive samples. The failure to detect *B. anthracis* by the 16S rRNA SNP-PCR assay in sample #4 is in line with our initial tests using massive excess of *B. cereus* DNA competing with *B. anthracis* detection (Figure S3; Table S9). Notably, the 16S rRNA SNP-PCR exhibited markedly, about three units, lower Ct values (23.6 ± 0.7 or 16.4 ± 0.0) than *dhp61* (26.2 ± 0.1 or 19.9 ± 0.1) or *PL3* (25.7 ± 0.0 or 19.5 ± 0.1) for samples #1 and #2, respectively. This result confirmed our preceding findings that the 16S rRNA SNP-PCR assay can reach a lower detection limit than the established assay as long as there is no large excess of other *B. cereus s.l.* DNA competing for amplification primers.

2.8. The New 16S rRNA SNP-PCR Assay also Functions as an RT-PCR Assay

We reasoned that the real-time 16S rRNA SNP-PCR assay targeting *B. anthracis* DNA might be converted into an RT-PCR assay targeting RNA in the form of 16S-BA-allele transcripts that harbor the *B. anthracis*-specific SNP. In order to test this, cells of *B. anthracis* Sterne or *B. cereus* 10987 were grown to exponential growth phase, inactivated, and total nucleic acids (including genomic DNA) were isolated alongside parallel preparations of DNA only. The one-step RT-PCR reaction was thus run with a mixture of genomic DNA and RNA, which can both be targeted by the assay. For comparison, the above-validated 16S rRNA real-time SNP-PCR was conducted in parallel with genomic DNA as the only template (no RT-reaction). When using identical samples, RT-PCR reactions (with templates consisting of total RNA and DNA) resulted in intensely lower Ct values than without reverse transcription (since only genomic DNA served as a template; Figure 4). Notably, differences in Ct values (RT-PCR vs. PCR) were in the range between 9 and 10 units. This translates to an about 1000-fold improvement using RT-PCR over DNA-only PCR. This result indicated that the 16S rRNA SNP-PCR assay functions both for DNA- and RNA-based (RT) PCR.

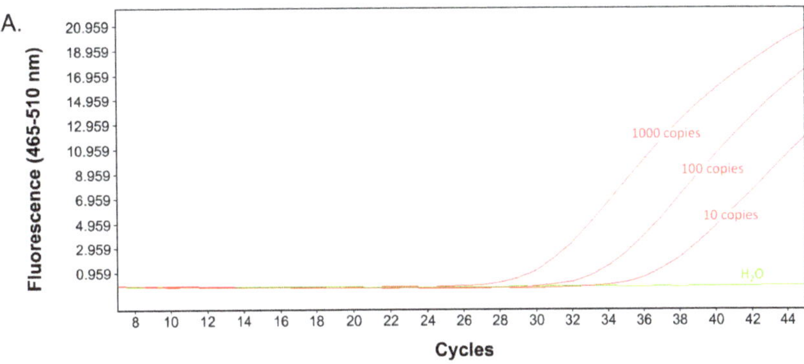

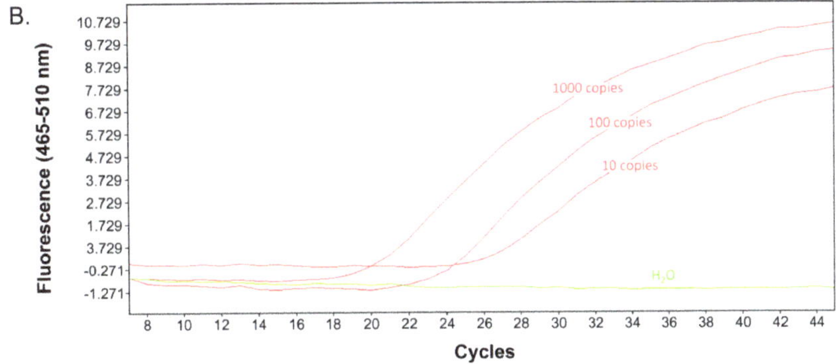

Figure 4. Comparison of the 16S rRNA SNP-PCR assay (DNA-only) with the RT-16S rRNA SNP-PCR assay (DNA+RNA). Total DNA or total DNA+RNA isolated from exponentially growing cells of *B. anthracis* or *B. cereus*, respectively, were used for PCR amplification of 16S-BA-allele DNA (**A**) or additionally after reverse transcription of 16S rRNA (ribosomal RNA) (**B**). Representative amplification curves (from *n* = 3 with similar results) are shown.

2.9. Linear Dynamic Range, Efficiency, and Limit of Detection of the B. anthracis 16S rRNA SNP RT-PCR Assay

To further characterize the RT-PCR, we determined the linear dynamic range and determined the LoD (Probit) of the 16S rRNA SNP RT-PCR using total RNA/DNA of *B. anthracis* Sterne (similar to DNA-only templates, see above). The RT-PCR was linear over a range from 10^0 to 10^8 template rRNA+DNA per reaction (Figure 5A; Table S15). The coefficient of determination (R^2) was 0.9982, and the efficacy of the RT-PCR was 1.92 (which is 92.3% of the theoretical optimum). Thus, the 16S rRNA SNP RT-PCR assay performed well over a wide 9 $\log_{10}$ concentration range of template RNA+DNA (higher template numbers than 1.5×10^8 were not tested).

The LoD for the 16S rRNA SNP RT-PCR assay as determined by probit analysis (Figure 5B; numerical data in Table S16) was 6.3 copies per reaction. This calculates to about 1.3 copies/µL with a probability of success of 95% with a confidence interval of 5.0–8.9 copies/assay. Thus, the RT-PCR reaction performed similarly well as the PCR reaction. Mindful of the about 3 $\log_{10}$ units higher number of 16S rRNAs in cells than genomes, detection of *B. anthracis* with the rRNA-directed RT-PCR is superior to the respective real-time PCR assay and all other *B. anthracis* PCR assays tested.

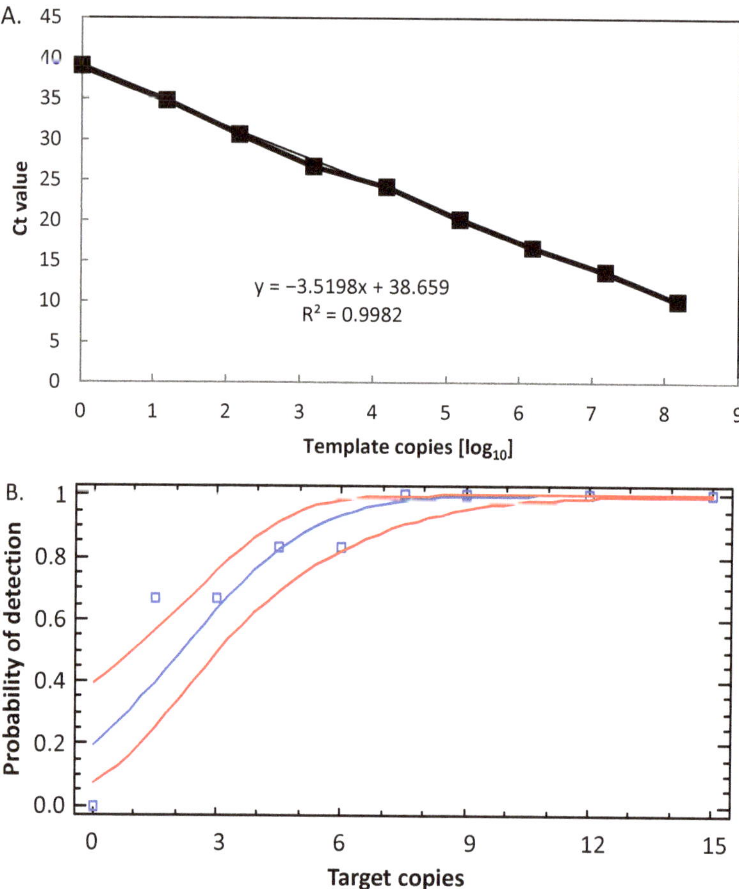

Figure 5. Linearity and LoD of the 16S rRNA SNP RT-PCR. Serial dilutions of RNA (with DNA) of *B. anthracis* strain Sterne were serially diluted 1:10, RT-PCR-tested and template copies plotted against Ct values (**A**). Indicated in the graph is the slope of the linear regression and the coefficients of determination (R^2). Individual data points represent average values from $n = 3 \times 3$ PCR-tests. Analytical sensitivity of the 16S rRNA SNP RT-PCR was determined by diluting samples from (**A**) to the indicated copies per reaction (numerical data in Table S16) and subjected to RT-PCR (12 replicates for each data point). To determine the LoD, probit analysis (plot of fitted model, blue squares, and line) was performed (as in Figure 2), and the lower and upper 95% confidence limits (red lines) were determined (**B**).

3. Materials and Methods

3.1. Bacterial Culture, Inactivation, and DNA Samples for Quality Assessment

B. anthracis strains and other Bacilli were cultivated at 37 °C on tryptic soy agar plates (TSA, Merck KGaA, Darmstadt, Germany). Bacteria comprising the negative panel (Table S1) were grown on appropriate agar media (with 10% CO_2 atmosphere where required) at 37 °C until colonies emerged. Risk group 3 (RG-3) *B. anthracis* strains were cultivated in the biosafety level 3 (BSL-3) facilities at the Bundeswehr Institute of Microbiology (IMB) and then chemically inactivated by resuspending a loop of colony material in aqueous peracetic acid solution (4% (*v/v*) Terralin PAA, Schülke & Mayr GmbH, Norderstedt, Germany) before further use [31]. RG-2 strains of endospore formers were inactivated likewise. All other bacterial cultures were inactivated by 70% (*v*/*v*) ethanol. Ring trial *B. anthracis* DNA samples published in [30] were obtained from Instant (Düsseldorf, Germany).

3.2. Isolation of DNA, RNA, and Nucleic Acid Quantification

Bacterial DNA and RNA were isolated using MasterPure™ Gram Positive DNA Purification kit (Lucigen, Middleton, WI, USA). For RNA (+DNA) isolation, RNase treatment was omitted. DNA and RNA concentrations were quantified using the Qubit dsDNA HS Assay or RNA HS Assay kits (ThermoFisher Scientific, Darmstadt, Germany) according to the manufacturer's protocols. DNA and RNA (+DNA) preparations were stored at $-20\,^{\circ}\mathrm{C}$ and $-80\,^{\circ}\mathrm{C}$, respectively, until further use.

3.3. Design and in Silico Bioinformatic Analysis of Primer and Probe DNA Sequences

All relevant DNA sequence data for oligonucleotide design were retrieved from public databases (NCBI). Primer and probe DNA oligonucleotides [25] were designed with Geneious Prime (version 2021.1.1; Biomatters, Auckland, New Zealand). In silico specificity analysis was performed by probing each primer and probing nucleotide sequences against the NCBI nt databases using BLASTN for short input sequences (Primer BLAST) [32]. The two amplification oligonucleotide primers target a consensus region within the 16S rRNA genes on the chromosome of *B. cereus s.l.* species (Table 1), including *B. anthracis*. The two oligonucleotide probes (Table 1) feature the centrally located discriminatory SNP (pos. 1110 in *B. anthracis* strain Ames Ancestor, NC_007530) [23,24]. These probes thus either match the allele unique for *B. anthracis* (named 16S-BA-allele; with an adenine, A at the SNP position) or the general 16S-BC-allele (guanine, G at the SNP position), respectively (the two alleles are depicted in Figure S1). Due to placement and length restrictions related to another non-discriminatory SNP (pos. 1119), each probe was amended with locked nucleic acids (LNA). LNA are modified nucleic acids in which the sugar is conformationally locked. This rigidity causes exceptional hybridization affinity through stable duplexes with DNA and RNA [33], eventually improving mismatch discrimination in SNP genotyping studies. Similar to unmodified ssDNA probes, the LNA-containing probes (Table 1) are susceptible to 5′-nuclease attack during PCR. LNA probes as well as primers were purchased from TIB MolBiol (Berlin, Germany).

3.4. Real-Time and Reverse Transcription PCR Conditions

All (pseudo) duplex real-time PCR amplifications were performed in reaction mixtures of a final volume of 20 μL containing 2 μL LightCycler® FastStart DNA Master HybProbe mix (Roche Diagnostics, Mannheim, Germany), 5 mM $MgCl_2$, 0.5 μM of each primer, 0.25 μM of 16S SNP BA probe, 0.75 μM of (dark) 16S SNP BC probe, and various quantities of template DNA template. All reactions were performed on a LightCycler 480 real-time PCR system fitted with color compensation (Roche Diagnostics, Mannheim, Germany). The optimized amplification conditions were 95 °C for 10 min, and then 45 consecutive cycles of first 15 s at 95 °C and then 20 s at 62 °C, followed by 20 s at 72 °C.

Reverse transcription PCR reaction mixtures contained 7.4 μL LightCycler® 480 RNA Master Hydrolysis Probes mix, 1.3 μL Activator, 1 μL Enhancer (Roche Diagnostics, Mannheim, Germany), 0.5 μM of each primer, 0.25 μM of 16S SNP BA probe, 0.75 μM of (Dark) 16S SNP BC probe, a variable volume of RNA and/or DNA template. Finally, nuclease-free water (Qiagen, Hilden, Germany) was added to a final volume of 20 μL. Using the LightCycler 480 real-time PCR system (Roche Diagnostics, Mannheim, Germany), reverse transcription was performed at 63 °C for 3 min followed by an activation step at 95 °C for 30 s and 45 cycles of 95 °C for 15 s, 62 °C for 20 s and 72 °C for 1 s.

A fluorescent signal 10-fold higher than the standard deviation of the mean baseline emission was counted as a positive detection. Samples were tested in triplicate (unless noted otherwise) and data recorded as Cycle thresholds (Ct) with Ct defined as the PCR cycle at which the fluorescent intensity raised above the threshold [34].

3.5. Droplet Digital PCR (ddPCR) and Reverse Transcription (RT) ddPCR

All DNA and RNA templates used for real-time and reverse transcription PCR were quantified by ddPCR and RT ddPCR, respectively. A 20 µL ddPCR reaction mixture consisted of 10 µL ddPCR Supermix for Probes (Bio-Rad Laboratories, Munich, Germany), 0.9 µM of each primer, 0.15 µM of each probe, and 5 µL of template DNA. RT-ddPCR reaction mixtures comprised of 5 µL One-Step RT-ddPCR Advanced Supermix for Probes (Bio-Rad, Munich, Germany), 2 µL of Reverse Transcriptase (Bio-Rad, Munich, Germany; final concentration 20 U/µL), 0.6 µL of DTT (Bio-Rad, Munich, Germany; final concentration 10 nM), 0.9 µM of each primer, 0.15 µM of each probe, and 5 µL of template RNA. Droplets were generated using a QX200 ddPCR droplet generator (Bio-Rad, Munich, Germany). PCR amplification for both assays was performed on the Mastercycler Gradient (Eppendorf, Hamburg, Germany) with the following conditions.

Initial reverse transcription was carried out at 48 °C for 60 min (only for RT-ddPCR). Enzyme activation at 95 °C for 10 min was followed by 40 cycles of denaturation at 94 °C for 30 s and annealing/extension at 58 °C for 1 min. Before the samples were cooled to 4 °C, a final enzyme inactivation was carried out at 98 °C for 10 min. The cooling and heating ramp rate was set to 2 °C/s for all steps. After PCR runs, droplets were analyzed using the QX100 Droplet Reader (Bio-Rad, Munich, Germany), and absolute target concentrations of each sample were calculated using Quantasoft Pro Software (Bio-Rad, Munich, Germany).

3.6. Generation of PCR Positive Controls from Reference Plasmids Harboring 16S-BA- or BC-Allele Fragments

Though we generally used genomic DNA from *B. cereus* or *B. anthracis*, respectively, for PCR testing and validation, generic positive control reference plasmids for either allele, the *B. anthracis*-specific 16S-BA-allele or the *B. cereus*-specific 16S-BC-allele were constructed. For this, a PCR-amplicon was generated from *B. anthracis* Ames DNA with primers 16S SNP F and 16S SNP R using Platinum™ Taq DNA Polymerase High Fidelity (ThermoFisher Scientific, Darmstadt, Germany). This DNA comprises a mixture of both alleles in a ratio of 4 to 7 [25]. The PCR-amplicon was analyzed on agarose gel electrophoresis, a band of the expected size (57 bp) cut from the gel and gel-purified using QIAquick Gel Extraction kit (QIAGEN, Hilden Germany). PCR products were ligated into pCR2.1 TOPO vector (ThermoFisher Scientific, Darmstadt, Germany) using TOPO TA Cloning kit (Thermo Scientific, Darmstadt, Germany) and transformed into One Shot TOP10 chemically competent cells (ThermoFisher Scientific, Darmstadt, Germany) according to the manufacturer's protocol. Several recombinant plasmids isolated from different clones were sequenced (Eurofins Genomics Germany, Ebersberg, Germany) in order to obtain plasmids harboring either the 16S-BA-allele or the 16S-BC-allele. From these plasmids, PCR products were generated using primers M13 F and M13 R, which contained the target region for the 16S rRNA SNP-PCR with either the 16S-BA- or BC-allele. After purification with QIAquick PCR purification kit (QIAGEN, Hilden Germany), PCR products were quantified using digital PCR and diluted as required.

3.7. Determination of the Specificity (Inclusivity/Exclusivity) of the B. anthracis 16S rRNA Allele Assay

PCR specificity for the 16S rRNA SNP assay was assessed by verifying the amplification of DNA containing or lacking respective markers. "Inclusivity" was evaluated by (exponential) amplification above threshold levels obtained with template DNA comprising the markers' sequences. Vice versa, "exclusivity" was confirmed by lack of amplification of genomic DNA from *B. cereus s.l.* strains reported to lack the particular 16S-BA-allele but also may harbor the alternative 16S-BC-allele or include no-template negative controls (NTC). Positive PCR results were further analyzed via agarose gel electrophoresis, demonstrating a single band with a molecular weight corresponding to the predicted size of the 16S rRNA SNP-PCR amplicon (note: this cannot differentiate between the two alternative SNP states in the 16S rRNA gene alleles).

3.8. Dynamic Linear Range, PCR Efficiency, and Limit of Detection

The dynamic linearity of the PCR assay was determined over a 9 $\log_{10}$ concentration range for DNA (real-time PCR) and RNA (RT-PCR) templates. Each dilution was assayed 6-fold, and analysis for linearity and PCR-efficiency (E) was performed from the plot of the Ct's versus the logarithm of the target concentrations [35]. The sensitivity of the PCR assay was expressed as the limit of detection (LoD) of 16S rRNA SNP genome or transcript copies. LoD was formally defined as the concentration permitting detection of the analyte at least 95% of the time. For this, DNA fragments comprising the 16S rRNA SNP were diluted to between 10 and 0 copies per reaction, subjected to real-time PCR with 12 replicates for each dilution step. Probit analysis (plot of the fitted model) was performed [36] using StatGraphics Centurion XVI.I (16.1.11; Statgraphics Technologies, The Plains, VA, USA) to determine the LoD by fitting template copies against the cumulative fractions of positive PCR observations and used for calculating the lower and upper 95% confidence limits. The LoD of the 16S rRNA SNP RT-PCR was determined likewise using samples with 0–15 rRNA copies per reaction (12 replicates for each dilution step).

4. Discussion

The use of SNPs as reliable markers for the identification of *B. anthracis* among its closest relatives of the *B. cereus* group is not a novel approach. This has previously been achieved with high specificity and sensitivity for nucleotide position 640 in the *plcR* gene [10] or at position 1050 in the *purA* gene [26], and diverse assays were thoroughly evaluated in [11]. Likewise, ribosomal gene sequences and intergenic transcribed spacers (ITS) between 16S and 23S rRNA genes have also been employed for *B. anthracis* identification in the past [37–40]. However, while these authors focused on the specific identification of *B. anthracis*, they neglected the potential of developing a sensitive assay making use of the multi-copy nature of their targets. An interesting exception is a study on fluorescent DNA-heteroduplex detection of *B. anthracis* [41]. Herein detection was preceded by general PCR-amplification of a fragment of the 16S rRNA gene region of *B. cereus s.l.* group strains containing a presumably specific SNP (pos. 980). This SNP, however, is neither specific for *B. anthracis* nor for the *B. cereus s.l.* group [24]. Anyway, Merrill et al. succeeded in establishing a LoD for their PCR of approximately 0.05 pg of purified *B. anthracis* genomic DNA (which can be calculated to represent 10–20 cell equivalents per reaction) [41]. This is higher than the LoD of about 1–2 cell equivalents per reaction found in our study. More importantly, Merrill et al. also took the effort to determine the detection limit of their presumably specific SNP in mixtures of 16S rRNA gene amplicons from *B. anthracis* and *B. cereus* [41]. The authors observed a detection limit of 1 out of 50 for *B. anthracis* DNA mixed with *B. cereus* DNA. They explained this limit as narrowed by methodological constraints and from competitive hybridization dynamics during probe annealing [41]. This finding can be compared with our results. The PCR assay developed here was able to detect at least one *B. anthracis* 16S-BA-allele target among 100 BC-allele targets (Figure S3 and Table S9). At higher alternative (16S-BC-allele) concentrations, these templates will outcompete the 16S-BA-allele for primer binding. Thus, the higher the fraction of 16S-BC-allele, the lower the relative amplification of 16S-BA-allele resulting in increasingly non-exponential amplification of the latter. In contrast, for a SNP in the DNA target *plcR* used for the differentiation of *B. anthracis* from *B. cereus*, a 20,000-fold excess of the alternate *B. cereus* allele did not preclude the detection of the *B. anthracis* allele [42]. With *B. cereus* spore counts in soils spanning a wide range of from 1×10^1 to 2.5×10^4 CFU per g soil [43], the *plcR* SNP-PCR should be able to detect *B. anthracis* in practically any sample. Here, the new 16S rRNA SNP-PCR on DNA as target molecule would fall short with only covering up to medium *B. cereus*-loaded soils. However, when targeting ribosomal RNA, the sensitivity (LoD) of the 16S rRNA SNP RT-PCR would be at least three orders of magnitude increased. Then, it should be possible to challenge the LoD values achieved by the *plcR* SNP-PCR (25 fg DNA or about 5 genome equivalents) [42].

A potential limitation of the multi-copy nature of the 16S-BA-allele may be the variable abundance of this allele in different *B. anthracis* strains. Previously, we could show that most *B. anthracis* strains harbor 3 (58.39%) 16S-BA-alleles. There are, however, also a number of isolates only possessing 2 (23.04%), 4 (17.10%), 5 (1.15%), and a single one with only 1 (0.31%) 16S-BA-alleles [25]. Thus, in most cases, this multi-copy gene allele can be harnessed nevertheless. A more typical multi-copy marker for detection of bacterial biothreat agents (and of other pathogens) constitute insertion sequence (IS) elements, which are widespread mobile genetic entities. For instance, in *Brucella* spp. *IS711* occurs in multiple genomic copies, and thus, the detection of this *IS711* is very sensitive. *B. melitensis* and *B. suis* contain seven complete copies, *B. abortus* carries six complete and one truncated *IS711* copies, *B. ovis*, *B. ceti*, and *B. pinnipedialis* even more than 20 copies [44]. Consequently, the lowest concentration of *Brucella* sp. DNA that could be detected was about ten times lower for *IS711* than, e.g., for single-copy genes *bcsp31* (*Brucella* cell surface 31 kDa protein) or *per* (perosamine synthetase), respectively [45]. Similarly, in *Coxiella burnetii*, the detection sensitivity of specific *IS1111* was compared to that of the single-copy *icd* gene (isocitrate dehydrogenase) [46]. While both PCRs for *icd* and *IS1111* had similar LoDs of 10.38 and 6.51, respectively, the sensitivity of *IS1111* was still superior because of its multiple-copy nature. Between 7 and 110 copies of this mobile element were found in various *C. burnetii* isolates [46].

The differences in threshold values (ΔCt = 9.96 $\pm$ 0.65) of identical samples obtained from (RT)-PCR using 16S-BA-allele DNA-only vs. 16S-BA-alleleDNA+RNA is enormous. There is an approximate factor of about 1000 ($2^{9.96}$) times more templates in the DNA+RNA sample than in the DNA-only sample. This factor favorably agrees with the numbers of genome copies and 16S rRNA transcripts in cells [17,18]. Similar to the work at hand, earlier work employed a combination of a DNA multi-copy marker and sensitive detection of rRNA transcript targets in *Mycobacterium ulcerans* [20]. The authors determined an LoD of six copies of the 16S rRNA transcript target sequence. For comparison, an LoD of two target copies of the high-copy insertion sequence element *IS2404*, which is present in 50–100 copies in different *M. ulcerans* strains, was calculated from parallel experiments [20]. Ribosomal RNA detection was also utilized for *Mycobacterium leprae* diagnosis by the same research team. Here, an LoD of three *M. leprae* target copies was achieved for a novel 16S rRNA RT-PCR assay; the same value as determined for the *M. leprae* specific multi-copy repetitive DNA target assayed in parallel [21]. At first glance, these values do not especially speak in favor of querying for 16S rRNA transcripts; however, one has to consider the high numbers of these molecules per cell in comparison to DNA markers (including the high-copy ones). Thus, the chance of capturing one of the more abundant rRNA molecules should be higher than that of the more limited DNA molecules. Indeed, this idea was explored, e.g., for *Escherichia coli*, *Enterococcus faecalis*, *Staphylococcus aureus*, *Clostridium perfringens*, and *Pseudomonas aeruginosa* by [19]. Comparative quantitative detection of these bacteria by RT-PCR (16S rRNA) vs. PCR (16S rRNA genes) revealed that the rRNA-detecting assay was from 64- to 1024-fold more sensitive than the one detecting DNA. Similarly, work on pathogenic spirochete *Leptospira* spp. found that 16S rRNA-based assays were at least 100-fold more sensitive than a DNA-based approach [47]. These authors also found that Leptospiral 16S rRNA molecules remain appreciably stable in blood. From this insight, the authors then highlighted the potential use of 16S RNA targets for the diagnosis of early infection. Nevertheless, potential limitations of this approach were also noted. Efficacy of the required reverse transcription reaction has to be considered, RNA molecules are notoriously less stable than other biomarkers, and their cellular abundance (and as a consequence, their detection) can be expected to be variable [47]. Finally, though not required for qualitative detection, absolute quantification of microbial cells based solely on enumeration of RNA molecules is complicated because of these variations in transcript numbers depending, e.g., on the growth phase [47]. However, the cell numbers determined by RT-PCR were similar when compared alongside standard methods such as cell counts, PCR, or fluorescence in situ hybridization (FISH) [48]. Yet, in certain instances, there might be an additional advantage of performing PCR on rRNA directly (via RT-PCR) instead of targeting DNA (including DNA of rRNA genes). Because DNA is more stable than RNA, DNA may originate from both live and dead bacterial cells. In contrast, rRNA molecules may be considered to be more closely associated with viable bacteria [49]. Though this might also be possible with the new PCR assays introduced in the work at hand, we chose to combine DNA and rRNA detection in a single test tube for the sake of simplicity (no troublesome DNase treatment of purified RNA required) and depth of detection.

5. Conclusions

In this work, we designed and validated a new PCR-based detection assay for the biothreat agent *B. anthracis*. This assay can be run as a real-time PCR with solely DNA as a template or as an RT-real-time version using both cellular nucleic acid pools (DNA and RNA) as a template. This assay was found to be highly species specific, yielding no false positives, and was sensitive with a LoD of about 0.6 copies/μL (DNA-only) and about 1.3 copies/μL (DNA+RNA). With the high abundance of 16S rRNA moieties in cells, this assay can be expected to facilitate the detection of *B. anthracis* by PCR. While standard PCR assays are well established for the identification of *B. anthracis* from pure culture, the exceptional sensitivity of the new 16S rRNA-based assay might excel in clinical and public health laboratories when detection of minute residues of the pathogen is required.

Supplementary Materials: The following are available online: https://www.mdpi.com/article/10.3390/ijms222212224/s1.

Author Contributions: Conceptualization, G.G. and P.B.; investigation, P.B., M.D.-T.N. and M.C.W.; methodology, M.D.-T.N. and P.B.; formal analysis and validation, P.B., M.D.-T.N. and G.G.; resources, G.G. and M.C.W.; data curation, P.B., M.D.-T.N. and M.C.W.; writing—original draft preparation, G.G. and P.B.; writing—review and editing, P.B., M.D.-T.N., M.C.W. and G.G.; visualization, M.D.-T.N., P.B. and G.G.; supervision and project administration, G.G. and P.B.; funding acquisition, G.G. All authors have read and agreed to the published version of the manuscript.

Funding: This research was funded by funds from the Medical Biological Defense Research Program of the Bundeswehr Joint Medical Service.

Institutional Review Board Statement: Not applicable.

Informed Consent Statement: Not applicable.

Data Availability Statement: Not applicable.

Acknowledgments: The authors thank Mandy Knüper (Bundeswehr Institute of Microbiology) for support with probit-analysis and the gift of DNA from spiked soil samples, Stefanie Gläser (Justus-Liebig-University Gießen, Germany), Erwin Märtlbauer (Ludwig-Maximilians-University, Munich, Germany) and Monika-Ehling Schultz (University of Veterinary Medicine, Vienna, Austria) for several *B. cereus* strains, as well as Paul Keim (Northern Arizona University, Flagstaff, AZ, USA) and Wolfgang Beyer (Hohenheim University, Stuttgart, Germany) for the gift of *B. anthracis* strains/DNA. Thanks are due to Rahime Terzioglu for technical assistance and Olfert Landt (TIB Molbiol, Berlin) for support in LNA-probe design.

Conflicts of Interest: The authors declare no conflict of interest. Opinions, interpretations, conclusions, and recommendations are those of the authors and are not necessarily endorsed by any governmental agency, department, or other institutions. The funders had no role in the design of the study; in the collection, analyses, or interpretation of data; in the writing of the manuscript, or in the decision to publish the results.

References

1. Turnbull, P.C.; World Health Organization. *Anthrax in Humans and Animals*; Turnbull, P., Ed.; WHO Press: Geneva, Switzerland, 2008.
2. Ehling-Schulz, M.; Lereclus, D.; Koehler, T.M. The *Bacillus cereus* group: *Bacillus* species with pathogenic potential. *Microbiol. Spectr.* **2019**, *7*. [CrossRef]
3. Mock, M.; Fouet, A. Anthrax. *Annu. Rev. Microbiol.* **2001**, *55*, 647–671. [CrossRef] [PubMed]

4. Jelacic, T.M.; Chabot, D.J.; Bozue, J.A.; Tobery, S.A.; West, M.W.; Moody, K.; Yang, D.; Oppenheim, J.J.; Friedlander, A.M. Exposure to *Bacillus anthracis* capsule results in suppression of human monocyte-derived dendritic cells. *Infect. Immun.* **2014**, *82*, 3405–3416. [CrossRef] [PubMed]

5. Patino-Navarrete, R.; Sanchis, V. Evolutionary processes and environmental factors underlying the genetic diversity and lifestyles of *Bacillus cereus* group bacteria. *Res. Microbiol.* **2017**, *168*, 309–318. [CrossRef]

6. Wielinga, P.R.; Hamidjaja, R.A.; Ågren, J.; Knutsson, R.; Segerman, B.; Fricker, M.; Ehling-Schulz, M.; de Groot, A.; Burton, J.; Brooks, T.; et al. A multiplex real-time PCR for identifying and differentiating *B. anthracis* virulent types. *Int. J. Food Microbiol.* **2011**, *145*, S137–S144. [CrossRef] [PubMed]

7. Ellerbrok, H.; Nattermann, H.; Ozel, M.; Beutin, L.; Appel, B.; Pauli, G. Rapid and sensitive identification of pathogenic and apathogenic *Bacillus anthracis* by real-time PCR. *FEMS Microbiol. Lett.* **2002**, *214*, 51–59. [CrossRef]

8. Ramisse, V.; Patra, G.; Garrigue, H.; Guesdon, J.L.; Mock, M. Identification and characterization of *Bacillus anthracis* by multiplex PCR analysis of sequences on plasmids pXO1 and pXO2 and chromosomal DNA. *FEMS Microbiol. Lett.* **1996**, *145*, 9–16. [CrossRef]

9. Antwerpen, M.H.; Zimmermann, P.; Bewley, K.; Frangoulidis, D.; Meyer, H. Real-time PCR system targeting a chromosomal marker specific for *Bacillus anthracis*. *Mol. Cell. Probes* **2008**, *22*, 313–315. [CrossRef]

10. Easterday, W.R.; Van Ert, M.N.; Simonson, T.S.; Wagner, D.M.; Kenefic, L.J.; Allender, C.J.; Keim, P. Use of single nucleotide polymorphisms in the *plcR* gene for specific identification of *Bacillus anthracis*. *J. Clin. Microbiol.* **2005**, *43*, 1995–1997. [CrossRef]

11. Ågren, J.; Hamidjaja, R.A.; Hansen, T.; Ruuls, R.; Thierry, S.; Vigre, H.; Janse, I.; Sundström, A.; Segerman, B.; Koene, M.; et al. In silico and in vitro evaluation of PCR-based assays for the detection of *Bacillus anthracis* chromosomal signature sequences. *Virulence* **2013**, *4*, 671–685. [CrossRef]

12. Straub, T.; Baird, C.; Bartholomew, R.A.; Colburn, H.; Seiner, D.; Victry, K.; Zhang, L.; Bruckner-Lea, C.J. Estimated copy number of *Bacillus anthracis* plasmids pXO1 and pXO2 using digital PCR. *J. Microbiol. Methods* **2013**, *92*, 9–10. [CrossRef]

13. Pena-Gonzalez, A.; Rodriguez-R, L.; Marston, C.K.; Gee, J.E.; Gulvik, C.A.; Kolton, C.B.; Saile, E.; Frace, M.; Hoffmaster, A.R.; Konstantinidis, K.T. Genomic characterization and copy number variation of *Bacillus anthracis* plasmids pXO1 and pXO2 in a historical collection of 412 strains. *mSystems* **2018**, *3*, e00065-18. [CrossRef]

14. Hansen, M.C.; Nielsen, A.K.; Molin, S.; Hammer, K.; Kilstrup, M. Changes in rRNA levels during stress invalidates results from mRNA blotting: Fluorescence in situ rRNA hybridization permits renormalization for estimation of cellular mRNA levels. *J. Bacteriol.* **2001**, *183*, 4747–4751. [CrossRef]

15. Bremer, H.; Dennis, P.P. Modulation of chemical composition and other parameters of the cell at different exponential growth rates. *EcoSal Plus* **2008**, *3*. [CrossRef] [PubMed]

16. Wiśniewski, J.R.; Rakus, D. Multi-enzyme digestion FASP and the 'Total Protein Approach'-based absolute quantification of the *Escherichia coli* proteome. *J. Proteom.* **2014**, *109*, 322–331. [CrossRef] [PubMed]

17. van Dijk-Salkinoja, M.S.; Planta, R.J. Rate of ribosome production in *Bacillus licheniformis*. *J. Bacteriol.* **1971**, *105*, 20–27. [CrossRef] [PubMed]

18. Barrera, A.; Pan, T. Interaction of the *Bacillus subtilis* RNase P with the 30S ribosomal subunit. *RNA* **2004**, *10*, 482–492. [CrossRef]

19. Matsuda, K.; Tsuji, H.; Asahara, T.; Kado, Y.; Nomoto, K. Sensitive quantitative detection of commensal bacteria by rRNA-targeted reverse transcription-PCR. *Appl. Environ. Microbiol.* **2007**, *73*, 32–39. [CrossRef]

20. Beissner, M.; Symank, D.; Phillips, R.O.; Amoako, Y.A.; Awua-Boateng, N.Y.; Sarfo, F.S.; Jansson, M.; Huber, K.L.; Herbinger, K.H.; Battke, F.; et al. Detection of viable *Mycobacterium ulcerans* in clinical samples by a novel combined 16S rRNA reverse transcriptase/IS2404 real-time qPCR assay. *PLoS Negl. Trop. Dis.* **2012**, *6*, e1756. [CrossRef]

21. Beissner, M.; Woestemeier, A.; Saar, M.; Badziklou, K.; Maman, I.; Amedifou, C.; Wagner, M.; Wiedemann, F.X.; Amekuse, K.; Kobara, B.; et al. Development of a combined RLEP/16S rRNA (RT) qPCR assay for the detection of viable *M. leprae* from nasal swab samples. *BMC Infect. Dis.* **2019**, *19*, 753. [CrossRef]

22. Ash, C.; Farrow, J.A.; Dorsch, M.; Stackebrandt, E.; Collins, M.D. Comparative analysis of *Bacillus anthracis*, *Bacillus cereus*, and related species on the basis of reverse transcriptase sequencing of 16S rRNA. *Int. J. Syst. Bacteriol.* **1991**, *41*, 343–346. [CrossRef] [PubMed]

23. Candelon, B.; Guilloux, K.; Ehrlich, S.D.; Sorokin, A. Two distinct types of rRNA operons in the *Bacillus cereus* group. *Microbiology* **2004**, *150 Pt 3*, 601–611. [CrossRef]

24. Hakovirta, J.R.; Prezioso, S.; Hodge, D.; Pillai, S.P.; Weigel, L.M. Identification and analysis of informative single nucleotide polymorphisms in 16S rRNA gene sequences of the *Bacillus cereus* group. *J. Clin. Microbiol.* **2016**, *54*, 2749–2756. [CrossRef] [PubMed]

25. Braun, P.; Zimmermann, F.; Walter, M.C.; Mantel, S.; Aistleitner, K.; Stürz, I.; Grass, G.; Stoecker, K. In-depth analysis of *Bacillus anthracis* 16S rRNA genes and transcripts reveals intra- and intergenomic diversity and facilitates anthrax detection. *bioRxiv* **2021**. [CrossRef]

26. Irenge, L.M.; Durant, J.F.; Tomaso, H.; Pilo, P.; Olsen, J.S.; Ramisse, V.; Mahillon, J.; Gala, J.L. Development and validation of a real-time quantitative PCR assay for rapid identification of *Bacillus anthracis* in environmental samples. *Appl. Microbiol. Biotechnol.* **2010**, *88*, 1179–1192. [CrossRef] [PubMed]

27. Sahl, J.W.; Pearson, T.; Okinaka, R.; Schupp, J.M.; Gillece, J.D.; Heaton, H.; Birdsell, D.; Hepp, C.; Fofanov, V.; Noseda, R.; et al. A *Bacillus anthracis* genome sequence from the Sverdlovsk 1979 autopsy specimens. *mBio* **2016**, *7*, e01501-16. [CrossRef]

28. Antwerpen, M.; Beyer, W.; Bassy, O.; Ortega-García, M.V.; Cabria-Ramos, J.C.; Grass, G.; Wölfel, R. Phylogenetic placement of isolates within the Trans-Eurasian clade A.Br.008/009 of *Bacillus anthracis. Microorganisms* **2019**, *7*, 689. [CrossRef]

29. Kämpfer, P.; Busse, H.J.; McInroy, J.A.; Glaeser, S.P. *Sphingomonas zeae* sp. nov., isolated from the stem of *Zea mays. Int. J. Syst. Evol. Microbiol.* **2015**, *65*, 2542–2548. [CrossRef]

30. Reischl, U.; Ehrenschwender, M.; Hiergeist, A.; Maaß, M.; Baier, M.; Frangoulidis, D.; Grass, G.; von Buttlar, H.; Scholz, H.; Fingerle, V.; et al. Bacterial and fungal genome detection PCR/NAT: Comprehensive discussion of the June 2018 distribution for external quality assessment of nucleic acid-based protocols in diagnostic medical microbiology by INSTAND e.V., GMS Z Forder Qualitatssich. *Med. Lab.* **2019**, *10*, 1869–4241.

31. Braun, P.; Grass, G.; Aceti, A.; Serrecchia, L.; Affuso, A.; Marino, L.; Grimaldi, S.; Pagano, S.; Hanczaruk, M.; Georgi, E.; et al. Microevolution of anthrax from a young ancestor (M.A.Y.A.) suggests a soil-borne life cycle of *Bacillus anthracis. PLoS ONE* **2015**, *10*, e0135346.

32. Altschul, S.F.; Gish, W.; Miller, W.; Myers, E.W.; Lipman, D.J. Basic local alignment search tool. *J. Mol. Biol.* **1990**, *215*, 403–410. [CrossRef]

33. Koshkin, A.A.; Singh, S.K.; Nielsen, P.; Rajwanshi, V.K.; Kumar, R.; Meldgaard, M.; Olsen, C.E.; Wengel, J. LNA (Locked Nucleic Acids): Synthesis of the adenine, cytosine, guanine, 5-methylcytosine, thymine and uracil bicyclonucleoside monomers, oligomerisation, and unprecedented nucleic acid recognition. *Tetrahedron* **1998**, *54*, 3607–3630. [CrossRef]

34. Perelle, S.; Dilasser, F.; Grout, J.; Fach, P. Detection by 5′-nuclease PCR of Shiga-toxin producing *Escherichia coli* O26, O55, O91, O103, O111, O113, O145 and O157:H7, associated with the world's most frequent clinical cases. *Mol. Cell. Probes* **2004**, *18*, 185–192. [CrossRef] [PubMed]

35. Stolovitzky, G.; Cecchi, G. Efficiency of DNA replication in the polymerase chain reaction. *Proc. Natl. Acad. Sci. USA* **1996**, *93*, 12947–12952. [CrossRef]

36. Forootan, A.; Sjöback, R.; Björkman, J.; Sjögreen, B.; Linz, L.; Kubista, M. Methods to determine limit of detection and limit of quantification in quantitative real-time PCR (qPCR). *Biomol. Detect. Quantif.* **2017**, *12*, 1–6. [CrossRef]

37. Nübel, U.; Schmidt, P.M.; Reiß, E.; Bier, F.; Beyer, W.; Naumann, D. Oligonucleotide microarray for identification of *Bacillus anthracis* based on intergenic transcribed spacers in ribosomal DNA. *FEMS Microbiol. Lett.* **2004**, *240*, 215–223. [CrossRef]

38. Daffonchio, D.; Raddadi, N.; Merabishvili, M.; Cherif, A.; Carmagnola, L.; Brusetti, L.; Rizzi, A.; Chanishvili, N.; Visca, P.; Sharp, R.; et al. Strategy for identification of *Bacillus cereus* and *Bacillus thuringiensis* strains closely related to *Bacillus anthracis. Appl. Environ. Microbiol.* **2006**, *72*, 1295–1301. [CrossRef]

39. Cherif, A.; Borin, S.; Rizzi, A.; Ouzari, H.; Boudabous, A.; Daffonchio, D. *Bacillus anthracis* diverges from related clades of the *Bacillus cereus* group in 16S-23S ribosomal DNA intergenic transcribed spacers containing tRNA genes. *Appl. Environ. Microbiol.* **2003**, *69*, 33–40. [CrossRef]

40. Hadjinicolaou, A.V.; Demetriou, V.L.; Hezka, J.; Beyer, W.; Hadfield, T.L.; Kostrikis, L.G. Use of molecular beacons and multi-allelic real-time PCR for detection of and discrimination between virulent *Bacillus anthracis* and other *Bacillus* isolates. *J. Microbiol. Methods* **2009**, *78*, 45–53. [CrossRef]

41. Merrill, L.; Richardson, J.; Kuske, C.R.; Dunbar, J. Fluorescent heteroduplex assay for monitoring *Bacillus anthracis* and close relatives in environmental samples. *Appl. Environ. Microbiol.* **2003**, *69*, 3317–3326. [CrossRef]

42. Easterday, W.R.; Van Ert, M.N.; Zanecki, S.; Keim, P. Specific detection of *Bacillus anthracis* using a TaqMan mismatch amplification mutation assay. *Biotechniques* **2005**, *38*, 731–735. [CrossRef]

43. Altayar, M.; Sutherland, A.D. *Bacillus cereus* is common in the environment but emetic toxin producing isolates are rare. *J. Appl. Microbiol.* **2006**, *100*, 7–14. [CrossRef] [PubMed]

44. Mancilla, M.; Ulloa, M.; López-Goñi, I.; Moriyón, I.; María Zárraga, A. Identification of new IS711 insertion sites in *Brucella abortus* field isolates. *BMC Microbiol.* **2011**, *11*, 176. [CrossRef] [PubMed]

45. Bounaadja, L.; Albert, D.; Chénais, B.; Hénault, S.; Zygmunt, M.S.; Poliak, S.; Garin-Bastuji, B. Real-time PCR for identification of *Brucella* spp.: A comparative study of IS711, bcsp31 and per target genes. *Veter- Microbiol.* **2009**, *137*, 156–164. [CrossRef]

46. Klee, S.R.; Tyczka, J.; Ellerbrok, H.; Franz, T.; Linke, S.; Baljer, G.; Appel, B. Highly sensitive real-time PCR for specific detection and quantification of *Coxiella burnetii. BMC Microbiol.* **2006**, *6*, 2. [CrossRef]

47. Backstedt, B.T.; Buyuktanir, O.; Lindow, J.; Wunder, E.A., Jr.; Reis, M.G.; Usmani-Brown, S.; Ledizet, M.; Ko, A.; Pal, U. Efficient detection of pathogenic Leptospires using 16S ribosomal RNA. *PLoS ONE* **2015**, *10*, e0128913. [CrossRef] [PubMed]

48. Kubota, H.; Tsuji, H.; Matsuda, K.; Kurakawa, T.; Asahara, T.; Nomoto, K. Detection of human intestinal catalase-negative, Gram-positive cocci by rRNA-targeted reverse transcription-PCR. *Appl. Environ. Microbiol.* **2010**, *76*, 5440–5451. [CrossRef] [PubMed]

49. Tsuji, H.; Matsuda, K.; Nomoto, K. Counting the countless: Bacterial quantification by targeting rRNA molecules to explore the human gut microbiota in health and disease. *Front. Microbiol.* **2018**, *9*, 1417. [CrossRef] [PubMed]

MDPI

St. Alban-Anlage 66

4052 Basel

Switzerland

Tel. +41 61 683 77 34

Fax +41 61 302 89 18

www.mdpi.com

International Journal of Molecular Sciences Editorial Office

E-mail: ijms@mdpi.com

www.mdpi.com/journal/ijms